Porphyrin-Based Compounds: Synthesis and Application

Porphyrin-Based Compounds: Synthesis and Application

Editors

Carlos J. P. Monteiro
M. Amparo F. Faustino
Carlos Serpa

Basel • Beijing • Wuhan • Barcelona • Belgrade • Novi Sad • Cluj • Manchester

Editors

Carlos J. P. Monteiro
University of Aveiro
Aveiro, Portugal

M. Amparo F. Faustino
University of Aveiro
Aveiro, Portugal

Carlos Serpa
University of Coimbra
Coimbra, Portugal

Editorial Office
MDPI
St. Alban-Anlage 66
4052 Basel, Switzerland

This is a reprint of articles from the Special Issue published online in the open access journal *Molecules* (ISSN 1420-3049) (available at: https://www.mdpi.com/journal/molecules/special_issues/Porphyrin_Based_Compounds).

For citation purposes, cite each article independently as indicated on the article page online and as indicated below:

Lastname, A.A.; Lastname, B.B. Article Title. *Journal Name* **Year**, *Volume Number*, Page Range.

ISBN 978-3-0365-9400-2 (Hbk)
ISBN 978-3-0365-9401-9 (PDF)
doi.org/10.3390/books978-3-0365-9401-9

Cover image courtesy of Carlos Jorge Pereira Monteiro

© 2023 by the authors. Articles in this book are Open Access and distributed under the Creative Commons Attribution (CC BY) license. The book as a whole is distributed by MDPI under the terms and conditions of the Creative Commons Attribution-NonCommercial-NoDerivs (CC BY-NC-ND) license.

Contents

About the Editors vii

Preface ix

Carlos J. P. Monteiro, M. Amparo F. Faustino and Carlos Serpa
Porphyrin-Based Compounds: Synthesis and Application
Reprinted from: *Molecules* **2023**, *28*, 7108, doi:10.3390/molecules28207108 1

Rui Liu, Jie Rong, Zhiyuan Wu, Masahiko Taniguchi, David F. Bocian, Dewey Holten and Jonathan S. Lindsey
Panchromatic Absorbers Tethered for Bioconjugation or Surface Attachment
Reprinted from: *Molecules* **2022**, *27*, 6501, doi:10.3390/molecules27196501 7

Sabrine Dridi, Jamel Eddine Khiari, Gabriele Magna, Manuela Stefanelli, Larisa Lvova, Federica Mandoj, et al.
Synthesis and Characterization of New-Type Soluble β-Substituted Zinc Phthalocyanine Derivative of Clofoctol
Reprinted from: *Molecules* **2023**, *28*, 4102, doi:10.3390/molecules28104102 27

Pol Torres, Marian Guillén, Marc Escribà, Joaquim Crusats and Albert Moyano
Synthesis of New Amino-Functionalized Porphyrins: Preliminary Study of Their Organophotocatalytic Activity
Reprinted from: *Molecules* **2023**, *28*, 1997, doi:10.3390/molecules28041997 37

Gabriela A. Corrêa, Susana L. H. Rebelo and Baltazar de Castro
Green Aromatic Epoxidation with an Iron Porphyrin Catalyst for One-Pot Functionalization of Renewable Xylene, Quinoline, and Acridine
Reprinted from: *Molecules* **2023**, *28*, 3940, doi:10.3390/molecules28093940 63

Marciana Pierina Uliana, Andréia da Cruz Rodrigues, Bruno Andrade Ono, Sebastião Pratavieira, Kleber Thiago de Oliveira and Cristina Kurachi
Photodynamic Inactivation of Microorganisms Using Semisynthetic Chlorophyll *a* Derivatives as Photosensitizers
Reprinted from: *Molecules* **2022**, *27*, 5769, doi:10.3390/molecules27185769 79

Sofia N. Sarabando, Cristina J. Dias, Cátia Vieira, Maria Bartolomeu, Maria G. P. M. S. Neves, Adelaide Almeida, et al.
Sulfonamide Porphyrins as Potent Photosensitizers against Multidrug-Resistant *Staphylococcus aureus* (MRSA): The Role of Co-Adjuvants
Reprinted from: *Molecules* **2023**, *28*, 2067, doi:10.3390/molecules28052067 93

Dragos Paul Mihai, Rica Boscencu, Gina Manda, Andreea Mihaela Burloiu, Georgiana Vasiliu, Ionela Victoria Neagoe, et al.
Interaction of Some Asymmetrical Porphyrins with U937 Cell Membranes–In Vitro and In Silico Studies
Reprinted from: *Molecules* **2023**, *28*, 1640, doi:10.3390/molecules28041640 111

Otávio Augusto Chaves, Bernardo A. Iglesias and Carlos Serpa
Biophysical Characterization of the Interaction between a Transport Human Plasma Protein and the 5,10,15,20-Tetra(pyridine-4-yl)porphyrin
Reprinted from: *Molecules* **2022**, *27*, 5341, doi:10.3390/molecules27165341 135

Caroline Ndung'U, Petia Bobadova-Parvanova, Daniel J. LaMaster, Dylan Goliber,
Frank R. Fronczek and Maria da Graça H. Vicente
8(*meso*)-Pyridyl-BODIPYs: Effects of 2,6-Substitution with Electron-Withdrawing Nitro, Chloro, and Methoxycarbonyl Groups
Reprinted from: *Molecules* **2023**, *28*, 4581, doi:10.3390/molecules28124581 149

Vitória Barbosa de Souza, Vinícius N. da Rocha, Paulo Cesar Piquini, Otávio Augusto Chaves and Bernardo A. Iglesias
Effects of Substituents on the Photophysical/Photobiological Properties of Mono-Substituted Corroles
Reprinted from: *Molecules* **2023**, *28*, 1385, doi:10.3390/molecules28031385 163

Mary-Ambre Carvalho, Khalissa Merahi, Julien Haumesser, Ana Mafalda Vaz Martins Pereira, Nathalie Parizel, Jean Weiss, et al.
Syntheses and Electrochemical and EPR Studies of Porphyrins Functionalized with Bulky Aromatic Amine Donors
Reprinted from: *Molecules* **2023**, *28*, 4405, doi:10.3390/molecules28114405 183

Pierpaolo Morgante and Roberto Peverati
Comparison of the Performance of Density Functional Methods for the Description of Spin States and Binding Energies of Porphyrins
Reprinted from: *Molecules* **2023**, *28*, 3487, doi:10.3390/molecules28083487 199

Ilaria Tomei, Beatrice Bonanni, Anna Sgarlata, Massimo Fanfoni, Roberto Martini,
Ilaria Di Filippo, et al.
Chiral Porphyrin Assemblies Investigated by a Modified Reflectance Anisotropy Spectroscopy Spectrometer
Reprinted from: *Molecules* **2023**, *28*, 3471, doi:10.3390/molecules28083471 219

Mateusz Pęgier, Krzysztof Kilian and Krystyna Pyrzynska
Increasing Reaction Rates of Water-Soluble Porphyrins for ^{64}Cu Radiopharmaceutical Labeling
Reprinted from: *Molecules* **2023**, *28*, 2350, doi:10.3390/molecules28052350 229

Francisco G. Moscoso, Carla Queirós, Paula González, Tânia Lopes-Costa, Ana M. G. Silva and Jose M. Pedrosa
Selective Determination of Glutathione Using a Highly Emissive Fluorescent Probe Based on a Pyrrolidine-Fused Chlorin
Reprinted from: *Molecules* **2023**, *28*, 568, doi:10.3390/molecules28020568 241

Rica Boscencu, Natalia Radulea, Gina Manda, Isabel Ferreira Machado, Radu Petre Socoteanu, Dumitru Lupuliasa, et al.
Porphyrin Macrocycles: General Properties and Theranostic Potential
Reprinted from: *Molecules* **2023**, *28*, 1149, doi:10.3390/molecules28031149 253

About the Editors

Carlos J. P. Monteiro

Carlos J. P. Monteiro is a researcher at the LAQV, University of Aveiro. He concluded his Ph.D. in Macromolecular Chemistry in 2012 (University of Coimbra). He was invited as a visiting research student at the Autonomous University of Barcelona, Spain (2009), and at Louisiana State University, USA (2011). Following his Ph.D., he served as a post-doctoral researcher between 2012 and 2014 at the University of Porto and University of Coimbra. He was a co-founder of Luzitin (2010) and participated in the medicinal chemistry discovery phase of the first Portuguese oncological drug candidate, Redaporfin, in clinical trials. He was awarded the Inventa Grand Prize (2011) for the best Portuguese patent in health and received the IUPAC Young Chemist Award at the 42nd IUPAC Congress in 2009. As a principal scientific contributor, he is the author of 29 papers in SCI journals and 2 book chapters, and the inventor of 3 patents, and has presented > 40 poster communications and 14 oral communications at scientific meetings/seminars. His papers have been cited more than 900 times in different journals. He acted as guest editor of six Special Issues in the fields of organic chemistry/ medicinal chemistry, photochemistry/photobiology, materials, and catalysis.

M. Amparo F. Faustino

Dr. M. Amparo F. Faustino is an accomplished associate professor in the Department of Chemistry at the University of Aveiro, and earned her Ph.D. in Chemistry with a specialization in organic synthesis in 1999. Her research focuses on its diverse applications, including Dye-Sensitized Solar Cells (DSSC), sensors, and photodynamic approaches. Dr. Faustino's scientific contributions are underscored by 3 patents, over 220 scientific papers in ISI journals, a H-index of 44, and more than 7000 citations. Beyond her academic achievements, Dr. Amparo Faustino plays important roles in scientific organizations. She currently holds the position of President of the Medicinal Chemistry and Biological Chemistry division of SPQ. Additionally, she serves as President of the Aveiro Regional Delegation of SPQ. Dr. Faustino is presently Past President of the European Photobiology Society (ESP) until 2025.

Carlos Serpa

Carlos Serpa, born in Lisbon in 1970, is presently an assistant professor in the Department of Chemistry at the University of Coimbra (Portugal). After the completion of his studies at the University of Coimbra (BSc Chemistry, 1995; MSc Physical Chemistry, 2000; PhD Photochemistry, 2004), Carlos Serpa took a one-year post-doc research position in the field of bioinorganic chemistry at Caltech (USA), under Harry B. Gray. He then earned two consecutive research positions within the national scientific and technological system (a Ciência 2007 Starting Grant and Investigator FCT Development Grant). His scientific interests cover a wide range of fundamental and applied research within the field of interactions between light and materials. His present research interests are focused on piezophotonic materials, materials chemistry for energy and health applications, and protein folding and aggregation. Carlos Serpa is co-author of over 60 papers in international peer-reviewed journals, over 185 conference proceedings, 3 book chapters, and 3 families of patents, one of them granted in 35 countries and licensed to a start-up company that emerged from his research. Carlos Serpa has been the principal investigator in R&D projects focused on piezophotonic materials and their applications in health sciences, and on nanostructured materials for application in renewable energies. In 2008, he was distinguished with the BES Innovation Prize. For many years, Carlos Serpa has contributed to teaching activities in graduate and post-graduate programs at the University of

Coimbra, as well as in students' supervision. Presently, Carlos Serpa serves as managing director of the Coimbra Laser Lab, a member of the inter-disciplinary network of European national laser infrastructures, Laserlab Europe. Carlos Serpa was one of the founders of the technological spin-off company LaserLeap Technologies, SA and presently serves as CEO of this company.

Preface

Porphyrins, metalloporphyrins, and their analogs belong to a group of macrocycles found abundantly in nature, where they play pivotal roles in various biological processes. For instance, chlorophyll, a magnesium–chlorin complex, is vital for plant light-harvesting, while the heme group, an iron–porphyrin complex, is responsible for oxygen binding and transport in animal cellular respiration. Bacteriochlorophylls are crucial in photosynthesis among phototropic bacteria, and cytochrome P450 enzymes, which oxidize and metabolize various substances, rely on porphyrinoid structures. Additionally, cyanocobalamin, a cobalt–corrin complex, serves as an enzyme co-factor in red blood cell formation. Consequently, these naturally occurring porphyrins are often referred to as "the Pigments of Life." Recognizing the significance of these natural porphyrinoids has inspired organic chemists to synthesize porphyrins and analogs in the laboratory. Efficient synthetic methods have been developed to create porphyrinoids with novel functionalities, electronic properties, and photophysical characteristics. This opens a wide array of applications, including their use in medicine (such as photodynamic therapy for cancer, antimicrobial photodynamic inactivation, medical imaging, and theragnostics), as catalysts, receptors in sensors, and dyes for solar cells. They also find utility in non-linear optics, molecular and supramolecular structures, and as components in materials and devices. In recent years, the multitude of applications for porphyrins has shifted the focus from purely academic to industrial processes. There is now a growing demand for the development of new synthetic methods that adhere to sustainable chemistry principles, avoiding hazardous solvents, reactants, and excessive energy consumption. It is of paramount importance to implement new, more selective, and environmentally friendly synthetic approaches.

Carlos J. P. Monteiro, M. Amparo F. Faustino, and Carlos Serpa
Editors

Editorial

Porphyrin-Based Compounds: Synthesis and Application

Carlos J. P. Monteiro [1,*], M. Amparo F. Faustino [1,*] and Carlos Serpa [2,*]

1 LAQV-Requimte and Department of Chemistry, University of Aveiro, 3810-193 Aveiro, Portugal
2 CQC-IMS, Department of Chemistry, University of Coimbra, Rua Larga, 3004-535 Coimbra, Portugal
* Correspondence: cmonteiro@ua.pt (C.J.P.M.); faustino@ua.pt (M.A.F.F.); serpasoa@ci.uc.pt (C.S.)

Porphyrin-based compounds are an attractive and versatile class of molecules that have attracted significant attention across different scientific disciplines [1]. These unique molecules, characterized by their distinctive macrocyclic structure featuring four pyrrole-type rings linked by methine bridges, have drawn the attention of chemists, biologists, and material scientists [2]. Porphyrins are not only notorious for their crucial roles in biological systems, serving as the core of heme in hemoglobin and chlorophyll in photosynthesis [3], but also for their remarkable capacity to undergo synthetic modification [4,5], leading to a wide range of functionalized derivatives. The synthesis and modification potential of porphyrins, combined with their extraordinary electronic and photophysical properties, has paved the way for an extensive possibility of applications, spanning from biomedical applications [6–8], to catalysis [9], sensors [10,11], energy conversion [12,13], and advanced materials [14]. The current Special Issue, entitled "Porphyrin-Based Compounds: Synthesis and Application", features 15 original research papers and one comprehensive review. These contributions are all dedicated to exploring the synthesis and functionalization of tetrapyrrolic macrocycles and their applications across several fields.

The importance of panchromatic absorbers and their potential applications motivated Bocian, Holten, Lindsey and co-workers [15] to develop an efficient synthetic procedure to obtain two panchromatic triads with absorption ranging from 350 to 700 nm and fluorescence emission from 733 to 743 nm. These triads comprise two perylene-monoimides connected to a porphyrin molecule through an ethyne unit. Additionally, these triads also contain a single anchoring group, either an alkynoic acid or an isophthalate unit for surface attachment, offering versatility in their use. The key steps in the synthesis involve the preparation of *trans*-AB-porphyrins, which are then coupled with the perylene-monoamide groups through copper-free Sonogashira-coupling reactions. The authors highlighted that these triads demonstrate significantly broader absorption spectra and enhanced fluorescence when compared to their individual components.

Paolesse and co-workers [16] reported a synthetic strategy to access zinc (II) phthalocyanine containing a 2-(2,4-dichloro-benzyl)-4-(1,1,3,3-tetramethyl-butyl)phenoxy group. They initiated this process from the Clofoctol derivative, a synthetic compound with antiviral activity against SARS-CoV-2. The synthesis involved the preparation of phthalonitrile precursors through nucleophilic aromatic nitro displacement of 2-(2,4-dichloro-benzyl)-4-(1,1,3,3-tetramethylbutyl)phenol (Clofoctol) with 4-nitrophthalonitrile. Subsequently, cyclotetramerization took place in the presence of zinc(II) salts to form the corresponding zinc(II) phthalocyanine complex, which was shown to be soluble in several organic solvents. Furthermore, the paper also explored the photochemical and electrochemical properties of the new compound and assessment as a solid-state sensing material in gravimetric chemical sensors, making it a promising material for gas sensing.

Moyano and co-workers [17] presented an efficient method for the synthesis of amino-functionalized porphyrins. These amino compounds were initially investigated as a new class of bifunctional catalysts for asymmetric organophotocatalysis. Two variants of amine-porphyrin hybrids were generated: one involving a cyclic secondary amine connected

Citation: Monteiro, C.J.P.; Faustino, M.A.F.; Serpa, C. Porphyrin-Based Compounds: Synthesis and Application. *Molecules* **2023**, *28*, 7108. https://doi.org/10.3390/molecules28207108

Received: 27 September 2023
Revised: 9 October 2023
Accepted: 12 October 2023
Published: 16 October 2023

Copyright: © 2023 by the authors. Licensee MDPI, Basel, Switzerland. This article is an open access article distributed under the terms and conditions of the Creative Commons Attribution (CC BY) license (https://creativecommons.org/licenses/by/4.0/).

to a β-pyrrolic position (referred to as Type A) and another linked to the *p*-position of a *meso* phenyl ring (referred to as Type B), along with their corresponding Cu(II) metalated derivatives. These synthetic steps involved the condensation or reductive amination reactions with appropriate chiral amines. Furthermore, an additional Type B bifunctional catalyst, 5,10,15-triphenyl-20-((S)-4-((pyrrolidine-2-carboxamido)methyl)phenyl)porphyrin, was achieved via the amidation of 5-(4-(aminomethyl)phenyl)-10,15,20-triphenylporphyrin with *N*-Boc-L-proline. The authors evaluated the potential use of Type A amine-porphyrin hybrids as asymmetric, bifunctional organophotocatalysts in a Diels–Alder cycloaddition reaction.

Rebelo and co-workers [18] contributed to the sustainable functionalization of renewable aromatic compounds. They reported one-pot oxidation reactions at room temperature, employing environmentally friendly H_2O_2 as an oxidant and ethanol as a solvent, along with an electron-withdrawing iron (III) porphyrin catalyst, used in a low catalyst loading (<2 mol%). The mechanistic aspects of these transformations were also investigated, offering insights into the reaction pathways. The investigation primarily centered on three inherently stable aromatic compounds: acridine, *o*-xylene, and quinoline. The results revealed an unconventional initial epoxidation of the aromatic ring catalyzed by this system. The study evidenced a distinctive preference for *o*-xylene oxidation, as it occurred exclusively on the aromatic ring and not on the methyl groups.

The synthesis and evaluation of porphyrin photosensitizers (PS) as antimicrobials have also been assessed by different authors in this Special Issue. In this context, Uliana and co-workers [19] elucidated the semisynthesis of PS derived from chlorophyll *a*, with several primary aliphatic amines (butylamine, hexylamine, and octylamine), to produce different derivatives and their utilization in antimicrobial photodynamic therapy (aPDT) toward different microorganisms such as *Staphylococcus aureus*, *Escherichia coli*, and *Candida albicans*. The modifications introduced in the porphyrin aimed to improve the solubility and amphiphilicity of the PSs to enhance their performance in aPDT. The purpurin-18 derivatives containing carboxylic acid groups were particularly effective against *S. aureus* and *C. albicans*, achieving significant inactivation. The length of the carbon chain in the PS influenced their performance, as longer chains led to reduced PS uptake by the microorganism and subsequently lower photoinactivation outcomes.

Monteiro, Faustino, and co-workers [20] published findings on the synthesis and antimicrobial action of a novel class of porphyrin-based PS incorporating sulfonamide groups and a zinc(II) complex. These compounds were evaluated to assess their effectiveness in aPDT against methicillin-resistant *Staphylococcus aureus* (MRSA) when combined with potassium iodide (KI) as a co-adjuvant. All examined compounds demonstrated the capacity to generate $^1O_2$, with the zinc(II) complex exhibiting the highest efficiency. The research also assessed the formation of iodine (I_2) in the presence of potassium iodide (KI) and each PS under white light irradiation. The results highlighted the effective generation of I_2 by all compounds, with the sulfonic acid derivative ($TPP(SO_3H)_4$) demonstrating the most notable efficiency. The photodynamic action of these PS was evaluated against MRSA. All compounds displayed significant photoinactivation of MRSA, with $TPP(SO_3H)_4$ emerging as the most effective. The zinc(II) complex, $ZnTPP(SO_2NHEt)_4$, also exhibited promising antimicrobial activity. These findings highlight the considerable potential of these PS, especially when used in combination with KI, in aPDT, addressing the pressing challenge of antimicrobial resistance.

Boscensu, Burloiu, Socoteanu and co-workers [21], investigated the effects exerted in vitro of three asymmetrical porphyrins, (5-(2-hydroxyphenyl)-10,15,20-tris(4-acetoxy-3-methoxyphenyl)porphyrin, 5-(2-hydroxyphenyl)-10,15,20-tris(4-acetoxy-3-methoxyphenyl) porphyrinatozinc(II), and 5-(2-hydroxyphenyl)-10,15,20-tris(4-acetoxy-3-methoxyphenyl) porphyrinatocopper(II)), on human U937 cell membranes. The porphyrins were studied for their impact on transmembrane potential and membrane anisotropy using fluorescent probes. The findings indicate that porphyrins induce cell membrane hyperpolarization and increase membrane anisotropy, suggesting enhanced rigidity. This could alter membrane protein interaction, impacting cellular function. Molecular docking simulations suggest that these derivatives may interact with membrane proteins such as SERCA2b, Slo1, and KATP channels, possibly modulating their activity. The research suggests that these asymmetrical porphyrins have notable effects on cell membranes and may hold promise for further study as potential agents for different biomedical applications. Further experimental validation is required to confirm these interactions and their impact on cellular homeostasis.

Chaves, Serpa and co-worker [22] explored the interaction between human serum albumin (HSA) and the non-charged synthetic porphyrin, 5,10,15,20-tetra(pyridin-4-yl)porphyrin (4-TPyP), evaluated via in vitro assays under physiological conditions using spectroscopic techniques (UV-vis, circular dichroism, steady-state, time-resolved, synchronous, and 3D-fluorescence) combined with in silico calculations via molecular docking. The UV-vis and steady-state fluorescence parameters indicated a ground-state association between HSA and 4-TPyP and the absence of any dynamic fluorescence quenching was confirmed by the same average fluorescence lifetime for HSA without and with 4-TPyP. Therefore, the Stern–Volmer quenching (K_{SV}) constant reflects the binding affinity, indicating a moderate interaction being spontaneous, enthalpically, and entropically driven. Binding produces only a very weak perturbation on the secondary structure of albumin. There is just one main binding site in HSA for 4-TPyP (n ≈ 1.0), probably in the subdomain IIA (site I), where the Trp-214 residue can be found. The microenvironment around this fluorophore seems not to be perturbed even with 4-TPyP interacting via hydrogen bonding and van der Waals forces with the amino acid residues in the subdomain IIA. To offer a molecular-level explanation of the binding HSA: 4-TPyP, molecular docking calculations were also carried out for the three main binding sites of albumin (subdomains IIA, IIIA, and IB).

BODIPY (boron-dipyrromethene) represents a class of synthetic dyes or fluorophores recognized for their tunable properties, such as absorption and emission wavelengths, achieved through chemical and structural modifications. In this context, Vicente and co-workers [23] undertook a study focused on synthesizing and characterizing a range of 8(*meso*)-pyridyl-BODIPY compounds featuring diverse substituents. The study encompassed the synthesis of BODIPY compounds incorporating 2-, 3-, or 4-pyridyl groups, along with functionalization involving nitro and chlorine groups at the 2,6-positions. Additionally, analogs with 2,6-methoxycarbonyl groups were synthesized. The study systematically investigated the structure and spectroscopic properties of these compounds using experimental and computational methods. Introducing 2,6-methoxycarbonyl groups significantly boosted fluorescence quantum yields of BODIPYs in polar organic solvents. Conversely, adding a single nitro group reduced fluorescence and caused shifts in absorption and emission bands to shorter wavelengths. Interestingly, incorporating a chloro substituent partially restored fluorescence and induced shifts to longer wavelengths. This research provides insights into how electron-withdrawing groups affect the photophysical properties of 8(*meso*)-pyridyl-BODIPYs, offering valuable information for potential applications.

Iglesias and co-workers [24] investigated the photophysical and photobiological properties of a series of *trans*-C_6F_5 *meso*-substituted corroles. These corroles have different groups attached to their *meso* positions (phenyl, naphthyl, 4-(hydroxy)phenyl, or 4-(thiomethyl)phenyl)) and are of interest for potential applications in PDT and other photo-related processes. The study involved characterizing these compounds using various techniques such as electrochemical methods and photophysical properties, including absorption and emission characteristics, as well as their behavior in different solvents. Theoretical calculations were performed to gain insights into their properties. The research suggests that these substituted corrole compounds show promise for various photoinduced processes and prefer binding to DNA in the minor grooves, as well as interacting with HSA.

The work reported by Ruhilman, Choua, Ruppert and co-workers [25] comprehensively investigated the synthesis and electronic properties of nickel(II) porphyrins with bulky nitrogen donors at the *meso* positions, prepared using Ullmann methodology or Buchwald–Hartwig amination reactions to establish new C-N bonds. They also conducted electrochemical studies, spectroelectrochemical measurements, electron paramagnetic resonance (EPR) studies, and density functional theory (DFT) calculations to gain insights into the electronic properties and delocalization of radical cations in these compounds. Electron paramagnetic resonance (EPR) and electron nuclear double resonance spectroscopy (ENDOR) were used to study the extent of delocalization of the generated radical cations. The results indicated that the radical cation distribution varied depending on the specific compound and the nature of the substituents. The study also explored the magnetic properties of some compounds, including the formation of biradicals and their exchange interactions. Additionally, DFT calculations supported the EPR spectroscopic data.

Peverati and Morgante [26] analyzed the performance of 250 electronic structure theory methods (including 240 density functional approximations) for the description of spin states and the binding properties of iron, manganese, and cobalt porphyrins. The assessment employed the Por21 database of high-level computational data. The results showed that the approximations failed to achieve the "chemical accuracy" target of 1.0 kcal/mol by a long margin. The best-performing methods achieved a mean unsigned error (MUE) < 15.0 kcal/mol, but the errors were at least twice as large for most methods. Semilocal functionals and global hybrid functionals with a low percentage of exact exchange were found to be the least problematic for spin states and binding energies, in agreement with the general knowledge in transition metal computational chemistry. These results reflect both an intrinsic difficulty of density functional calculations on metalloporphyrin and the difficulties in obtaining reliable reference energies and experimental results.

Golleti and co-workers [27] highlighted the importance of chirality in organic materials, using Reflectance Anisotropy Spectroscopy (RAS) to investigate porphyrins and porphyrin-related compounds, in various experimental conditions, including ultra-high vacuum-controlled atmospheres and liquid. The study introduced a technical enhancement to the RAS spectrometer called CD-RAS (circular dichroism RAS), which allows for the measurement of circular dichroism (CD) in samples under right- and left-circularly polarized light in transmission mode, offering flexibility and potential applications in various experimental setups. The research emphasizes the significance of chirality in organic materials and the role of porphyrins in exploring this phenomenon.

Exploring novel compounds and synthetic pathways for medical applications led Pęgier and co-workers [28] to optimize the synthesis of copper complexes with various water-soluble porphyrins (5,10,15,20-tetrakis(1-methylpyridinium-4-yl)porphyrin tetratosylate (TMPyP), 5,10,15,20-tetrakis(4-sulfonatophenyl)porphyrin (TSPP), and also 5,10,15,20-tetrakis(4-carboxyphenyl)porphyrin (TCPP)) for potential use in radiopharmaceuticals, with a particular emphasis on applications in positron emission tomography (PET). Efforts were focused on developing the fastest possible method that would meet the requirements of labeling with short-lived copper isotopes used in PET. Two optimized methods were applied for the synthesis of the ^{64}Cu–porphyrin complex: (i) using ascorbic acid (AA) as a reducing agent for accelerated room temperature complexation; and (ii) utilizing microwave-assisted synthesis at 140 °C for 1–2 min. Both methods are efficient for application in radiopharmaceutical synthesis as they are one-step, fast, and avoid the use of toxic or harsh chemicals requiring separation.

Moscoso, Pedrosa and co-workers [29] introduced an innovative colorimetric and fluorescent probe designed for the detection of glutathione (GSH) within a pH 7.4 phosphate buffer solution. This probe is built upon a highly luminescent porphyrin analog, specifically a carboxylated pyrrolidine-fused chlorin (TCPC). The probe runs on the principle of forming a TCPC-Hg^{2+} complex, where TCPC binds to mercury ions (Hg^{2+}), resulting in a notable reduction in its fluorescence output. However, when GSH is introduced into the system, it competes with Hg^{2+} for binding to TCPC, leading to an increase in fluorescence intensity. This fluorescence enhancement serves as the basis for GSH detection. The sensitivity of this detection method depends on the concentration ratio of TCPC to Hg^{2+}, with a remarkable achievement of a detection limit as low as 40 nM under specific ratios. The study further highlights the selectivity of the sensor, as it demonstrates the ability to discriminate against potential interferents, including various metal ions and cysteine.

In this Special Issue, an important contribution is a comprehensive review authored by Boscescu, Socoteanu, Burloi, Ferreira and co-workers [30] focusing on porphyrin macrocycles and their potential in cancer diagnosis and therapy. The paper shed light on photodynamic therapy (PDT) as a non-invasive therapy with minimal side effects compared to conventional cancer treatments. The properties of porphyrin macrocycles, including their absorption and emission spectra, were discussed in detail. The authors also presented a critical overview of the main commercial PS, followed by short descriptions of some strategies approached in the development of third-generation PS. The paper discussed challenges in solubility and molecular aggregation for tetrapyrrolic compounds, emphasizing efforts to enhance their clinical performance through modifications and functionalization. Strategies for functionalizing porphyrinic macrocycles with bioactive molecule fragments such as amino acids, peptides, and sugars to optimize the biodistribution of PS were also comprehensively explored and discussed.

To sum up, the Special Issue entitled "Porphyrin-Based Compounds: Synthesis and Application" provides a current perspective on the synthesis and modification of porphyrins, metalloporphyrins, and related compounds. Considering the challenges in this exciting field, this edition not only enhances our understanding of synthetic methodologies but also unveils a wide range of impactful applications. These applications cover critical domains such as biomedical applications (including radiopharmaceuticals, DNA and protein interactions, cancer therapeutics, and microbial inactivation) as well as catalysis, sensors, advanced materials, and computational studies. The editors wish to thank the invited authors for their engaging and insightful contributions. They anticipate that the joint insights presented in this Special Issue will have a significant impact on the field, enriching research, development, and knowledge concerning the applications of porphyrins and their analogs.

Author Contributions: Conceptualization, C.J.P.M., M.A.F.F. and C.S.; writing—original draft preparation, C.J.P.M., M.A.F.F. and C.S.; writing—review and editing, C.J.P.M., M.A.F.F. and C.S. All authors have read and agreed to the published version of the manuscript.

Funding: CJPM and MAFF thank the University of Aveiro and FCT/MCT for the financial support provided to LAQV-REQUIMTE (UIDB/50006/2020 and UIDP/50006/2020), through national funds (OE), and where applicable, co-financed by the FEDER-Operational Thematic Program for Competitiveness and Internationalization-COMPETE 2020, within the PT2020 Partnership Agreement. CS acknowledge funding support from FCT/MCT through projects UID/QUI/00313/2020, UIDB/00285/2020 and PTDC/QUI-OUT/0303/2021, and the European Union through H2020-INFRAIA-2018 under grant agreement number 871124 Laserlab-Europe.

Acknowledgments: We would like to thank all the authors for their contributions to this Special Issue and the reviewers for their careful work.

Conflicts of Interest: The authors declare no conflict of interest.

References

1. Milgrom, L.R. *The Colours of Life: An Introduction to the Chemistry of Porphyrins and Related Compounds*; Oxford University Press: New York, NY, USA, 1997.
2. Park, J.M.; Hong, K.-I.; Lee, H.; Jang, W.-D. Bioinspired Applications of Porphyrin Derivatives. *Acc. Chem. Res.* **2021**, *54*, 2249–2260. [CrossRef] [PubMed]
3. Battersby, A.R. Tetrapyrroles: The pigments of life. *Nat. Prod. Rep.* **2000**, *17*, 507–526. [CrossRef] [PubMed]
4. Moura, N.M.; Monteiro, C.J.P.; Tomé, A.C.; Neves, M.G.P.M.S.; Cavaleiro, J.A. Synthesis of chlorins and bacteriochlorins from cycloaddition reactions with porphyrins. *Arkivoc* **2022**, *2022*, 54–98. [CrossRef]
5. Pereira, M.M.; Monteiro, C.J.P.; Peixoto, A.F. Meso-substituted porphyrin synthesis from monopyrrole: An overview. In *Targets in Heterocyclic Systems-Chemistry and Properties*; Attanasi, O.A., Spinelli, D., Eds.; Italian Society of Chemistry: Rome, Italy, 2009; Volume 12, pp. 258–278.
6. Pham, T.C.; Nguyen, V.N.; Choi, Y.; Lee, S.; Yoon, J. Recent Strategies to Develop Innovative Photosensitizers for Enhanced Photodynamic Therapy. *Chem. Rev.* **2021**, *121*, 13454–13619. [CrossRef]
7. Monteiro, C.J.P.; Neves, M.G.P.M.S.; Nativi, C.; Almeida, A.; Faustino, M.A.F. Porphyrin Photosensitizers Grafted in Cellulose Supports: A Review. *Int. J. Mol. Sci.* **2023**, *24*, 3475. [CrossRef]

8. Vallejo, M.C.S.; Moura, N.M.M.; Gomes, A.; Joaquinito, A.S.M.; Faustino, M.A.F.; Almeida, A.; Goncalves, I.; Serra, V.V.; Neves, M. The Role of Porphyrinoid Photosensitizers for Skin Wound Healing. *Int. J. Mol. Sci.* **2021**, *22*, 43. [CrossRef]
9. Barona-Castaño, J.C.; Carmona-Vargas, C.C.; Brocksom, T.J.; De Oliveira, K.T. Porphyrins as Catalysts in Scalable Organic Reactions. *Molecules* **2016**, *21*, 310. [CrossRef]
10. Paolesse, R.; Nardis, S.; Monti, D.; Stefanelli, M.; Di Natale, C. Porphyrinoids for Chemical Sensor Applications. *Chem. Rev.* **2017**, *117*, 2517–2583. [CrossRef]
11. Norvaiša, K.; Kielmann, M.; Senge, M.O. Porphyrins as Colorimetric and Photometric Biosensors in Modern Bioanalytical Systems. *ChemBioChem* **2020**, *21*, 1793–1807. [CrossRef]
12. Mahmood, A.; Hu, J.-Y.; Xiao, B.; Tang, A.; Wang, X.; Zhou, E. Recent progress in porphyrin-based materials for organic solar cells. *J. Mater. Chem. A* **2018**, *6*, 16769–16797. [CrossRef]
13. Monteiro, C.J.P.; Jesus, P.; Davies, M.L.; Ferreira, D.; Arnaut, L.G.; Gallardo, I.; Pereira, M.M.; Serpa, C. Control of the distance between porphyrin sensitizers and the TiO2 surface in solar cells by designed anchoring groups. *J. Mol. Struct.* **2019**, *1196*, 444–454. [CrossRef]
14. Tanaka, T.; Osuka, A. Conjugated porphyrin arrays: Synthesis, properties and applications for functional materials. *Chem. Soc. Rev.* **2015**, *44*, 943–969. [CrossRef] [PubMed]
15. Liu, R.; Rong, J.; Wu, Z.; Taniguchi, M.; Bocian, D.F.; Holten, D.; Lindsey, J.S. Panchromatic Absorbers Tethered for Bioconjugation or Surface Attachment. *Molecules* **2022**, *27*, 6501. [CrossRef] [PubMed]
16. Dridi, S.; Khiari, J.E.; Magna, G.; Stefanelli, M.; Lvova, L.; Mandoj, F.; Khezami, K.; Durmuş, M.; Di Natale, C.; Paolesse, R. Synthesis and Characterization of New-Type Soluble β-Substituted Zinc Phthalocyanine Derivative of Clofoctol. *Molecules* **2023**, *28*, 4102. [CrossRef]
17. Torres, P.; Guillén, M.; Escribà, M.; Crusats, J.; Moyano, A. Synthesis of New Amino-Functionalized Porphyrins:Preliminary Study of Their Organophotocatalytic Activity. *Molecules* **2023**, *28*, 1997. [CrossRef]
18. Corrêa, G.A.; Rebelo, S.L.H.; de Castro, B. Green Aromatic Epoxidation with an Iron Porphyrin Catalyst for One-Pot Functionalization of Renewable Xylene, Quinoline, and Acridine. *Molecules* **2023**, *28*, 3940. [CrossRef]
19. Uliana, M.P.; da Cruz Rodrigues, A.; Ono, B.A.; Pratavieira, S.; de Oliveira, K.T.; Kurachi, C. Photodynamic Inactivation of Microorganisms Using Semisynthetic Chlorophyll a Derivatives as Photosensitizers. *Molecules* **2022**, *27*, 5769. [CrossRef]
20. Sarabando, S.N.; Dias, C.; Vieira, C.; Bartolomeu, M.; Neves, M.G.P.M.S.; Almeida, A.; Monteiro, C.J.P.; Faustino, M.A.F. Sulfonamide Porphyrins as Potent Photosensitizers against Multidrug-Resistant Staphylococcus aureus (MRSA): The Role of Co-Adjuvants. *Molecules* **2023**, *28*, 2067. [CrossRef]
21. Mihai, D.P.; Boscencu, R.; Manda, G.; Burloiu, A.M.; Vasiliu, G.; Neagoe, I.V.; Socoteanu, R.P.; Lupuliasa, D. Interaction of Some Asymmetrical Porphyrins with U937 Cell Membranes-In Vitro and In Silico Studies. *Molecules* **2023**, *28*, 1640. [CrossRef]
22. Chaves, O.A.; Iglesias, B.A.; Serpa, C. Biophysical Characterization of the Interaction between a Transport Human Plasma Protein and the 5,10,15,20-Tetra(pyridine-4-yl)porphyrin. *Molecules* **2022**, *27*, 5341. [CrossRef]
23. Ndung'U, C.; Bobadova-Parvanova, P.; LaMaster, D.J.; Goliber, D.; Fronczek, F.R.; Vicente, M.d.G.H. 8(meso)-Pyridyl-BODIPYs: Effects of 2,6-Substitution with Electron-Withdrawing Nitro, Chloro, and Methoxycarbonyl Groups. *Molecules* **2023**, *28*, 4581. [CrossRef] [PubMed]
24. de Souza, V.B.; da Rocha, V.N.; Piquini, P.C.; Chaves, O.A.; Iglesias, B.A. Effects of Substituents on the Photophysical/Photobiological Properties of Mono-Substituted Corroles. *Molecules* **2023**, *28*, 1385. [CrossRef] [PubMed]
25. Carvalho, M.-A.; Merahi, K.; Haumesser, J.; Pereira, A.M.V.M.; Parizel, N.; Weiss, J.; Orio, M.; Maurel, V.; Ruhlmann, L.; Choua, S.; et al. Syntheses and Electrochemical and EPR Studies of Porphyrins Functionalized with Bulky Aromatic Amine Donors. *Molecules* **2023**, *28*, 4405. [CrossRef]
26. Morgante, P.; Peverati, R. Comparison of the Performance of Density Functional Methods for the Description of Spin States and Binding Energies of Porphyrins. *Molecules* **2023**, *28*, 3487. [CrossRef] [PubMed]
27. Tomei, I.; Bonanni, B.; Sgarlata, A.; Fanfoni, M.; Martini, R.; Di Filippo, I.; Magna, G.; Stefanelli, M.; Monti, D.; Paolesse, R.; et al. Chiral Porphyrin Assemblies Investigated by a Modified Reflectance Anisotropy Spectroscopy Spectrometer. *Molecules* **2023**, *28*, 3471. [CrossRef] [PubMed]
28. Pęgier, M.; Kilian, K.; Pyrzynska, K. Increasing Reaction Rates of Water-Soluble Porphyrins for ^{64}Cu Radiopharmaceutical Labeling. *Molecules* **2023**, *28*, 2350. [CrossRef]
29. Moscoso, F.G.; Queirós, C.; González, P.; Lopes-Costa, T.; Silva, A.M.G.; Pedrosa, J.M. Selective Determination of Glutathione Using a Highly Emissive Fluorescent Probe Based on a Pyrrolidine-Fused Chlorin. *Molecules* **2023**, *28*, 568. [CrossRef]
30. Boscencu, R.; Radulea, N.; Manda, G.; Machado, I.F.; Socoteanu, R.P.; Lupuliasa, D.; Burloiu, A.M.; Mihai, D.P.; Ferreira, L.F.V. Porphyrin Macrocycles: General Properties and Theranostic Potential. *Molecules* **2023**, *28*, 1149. [CrossRef]

Disclaimer/Publisher's Note: The statements, opinions and data contained in all publications are solely those of the individual author(s) and contributor(s) and not of MDPI and/or the editor(s). MDPI and/or the editor(s) disclaim responsibility for any injury to people or property resulting from any ideas, methods, instructions or products referred to in the content.

Article

Panchromatic Absorbers Tethered for Bioconjugation or Surface Attachment

Rui Liu [1,†], Jie Rong [1,†], Zhiyuan Wu [1,†], Masahiko Taniguchi [1], David F. Bocian [2,*], Dewey Holten [3,*] and Jonathan S. Lindsey [1,*]

1 Department of Chemistry, North Carolina State University, Raleigh, NC 27695-8204, USA
2 Department of Chemistry, University of California, Riverside, CA 92521-0403, USA
3 Department of Chemistry, Washington University, St. Louis, MO 63130-4889, USA
* Correspondence: david.bocian@ucr.edu (D.F.B.); holten@wustl.edu (D.H.); jlindsey@ncsu.edu (J.S.L.); Tel.: +1-919-515-6406 (J.S.L.)
† These authors contributed equally to this work.

Abstract: The syntheses of two triads are reported. Each triad is composed of two perylene-monoimides linked to a porphyrin via an ethyne unit, which bridges the perylene 9-position and a porphyrin 5- or 15-position. Each triad also contains a single tether composed of an alkynoic acid or an isophthalate unit. Each triad provides panchromatic absorption (350–700 nm) with fluorescence emission in the near-infrared region (733 or 743 nm; fluorescence quantum yield ~0.2). The syntheses rely on the preparation of *trans*-AB-porphyrins bearing one site for tether attachment (A), an aryl group (B), and two open meso-positions. The AB-porphyrins were prepared by the condensation of a 1,9-diformyldipyrromethane and a dipyrromethane. The installation of the two perylene-monoimide groups was achieved upon the 5,15-dibromination of the porphyrin and the subsequent copper-free Sonogashira coupling, which was accomplished before or after the attachment of the tether. The syntheses provide relatively straightforward access to a panchromatic absorber for use in bioconjugation or surface-attachment processes.

Keywords: array; artificial photosynthesis; building block; copper-free Sonogashira coupling; ethyne; perylene; porphyrin; solar

1. Introduction

Chromophores that absorb across the visible spectrum (400–700 nm) are of essential importance for diverse studies in the photosciences. The chlorophylls of plant photosynthesis exhibit strong absorption in the violet and red regions, with relatively weak absorption across the rest of the visible region [1]. Natural photosynthetic systems use carotenoids and/or other accessory chromophores in complementary fashion to fill the blue–orange region of the solar spectrum [2]. In this approach, the distinct absorption of a given accessory pigment is followed by excited-state energy transfer among a set of pigments, thereby increasing the wavelength expanse of absorption beyond that of chlorophyll alone. Work that began some 45 years ago [3] led to the realization in the early 1990s that the absorption spectrum of porphyrins could be substantially broadened upon the conjugation of ethynyl groups to the macrocycle [4–7]. Penetrating studies thereafter by the groups of Anderson [8] and Therien [9–21] chiefly focused on joining porphyrins via butadiyne and ethyne linkers, respectively, eliciting fascinating spectroscopic features of the resulting arrays. The joining of porphyrins and other chromophores via ethynyl linkers quickly became a prominent molecular motif [22–25], a topic that has been reviewed [26,27].

In a quite separate research thread, we had prepared and characterized a large number of tetrapyrrole-perylene constructs for studies of excited-state energy transfer as part of a program in artificial photosynthesis [28–40]. In such constructs, we had employed long linkers such as phenyl-ethynyl-phenyl to separate the tetrapyrrole and perylene or ethynylphenyl joining the pigments in the manner perylene-ethynylphenyl-tetrapyrrole. As an intended "last molecular design" in this area [41], we omitted any phenyl groups and prepared a perylene-ethynyl-porphyrin dyad (**0**). The dyad exhibited such broad and unusual absorption across the visible region that the characteristic Soret band, a hallmark of porphyrins, was almost unrecognizable (Figure 1) [41].

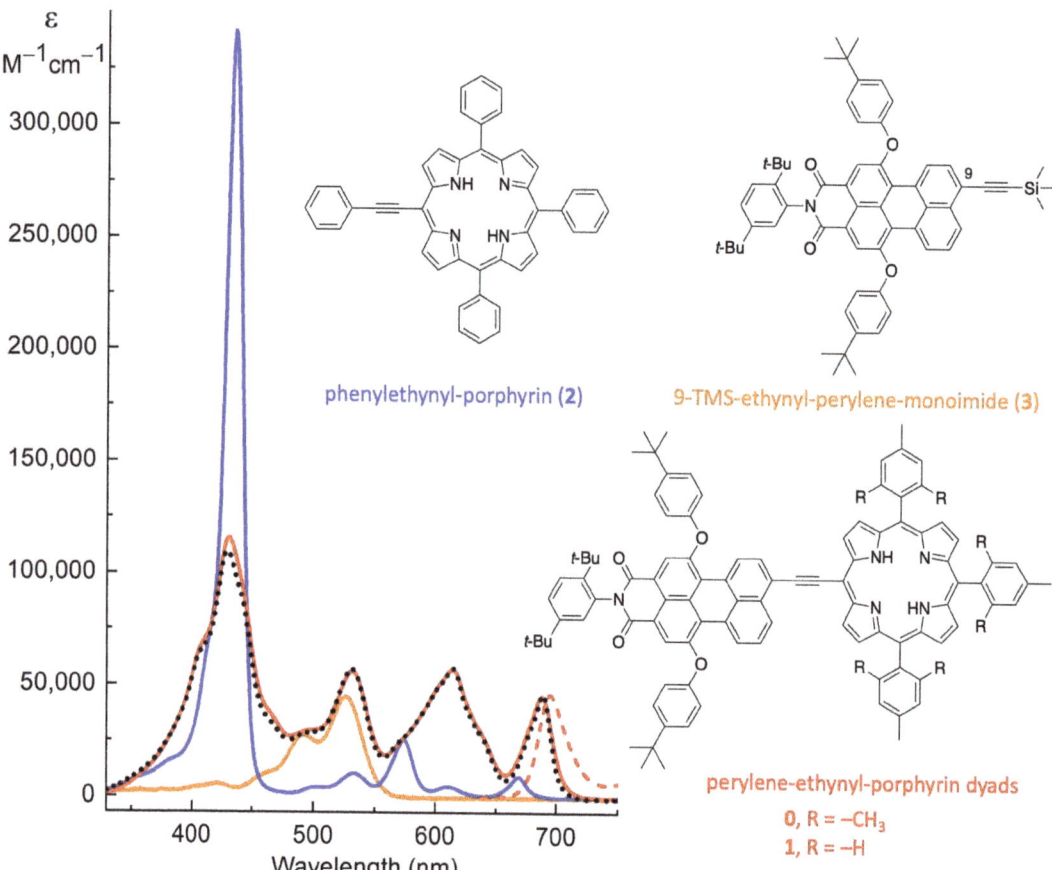

Figure 1. Absorption spectra of a perylene-ethynyl-porphyrin dyad (**1**, solid red line), to be compared with phenylethynyl-porphyrin (**2**, solid blue line) and perylene-ethyne (**3**, solid orange line) constituents (in toluene at room temperature). The fluorescence spectrum of **1** also is shown (dashed red line; λ_{em} 695 nm) [42]. The absorption spectrum of the perylene-ethynyl-porphyrin dyad **0** (dotted black trace [41]) is essentially identical with that of analogous dyad **1** (solid red line).

For exacting comparison, we then prepared an analogous perylene-ethynyl-porphyrin dyad (**1**) [42], which differed from the initial dyad (**0**) in the nature of the non-linking substituents (*p*-tolyl versus mesityl groups) on the porphyrin, and also the corresponding ethynyl-porphyrin (**2**) [43,44] and ethynyl-perylene (**3**) [40] benchmark compounds bearing nearly identical substituents. The change from mesityl to *p*-tolyl substituents was inconsequential, as shown in the overlay of the spectra of dyads **0** and **1**. While the absorption spectrum of each dyad resembles that of a strongly potentiated perylene and diminished porphyrin (Figure 1), the fluorescence emission spectrum resembles that of a porphyrin, albeit one shifted bathochromically (~40 nm) and with substantially increased intensity. Indeed, the fluorescence quantum yield (Φ_f) of dyad **1** is 0.38 instead of 0.14 for the phenylethynyl-substituted porphyrin **2** [42].

This unexpected finding in 2013 [41] did not close the experimental program but instead prompted an additional decade of research [45–50] to establish the physical basis for the origin of panchromaticity. An ideal panchromatic absorber was found to be a porphyrin bearing two perylene-monoimides each joined via a single ethyne unit bridging the perylene 9-position and a porphyrin meso-position (**4**, Chart 1). The absorption and fluorescence spectra of **4** are shown in Figure 2, upper panel. Conversion of the free base porphyrin to the zinc chelate (**Zn4**) alters the spectral features (Figure 2, lower panel). The free base porphyrin triad also has been incorporated as a crossbar unit in a pentad array for solar light capture and charge separation [48].

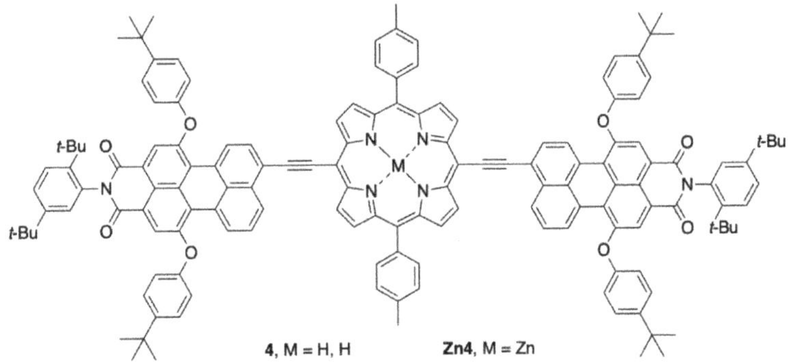

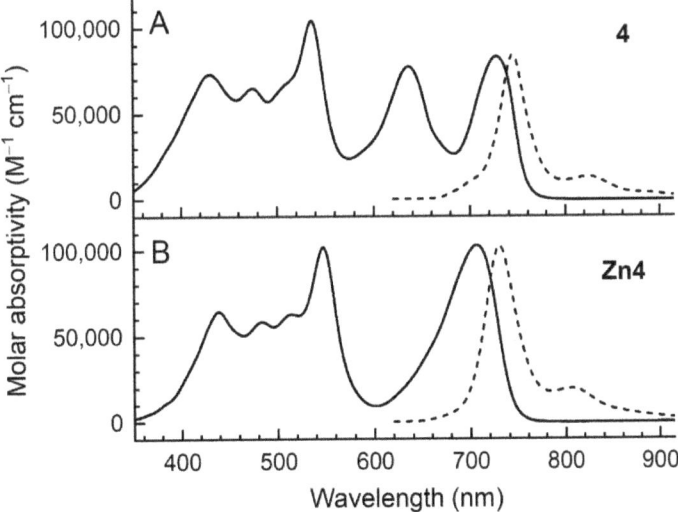

Chart 1. Structure of panchromatic triad (**4**) and the zinc chelate **Zn4**.

Figure 2. Absorption spectra (solid line) and fluorescence spectra (dashed line) of triads in toluene at room temperature. The molar absorption coefficient values for free base porphyrin triad **4** are as follows: 431 nm (73,200 $M^{-1}cm^{-1}$), 537 nm (104,000 $M^{-1}cm^{-1}$), 637 nm (77,400 $M^{-1}cm^{-1}$), and 728 nm (83,200 $M^{-1}cm^{-1}$) [42,45]. The molar absorption coefficient values for zinc porphyrin triad **Zn4** are as follows: 439 nm (64,700 $M^{-1}cm^{-1}$), 548 nm (102,000 $M^{-1}cm^{-1}$), and 707 nm (103,000 $M^{-1}cm^{-1}$) [45]. The Φ_f value of **4** is 0.26, whereas that of **Zn4** is 0.30 [45].

In this paper, we describe the extension of the panchromatic triad for studies in light-harvesting. One design incorporates a single carboxylic acid for bioconjugation, such as in biohybrid assemblies wherein additional solar light capture is advantageous. A second design incorporates an isophthalic acid terminal unit for attachment to surfaces, such as metal oxides for studies of photoinduced charge injection. The syntheses rely on established rational methods and provide relatively direct access to lipophilic panchromatic triads each bearing a single attachment handle.

2. Results and Discussion

2.1. Synthesis of a Bioconjugatable Panchromatic Triad

We sought to place a hexanoic acid chain on the central porphyrin for bioconjugation purposes. A *meso*-dibromo-substituted porphyrin building block was prepared for the construction of the bioconjugatable panchromatic triad. The core porphyrin contains an AB pattern of substituents with two open meso-positions and was prepared in a standard manner [51] from two dipyrromethanes (Scheme 1). The dibutyltin-complexed 1,9-diformyl-5-phenyldipyrromethane **5** [52] was treated with propylamine [51] and then reacted with 1,9-diunsubstituted 5-(4-bromophenyl)dipyrromethane **6** [53] in the presence of Zn(OAc)$_2$ to afford the *trans*-AB zinc porphyrin **7**, which bears a bromine atom at a site for the introduction of a bioconjugatable tether. The bromo-porphyrin **7** thereafter reacted with ethynyl tether **8** [54] via a Sonogashira coupling reaction to form the bioconjugatable zinc porphyrin **9**. The Pd-mediated Sonogashira reaction [55] was carried out under copper-free conditions [56,57] using a solvent mixture of toluene and triethylamine (TEA). The bromination [58] of porphyrin **9** using *N*-bromosuccinimide (NBS) at 0 °C yielded the dibromoporphyrin building block **10**. The next step entailed attachment of two perylene-ethyne units.

Several points warrant comment concerning the choice of ethynyl-perylene for coupling with the dibromoporphyrin. The chosen perylene (**11**) bears two aryloxy groups (one in each bay region) and a 2,6-diisopropylphenyl group at the *N*-imide site, which together impart structural features that enable good solubility of the perylene in hydrocarbon solvents [46]. The ethyne is located at the perylene 9-position, a site of considerable electron density in the frontier molecular orbitals and a site known to afford substantial electronic communication upon covalent attachment to the porphyrin [42,45–50]. The ethynyl-perylene-monoimide **11** is an advanced functional dye that has emerged from extensive studies over several decades beginning with the pioneering work of Langhals [59–65]. A key distinction between **11** and many other rylene dyes [66–68] is the presence of a single imide versus two imides. The presence of two imides causes the resulting dye to be a good photooxidant, whereas the mono-imide is more electroneutral. The synthesis of **11** [46] follows methods [39,69–71] established earlier using a 2,5-di-*tert*-butylphenyl group at the *N*-imide site. The 2,6-diisopropylphenyl group of **11** is superior in not giving rise to stereoisomers as occur with the 2,5-di-*tert*-butylphenyl group (as in **1**) [60,61,65].

The absorption spectra of ethynyl-perylene-monoimides bearing the 2,6-diisopropylphenyl group (**11**) or the 2,5-di-*tert*-butylphenyl (**11-tBu** [70,71]) are nearly identical, as shown in Figure 3. The similarity of spectra upon use of either solubilization motif is consistent with observation that the 2,6-diisopropyl *versus* 2,5-di-*tert*-butyl group has insignificant effects on perylene photophysics [40]. The similar spectra and photophysics are attributed to the presence of a node at the perylene-imide nitrogen atom in both the highest occupied molecular orbital (HOMO) and the lowest unoccupied molecular orbital (LUMO) [61]. Thus, the changes in porphyrin substituents from mesityl to *p*-tolyl groups, and perylene *N*-imide from a 2,5-di-*tert*-butylphenyl to a 2,6-diisopropylphenyl group, facilitate synthesis and chemical characterization but have hardly any effects on spectral and photophysical properties.

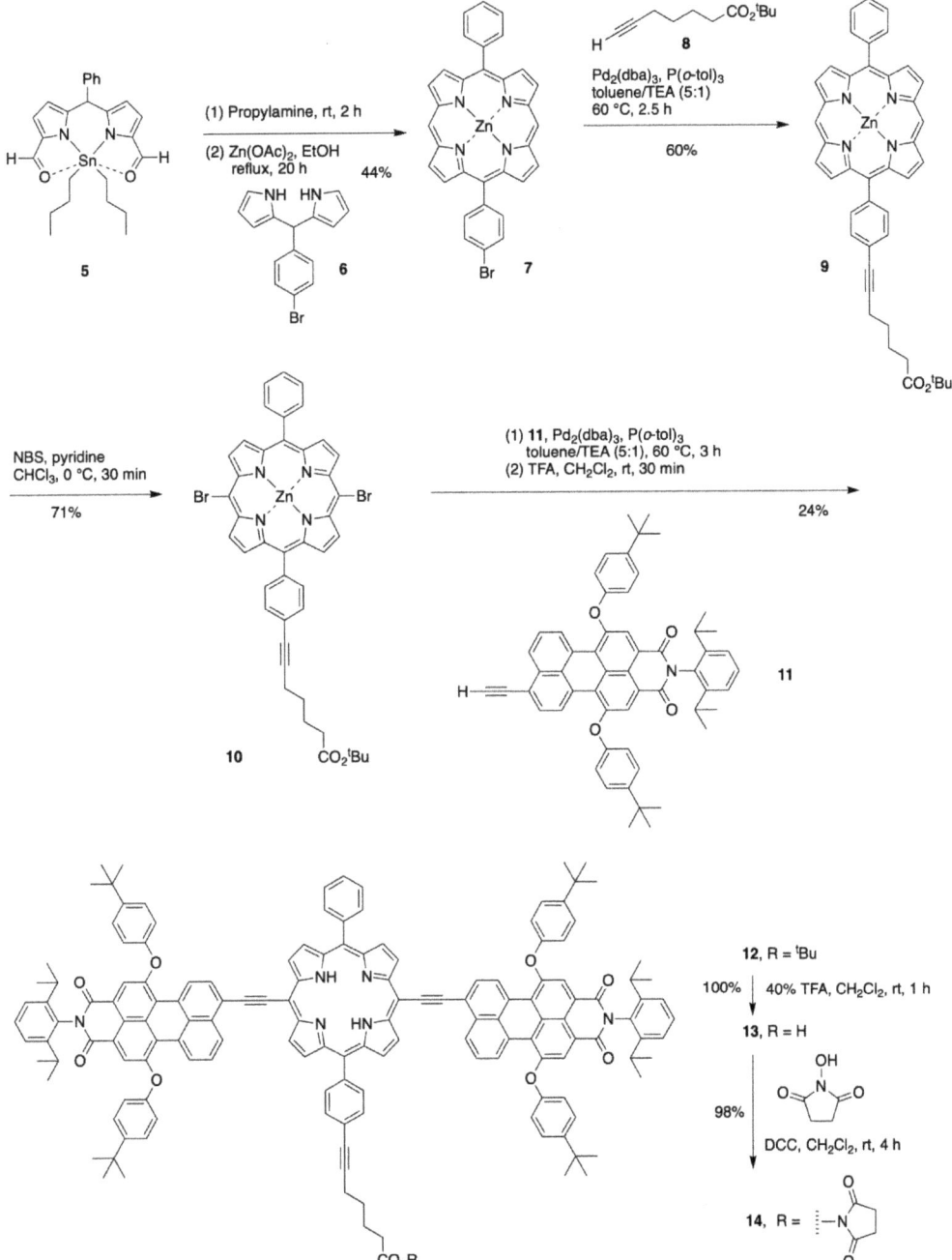

Scheme 1. Synthesis of the bioconjugatable panchromatic absorber.

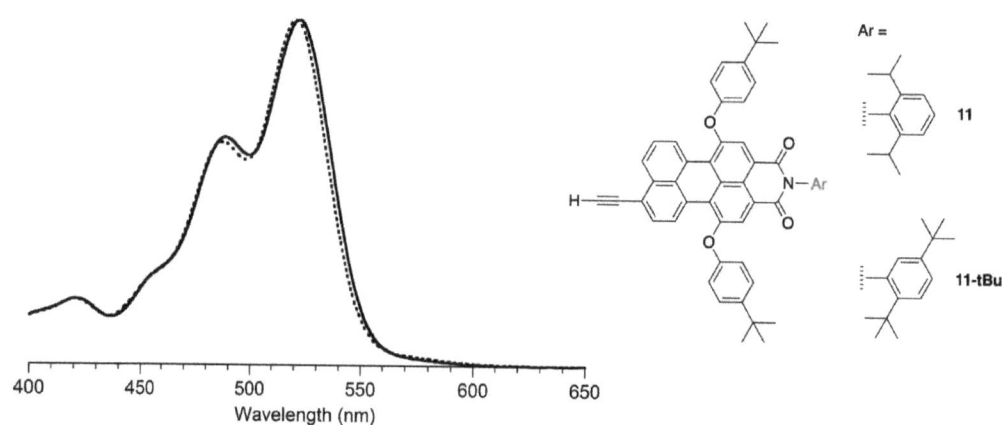

Figure 3. Absorption spectra (solid lines) in toluene at room temperature of ethynyl-perylene **11** (λ_{abs} 522 nm, $\varepsilon_{522\,nm}$ = 40,500 M^{-1}cm^{-1}, solid line) and ethynyl-perylene **11-tBu** (λ_{abs} 520 nm, dashed line).

The Sonogashira coupling [55] reaction of ethynyl-perylene **11** and dibromoporphyrin **10** was carried out under copper-free [56,57] Pd-mediated conditions (as in the reaction of **7** and **8**). Such a reaction of an ethynylphenyl-chromophore and a meso-bromoporphyrin is facile and has been carried out under nearly identical conditions to form porphyrin-phenylethynyl-porphyrin dyads [44]. The rationale for copper-free conditions in the synthesis was to avoid any transmetalation of the zinc porphyrin given (1) the avidity of porphyrins for Cu(II); (2) the very short-lived singdoublet excited-state lifetime of copper porphyrins (marked by the absence of fluorescence); and (3) the subsequent difficulty of removing copper from porphyrins, unlike the facile removal of zinc achieved upon treatment with mild acid. Thus, the subsequent acid-mediated demetalation of zinc [42] with trifluoroacetic acid (TFA) afforded the target panchromatic triad **12**. Purification was achieved via a three-chromatography sequence that includes use of size-exclusion chromatography (SEC) [42,72]. Removal of the *tert*-butyl protecting group using 40% TFA [73] in CH$_2$Cl$_2$ gave the carboxy-triad **13** (Scheme 1). For bioconjugation purposes, **13** was further transformed to an *N*-hydroxysuccinimide ester **14** via reaction [74] with *N*-hydroxysuccinimide in the presence of *N,N*-dicyclohexylcarbodiimide (DCC).

2.2. Synthesis of a Panchromatic Triad for Surface Attachment

We previously prepared a set of tetrapyrrole macrocycles (**15–17**) bearing an isophthalic acid tether [75] for surface attachment (Chart 2) [76]. The present work extends this design motif.

The synthesis of the core porphyrin follows that shown for the bioconjugatable triad. Thus, the 1,9-formyldipyrromethane **18** was reacted with propylamine [51] to form the bis(imine), which was then treated with the complementary dipyrromethane **19** [77] to afford the zinc porphyrin **20** in 38% yield (Scheme 2). The bromination [58] of zinc porphyrin **20** afforded the corresponding dibromo zinc porphyrin **21** in 75% yield. Sonogashira coupling [55] of the dibromo zinc porphyrin **21** and ethynyl-perylene **11** under copper-free conditions [56,57] afforded the triad **22** bearing two perylenes and one zinc porphyrin in 90% yield (~200 mg). The reaction progress was monitored by analytical SEC [72], as has been done previously with other panchromatic arrays [42,46]. The analytical SEC traces with absorption spectral determination show the starting materials (Figure 4, panel a), crude mixture after reaction for several hours (panel b), and the purified triad **22** (panel c) following preparative purification using the three-column chromatography process. The cleavage of the trimethylsilyl (TMS) group [48] of **22** afforded triad **23** bearing a free ethynyl group in 93% yield.

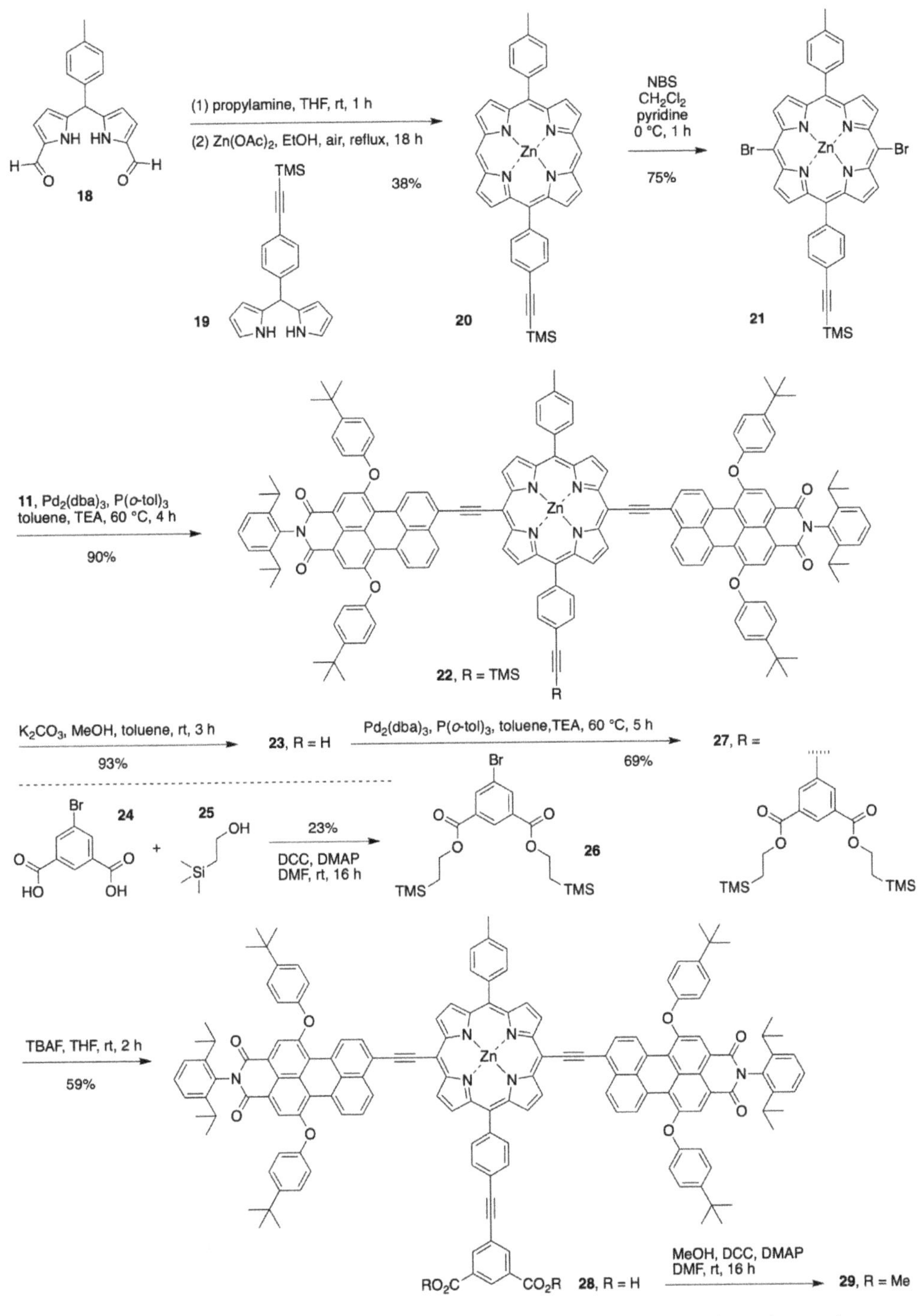

Scheme 2. Synthesis of a perylene-porphyrin-perylene triad (**28**) for surface attachment.

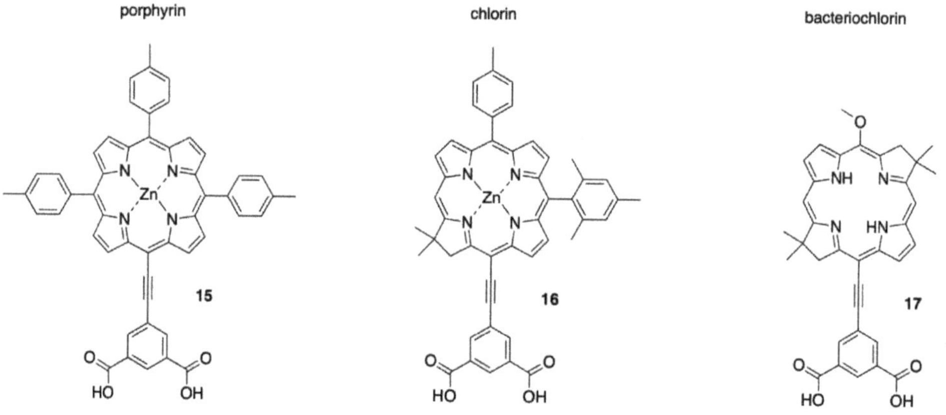

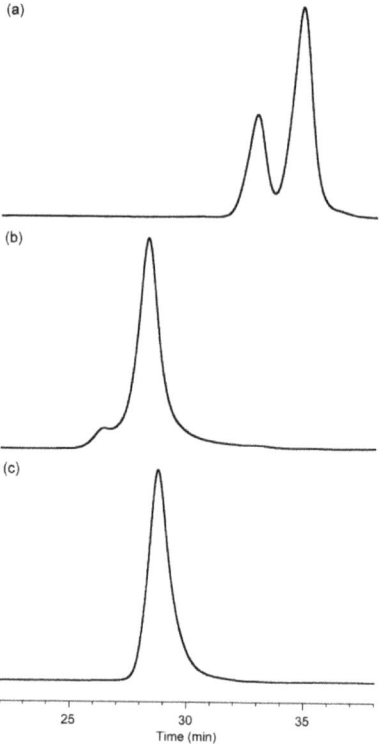

Chart 2. Tetrapyrrole dyes bearing an isophthalate tether.

Figure 4. Analytical SEC traces (λ_{det} = 550 nm) for the synthesis of a triad array **22**. (**a**): Mixture of starting materials **11** and **21**. (**b**): Crude mixture after reaction for 4 h. (**c**): Purified triad **22**.

The surface-attachment motif was prepared through the DCC-mediated condensation of 5-bromoisophthalic acid (**24**) and 2-(trimethylsilyl)ethanol (**25**) in N,N-dimethylformamide (DMF) to give the protected 5-bromo-isophthalate **26** in 23% yield (Scheme 2). The Sonogashira coupling [55] of ethynyl-triad **23** and isophthalate **26** under copper-free conditions [56,57] afforded the protected tethered triad **27** in 69% yield. The removal of the

2-trimethylsilylethyl group [76] of **27** upon treatment with tetrabutylammonium fluoride (TBAF) in tetrahydrofuran (THF) afforded isophthalate-tethered triad **28** in 59% yield. While protected triad **27** was fully characterized, all efforts to characterize **28** by NMR spectroscopy and mass spectrometry were unsuccessful. For confirmation purposes, a small portion of **28** was treated with methanol in the presence of DCC and and 4-dimethylaminopyridine (DMAP) to form the corresponding dimethyl ester **29**, which gave the expected mass peak upon matrix-assisted laser-desorption ionization mass spectrometry (MALDI-MS) analysis.

A tethered porphyrin analogue lacking the two perylene-ethynyl groups was prepared. The cleavage of the TMS group [48] of zinc porphyrin **20** afforded zinc porphyrin **30** bearing an ethynyl group in 91% yield (Scheme 3). The Sonogashira coupling [55] of **30** and isophthalate **26** under copper-free conditions [56,57] afforded the protected tethered porphyrin **31** in 45% yield. The removal of the 2-trimethylsilylethyl groups [76] of **31** upon treatment with TBAF afforded isophthalate-tethered porphyrin **32** in 78% yield.

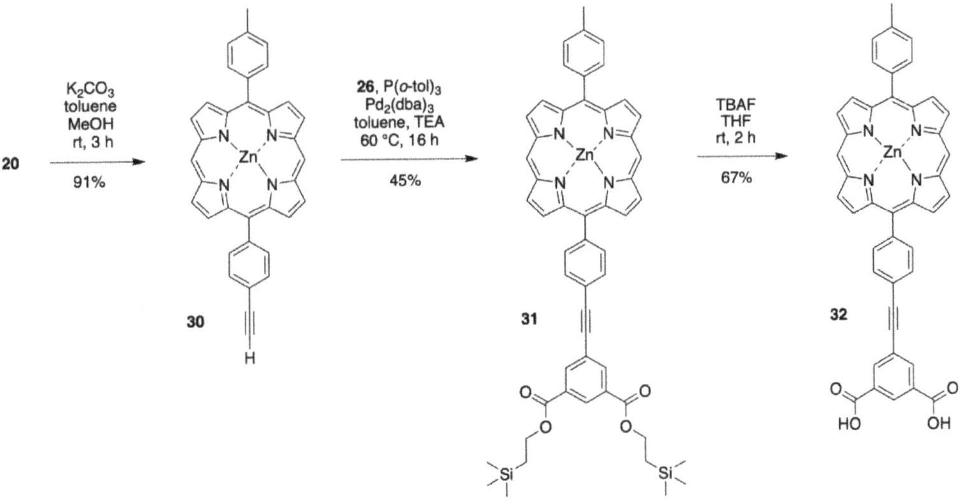

Scheme 3. Synthesis of tethered porphyrin **32** lacking perylenes.

2.3. Chemical Characterization

The triads were generally characterized by ^{1}H NMR spectroscopy, absorption spectroscopy, and MALDI-MS analysis. Limited solubility precluded the collection of ^{13}C NMR spectra for a number of the triads as well as other compounds. Accurate mass data were obtained by electrospray ionization mass spectrometry (ESI-MS) where possible. The absorption and fluorescence spectra of two tethered triads are shown in Figure 5. The carboxy-triad **13** contains a free base porphyrin, whereas the isophthalate-triad **27** contains a zinc porphyrin. The absorption spectrum of carboxy-triad **13** (panel a) is nearly identical to that of the untethered triad **4** shown in Figure 2. The absorption spectrum of isophthalate-triad **27** (panel b) is nearly identical to that of the untethered triad **Zn4** shown in Figure 2. For comparison purposes, the absorption and fluorescence spectra of a benchmark perylene (**33**) [46] and the *trans*-AB zinc porphyrin **31** are shown in panels c and d, respectively. Perylene **33** includes a phenyl group at the terminus of the ethyne and a 2,6-diisopropylphenyl substituent at the *N*-imide position (Chart 3) and is displayed correctly in the original report of synthesis and characterization [46], but is shown incorrectly with the 2,5-di-*tert*-butylphenyl substituent at the *N*-imide position in a subsequent report [50]. Molar absorption coefficient values are reported for triads **13**, **23**, **27**, and **28**, as well as for benchmark ethynyl-perylene-monoimide **11**. The Φ_f values measured in toluene for triad

13, triad **27**, perylene **33**, and porphyrin **31** are 0.21, 0.24, 0.94 [50], and 0.011, respectively. For comparison, the Φ_f values for free base porphyrin triad **4** and zinc porphyrin triad **Zn4** are 0.26 and 0.30, respectively [45]. The fluorescence emission of each triad exhibits (as expected) the general spectral features of a porphyrin but with an enhanced quantum yield. ^{1}H and ^{13}C{^{1}H} NMR spectra, mass spectra, and absorption spectra (where available) for new compounds are provided in the Supplementary Materials.

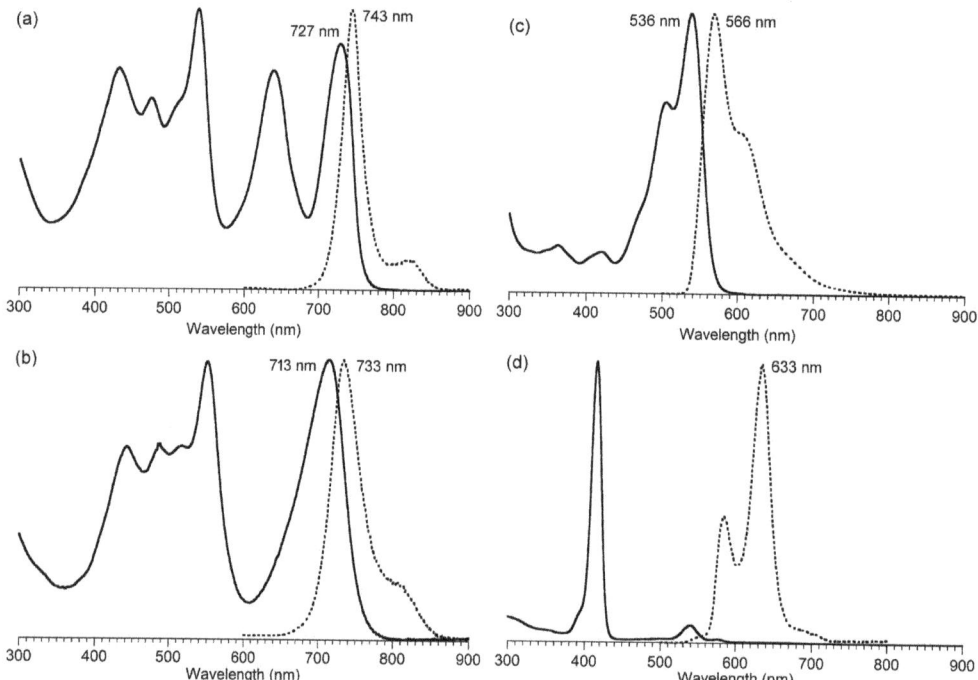

Figure 5. Absorption spectra (solid lines) and emission spectra (dashed lines) in toluene at room temperature (normalized displays). (**a**): free base porphyrin carboxy-triad **13** shows λ_{abs} 537 nm, $\varepsilon_{537\,nm}$ = 67,800 M^{-1}cm^{-1}; fluorescence λ_{em} 743 nm upon λ_{ex} = 521 nm). (**b**): zinc porphyrin isophthalate-triad **27** shows λ_{abs} 548 nm, $\varepsilon_{548\,nm}$ = 99,800 M^{-1}cm^{-1}; fluorescence λ_{em} 733 nm upon λ_{ex} = 549 nm). (**c**): perylene **33** (fluorescence λ_{em} 566 nm upon λ_{ex} ~500 nm). (**d**): *trans*-AB zinc porphyrin **31** (fluorescence λ_{em} 633 nm upon λ_{ex} = 415 nm).

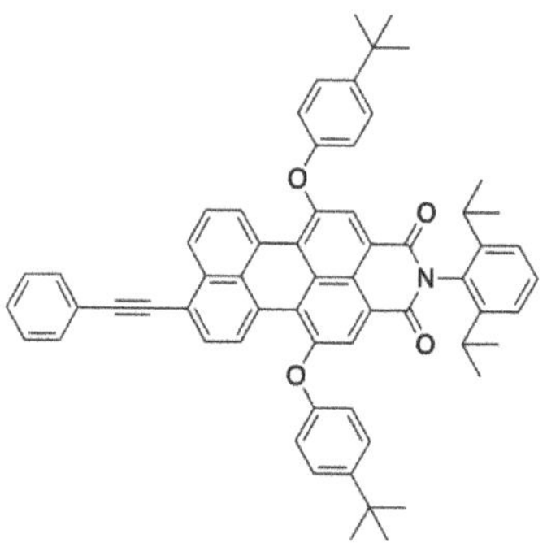

Chart 3. Benchmark perylene.

3. Materials and Methods

3.1. General Methods

All chemicals obtained commercially were used as received unless noted otherwise. Reagent-grade solvents (CH_2Cl_2, hexanes, methanol, toluene, ethyl acetate) and HPLC-grade solvents (toluene, CH_2Cl_2, hexanes) were used as received. THF was freshly distilled from sodium/benzophenone ketyl and used immediately. MALDI-MS was performed with the matrix 1,4-bis(5-phenyl-2-oxazol-2-yl)benzene (POPOP) [78] or α-cyano-4-hydroxycinnamic acid (α-CHCA) as noted. ESI-MS data were obtained in positive-ion mode unless noted otherwise. Known building blocks **5** [52], **6** [53], **8** [54], **11** [46], **18** [51], and **19** [77] were prepared via reported methods.

3.2. Purification Following Sonogashira Coupling Reactions

Following a three-chromatography procedure [42,70], reaction mixtures of arrays were first chromatographed with adsorption column chromatography (flash silica, Baker) to remove catalysts and ligands from the coupling reaction. Then, preparative-scale size exclusion chromatography (SEC) was performed using BioRad Bio-Beads S-X1. A preparative-scale glass column (4.3 × 53 cm) was packed using BioRad Bio-Beads S-X1 in HPLC grade toluene. The chromatography was performed with gravity flow (~0.2 mL/min). Thereafter, a subsequent adsorption column chromatography (flash silica, Baker) procedure was performed (with HPLC-grade CH_2Cl_2 and hexanes unless noted otherwise) to remove material that may have leached from the SEC resin.

The preparative purification procedure is generally most effective when the reaction affords a change in size; e.g., in instances where the product is substantially larger than the starting materials. Such is the case of forming triads via Sonogashira coupling procedures as described herein. The purification protocol is applicable to both zinc and free base porphyrins. Here, all Sonogashira coupling reactions were carried out under anaerobic, copper-free conditions [56,57] using zinc porphyrins in relatively dilute solution, as required for homogeneous solubilization of each porphyrin reactant.

Analytical-scale SEC was performed to monitor the purification of arrays [42,56]. Analytical SEC columns (styrene-divinylbenzene copolymer) were purchased from Polymer Laboratories. Analytical SEC was performed with a Hewlett-Packard 1100 HPLC using PLgel 100 Å, Plgel 500 Å, and PLgel 1000 Å columns (each ~30-cm in length) in series, eluting with toluene (flow rate = 1.0 mL/min). Sample detection was achieved by absorption spectroscopy using a diode array detector with quantitation at 422, 521, 638, and 726 nm (±8 nm band width), which best captures the peaks of the arrays. In other cases, analytical SEC was performed using PLgel 50 Å, PLgel 100 Å × 2, and PLgel 500 Å columns (each ~30-cm in length) in series, eluting with THF (flow rate = 0.8 mL/min) at room temperature. Sample detection was achieved by absorption spectroscopy using a diode array detector with quantitation at 440, 488, 515, 550, and 710 nm.

3.3. Synthesis and Characterization

Zinc(II)-5-(4-Bromophenyl)-15-phenylporphyrin (**7**). Following a general procedure [51], a solution of **5** (126 mg, 247 µmol) in propylamine (0.4 mL, 5 mmol) was stirred at room temperature for 2 h. Then the mixture was concentrated and dried at high vacuum for 5 min. The resulting solid was dissolved in ethanol (30 mL) and then treated with **6** (75.0 mg, 250 µmol) and zinc acetate (0.52 g, 25 mmol). The mixture was refluxed at 90 °C open to the air for 20 h. The mixture was allowed to cool to room temperature and then was concentrated. Column chromatography [silica, hexanes/CH_2Cl_2 (1:2) to CH_2Cl_2] afforded a pink solid (67 mg, 45%): ^{1}H NMR (THF-d_8, 400 MHz) δ 10.29 (s, 2H), 9.44–9.41 (m, 4H), 9.04–9.02 (m, 4H), 8.26–8.24 (m, 2H), 8.18–8.16 (m, 2H), 7.98–7.96 (m, 2H), 7.80–7.78 (m, 3H); ^{13}C{^{1}H} NMR (THF-d_8, 100 MHz) δ 150.7, 149.8, 136.4, 134.8, 132.0, 131.7, 131.5, 129.7, 127.3, 126.5, 105.8; MALDI-MS (POPOP) obsd 603.8, calcd 603.0 [M + H]$^+$; ESI-MS obsd 602.0089, calcd 602.0079 (M$^+$), M = $C_{32}H_{19}BrN_4Zn$; λ_{abs} (toluene) 413, 538, 573 nm.

Zinc(II)-5-[4-(7-tert-Butoxy-7-oxohept-1-ynyl)phenyl]-15-phenylporphyrin (**9**). Following a general procedure [56,57], a mixture of toluene/triethylamine (5:1, *v*:*v*) was deaerated with a continuous steam of argon for 1 h. A mixture of zinc porphyrin **7** (22 mg, 36 µmol), P(*o*-tol)$_3$ (14 mg, 46 µmol), and Pd$_2$(dba)$_3$ (5 mg, 6 µmol) was placed into a 100 mL Schlenk flask and evacuated under high vacuum for 10 min. The flask was refilled with argon thereafter, and the procedure was repeated three times. Then the degassed solvent (10 mL) was added to the flask, whereupon three freeze–pump–thaw cycles were performed. The mixture was allowed to warm to temperature. Ethynyl coupling partner **8** (19 mg, 0.10 mmol) was added dropwise to the mixture under a continuous stream of argon. The mixture was then stirred for 3 h at 60 °C. The mixture was allowed to cool to room temperature and then was quenched by the addition of water. The mixture was extracted with CH$_2$Cl$_2$. The organic layer was washed with water, dried (Na$_2$SO$_4$), concentrated and chromatographed [silica, hexanes/CH$_2$Cl$_2$ (1:2) to CH$_2$Cl$_2$] to afford a dark purple solid (47 mg, 60%): ^{1}H NMR (THF-d_8, 400 MHz) δ 10.46 (s, 2H), 9.61–9.58 (m, 4H), 9.24–9.21 (m, 4H), 8.45–8.37 (m, 4H), 8.01–7.96 (m, 5H), 2.81 (t, *J* = 6.8 Hz, 2 H), 2.55 (t, *J* = 6.8 Hz, 2H), 2.08 (q, *J* = 7.2 Hz, 2H), 2.00–1.95 (m, 2H), 1.69 (s, 9H); ^{13}C{^{1}H} NMR (THF-d_8, 100 MHz) δ 187.5, 172.0, 150.1, 149.9, 149.8, 149.7, 143.6, 143.0, 142.4, 135.6, 134.9, 134.8, 131.9, 131.7, 131.6, 131.5, 130.2, 129.7, 128.9, 128.4, 127.3, 126.5, 125.8, 123.5, 119.7, 118.8, 105.8, 90.6, 81.1, 79.3, 34.8, 28.6, 19.2; MALDI-MS (POPOP) obsd 705.2, calcd 705.2 [M + H]$^+$; ESI-MS obsd 704.2134, calcd 704.2124 (M$^+$), M = C$_{43}$H$_{36}$N$_4$O$_2$Zn; λ_{abs} (toluene) 413, 539, 573 nm.

Zinc(II)-10,20-Dibromo-5-[4-(7-tert-butoxy-7-oxohept-1-ynyl)phenyl]-15-phenylporphyrin (**10**). Following a general procedure [58], a solution of **9** (47 mg, 67 µmol) in CHCl$_3$ (23 mL) was stirred at 0 °C in ice bath, then treated with NBS (38 mg, 21 µmol). The mixture was stirred at 0 °C for 30 min, whereupon acetone was added to quench the reaction. Then the mixture was washed with water and extracted with CH$_2$Cl$_2$. The organic extract was dried (Na$_2$SO$_4$), concentrated, and chromatographed [silica, CH$_2$Cl$_2$] to afford a green solid (41 mg, 71%): ^{1}H NMR (THF-d_8, 400 MHz) δ 9.69–9.67 (m, 4H), 8.90–8.87 (m, 4H), 8.19 (d, *J* = 6.4 Hz, 2H), 8.13 (d, *J* = 8.0 Hz, 2H), 7.83–7.79 (m, 5H), 2.64 (t, *J* = 6.8 Hz, 2H), 2.38 (t, *J* = 8.0 Hz, 2H), 1.91 (p, *J* = 6.8 Hz, 2H), 1.82 (p, *J* = 7.2 Hz, 2H), 1.52 (s, 9H); ^{13}C{^{1}H} NMR (THF-d_8, 100 MHz) δ 172.0, 151.0, 150.7, 150.2, 142.9, 142.2, 134.7, 134.6, 133.2, 133.0, 132.9, 132.7, 129.6, 127.7, 126.6, 123.9, 122.3, 121.5, 104.6, 90.9, 80.9, 79.3, 34.8, 28.6, 27.6, 19.2; ESI-MS obsd 860.0335, calcd 860.03245 (M$^+$), M = C$_{43}$H$_{34}$N$_4$O$_2$Br$_2$Zn; λ_{abs} (toluene) 429, 539, 598 nm.

5-[4-(7-tert-Butoxy-7-oxohept-1-ynyl)phenyl]-10,20-bis[2-(3,4-(N-(2,6-diisopropylphenyl) iminodicarbonyl)-1,6-bis(4-tert-butylphenoxy)perylen-9-yl)ethynyl]-15-phenylporphyrin (**12**). Following a standard procedure [44,46,56,57], a mixture of zinc dibromoporphyrin **10** (3.0 mg, 3.5 µmol), ethynyl-perylene **11** (6.0 mg, 7.5 µmol), P(*o*-tol)$_3$ (2.2 mg, 7.2 µmol), and Pd$_2$(dba)$_3$ (0.8 mg, 1.0 µmol) in degassed toluene/triethylamine (2.2 mL, 5:1, *v*:*v*) was stirred at 60 °C for 3 h. The mixture was allowed to cool at room temperature. The mixture was then washed with water and extracted with CH$_2$Cl$_2$. The organic extract was dried (Na$_2$SO$_4$) and concentrated. The resulting solid was dissolved in CH$_2$Cl$_2$ (2.0 mL) and treated with TFA (14 µL). The mixture was stirred at room temperature for 1 h, whereupon excess triethylamine was added to quench the reaction. The solution was then washed with water, dried (Na$_2$SO$_4$), concentrated, and chromatographed using the standard three-chromatography procedure to afford a black solid (2.3 mg, 29%): ^{1}H NMR (CDCl$_3$, 300 MHz) δ 9.78–9.73 (m, 4H), 9.31 (d, *J* = 7.8 Hz, 2H), 9.20 (d, *J* = 8.1 Hz, 2H), 9.06 (d, *J* = 8.1 Hz, 2H), 8.89–8.83 (m, 4H), 8.34 (s, 2H), 8.26–8.04 (m, 8H), 7.91 (d, *J* = 8.1 Hz, 2H), 7.84–7.80 (m, 3H), 7.74–7.68 (m, 2H), 7.46–7.39 (m, 10H), 7.32 (d, *J* = 7.8 Hz, 4H), 7.08 (d, *J* = 9.0 Hz, 4H), 6.98–6.95 (m, 4H), 2.77 (q, *J* = 6.6 Hz, 4H), 2.63 (t, *J* = 6.8 Hz, 2H), 2.39 (t, *J* = 7.2 Hz, 2H), 1.92 (q, *J* = 7.2 Hz, 2H), 1.83 (q, *J* = 6.6 Hz, 2H), 1.52 (s, 9H), 1.34 (s, 36H), 1.18 (d, *J* = 6.6 Hz, 24H), −2.01 (br, 2H); MALDI-MS (POPOP) obsd 2242.7, calc 2241.0 [M + H]$^+$, M = C$_{155}$H$_{136}$N$_6$O$_{10}$; λ_{abs} (toluene) 431, 476, 537, 638, 727 nm.

5-[4-(7-Hydroxy-7-oxohept-1-ynyl)phenyl]-10,20-bis[2-(3,4-(N-(2,6-diisopropylphenyl)iminodicarbonyl) -1,6-bis(4-tert-butylphenoxy)perylen-9-yl)ethynyl]-15-phenylporphyrin (**13**). A solution of the *tert*-butyl protected triad **12** (2.3 mg, 1.0 µmol) in CH$_2$Cl$_2$ (1.2 mL) was treated with TFA (0.8 mL). The mixture was stirred at room temperature for 1 h, whereupon excess triethylamine was added to quench the reaction. The mixture was then washed with water, dried (Na$_2$SO$_4$), and concentrated to afford a black solid (2.2 mg, 100%): ^{1}H NMR (CDCl$_3$, 300 MHz) δ 9.73–9.66 (m, 4H), 9.24 (d, *J* = 8.1 Hz, 2H), 9.18 (d, *J* = 8.4 Hz, 2H), 8.99 (d, *J* = 8.4 Hz, 2H), 8.83–8.78 (m, 4H), 8.32–8.28 (m, 2H), 8.16–8.07 (m, 8H), 7.96 (d, *J* = 8.4 Hz, 2H), 7.83–7.76 (m, 3H), 7.66–7.61 (m, 2H), 7.43–7.38 (m, 10H), 7.31 (d, *J* = 7.5 Hz, 4H), 7.06 (d, *J* = 9.0 Hz, 4H), 6.99 (d, *J* = 9.0 Hz, 4H), 2.77 (q, *J* = 6.6 Hz, 4H), 2.65 (t, *J* = 6.6 Hz, 2H), 2.41 (t, *J* = 7.5 Hz, 2H), 1.86 (q, *J* = 7.2 Hz, 2H), 1.83 (q, *J* = 6.9 Hz, 2H), 1.33 (s, 36H), 1.17 (d, *J* = 7.2 Hz, 24H), −1.97 (br, 2H), a signal due to the carboxylic acid was not observed; MALDI-MS (POPOP) obsd 2187.9, calc 2185.0 (M$^+$), M = C$_{151}$H$_{128}$N$_6$O$_{10}$; λ$_{abs}$ (toluene) 430, 476, 537, 638, 727 nm, ε$_{537\ nm}$ = 67,800 M^{-1}cm^{-1}, ε$_{727\ nm}$ = 56,100 M^{-1}cm^{-1}; λ$_{em}$ (toluene, λ$_{ex}$ = 521 nm) 743 nm.

5-[4-(7-(N-succinimidooxy)-7-oxohept-1-ynyl)phenyl]-10,20-bis[2-(3,4-(N-(2,6-diisopropylphenyl) iminodicarbonyl)-1,6-bis(4-tert-butylphenoxy)perylen-9-yl)ethynyl]-15-phenylporphyrin (**14**). Following a general procedure [74], a solution of carboxy-triad **13** (5.5 mg, 2.5 µmol), *N*-hydroxysuccinimide (2.8 mg, 25 µmol), and DCC (5.0 mg, 25 µmol) in CH$_2$Cl$_2$ (0.25 mL) was stirred under argon at room temperature for 4 h. Then the mixture was washed with water, dried (Na$_2$SO$_4$), concentrated, and chromatographed [silica, CH$_2$Cl$_2$] to afford a black solid (5.6 mg, 98%): ^{1}H NMR (CDCl$_3$, 300 MHz) δ 9.78 (d, *J* = 4.5 Hz, 4H), 9.47–9.42 (m, 4H), 9.10 (d, *J* = 8.4, 2H), 8.89–8.87 (m, 4H), 8.36 (d, *J* = 4.2 Hz, 4H), 8.31 (d, *J* = 8.4 Hz, 2H), 8.13 (d, *J* = 8.1 Hz, 2H), 7.86–7.80 (m, 7H), 7.46–7.43 (m, 10H), 7.31 (d, *J* = 7.8 Hz, 4H), 7.14–1.09 (m, 8H), 2.87 (s, 4H), 2.84–2.71 (m, 6H), 2.68 (t, *J* = 6.6 Hz, 2H), 2.17–2.06 (m, 2H), 1.95–1.85 (m, 4H), 1.35 (s, 36H), 1.17 (d, *J* = 6.6 Hz, 24H), −1.78 (br, 2H); MALDI-MS (POPOP) obsd 2283.5, calc 2283.0 (M$^+$), M = C$_{155}$H$_{131}$N$_7$O$_{12}$.

Zinc(II) 15-p-Tolyl-5-[4-(2-(trimethylsilyl)ethynyl)phenyl]porphyrin (**20**). Following a reported method [51], a solution of diformyldipyrromethane **18** (438 mg, 1.5 mmol) in THF (5 mL) was treated with propylamine (1.8 g, 30 mmol) at room temperature for 1 h. The solution was concentrated to dryness. The resulting solid was dissolved in ethanol (150 mL) and treated with dipyrromethane **19** (480 mg, 1.5 mmol) and Zn(OAc)$_2$ (2.8 g, 15 mmol). The mixture was refluxed open to the air for 18 h. The reaction mixture was allowed to cool to room temperature, then concentrated to dryness and purified by chromatography [silica, hexanes/CH$_2$Cl$_2$ (4:1 to 1:1)] to afford a purple solid (351 mg, 38%): ^{1}H NMR (600 MHz, CDCl$_3$) δ 10.28 (s, 2H), 9.41 (d, *J* = 4.4 Hz, 4H), 9.16 (d, *J* = 4.4 Hz, 2H), 9.08 (d, *J* = 4.4 Hz, 2H), 8.19 (d, *J* = 7.7 Hz, 2H), 8.14 (d, *J* = 7.4 Hz, 2H), 7.91 (d, *J* = 7.7 Hz, 2H), 7.61 (d, *J* = 7.4 Hz, 2H), 2.75 (s, 3H), 0.41 (s, 9H); ^{13}C{^{1}H} NMR (151 MHz, CDCl$_3$) δ 150.2, 149.6, 149.4, 149.3, 142.9, 137.1, 134.5, 134.4, 132.6, 132.0, 131.8, 131.6, 130.2, 127.3, 122.2, 120.4, 118.9, 106.2, 21.4; MALDI-MS (α-CHCA) obsd 634.2, calcd 634.2 (M$^+$), M = C$_{38}$H$_{30}$N$_4$SiZn; λ$_{abs}$ (toluene) 414, 540 nm.

Zinc(II) 5,15-Dibromo-20-p-tolyl-10-[4-(2-(trimethylsilyl)ethynyl)phenyl]porphyrin (**21**). Following a standard bromination method [58], a solution of zinc porphyrin **20** (127 mg, 0.20 mmol) in CH$_2$Cl$_2$ (70 mL) containing pyridine (1.4 mL) was treated with NBS (85 mg, 0.48 mmol) at 0 °C for 1 h. The reaction mixture was quenched by the addition of acetone (3.0 mL). The mixture was washed with water (80 mL) and brine (80 mL), dried, concentrated to dryness, and purified by chromatography [silica, hexanes/CH$_2$Cl$_2$ (10:1 to 3:1)] to afford a greenish purple solid (119 mg, 75%): ^{1}H NMR (600 MHz, CDCl$_3$) δ 9.68–9.62 (m, 4H), 8.89 (d, *J* = 4.6 Hz, 2H), 8.82 (d, *J* = 4.6 Hz, 2H), 8.09 (d, *J* = 7.9 Hz, 2H), 8.02 (d, *J* = 7.5 Hz, 2H), 7.86 (d, *J* = 7.9 Hz, 2H), 7.55 (d, *J* = 7.4 Hz, 2H), 2.72 (s, 3H), 0.39 (s, 9H); MALDI-MS (α-CHCA) obsd 790.2, calcd 790.0 (M$^+$), M = C$_{38}$H$_{28}$Br$_2$N$_4$SiZn; λ$_{abs}$ (toluene) 432, 560, 600 nm.

Zinc(II) 10,20-Bis[2-(3,4-(N-(2,6-diisopropylphenyl)iminodicarbonyl)-1,6-bis(4-tert-butylphenoxy) perylen-9-yl)ethynyl]-5-p-tolyl-15-(4-(2-trimethylsilylethynyl)phenyl)porphyrin (**22**). Following a standard procedure [44,46,56,57], a Schlenk flask containing zinc porphyrin **21** (79 mg, 0.10 mmol), ethynyl-perylene **11** (192 mg, 0.24 mmol), and tri(*o*-tolyl)phosphine (73 mg,

0.24 mmol) was flushed with argon, treated with degassed toluene/TEA (50 mL, 5:1, *v*:*v*), and subjected to three freeze–pump–thaw cycles. The mixture was then treated with Pd$_2$(dba)$_3$ (27 mg, 0.030 μmol) and subjected to two additional freeze–pump–thaw cycles. The resulting mixture was heated to 60 °C for 4 h. The crude mixture was purified following the standard three-chromatography procedure [silica (hexanes/CH$_2$Cl$_2$ (2:1 to 1:1)), SEC (toluene), silica (hexanes to hexanes/CH$_2$Cl$_2$ (1:1))] to afford a purple solid (161 mg, 72%): ^{1}H NMR (600 MHz, CDCl$_3$) δ 9.80–9.69 (m, 4H), 9.11–8.98 (m, 4H), 8.87 (d, J = 4.2 Hz, 2H), 8.81 (d, J = 4.3 Hz, 2H), 8.58 (s, 2H), 8.26 (s, 4H), 8.06–7.99 (m, 2H), 7.99–7.93 (m, 2H), 7.91–7.85 (m, 4H), 7.82 (br, 2H), 7.58 (d, J = 7.2 Hz, 2H), 7.47–7.38 (m, 8H), 7.31 (d, J = 8.1 Hz, 4H), 7.27 (s, 2H), 6.97 (d, J = 8.3 Hz, 4H), 6.49 (s, 4H), 2.82–2.72 (m, 7H), 1.35 (s, 18H), 1.30 (s, 18H), 1.23–1.13 (m, 24H), 0.43 (s, 9H); MALDI-MS (α-CHCA) obsd 2232.7, calcd 2232.9 (M$^+$), M = C$_{150}$H$_{128}$N$_6$O$_8$SiZn; λ$_{abs}$ (toluene) 440, 488, 515, 549, 706 nm.

Zinc(II) 10,20-Bis[2-(3,4-(N-(2,6-diisopropylphenyl)iminodicarbonyl)-1,6-bis(4-tert-butylphenoxy)perylen-9-yl)ethynyl]-5-p-tolyl-15-(4-ethynylphenyl)porphyrin (**23**). A solution of triad **22** (197 mg, 88 μmol) in toluene (75 mL) and methanol (75 mL) was treated with K$_2$CO$_3$ (1.21 g, 8.8 mmol) at room temperature for 3 h. The reaction mixture was poured into water (300 mL) and extracted with CH$_2$Cl$_2$ (200 mL × 2). The combined organic extract was washed with brine (200 mL), dried, and concentrated to dryness to afford a dark red solid (178 mg, 93%): ^{1}H NMR (600 MHz, CDCl$_3$) δ 9.81 (d, J = 4.4 Hz, 4H), 9.24 (s, 2H), 9.12 (d, J = 8.0 Hz, 2H), 8.96 (d, J = 4.4 Hz, 2H), 8.88 (d, J = 4.4 Hz, 2H), 8.32 (s, 4H), 8.16 (d, J = 8.0 Hz, 2H), 8.06 (d, J = 7.5 Hz, 2H), 8.03 (s, 2H), 7.99 (d, J = 7.2 Hz, 2H), 7.93 (d, J = 7.6 Hz, 2H), 7.65–7.59 (m, 4H), 7.46–7.42 (m, 6H), 7.34 (d, J = 8.6 Hz, 4H), 7.31 (d, J = 8.0 Hz, 4H), 7.05 (d, J = 8.5 Hz, 4H), 6.73 (s, 4H), 3.37 (s, 1H), 2.80–2.74 (m, 7H), 1.35 (s, 18H), 1.32 (s, 18H), 1.21–1.13 (m, 24H); MALDI-MS (α-CHCA) obsd 2160.9 (M$^+$), calcd 2160.8 (M$^+$), M = C$_{147}$H$_{120}$N$_6$O$_8$Zn; λ$_{abs}$ (toluene) 440, 488, 515, 548, 706 nm, ε$_{548\,nm}$ = 103,000 M^{-1}cm^{-1}.

Bis(2-(trimethylsilyl)ethyl) 5-bromoisophthalate (**26**). A solution of 5-bromoisophthalic acid (**24**, 490 mg, 2.0 mmol) and 2-(trimethylsilyl)ethanol (**25**, 497 mg, 4.2 mmol) in DMF (12 mL) was treated with DCC (866 mg, 4.2 mmol) and DMAP (489 mg, 4.0 mmol) at room temperature for 16 h. The reaction mixture was diluted with ethyl acetate (80 mL) and filtered. The filtrate was washed with water (50 mL) and a saturated NH$_4$Cl aqueous solution (50 mL), dried, concentrated to dryness, and then purified by chromatography [silica, hexanes to hexanes/CH$_2$Cl$_2$ (10:1)] to afford a white amorphous solid (204 mg, 23%): ^{1}H NMR (600 MHz, CDCl$_3$) δ 8.59 (t, J = 1.5 Hz, 1H), 8.33 (d, J = 1.5 Hz, 2H), 4.47–4.42 (m, 4H), 1.18–1.13 (m, 4H), 0.09 (s, 18H); ^{13}C{^{1}H} NMR (151 MHz, CDCl$_3$) δ 166.2, 137.9, 134.3, 130.5, 123.9, 65.6, 18.9, 1.5; ESI-MS (negative-ion mode) obsd 442.92573, calcd 443.0715 [M − H]$^-$, M = C$_{18}$H$_{29}$BrO$_4$Si$_2$.

Zinc(II) 10,20-Bis[2-(3,4-(N-(2,6-diisopropylphenyl)iminodicarbonyl)-1,6-bis(4-tert-butylphenoxy)perylen-9-yl)ethynyl]-5-(4-(2-(3,5-bis((2-(trimethylsilyl)ethoxy)carbonyl)phenyl)ethynyl)phenyl)-15-p-tolylporphyrin (**27**). Following a standard procedure [56,57], a Schlenk flask containing zinc porphyrin triad **23** (51.9 mg, 24 μmol), bromoisophthalate **26** (16.0 mg, 36 μmol), and tri(*o*-tolyl)phosphine (11.0 mg, 36 μmol) was flushed with argon, treated with degassed toluene/TEA (12 mL, 5:1, *v*:*v*) and subjected to three freeze–pump–thaw cycles, and then treated with Pd$_2$(dba)$_3$ (6.6 mg, 7.2 μmol) and subjected to two additional freeze–pump–thaw cycles. The resulting mixture was heated to 60 °C for 5 h. The crude mixture was purified following the standard three-chromatography procedure [silica (hexanes/CH$_2$Cl$_2$ (2:1 to 1:1)), SEC (toluene), silica (hexanes to hexanes/CH$_2$Cl$_2$ (1:1))] to afford a dark red solid (42.0 mg, 69%): ^{1}H NMR (600 MHz, CDCl$_3$) δ 9.85–9.75 (m, 4H), 9.21–9.14 (m, 2H), 9.11 (d, J = 8.1 Hz, 2H), 8.99–8.94 (m, 2H), 8.91–8.87 (m, 2H), 8.67 (s, 2H), 8.51–8.43 (m, 3H), 8.30 (s, 4H), 8.16–8.11 (m, 2H), 8.10–8.06 (m, 2H), 8.03–7.99 (m, 2H), 7.98–7.94 (m, 2H), 7.61 (d, J = 7.4 Hz, 2H), 7.59–7.54 (m, 2H), 7.47–7.40 (m, 6H), 7.35–7.28 (m, 8H), 7.02 (d, J = 8.3 Hz, 4H), 6.69 (s, 4H), 4.53–4.46 (m, 4H), 2.82–2.72 (m, 7H), 1.34 (s, 18H), 1.31 (s, 18H), 1.22–1.12 (m, 28H), 0.14 (s, 18H); MALDI-MS (α-CHCA) obsd 2525.8, calcd 2525.0 (M$^+$), M = C$_{165}$H$_{148}$N$_6$O$_{12}$Si$_2$Zn; λ$_{abs}$ (toluene) 441, 486, 516, 550, 713 nm, ε$_{549\,nm}$ = 99,800 M^{-1}cm^{-1}; λ$_{em}$ (toluene, λ$_{exc}$ = 549 nm) 733 nm.

Zinc(II) 10,20-Bis[2-(3,4-(N-(2,6-diisopropylphenyl)iminodicarbonyl)-1,6-bis(4-tert-butylphenoxy) perylen-9-yl)ethynyl]-5-(4-(2-(3,5-dicarboxyphenyl)ethynyl)phenyl)-15-p-tolylporphyrin (**28**). Following a reported method [76], a solution of **27** (30 mg, 12 μmol) in THF (30 mL) was treated with a tetrabutylammonium fluoride solution (50 μL, 1.0 M in THF) at room temperature for 2 h. The reaction mixture was diluted with CH_2Cl_2 (100 mL) and then washed with water (100 mL × 2) and brine (100 mL). The organic extract was dried, concentrated to dryness, and purified by chromatography [silica, CH_2Cl_2 to CH_2Cl_2/MeOH (5:1)] to afford a purple solid (16 mg, 59%): λ_{abs} (toluene) 454, 515, 756 nm, $\varepsilon_{454\,nm}$ = 44,400 $M^{-1}cm^{-1}$ (broad spectrum).

10,20-Bis[2-(3,4-(N-(2,6-diisopropylphenyl)iminodicarbonyl)-1,6-bis(4-tert-butylphenoxy) perylen-9-yl)ethynyl]-5-(4-(2-(3,5-dimethoxycarbonylphenyl)ethynyl)phenyl)-15-p-tolylporphyrin (**29**). A solution of **28** (2.0 mg, 0.86 μmol) and methanol (100 μL) in DMF (900 μL) was treated with DCC (1.0 mg, 4.8 μmol) and DMAP (1 mg, 8 μmol) at room temperature for 16 h. The reaction mixture was diluted with CH_2Cl_2 (20 mL) and then filtered. The filtrate was washed with water (20 mL) and a saturated NH_4Cl aqueous solution (20 mL), dried, concentrated to dryness, and then dissolved in CH_2Cl_2 (500 μL). The resulting solution was used for MALDI-MS characterization without further purification: MALDI-MS (α-CHCA) obsd 2352.6, calcd 2352.9 (M^+), M = $C_{157}H_{128}N_6O_{12}Zn$.

Zinc(II) 5-(4-Ethynyl)phenyl-15-p-tolylporphyrin (**30**). A solution of porphyrin **20** (63.6 mg, 0.10 mmol) in toluene (30 mL) and methanol (30 mL) was treated with K_2CO_3 (1.38 g, 10 mmol) at room temperature for 3 h. The reaction mixture was poured into water (120 mL) and extracted with CH_2Cl_2 (100 mL × 2). The combined organic extract was washed with brine (100 mL), dried, and then concentrated to dryness to afford a purple solid (51.1 mg, 91%): ^{1}H NMR (600 MHz, $CDCl_3$) δ 10.32 (s, 2H), 9.46–9.41 (m, 4H), 9.18 (d, *J* = 4.4 Hz, 2H), 9.11 (d, *J* = 4.4 Hz, 2H), 8.23 (d, *J* = 7.8 Hz, 2H), 8.15 (d, *J* = 7.4 Hz, 2H), 7.94 (d, *J* = 7.7 Hz, 2H), 7.61 (d, *J* = 7.4 Hz, 2H), 3.34 (s, 1H), 2.75 (s, 3H); MALDI-MS (α-CHCA) obsd 562.2, calcd 562.1 (M^+), M = $C_{35}H_{22}N_4Zn$; λ_{abs} (toluene) 414, 539, 576 nm.

Zinc(II) 5-(4-(2-(3,5-Bis((2-(trimethylsilyl)ethoxy)carbonyl)phenyl)ethynyl)phenyl)-15-p-tolylporphyrin (**31**). Following a standard procedure [56,57], a Schlenk flask containing zinc porphyrin **30** (37.9 mg, 65 μmol), bromoisophthalate **26** (44.0 mg, 98 μmol), and tri(*o*-tolyl)phosphine (29.8 mg, 98 μmol) was flushed with argon, treated with degassed toluene/TEA (35 mL, 5:1, *v:v*), and subjected to three freeze–pump–thaw cycles, and then treated with $Pd_2(dba)_3$ (18.3 mg, 20 μmol) and subjected to two additional freeze–pump–thaw cycles. The resulting mixture was heated to 60 °C for 16 h. The crude mixture was concentrated to dryness and then purified by chromatography [silica, hexanes/CH_2Cl_2 (1:1) to CH_2Cl_2] to afford a purple solid (22.4 mg, 45%): ^{1}H NMR (600 MHz, $CDCl_3$) δ 10.34 (s, 2H), 9.47 (d, *J* = 4.4 Hz, 2H), 9.45 (d, *J* = 4.4 Hz, 2H), 9.19 (d, *J* = 4.4 Hz, 2H), 9.16 (d, *J* = 4.4 Hz, 2H), 8.69 (t, *J* = 1.6 Hz, 1H), 8.52 (d, *J* = 1.6 Hz, 2H), 8.29 (d, *J* = 7.9 Hz, 2H), 8.16 (d, *J* = 7.6 Hz, 2H), 8.00 (d, *J* = 7.9 Hz, 2H), 7.61 (d, *J* = 7.6 Hz, 2H), 4.55–4.48 (m, 4H), 2.75 (s, 3H), 1.25–1.20 (m, 4H), 0.14 (s, 18H); MALDI-MS (α-CHCA) obsd 926.6, calcd 926.3 (M^+), M = $C_{53}H_{50}N_4O_4Si_2Zn$; λ_{abs} (toluene) 415, 540, 575 nm; λ_{em} (λ_{exc} = 415 nm) 584, 633 nm.

Zinc(II) 5-(4-(2-(3,5-Dicarboxyphenyl)ethynyl)phenyl)-15-p-tolylporphyrin (**32**). Following a reported method [76], a solution of porphyrin **31** (18.5 mg, 20 μmol) in THF (6.0 mL) was treated with a tetrabutylammonium fluoride solution (200 μL, 1.0 M in THF) at room temperature for 2 h. The reaction mixture was diluted with CH_2Cl_2 (20 mL) and then washed with water (20 mL × 2) and brine (20 mL). The organic extract was dried, concentrated to dryness, and purified by chromatography [silica, CH_2Cl_2/MeOH (20:1 to 1:1) to afford a purple solid (9.8 mg, 67%): ^{1}H NMR (600 MHz, dimethylsulfoxide-d_6) δ 10.38 (s, 2H), 9.54 (d, *J* = 4.4 Hz, 2H), 9.50 (d, *J* = 4.3 Hz, 2H), 9.01 (d, *J* = 4.3 Hz, 2H), 8.97 (d, *J* = 4.4 Hz, 2H), 8.27 (d, *J* = 7.7 Hz, 2H), 8.23 (d, *J* = 7.4 Hz, 1H), 8.15–8.09 (m, 4H), 8.04 (d, *J* = 7.6 Hz, 2H), 7.66 (d, *J* = 7.5 Hz, 2H), 2.71 (s, 3H); ESI-MS (negative-ion mode) obsd 725.1157, calcd 725.1173 $[M - H]^-$; ESI-MS obsd 727.1301, calcd 727.1318 $[M + H]^+$, M = $C_{43}H_{26}N_4O_4Zn$; λ_{abs} (dimethylsulfoxide) 419, 548, 587 nm.

3.4. Fluorescence Yield Determinations

The Φ_f values were determined in the standard manner by a comparison of integrated spectra (corrected for instrument sensitivity) with a known fluorophore of similar absorption and fluorescence spectral features. For **13**, the standard was dyad **1** ($\Phi_f = 0.38$) [42]. For **27**, the standard was triad **4** ($\Phi_f = 0.41$) [42]. For **31**, the standard was *meso*-tetraphenylporphyrin ($\Phi_f = 0.070$) [79]. The resulting Φ_f values for **13**, **27**, and **31** are 0.21, 0.24, and 0.011, respectively.

4. Outlook

Building block routes have been established for the preparation of triads comprised of two perylene-monoimides, one porphyrin, and a single tether. All of this work has emanated from the unexpected observation a decade ago that a perylene-ethynyl-porphyrin (Dyad **0**) exhibits an absorption spectrum essentially lacking a characteristic porphyrin Soret band [41], as shown in Figure 1 [45–47,49,50]. The panchromatic absorption provided by the perylene-ethynyl-porphyrin construct differs profoundly from that of the constituent parts. Among all other chromophore-tetrapyrrole constructs subsequently examined [42,45–50], including the exploration of the type and number of chromophores, the attachment site on the tetrapyrrole, and the composition of the tetrapyrrole, the linear (i.e., *trans*) arrangement of perylene-ethynyl-porphyrin-ethynyl-perylene with attachment at the porphyrin meso-positions has proven superior for panchromaticity and photophysical features. The triads described herein provide absorption across the 350–750 nm region and fluorescence in the near-infrared region. Such spectral features closely resemble those of triads prepared previously that lack tethers. The triads described herein are hydrophobic and may be best suited for use in membraneous assemblies and other lipophilic environments. For perspective, a phenazinyl-ethynyl-porphyrin that bears a benzoic acid tether has been prepared [80]. Triads **13** and **27** provide broader spectral coverage but also are substantially larger. The building block chemistry described herein should enable the preparation of a variety of porphyrin constructs with a range of tethers for studies of panchromatic absorbers in diverse applications.

Supplementary Materials: The following supporting information can be downloaded at: https://www.mdpi.com/article/10.3390/molecules27196501/s1, ^{1}H and ^{13}C{^{1}H} NMR spectra, mass spectra, and absorption spectra (where available) for new compounds, comprising 59 pages.

Author Contributions: R.L. and J.R. carried out the synthesis of triads **13** and **27**, respectively, and companion compounds. Z.W. carried out analysis. M.T. performed spectral comparisons. Z.W. and J.S.L. wrote the paper. D.F.B., D.H. and J.S.L. designed the compounds. All authors have read and agreed to the published version of the manuscript.

Funding: This work was funded by the Division of Chemical Sciences, Geosciences, and Biosciences, Office of Basic Energy Sciences of the U.S. Department of Energy (DE-FG02-05ER15661). Compound characterization was performed in part by the Molecular Education, Technology and Research Innovation Center (METRIC) at NC State University, which is supported by the State of North Carolina.

Institutional Review Board Statement: Not applicable.

Informed Consent Statement: Not applicable.

Data Availability Statement: All data are contained within the paper and Supplementary Materials.

Conflicts of Interest: The authors declare no competing financial interest.

Sample Availability: Samples have generally been consumed during the course of research.

References

1. Scheer, H. An Overview of Chlorophylls and Bacteriochlorophylls: Biochemistry, Biophysics, Functions and Applications. In *Chlorophylls and Bacteriochlorophylls: Biochemistry, Biophysics, Functions and Applications*; Scheer, H., Grimm, B., Porra, R.J., Rüdiger, W., Scheer, H., Eds.; Springer: Dordrecht, The Netherlands, 2006; Volume 25, pp. 1–26.
2. Blankenship, R.E. *Molecular Mechanisms of Photosynthesis*; Wiley-Blackwell: Chichester, UK, 2014.

3. Arnold, D.P.; Johnson, A.W.; Mahendran, M. Some Reactions of *meso*-Formyloctaethylporphyrin. *J. Chem. Soc. Perkin Trans. I* **1978**, 366–370. [CrossRef]
4. Arnold, D.P.; Nitschinsk, L.J. Porphyrin Dimers Linked by Conjugated Butadiynes. *Tetrahedron* **1992**, *48*, 8781–8792. [CrossRef]
5. Proess, G.; Pankert, D.; Hevesi, L. Synthesis of meso-Tetraalkynyl Porphyrins Using 1-Seleno-2-alkynyl Cation Precursors. *Tetrahedron Lett.* **1992**, *33*, 269–272. [CrossRef]
6. Anderson, H.L. Meso-Alkynyl Porphyrins. *Tetrahedron Lett.* **1992**, *33*, 1101–1104. [CrossRef]
7. Lin, V.S.-Y.; DiMagno, S.G.; Therien, M.J. Highly Conjugated, Acetylenyl Bridged Porphyrins: New Models for Light-Harvesting Antenna Systems. *Science* **1994**, *264*, 1105–1111. [CrossRef] [PubMed]
8. Rickhaus, M.; Jentzsch, A.V.; Tejerina, L.; Grübner, I.; Jirasek, M.; Claridge, T.D.W.; Anderson, H.L. Single-Acetylene Linked Porphyrin Nanorings. *J. Am. Chem. Soc.* **2017**, *139*, 16502–16505. [CrossRef]
9. Angiolillo, P.J.; Lin, V.S.-Y.; Vanderkooi, J.M.; Therien, M.J. EPR Spectroscopy and Photophysics of the Lowest Photoactivated Triplet State of a Series of Highly Conjugated (Porphinato)Zn Arrays. *J. Am. Chem. Soc.* **1995**, *117*, 12514–12527. [CrossRef]
10. Lin, V.S.-Y.; Therien, M.J. The Role of Porphyrin-to-Porphyrin Linkage Topology in the Extensive Modulation of the Absorptive and Emissive Properties of a Series of Ethynyl- and Butadiynyl-Bridged Bis- and Tris(porphinato)zinc Chromophores. *Chem. Eur. J.* **1995**, *1*, 645–651. [CrossRef]
11. Kumble, R.; Palese, S.; Lin, V.S.-Y.; Therien, M.J.; Hochstrasser, R.M. Ultrafast Dynamics of Highly Conjugated Porphyrin Arrays. *J. Am. Chem. Soc.* **1998**, *120*, 11489–11498. [CrossRef]
12. Shediac, R.; Gray, M.H.B.; Uyeda, H.T.; Johnson, R.C.; Hupp, J.T.; Angiolillo, P.J.; Therien, M.J. Singlet and Triplet Excited States of Emissive, Conjugated Bis(porphyrin) Compounds Probed by Optical and EPR Spectroscopic Methods. *J. Am. Chem. Soc.* **2000**, *122*, 7017–7033. [CrossRef]
13. Fletcher, J.T.; Therien, M.J. Strongly Coupled Porphyrin Arrays Featuring Both π-Cofacial and Linear-π-Conjugative Interactions. *Inorg. Chem.* **2002**, *41*, 331–341. [CrossRef] [PubMed]
14. Rubtsov, I.V.; Susumu, K.; Rubtsov, G.I.; Therien, M.J. Ultrafast Singlet Excited-State Polarization in Electronically Asymmetric Ethyne-Bridged Bis[(porphinato)zinc(II)] Complexes. *J. Am. Chem. Soc.* **2003**, *125*, 2687–2696. [CrossRef] [PubMed]
15. Duncan, T.V.; Susumu, K.; Sinks, L.E.; Therien, M.J. Exceptional Near-Infrared Fluorescence Quantum Yields and Excited-State Absorptivity of Highly Conjugated Porphyrin Arrays. *J. Am. Chem. Soc.* **2006**, *128*, 9000–9001. [CrossRef] [PubMed]
16. Duncan, T.V.; Ishizuka, T.; Therien, M.J. Molecular Engineering of Intensely Near-Infrared Absorbing Excited States in Highly Conjugated Oligo(porphinato)zinc–(Polypyridyl)metal(II) Supermolecules. *J. Am. Chem. Soc.* **2007**, *129*, 9691–9703. [CrossRef]
17. Fisher, J.A.N.; Susumu, K.; Therien, M.J.; Yodh, A.G. One- and Two-Photon Absorption of Highly Conjugated Multiporphyrin Systems in the Two-Photon Soret Transition Region. *J. Chem. Phys.* **2009**, *130*, 134506. [CrossRef]
18. Duncan, T.V.; Frail, P.R.; Miloradovic, I.R.; Therien, M.J. Excitation of Highly Conjugated (Porphinato)palladium(II) and (Porphinato)platinum(II) Oligomers Produces Long-Lived, Triplet States at Unit Quantum Yield That Absorb Strongly over Broad Spectral Domains of the NIR. *J. Phys. Chem. B* **2010**, *114*, 14696–14702. [CrossRef] [PubMed]
19. Singh-Rachford, T.N.; Nayak, A.; Muro-Small, M.L.; Goeb, S.; Therien, M.J.; Castellano, F.N. Supermolecular-Chromophore-Sensitized Near-Infrared-to-Visible Photon Upconversion. *J. Am. Chem. Soc.* **2010**, *132*, 14203–14211. [CrossRef] [PubMed]
20. Susumu, K.; Therien, M.J. Design of Diethynyl Porphyrin Derivatives with High Near Infrared Fluorescence Quantum Yields. *J. Porphyr. Phthalocyanines* **2015**, *19*, 205–218. [CrossRef]
21. Viere, E.J.; Qi, W.; Stanton, I.N.; Zhang, P.; Therien, M.J. Driving High Quantum Yield NIR Emission through Proquinoidal Linkage Motifs in Conjugated Supermolecular Arrays. *Chem. Sci.* **2020**, *11*, 8095–8104.
22. Milgrom, L.R.; Yahioglu, G.; Bruce, D.W.; Morrone, S.; Henari, F.Z.; Blau, W.J. Mesogenic Zinc(II) Complexes of 5,10,15,20-Tetraarylethynyl-Substituted Porphyrins. *Adv. Mater.* **1997**, *9*, 313–316. [CrossRef]
23. Sutton, J.M.; Boyle, R.W. First Synthesis of Porphyrin–Phthalocyanine Heterodimers With a Direct Ethynyl Linkage. *Chem. Commun.* **2001**, 2014–2015. [CrossRef] [PubMed]
24. Pereira, A.M.V.M.; Soares, A.R.M.; Calvete, M.J.F.; de la Torre, G. Recent Developments in the Synthesis of Homo- And Heteroarrays of Porphyrins and Phthalocyanines. *J. Porphyr. Phthalocyanines* **2009**, *13*, 419–428. [CrossRef]
25. Ke, H.; Li, W.; Zhang, T.; Zhu, X.; Tam, H.-L.; Hou, X.; Kwong, D.W.J.; Wong, W.-K. Acetylene Bridged Porphyrin–Monophthalocyaninato Ytterbium(III) Hybrids with Strong Two-Photon Absorption and High Singlet Oxygen Quantum Yield. *Dalton Trans.* **2012**, *41*, 4536–4543. [CrossRef] [PubMed]
26. Anderson, H.L. Building Molecular Wires From the Colours of Life: Conjugated Porphyrin Oligomers. *Chem. Commun.* **1999**, 2323–2330. [CrossRef]
27. Maretina, I.A. Porphyrin–Ethynyl Arrays: Synthesis, Design, and Application. *Russ. J. Gen. Chem.* **2009**, *79*, 1544–1581. [CrossRef]
28. Miller, M.A.; Lammi, R.K.; Prathapan, S.; Holten, D.; Lindsey, J.S. A Tightly Coupled Linear Array of Perylene, Bis(Porphyrin), and Phthalocyanine Units that Functions as a Photoinduced Energy-Transfer Cascade. *J. Org. Chem.* **2000**, *65*, 6634–6649. [CrossRef]
29. Prathapan, S.; Yang, S.I.; Seth, J.; Miller, M.A.; Bocian, D.F.; Holten, D.; Lindsey, J.S. Synthesis and Excited-State Photodynamics of Perylene-Porphyrin Dyads. 1. Parallel Energy and Charge Transfer via a Diphenylethyne Linker. *J. Phys. Chem. B* **2001**, *105*, 8237–8248. [CrossRef]
30. Yang, S.I.; Prathapan, S.; Miller, M.A.; Seth, J.; Bocian, D.F.; Lindsey, J.S.; Holten, D. Synthesis and Excited-State Photodynamics in Perylene-Porphyrin Dyads 2. Effects of Porphyrin Metalation State on the Energy-Transfer, Charge-Transfer, and Deactivation Channels. *J. Phys. Chem. B* **2001**, *105*, 8249–8258. [CrossRef]

31. Yang, S.I.; Lammi, R.K.; Prathapan, S.; Miller, M.A.; Seth, J.; Diers, J.R.; Bocian, D.F.; Lindsey, J.S.; Holten, D. Synthesis and Excited-State Photodynamics of Perylene–Porphyrin Dyads Part 3. Effects of Perylene, Linker, and Connectivity on Ultrafast Energy Transfer. *J. Mater. Chem.* **2001**, *11*, 2420–2430. [CrossRef]
32. Ambroise, A.; Kirmaier, C.; Wagner, R.W.; Loewe, R.S.; Bocian, D.F.; Holten, D.; Lindsey, J.S. Weakly Coupled Molecular Photonic Wires: Synthesis and Excited-State Energy-Transfer Dynamics. *J. Org. Chem.* **2002**, *67*, 3811–3826. [CrossRef]
33. Kirmaier, C.; Yang, S.I.; Prathapan, S.; Miller, M.A.; Diers, J.R.; Bocian, D.F.; Lindsey, J.S.; Holten, D. Synthesis and Excited-State Photodynamics in Perylene-Porphyrin Dyads. 4. Ultrafast Charge Separation and Charge Recombination Between Tightly Coupled Units in Polar Media. *Res. Chem. Intermed.* **2002**, *28*, 719–740. [CrossRef]
34. Tomizaki, K.-Y.; Loewe, R.S.; Kirmaier, C.; Schwartz, J.K.; Retsek, J.L.; Bocian, D.F.; Holten, D.; Lindsey, J.S. Synthesis and Photophysical Properties of Light-Harvesting Arrays Comprised of a Porphyrin Bearing Multiple Perylene-Monoimide Accessory Pigments. *J. Org. Chem.* **2002**, *67*, 6519–6534. [CrossRef] [PubMed]
35. Loewe, R.S.; Tomizaki, K.-Y.; Youngblood, W.J.; Bo, Z.; Lindsey, J.S. Synthesis of Perylene–Porphyrin Building Blocks and Rod-Like Oligomers for Light-Harvesting Applications. *J. Mater. Chem.* **2002**, *12*, 3438–3451. [CrossRef]
36. Loewe, R.S.; Tomizaki, K.-Y.; Chevalier, F.; Lindsey, J.S. Synthesis of Perylene–Porphyrin Dyads for Light-Harvesting Studies. *J. Porphyr. Phthalocyanines* **2002**, *6*, 626–642. [CrossRef]
37. Muthukumaran, K.; Loewe, R.S.; Kirmaier, C.; Hinden, E.; Schwartz, J.K.; Sazanovich, I.V.; Diers, J.R.; Bocian, D.F.; Holten, D.; Lindsey, J.S. Synthesis and Excited-State Photodynamics of A Perylene-Monoimide-Oxochlorin Dyad. A Light-Harvesting Array. *J. Phys. Chem. B* **2003**, *107*, 3431–3442. [CrossRef]
38. Kirmaier, C.; Hinden, E.; Schwartz, J.K.; Sazanovich, I.V.; Muthukumaran, K.; Taniguchi, M.; Bocian, D.F.; Lindsey, J.S.; Holten, D. Synthesis and Excited-State Photodynamics of Perylene-Bis(imide)-Oxochlorin Dyads. A Charge-Separation Motif. *J. Phys. Chem. B* **2003**, *107*, 3443–3454. [CrossRef]
39. Tomizaki, K.-Y.; Thamyongkit, P.; Loewe, R.S.; Lindsey, J.S. Practical Synthesis of Perylene-Monoimide Building Blocks That Possess Features Appropriate for Use in Porphyrin-Based Light-Harvesting Arrays. *Tetrahedron* **2003**, *59*, 1191–1207. [CrossRef]
40. Kirmaier, C.; Song, H.-E.; Yang, E.K.; Schwartz, J.K.; Hindin, E.; Diers, J.R.; Loewe, R.S.; Tomizaki, K.-Y.; Chevalier, F.; Ramos, L.; et al. Excited-State Photodynamics of Perylene–Porphyrin Dyads. 5. Tuning Light-Harvesting Characteristics via Perylene Substituents, Connection Motif, and 3-Dimensional Architecture. *J. Phys. Chem. B* **2010**, *114*, 14249–14264. [CrossRef]
41. Wang, J.; Yang, E.; Diers, J.R.; Niedzwiedzki, D.M.; Kirmaier, C.; Bocian, D.F.; Lindsey, J.S.; Holten, D. Distinct Photophysical and Electronic Characteristics of Strongly Coupled Dyads Containing a Perylene Accessory Pigment and a Porphyrin, Chlorin, or Bacteriochlorin. *J. Phys. Chem. B* **2013**, *117*, 9288–9304. [CrossRef]
42. Alexy, E.J.; Yuen, J.M.; Chandrashaker, V.; Diers, J.R.; Kirmaier, C.; Bocian, D.F.; Holten, D.; Lindsey, J.S. Panchromatic Absorbers for Solar Light-Harvesting. *Chem. Commun.* **2014**, *50*, 14512–14515. [CrossRef]
43. Muthiah, C.; Kee, H.L.; Diers, J.R.; Fan, D.; Ptaszek, M.; Bocian, D.F.; Holten, D.; Lindsey, J.S. Synthesis and Excited-State Photodynamics of a Chlorin–Bacteriochlorin Dyad–Through-Space Versus Through-Bond Energy Transfer in Tetrapyrrole Arrays. *Photochem. Photobiol.* **2008**, *84*, 786–801. [CrossRef] [PubMed]
44. Tomizaki, K.-Y.; Lysenko, A.B.; Taniguchi, M.; Lindsey, J.S. Synthesis of Phenylethyne-linked Porphyrin Dyads. *Tetrahedron* **2004**, *60*, 2011–2023. [CrossRef]
45. Amanpour, J.; Hu, G.; Alexy, E.J.; Mandal, A.K.; Kang, H.S.; Yuen, J.M.; Diers, J.R.; Bocian, D.F.; Lindsey, J.S.; Holten, D. Tuning the Electronic Structure and Properties of Perylene–Porphyrin–Perylene Panchromatic Absorbers. *J. Phys. Chem. A* **2016**, *120*, 7434–7450. [CrossRef]
46. Hu, G.; Liu, R.; Alexy, E.J.; Mandal, A.K.; Bocian, D.F.; Holten, D.; Lindsey, J.S. Panchromatic Chromophore–Tetrapyrrole Light-Harvesting Arrays Constructed from Bodipy, Perylene, Terrylene, Porphyrin, Chlorin, and Bacteriochlorin Building Blocks. *New J. Chem.* **2016**, *40*, 8032–8052. [CrossRef]
47. Mandal, A.K.; Diers, J.R.; Niedzwiedzki, D.M.; Hu, G.; Liu, R.; Alexy, E.J.; Lindsey, J.S.; Bocian, D.F.; Holten, D. Tailoring Panchromatic Absorption and Excited-State Dynamics of Tetrapyrrole–Chromophore (Bodipy, Rylene) Arrays—Interplay of Orbital Mixing and Configuration Interaction. *J. Am. Chem. Soc.* **2017**, *139*, 17547–17564. [CrossRef] [PubMed]
48. Hu, G.; Kang, H.S.; Mandal, A.K.; Roy, A.; Kirmaier, C.; Bocian, D.F.; Holten, D.; Lindsey, J.S. Synthesis of Arrays Containing Porphyrin, Chlorin, and Perylene-imide Constituents for Panchromatic Light-Harvesting and Charge Separation. *RSC Adv.* **2018**, *8*, 23854–23874. [CrossRef]
49. Yuen, J.; Diers, J.R.; Alexy, E.J.; Roy, A.; Mandal, A.K.; Kang, H.S.; Niedzwiedzki, D.M.; Kirmaier, C.; Lindsey, J.S.; Bocian, D.F.; et al. Origin of Panchromaticity in Multichromophore–Tetrapyrrole Arrays. *J. Phys. Chem. A* **2018**, *122*, 7181–7201. [CrossRef]
50. Rong, J.; Magdaong, N.C.M.; Taniguchi, M.; Diers, J.R.; Niedzwiedzki, D.M.; Kirmaier, C.; Lindsey, J.S.; Bocian, D.F.; Holten, D. Electronic Structure and Excited-State Dynamics of Rylene–Tetrapyrrole Panchromatic Absorbers. *J. Phys. Chem. A* **2021**, *125*, 7900–7919. [CrossRef]
51. Taniguchi, M.; Balakumar, A.; Fan, D.; McDowell, B.E.; Lindsey, J.S. Imine-Substituted Dipyrromethanes in the Synthesis of Porphyrins Bearing One or Two *Meso* Substituents. *J. Porphyr. Phthalocyanines* **2005**, *9*, 554–574. [CrossRef]

52. Tamaru, S.-I.; Yu, L.; Youngblood, W.J.; Muthukumaran, K.; Taniguchi, M.; Lindsey, J.S. A Tin-Complexation Strategy for Use with Diverse Acylation Methods in the Preparation of 1,9-Diacyldipyrromethanes. *J. Org. Chem.* **2004**, *69*, 765–777. [CrossRef]
53. Borbas, K.E.; Mroz, P.; Hamblin, M.R.; Lindsey, J.S. Bioconjugatable Porphyrins Bearing a Compact Swallowtail Motif for Water Solubility. *Bioconjugate Chem.* **2006**, *17*, 638–653. [CrossRef] [PubMed]
54. Liu, R.; Liu, M.; Hood, D.; Chen, C.-Y.; MacNevin, C.J.; Holten, D.; Lindsey, J.S. Chlorophyll-Inspired Red-Region Fluorophores: Building Block Syntheses and Studies in Aqueous Media. *Molecules* **2018**, *23*, 130. [CrossRef] [PubMed]
55. Chinchilla, R.; Nájera, C. The Sonogashira Reaction: A Booming Methodology in Synthetic Organic Chemistry. *Chem. Rev.* **2007**, *107*, 874–922. [CrossRef]
56. Wagner, R.W.; Johnson, T.E.; Li, F.; Lindsey, J.S. Synthesis of Ethyne-Linked or Butadiyne-Linked Porphyrin Arrays Using Mild, Copper-Free, Pd-Mediated Coupling Reactions. *J. Org. Chem.* **1995**, *60*, 5266–5273. [CrossRef]
57. Wagner, R.W.; Ciringh, Y.; Clausen, C.; Lindsey, J.S. Investigation and Refinement of Palladium-Coupling Conditions for the Synthesis of Diarylethyne-Linked Multiporphyrin Arrays. *Chem. Mater.* **1999**, *11*, 2974–2983. [CrossRef]
58. Schmidt, I.; Jiao, J.; Thamyongkit, P.; Sharada, D.S.; Bocian, D.F.; Lindsey, J.S. Investigation of Stepwise Covalent Synthesis on a Surface Yielding Porphyrin-Based Multicomponent Architectures. *J. Org. Chem.* **2006**, *71*, 3033–3050. [CrossRef]
59. Rademacher, A.; Märkle, S.; Langhals, H. Lösliche Perylen-Fluoreszenzfarbstoffe mit hoher Photostabilität. *Chem. Ber.* **1982**, *115*, 2927–2934. [CrossRef]
60. Langhals, H. Synthese von hochreinen Perylen-Fluoreszenzfarbstoffen in großen Mengen–gezielte Darstellung von Atrop-Isomeren. *Chem. Ber.* **1985**, *118*, 4641–4645. [CrossRef]
61. Langhals, H.; Demmig, S.; Huber, H. Rotational Barriers in Perylene Fluorescent Dyes. *Spectrochim. Acta* **1988**, *44A*, 1189–1193. [CrossRef]
62. Demmig, S.; Langhals, H. Leichtlösliche, lichtechte Perylen-Fluoreszenzfarbstoffe. *Chem. Ber.* **1988**, *121*, 225–230. [CrossRef]
63. Ebeid, E.-Z.M.; El-Daly, S.A.; Langhals, H. Emission Characteristics and Photostability of N,N'-Bis(2,5-di-*tert*-butylphenyl)-3,4:9,10-perylenebis(dicarboximide). *J. Phys. Chem.* **1988**, *92*, 4565–4568. [CrossRef]
64. Feiler, L.; Langhals, H.; Polborn, K. Synthesis of Perylene-3,4-dicarboximides—Novel Highly Photostable Fluorescent Dyes. *Liebigs Ann.* **1995**, *1995*, 1229–1244. [CrossRef]
65. Langhals, H. Cyclic Carboxylic Imide Structures as Structure Elements of High Stability. Novel Developments in Perylene Dye Chemistry. *Heterocycles* **1995**, *40*, 477–500. [CrossRef]
66. Würthner, F. Perylene Bisimide Dyes as Versatile Building Blocks for Functional Supramolecular Architectures. *Chem. Commun.* **2004**, 1564–1579. [CrossRef] [PubMed]
67. Langhals, H. Control of the Interactions in Multichromophores: Novel Concepts. Perylene Bis-imides as Components for Larger Functional Units. *Helv. Chim. Acta* **2005**, *88*, 1309–1343. [CrossRef]
68. Weil, T.; Vosch, T.; Hofkens, J.; Peneva, K.; Müllen, K. The Rylene Colorant Family—Tailored Nanoemitters for Photonics Research and Applications. *Angew. Chem. Int. Ed.* **2010**, *49*, 9068–9093. [CrossRef]
69. Gosztola, D.; Niemczyk, M.P.; Wasielewski, M.R. Picosecond Molecular Switch Based on Bidirectional Inhibition of Photoinduced Electron Transfer Using Photogenerated Electric Fields. *J. Am. Chem. Soc.* **1998**, *120*, 5118–5119. [CrossRef]
70. Odobel, F.; Séverac, M.; Pellegrin, Y.; Blart, E.; Fosse, C.; Cannizzo, C.; Mayer, C.R.; Elliott, K.J.; Harriman, A. Coupled Sensitizer–Catalyst Dyads: Electron-Transfer Reactions in a Perylene–Polyoxometalate Conjugate. *Chem. Eur. J.* **2009**, *15*, 3130–3138. [CrossRef] [PubMed]
71. Boixel, J.; Blart, E.; Pellegrin, Y.; Odobel, F.; Perin, N.; Chiorboli, C.; Fracasso, S.; Ravaglia, M.; Scandola, F. Hole-Transfer Dyads and Triads Based on Perylene Monoimide, Quaterthiophene, and Extended Tetrathiafulvalene. *Chem. Eur. J.* **2010**, *16*, 9140–9153. [CrossRef]
72. Wagner, R.W.; Johnson, T.E.; Lindsey, J.S. Soluble Synthetic Multiporphyrin Arrays. 1. Modular Design and Synthesis. *J. Am. Chem. Soc.* **1996**, *118*, 11166–11180. [CrossRef]
73. Gorrea, E.; Carbajo, D.; Gutiérrez-Abad, R.; Illa, O.; Branchadell, V.; Royo, M.; Ortuño, R.M. Searching for New Cell-Penetrating Agents: Hybrid Cyclobutane–Proline γ,γ-Peptides. *Org. Biomol. Chem.* **2012**, *10*, 4050–4057. [CrossRef] [PubMed]
74. Reddy, K.R.; Jiang, J.; Krayer, M.; Harris, M.A.; Springer, J.W.; Yang, E.; Jiao, J.; Niedzwiedzki, D.M.; Pandithavidana, D.; Parkes-Loach, P.S.; et al. A Palette of Lipophilic Bioconjugatable Bacteriochlorins for Construction of Biohybrid Light-Harvesting Architectures. *Chem. Sci.* **2013**, *4*, 2036–2053. [CrossRef]
75. Morisue, M.; Haruta, N.; Kalita, D.; Kobuke, Y. Efficient Charge Injection from the S_2 Photoexcited State of Special-Pair Mimic Porphyrin Assemblies Anchored on a Titanium-Modified ITO Anode. *Chem. Eur. J.* **2006**, *12*, 8123–8135. [CrossRef] [PubMed]
76. Muthiah, C.; Taniguchi, M.; Kim, H.-J.; Schmidt, I.; Kee, H.L.; Holten, D.; Bocian, D.F.; Lindsey, J.S. Synthesis and Photophysical Characterization of Porphyrin, Chlorin and Bacteriochlorin Molecules Bearing Tethers for Surface Attachment. *Photochem. Photobiol.* **2007**, *83*, 1513–1528. [CrossRef] [PubMed]
77. Lee, C.-H.; Lindsey, J.S. One-Flask Synthesis of *Meso*-Substituted Dipyrromethanes and Their Application in the Synthesis of *Trans*-Substituted Porphyrin Building Blocks. *Tetrahedron* **1994**, *50*, 11427–11440. [CrossRef]
78. Srinivasan, N.; Haney, C.A.; Lindsey, J.S.; Zhang, W.; Chait, B.T. Investigation of MALDI-TOF Mass Spectrometry of Diverse Synthetic Metalloporphyrins, Phthalocyanines, and Multiporphyrin Arrays. *J. Porphyr. Phthalocyanines* **1999**, *3*, 283–291. [CrossRef]

79. Taniguchi, M.; Lindsey, J.S.; Bocian, D.F.; Holten, D. Comprehensive Review of Photophysical Parameters (ε, Φ_f, τ_S) of Tetraphenylporphyrin (H$_2$TPP) and Zinc Tetraphenylporphyrin (ZnTPP)–Critical Benchmark Molecules in Photochemistry and Photosynthesis. *J. Photochem. Photobiol. C Photochem. Rev.* **2021**, *46*, 100401. [CrossRef]
80. Lee, S.H.; Matula, A.J.; Hu, G.; Troiano, J.L.; Karpovich, C.J.; Crabtree, R.H.; Batista, V.S.; Brudvig, G.W. Strongly Coupled Phenazine−Porphyrin Dyads: Light-Harvesting Molecular Assemblies with Broad Absorption Coverage. *ACS Appl. Mater. Interfaces* **2019**, *11*, 8000–8008. [CrossRef]

Article

Synthesis and Characterization of New-Type Soluble β-Substituted Zinc Phthalocyanine Derivative of Clofoctol

Sabrine Dridi [1], Jamel Eddine Khiari [2], Gabriele Magna [3], Manuela Stefanelli [3], Larisa Lvova [3], Federica Mandoj [3], Khaoula Khezami [4,5], Mahmut Durmuş [5], Corrado Di Natale [6] and Roberto Paolesse [3,*]

[1] Experimental Sciences and Supramolecular Chemistry, Laboratory of Didactic Research, Higher Institute of Education and Continuing Training (ISEFC), University of Tunis El Manar, Tunis 1002, Tunisia; sabrinegandour@gmail.com
[2] Experimental Sciences and Supramolecular Chemistry, Laboratory of Didactic Research, Higher Institute of Education and Continuing Training (ISEFC), University of Carthage, Tunis 1054, Tunisia; jamelkhiari@yahoo.fr
[3] Department of Chemical Science and Technologies, University of Rome Tor Vergata, 00133 Rome, Italy; gabriele.magna@uniroma2.it (G.M.); manuela.stefanelli@uniroma2.it (M.S.); larisa.lvova@uniroma2.it (L.L.); federica.mandoj@uniroma2.it (F.M.)
[4] Department of Chemistry, Faculty of Engineering and Natural Sciences, Istinye University, 34396 Istanbul, Turkey; khaoula@gtu.edu.tr
[5] Department of Chemistry, Gebze Technical University, 41400 Kocaeli, Turkey; durmus@gtu.edu.tr
[6] Department of Electronic Engineering, University of Rome Tor Vergata, 00133 Rome, Italy; dinatale@uniroma2.it
* Correspondence: roberto.paolesse@uniroma2.it

Abstract: In this work, we have described the synthesis and characterization of novel zinc (II) phthalocyanine bearing four 2-(2,4-dichloro-benzyl)-4-(1,1,3,3-tetramethyl-butyl)-phenoxy substituents on the peripheral positions. The compound was characterized by elemental analysis and different spectroscopic techniques, such as FT-IR, ^{1}H NMR, MALDI-TOF, and UV-Vis. The Zn (II) phthalocyanine shows excellent solubility in organic solvents such as dichloromethane (DCM), n-hexane, chloroform, tetrahydrofuran (THF), and toluene. Photochemical and electrochemical characterizations of the complex were performed by UV-Vis, fluorescence spectroscopy, and cyclic voltammetry. Its good solubility allows a direct deposition of this compound as film, which has been tested as a solid-state sensing material in gravimetric chemical sensors for gas detection, and the obtained results indicate its potential for qualitative discrimination and quantitative assessment of various volatile organic compounds, among them methanol, n-hexane, triethylamine (TEA), toluene and DCM, in a wide concentration range.

Keywords: phthalocyanines; solubility; gas sensors; quartz microbalances

1. Introduction

Phthalocyanines (Pcs) are synthetic tetrapyrrolic macrocycles that have been extensively studied due to their unique properties, such as their aromatic 18 π-electronic structure, planarity, high symmetry, good thermal stability, and efficient light absorption [1,2]. Pcs and their metal complexes (MPcs) show green and blue colors, leading to their first use as dyes and pigments [3]. They have also been exploited in sensors, electronic displays, non-linear optical devices, solar cells, semiconductors, data storage systems, catalysts, and photodynamic therapy (PDT) [4]. One relevant requirement for Pcs applications is their solubility, which can allow their purification and consequent exploitation in target devices. However, unsubstituted phthalocyanines have very low solubility in common organic solvents [5], so a decisive ambition of Pc chemistry is to obtain soluble derivatives. This aim can be achieved by introducing appropriate functional groups at the peripheral positions of the phthalocyanine ring and selecting the metal ion coordinated at the inner core [6]. Therefore, the modification of π-conjugated system size by alternating the type, number, and positions of the substituents on the phthalocyanine frameworks enhances their solubility and affect their aggregation in solutions [7], which is generally related to intermolecular interactions such as π—π stacking, hydrogen–bond, acid–base interactions, and donor–acceptor interactions [8,9]. In this respect, it is possible to design Pcs that have high solubility by introducing bulky groups either on peripheral or non-peripheral positions [10]. Avoiding aggregation leads to soluble phthalocyanines, making them good candidates for particular applications.

In this paper, we report the synthesis of zinc phthalocyanine containing 2-(2,4-dichloro-benzyl)-4-(1,1,3,3-tetramethyl-butyl)-phenoxy moiety starting from the Clofoctol derivative, which is a synthetic compound used in the treatment of bacterial respiratory infections and exhibits antiviral activity against SARS-CoV-2 (see **3** and **4** in Scheme 1) [11]. Solubility, aggregation, fluorescence, electrochemical, and optical properties of the complex were determined, and gravimetric sensors investigated its possible use as a sensing material. In particular, two sensors were produced by depositing both titled and Ni phthalocyanines onto Quartz Microbalances (QMBs), and a comparison between their chemical sensitivities toward volatile organic compound (VOC) vapors was made.

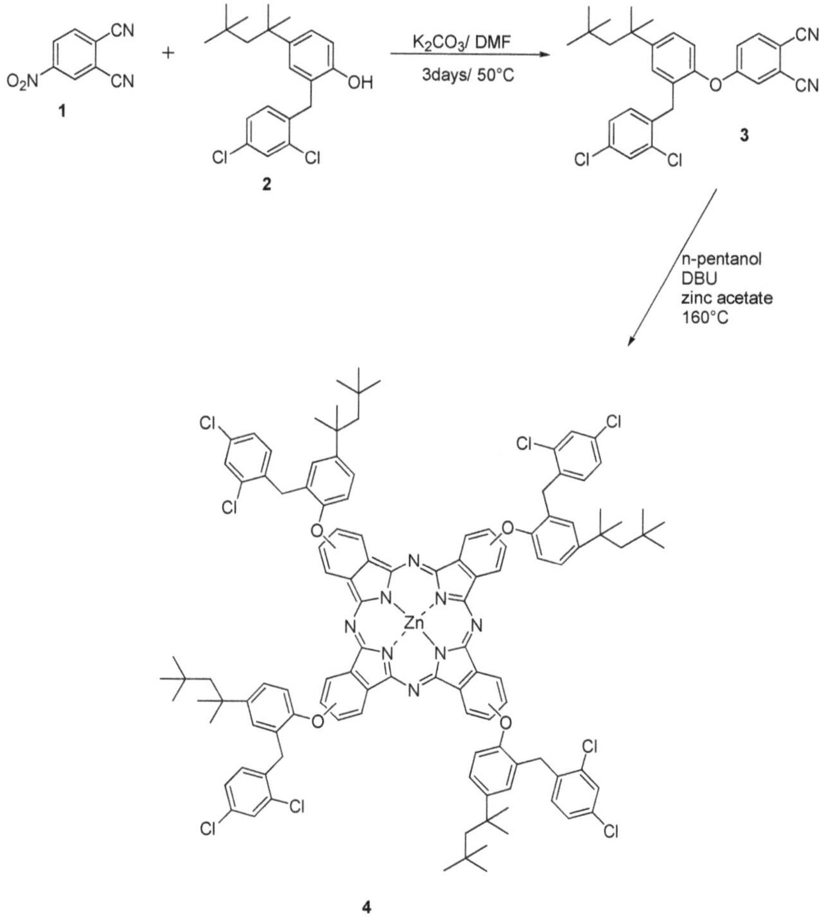

Scheme 1. Synthetic route of complex **4**.

2. Results

2.1. Synthesis and Characterization

The procedure for the synthesis of **3** and **4** is given in Scheme 1. The preparation of phthalonitrile **3** was performed by nucleophilic aromatic nitro displacement of 2-(2,4-dichloro-benzyl)-4-(1,1,3,3-tetramethyl-butyl) phenol (**2**) with 4-nitrophthalonitrile in the presence of K_2CO_3 in dry DMF [12]. Following the usual protocol for the synthesis of tetrasubstituted Pcs, **4** was achieved in dry pentanol by cyclotetramerization reaction of phthalonitrile **3** in the presence of anhydrous $Zn(OAc)_2$ and 1,8-diazabicyclo[5.4.0]undec-7-ene (DBU) base [13]. The symmetric Pc derivative **4** was easily purified by column chromatography, with a satisfying yield of 58%.

2.2. Spectroscopic Characterization

The structures of compounds **3** and **4** were checked by employing elemental analysis and common spectroscopic techniques such as FT-IR, ^{1}H NMR, ^{13}C NMR, MALDI-TOF, and UV-Vis. According to FT-IR spectral data, **3** showed its vibration band at 2229 cm^{-1}, indicating the presence of the -C≡N group (Figure S1). The ^{1}H NMR spectrum of **3** proves the disappearance of the OH signal of the starting phenol, with the appearance of further aromatic protons confirming the occurrence of the substitution (Figure S1). The structure of compound **3** was identified within the reflectron mode by matrix-assisted laser desorption/ionization time-of-flight mass (MALDI-TOF) spectrometry using 2,5-dihydroxybenzoic acid (DHB) as matrix. The presence of potassium ion K$^+$ was detected in the molecular ion peak at 530.42 as $[M + K]^+$ (Figure S2). All spectral data collected support the proposed structure of **3**.

The formation of **4** was confirmed in the FT-IR spectra by the disappearance of the sharp C≡N vibration of precursor **3** at 2238 cm^{-1} (Figure S1). The ^{1}H NMR spectrum of **4** suffers from signal broadening, probably due to aggregation at the concentration for NMR measurements, although the pattern is similar to that of compound **3** [14]. The MALDI-TOF spectrum indicated that the molecular weight value of this phthalocyanine is in accordance with the suggested molecule. The molecular ion peak of compound **4** was observed at 2032 as $[M+H]^+$ (Figure S3).

2.3. UV-Vis Characterization

UV-Vis absorption spectroscopy is the most effective technique for proving the formation of Pcs. Pcs generally display two strong absorption bands in their ground-state absorption spectra. The first is the B band observed between 300–500 nm in the UV region, and the other is the Q band placed at 600–750 nm in the visible region, showing a more intense absorption [15]. Figure 1a shows the UV-Vis absorption spectra of **4** in different solvents. In the UV-Vis absorption spectra of **4**, an intense Q band absorption was observed at 682 nm, while the B band absorption was observed around 354 nm. Metallophthalocyanines that have D4h symmetry exhibit only an intense Q band absorption in their UV-Vis spectra [16]. The synthesized phthalocyanine **4** showed good solubility in most of the common organic solvents, such as toluene, THF, DCM, chloroform, and hexane, although in this latter solvent, the UV-Vis spectrum of **4** shows a broadening of bands with a relative lowering of the main Q-band, suggesting the presence of macrocycle aggregation.

UV-Vis absorption spectra in DCM (see spectra in Figure 1b) at different concentrations show that the main Q-band of **4** has a high molar extinction coefficient of 3.14×10^5 M^{-1}cm^{-1} at 683 nm in DCM, with good linearity observed for the Lambert–Beer law, confirming the absence of aggregation ($R^2 = 99.79$).

As a further confirmation of the solubility of the macrocycle, fluorescence measurements have been recorded in DCM. Figure 2 shows the fluorescence emission and excitation spectra for **4**. In the case of emission spectra, excitation was set at 610 nm, corresponding to the main Q-band peak, and emission was recorded in the 650–800 nm range. On the other hand, in the case of excitation spectra, emission was recorded at 756 nm scanning excitation wavelength in the 500–710 nm interval. Complex **4** is fluorescent, with the excitation spectrum superimposable to the absorption spectrum for **4** in DCM. Furthermore, the fluorophore showed no photodegradation during the experiments due to light excitation.

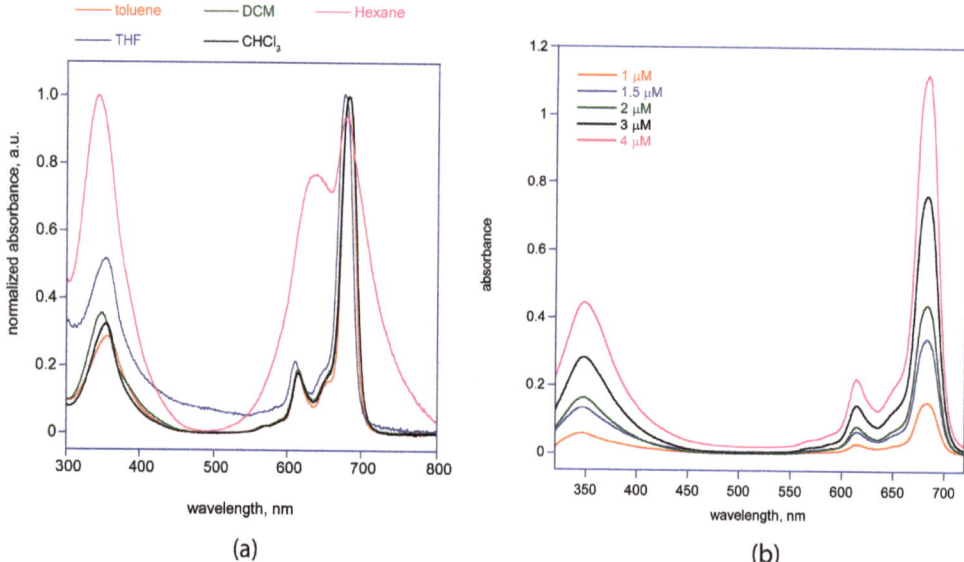

Figure 1. (**a**) UV–Vis spectra of **4** in different solvents. (**b**) **4** in DCM at different concentrations.

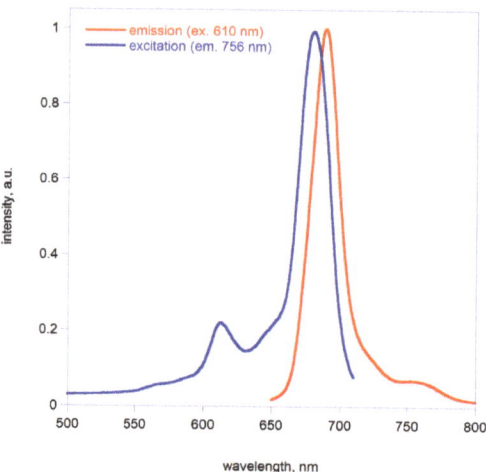

Figure 2. Excitation and emission spectra of **4** in DCM.

2.4. Electrochemical Studies

The investigation of the electrochemical behavior of the newly synthesized complex **4** is important to predict the possible usage in different technological areas. To characterize the degree of electronic interactions among the Zn(II)Pc and 2-(2,4-dichloro-benzyl)-4-(1,1,3,3-tetramethyl-butyl)-phenoxy peripheral substituents in complex **4**, we investigated the electrochemical properties of this macrocycle by cyclic voltammetry (CV) technique. The CV of **4** was recorded in the potential range from −2.0 to 2.0 V, and the details on complex **4** redox behavior are summarized in Figure 3.

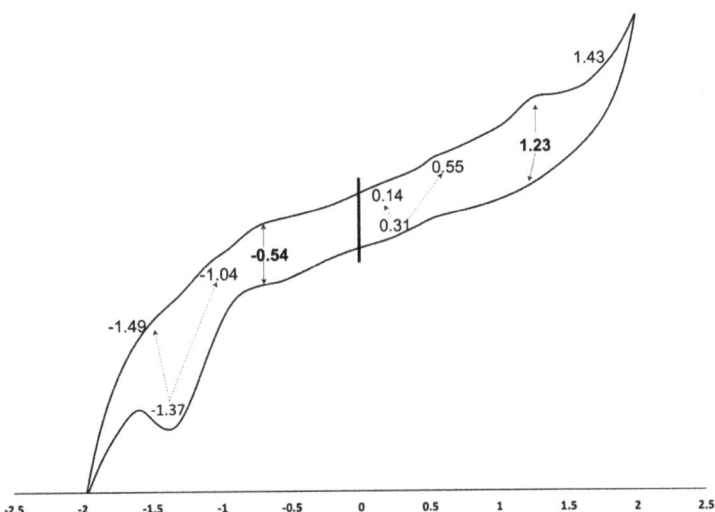

Figure 3. Cyclic voltammograms of 5×10^{-5} M of complex 4 in DCM and 0.1 M TBAClO$_4$ solution.

In the cathodic region, Zn(II)Pc 4 undergoes one well-defined one-electron process at −0.54 V and one split two-electron reduction at −1.37 V followed by two small back-way oxidation waves at −1.49 V and at −1.04 V, respectively. In the anodic region, the oxidation waves at 0.43 V and 1.23 V are well seen and correspond to the data on Zn(II)Pc electrochemical behavior reported in the literature [17–19]. Thus, for instance, for octa-3-hydroxypropylthio substituted Zn(II) phthalocyanine, the first oxidation wave was registered at 0.49 V, and the second one was observed at 0.98 V, and the processes have been assigned to the metal- and ring-based processes [1]. An additional irreversible small oxidation peak at 1.43 V and the split oxidations at 0.11 V and 0.55 V reflect the complex 4 redox behavior registered in the cathodic region and may be assigned to the influence of the peripheral substituents on the electron transfer to the phthalocyanine center.

The splitting of redox processes indicates that the peripheral substituents of complex 4 behave as equivalent but not interacting reaction centers. The HOMO-LUMO gap of compound 4 evaluated as the potential difference between the first oxidation and first reduction corresponds to the high value of 2.6 V, which is indicative of the high inter-molecular electron transfer abilities and potential of the developed compound as a sensing semiconductor material for both electron-donating and electron-withdrawing analytes [20,21]. Zn(II)Pc 4 gives a ring-based redox process only due to the electro-inactive nature of the Zn(II) metal coordinated to the macrocycle, as the energy level of the d orbital of the zinc is placed out of the HOMO-LUMO energy levels [22]. This result may also be due to the smaller atomic radius of Zn(II) metal compared to other metal atomic radii such as Co and Mn [17].

2.5. Gas Sensing Properties

The detection of VOCs vapors plays an important role in various fields, such as agriculture, medicine, and the food industry, since they are widely used and may provoke respiratory difficulties and irritation effects [23]. As gas-sensing materials, phthalocyanines have revealed high sensitivity to large sets of organic vapors [24,25], even if solubility and limited sensitivity often hinder their implementation in broader applications. The evaluation of the sensing properties of phthalocyanine for the detection of VOCs has received a lot of interest. Therefore, many types of research have been conducted on the sensing performance of Pc-based devices and the effect of environmental conditions on their performances [26–28]. Unlike the usual methods based on conductometric devices, in the present study, we deposited the sensing material onto a gravimetric sensor, such as Quartz Microbalances (QMBs). The sensing performances of **4** were then explored by depositing 15 kHz of material onto each of the two electrodes of a QMB. Subsequently, the sensor has been exposed to different VOCs, utilizing a series of mass flow controllers.

Several concentrations have been obtained from liquid samples in bubblers diluting the saturated vapors of the analytes with pure nitrogen in order to obtain the desired concentrations. By using this approach, it is possible to dilute vapors in the headspace of bubblers up to 1:100. Saturated vapors at 20 °C were calculated accordingly to Antoine equations (parameters can be found at https://webbook.nist.gov/chemistry/, accessed on 1 March 2023). As probe analytes, we selected dichloromethane, ethanol, n-hexane, triethylamine (TEA), toluene, and water due to the different interactions that may occur with receptors in the film. Usually, the Linear Solvation Equation Relationship (LSER) reports five parameters that account for electron donor, electron acceptor, dipole–dipole, dipole–"induced dipole", and "induced dipole"–"induced dipole" interactions. The sensor exhibited high variations in oscillation frequency toward most of the volatile compounds tested, especially toluene and water (see Figure 4a). This first outcome suggests that the macrocycle preferentially interacts with guest molecules by π–π interactions (toluene) or π–hydrogen bonds (water), likely due to the additional aromatic units conferred by the four clofoctol appended units. At the same time, this functional unit is supposed to favor dispersion interactions, such as in the case of hexane. Finally, the presence of Zn(II) metal facilitates the coordination with molecules bearing electron-donating groups, as in the case of amines. Considering the sensitivity as the first derivative of the response curves, it is possible to extrapolate the selective pattern of the sensor. Remarkably, if we compare the sensitivity of **4** Pc with other phthalocyanines previously studied in similar experiments by our group, such as a NiPc derivative [25] (1,4,8,11,15,18,22,25-octabutoxy-Ni(II)-phthalocyanine), we can observe how the sensitivity of the novel synthesized material is from four to ten times higher (see Figure 4b). This outcome suggests that this new phthalocyanine derivative can be extremely useful in the sensor field firstly because it made easy the deposition of films onto QMBs or transducer surface directly from a chloroform solution thanks to the good solubility of this macrocycle. Furthermore, this class of phthalocyanine derivatives showed to be a receptor with improved sensitivity toward a large class of compounds, and a potentially excellent porphyrinoid-based material for expanding the cross-sensor selectivity in electronic nose platforms, for example.

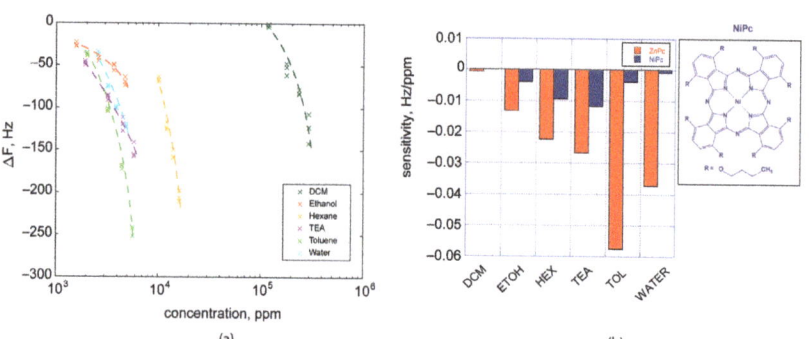

Figure 4. (**a**) Characteristic curves of QMB sensor based on **4**. Data are reported on a semilogarithmic scale due to the wide range of VOC concentrations. (**b**) Sensitivities calculated as the first derivative of linear fitting equations of data in the range of concentration measured. In the case of NiPc data, and experimental details were previously reported in [25].

3. Materials and Methods

3.1. Equipment

All reagents were purchased from Merck or Aldrich and used without further purification. 4-Nitrophthalonitrile was synthesized and purified according to the protocol mentioned in the literature [29]. Volatile Organic Compounds (VOC) were reagent grade from Sigma Aldrich and used as received. Fluorescence experiments were carried out by SHIMADZU RF-1501 spectrofluorimeter with automatized internal working settings. UV-Vis spectra were performed by an Agilent Cary 60 spectrophotometer.

Voltammetric measurements were performed by PalmSens3 potentiostat/galvanostat (PALMSENS BV, Utrecht, The Netherlands) on disk Pt WE (2.5 mm diameter) versus SCE reference electrode (Amel, Milan, Italy) and Pt wire counter electrode (Sigma-Aldrich, Darmstadt, Germany) in 5.0×10^{-5} M solution of 4 in dichloromethane containing 0.1 M tetrabutylammonium perchlorate salt as background electrolyte. The CVs were registered in the −2.0 to +2.0 V range with a scan rate of 100 mV/s. The FT-IR spectra were recorded on a Perkin Elmer 1600 FT-IR spectrophotometer using KBr pellets. ^{1}H-NMR spectra were recorded on a Bruker 700 MHz spectrometer in CDCl$_3$ and chemical shifts were reported (δ) relative to Me$_4$Si as internal standard. Mass spectra of the compounds were measured on a Micromass Quattro LC/ULTIMA LC-MS/MS spectrometer and MALDI-MS in dihydroxybenzoic acid as MALDI matrix using nitrogen laser accumulating 50 laser shots using Bruker Microflex LT MALDI-TOF mass spectrometer Bremen, Germany. The elemental analyses were performed on a Costech ECS 4010 instrument. FT-IR, MALDI-MS, ^{1}H NMR, and ^{13}C-NMR spectra of compounds 3 and 4 are reported in the Supporting Information (Figures S1–S5).

Analytes were delivered in the vapor phase by a gas flow system, including MKS mass flow meter and Channel Readout mass controllers.

3.2. Sensors Preparation and Characterization

Quartz microbalance (QMBs) were AT-cut quartzes oscillating at 20 MHz (KVG GmbH). In QMBs sensors, the resonance frequency is inversely proportional to the mass adsorbed onto the electrode surface. QMB utilized has a nominal and experimental mass sensitivity of about 7 Hz/ng [30]. The drop casting technique was used for sensing layer deposition, ensuring a total frequency shift of approximately 30 kHz. An electronic oscillator circuit acquired the fundamental oscillation frequency of the quartz each second. The sensor was tested with vapors of different volatile compounds, pouring liquid samples in bubblers at a constant temperature of 303 K. The VOC concentration can be estimated using the Antoine law; parameters of the volatile compound are available in the NIST database (available at www.nist.gov/webbook, accessed on 1 March 2023). The measurement protocol has an initial baseline phase under nitrogen for acquiring the baseline frequencies of the sensor, an exposure phase during which a known concentration of analyte vapor is fluxed in the measurement chamber containing the sensor, and a final desorption phase where nitrogen flow allows the desorption of analyte molecules from the sensing films. The frequency variations between the end of the measurement phase and the baseline value were chosen as sensor response features. Each volatile compound was measured at four different concentrations, and each concentration was randomly measured three times.

3.3. Synthesis

3.3.1. 4-[2-(2,4-Dichloro-benzyl)-4-(1,1,3,3-tetramethyl-butyl)-phenoxy]-phthalonitrile (3)

Clofoctol 2 (0.4 g, 1.15 mmol) was dissolved in 10 mL dry DMF. An excess of anhydrous K$_2$CO$_3$ was added to the mixture. Then 4-nitrophthalonitrile 1 (0.2 g, 1.15 mmol) was added to the mixture. The reaction mixture was continually stirred powerfully under the N$_2$ atmosphere at 50 °C for 3 days. After the reaction mixture was cooled down at room temperature, it was poured into 100 mL of ice-water media. A creamy white precipitate was formed; it was filtered and then washed with water until the filtrate became neutral. The white product was recrystallized from methanol, filtered, and finally dried under vacuum forming a white purified powder. Yield 0.32 g (57%), mp 211–213 °C; FT-IR ν_{max}/cm^{-1} (KBr): 3042 (Ar–H), 2229 (C≡N), 1565, 1469, 1251 (Ar–O–C), 1097, 956, 834, 823; ^{1}H NMR (700 MHz, CDCl$_3$, δ ppm): 7.67 (d, J = 8.7 Hz, 1H), 7.39–7.35 (m, 1H), 7.30 (d, J = 9.4 Hz, 2H), 7.15–7.08 (m, 2H), 7.07 (s, 1H), 6.96 (d, J = 8.2 Hz, 1H), 6.92 (d, J = 8.4 Hz, 1H), 3.97 (s,

2H), 1.75 (s, 2H), 1.40 (s, 6H), 0.76 (s, 9H). ^{13}C-NMR (CDCl$_3$, 125 MHz, δ ppm): 31.11, 32.68, 33.50, 38.56, 56.87, 108.62, 114.20, 117.23, 120.17, 127.79, 129.38, 130.93, 131.30, 132.49, 134.68, 135.68, 135.74, 148.82, 161.52. Elemental analysis C$_{29}$H$_{28}$Cl$_2$N$_2$O: 70.87 C%, 5.74 H%, 5.70 N%; found: 70.96 C%, 5.73 H%, 5.54 N%. MALDI-TOF (+) (m/z): Calc. for C$_{29}$H$_{28}$Cl$_2$N$_2$O 491.46; found 530.42 [M + K]$^+$.

3.3.2. 2(3),9(10),16(17),23(24)-Tetrakis 2-(2,4-dichloro-benzyl)-4-(1,1,3,3-tetramethyl-butyl)-phenoxy zinc (II) phthalocyanine (4)

Compound 3 (0.1 g, 0.2 mmol) and few drops of 1,8-diazabicyclo [5.4.0] undec-7-ene (DBU) in dry pentanol (3.00 mL) with anhydrous Zn(OAc)$_2$ (0.018 g, 0.1 mmol) were added in round bottom flask and heated with efficient stirring at 160 °C under inert N$_2$ atmosphere for 16 h. The green–blue color was examined during the progress of time. The observed green–blue product was cooled down to room temperature when the mixture was diluted with methanol. The obtained crude product was washed several times with methanol, water to eliminate the excessive impurities. Finally, the product was purified by silica gel chromatography (Ethyl acetate-hexane) and crystallized from methanol to obtain the target macrocycle. Yield 60 mg (58%); mp > 300 °C; C$_{116}$H$_{112}$Cl$_8$N$_8$O$_4$Zn; FT-IR ν_{max}/cm^{-1} (KBr): 3020, 1553, 1450, 1250, 1284, 1104, 925, 830, 760; ^{1}H NMR (700 MHz, CDCl$_3$, δ ppm): 7.58–6.97 (m, 36H), 4.30–4.05 (m, 8H), 1.84–1.70 (m, 8H), 1.50–1.34 (m, 24H), 0.93–0.67 (m, 36H). Elemental analysis C$_{116}$H$_{112}$Cl$_8$N$_8$O$_4$Zn: 68.59 C%, 5.56 H%, 5.52 N%; found: 68.56 C%, 5.53 H%, 5.54 N%. MALDI-TOF (+) (m/z): Calc. C$_{116}$H$_{112}$Cl$_8$N$_8$O$_4$Zn 2031.20; found 2032.82 [M+H]$^+$.

4. Conclusions

The main goal of this study was to investigate the possible use of novel soluble zinc(II) phthalocyanine **4** as a sensing material. The major advantage of this compound is its solubility in organic solvents and the absence of aggregation in the concentration range studied. Zinc(II) phthalocyanine **4** shows fluorescence properties that should be used, in turn, to investigate the sensing properties of this fluorophore in solution. This property indicates that the compound can potentially be used in medical applications. The electrochemical properties of the proposed compound were also investigated because of its possible potential usage in electro-catalysis, electro-sensing, and electrochromic devices.

Supplementary Materials: The following supporting information can be downloaded at: https://www.mdpi.com/article/10.3390/molecules28104102/s1, Figure S1: FT-IR spectrum of compounds **3** and **4**; Figure S2: MALDI-TOF mass spectrum of compound **3**; Figure S3: MALDI-TOF mass spectrum of compound **4**; Figure S4: ^{1}H NMR spectrum of **3** in CDCl$_3$, at 298 K; Figure S5: ^{13}C NMR spectrum of **3** in CDCl$_3$, at 298 K.

Author Contributions: Conceptualization, S.D., G.M. and R.P.; methodology, G.M., M.S. and F.M.; formal analysis, K.K., L.L. and F.M.; investigation, S.D., L.L. and G.M.; resources, S.D. and R.P.; data curation, K.K. and G.M.; writing—original draft preparation, S.D., G.M. and R.P.; writing—review and editing, J.E.K., M.D., G.M., L.L., M.S., F.M., C.D.N. and R.P.; supervision, J.E.K., M.D., C.D.N. and R.P.; funding acquisition, S.D. All authors have read and agreed to the published version of the manuscript.

Funding: This research was funded by the "Ministry of Higher Education and Scientific Research-Tunisia".

Institutional Review Board Statement: Not applicable.

Informed Consent Statement: Not applicable.

Data Availability Statement: Data available upon request.

Conflicts of Interest: The authors declare no conflict of interest.

Sample Availability: Sample of compound **4** available from the authors upon request.

References

1. Zheng, B.D.; He, Q.X.; Li, X.; Yoon, J.; Huang, J.D. Phthalocyanines as contrast agents for photothermal therapy. *Coord. Chem. Rev.* **2021**, *426*, 213548. [CrossRef]
2. Demirbaş, Ü.; Akçay, H.T.; Koca, A.; Kantekin, H. Synthesis, characterization and investigation of electrochemical and spectro-electrochemical properties of peripherally tetra 4-phenylthiazole-2-thiol substituted metal-free, zinc(II), copper(II) and cobalt(II) phthalocyanines. *J. Mol. Struct.* **2017**, *1141*, 643–649. [CrossRef]
3. Odabaş, Z.; Orman, E.B.; Durmuş, M.; Dumludağ, F.; Özkaya, A.R.; Bulut, M. Novel alpha-7-oxy-4-(4-methoxyphenyl)-8-methylcoumarin substituted metal-free, Co(II) and Zn(II) phthalocyanines: Photochemistry, photophysics, conductance and electrochemistry. *Dye. Pigment.* **2012**, *95*, 540–552. [CrossRef]
4. Acar, E.T.; Tabakoglu, T.A.; Atilla, D.; Yuksel, F.; Atun, G. Synthesis, electrochemistry and electrocatalytic activity of cobalt phthalocyanine complexes–effects of substituents for oxygen reduction reaction. *Polyhedron* **2018**, *152*, 114–124. [CrossRef]
5. Ertem, B.; Yalazan, H.; Güngör, Ö.; Sarkı, G.; Durmuş, M.; Saka, E.T.; Kantekin, H. Synthesis, structural characterization, and investigation on photophysical and photochemical features of new metallophthalocyanines. *J. Lumin.* **2018**, *204*, 464–471. [CrossRef]
6. Günsel, A.; Kobyaoğlu, A.; Bilgicli, A.T.; Tüzün, B.; Tosun, B.; Arabaci, G.; Yarasir, M.N. Novel biologically active metallophthalocyanines as promising antioxidant-antibacterial agents: Synthesis, characterization and computational properties. *J. Mol. Struct.* **2020**, *1200*, 127127. [CrossRef]
7. Kırbaç, E.; Erdoğmuş, A. New non-peripherally substituted zinc phthalocyanines; synthesis, and comparative photophysico-chemical properties. *J. Mol. Struct.* **2020**, *1202*, 127392. [CrossRef]
8. Bilgiçli, A.T.; Kandemir, T.; Tüzün, B.; Arıduru, R.; Günsel, A.; Abak, Ç.; Yarasir, M.N.; Arabaci, G. Octa-substituted Zinc(II), Cu(II), and Co(II) phthalocyanines with 1-(4-hydroxyphenyl) propane-1-one: Synthesis, sensitive protonation behaviors, Ag(I) induced H-type aggregation properties, antibacterial–antioxidant activity, and molecular docking studies. *Appl. Organomet. Chem.* **2021**, *35*, e6353. [CrossRef]
9. Bian, Y.; Chen, J.; Xu, S.; Zhou, Y.; Zhu, L.; Xiang, Y.; Xia, D. The effect of a hydrogen bond on the supramolecular self-aggregation mode and the extent of metal-free benzoxazole-substituted phthalocyanines. *New J. Chem.* **2015**, *39*, 5750–5758. [CrossRef]
10. Gonzalez, A.C.; Damas, L.; Aroso, R.T.; Tome, V.A.; Dias, L.D.; Pina, J.; Carrilho, R.M.B.; Pereira, M.M. Monoterpene-based metallophthalocyanines: Sustainable synthetic approaches and photophysical studies. *J. Porphyr. Phthalocyanines* **2020**, *24*, 947–958. [CrossRef]
11. Belouzard, S.; Machelart, A.; Sencio, V.; Vausselin, T.; Hoffmann, E.; Deboosere, N.; Rouillé, Y.; Desmarets, L.; Se'ron, K.; Danneels, A.; et al. Clofoctol inhibits SARS-CoV-2 replication and reduces lung pathology in mice. *PLoS Pathog.* **2022**, *18*, e1010498. [CrossRef]
12. Demirbaş, Ü.; Kobak, R.Z.U.; Barut, B.; Bayrak, R.; Koca, A.; Kantekin, H. Synthesis and electrochemical characterization of tetra-(5-chloro-2-(2, 4-dichlorophenoxy) phenol) substituted Ni (II), Fe (II) and Cu (II) metallophthalocyanines. *Synth. Met.* **2016**, *215*, 7–13. [CrossRef]
13. Breloy, L.; Yavuz, O.; Yilmaz, I.; Yagci, Y.; Versace, D.L. Design, synthesis and use of phthalocyanines as a new class of visible-light photoinitiators for free-radical and cationic polymerizations. *Polym. Chem.* **2021**, *12*, 4291–4316. [CrossRef]
14. Bayrak, R.; Akçay, H.T.; Pişkin, M.; Durmuş, M.; Değirmencioğlu, İ. Azine-bridged binuclear metallophthalocyanines functioning photophysical and photochemical-responsive. *Dye. Pigment.* **2012**, *95*, 330–337. [CrossRef]
15. Amitha, G.S.; Yoosuf Ameen, M.; Sivaji Reddy, V.; Vasudevan, S. Synthesis of peripherally tetra substituted neutral azophenoxy zinc phthalocyanine and its application in bulk hetero junction solar cells. *J. Mol. Struct.* **2019**, *1185*, 425–431. [CrossRef]
16. Sevim, A.M.; Yüzeroğlu, M.; Gül, A. Novel metallophthalocyanines with bulky 4-[3, 4-bis (benzyloxy) benzylidene] aminophenoxy substituents. *Mon. Chem. Chem. Mon.* **2020**, *151*, 1059–1068. [CrossRef]
17. Kabay, N.; Baygu, Y.; Metin, A.K.; İzzet, K.A.R.A.; EsraNur, K.A.Y.A.; Durmu, Ş.M.; Yaşar, G.Ö.K. Novel nonperipheral octa-3-hydroxypropylthio substituted metallo-phthalocyanines: Synthesis, characterization, and investigation of their electrochemical, photochemical and computational properties. *Turk. J. Chem.* **2021**, *45*, 143. [CrossRef]
18. Nas, A.; Biyiklioglu, Z.; Fandaklı, S.; Sarkı, G.; Yalazan, H.; Kantekin, H. Tetra (3-(1, 5-diphenyl, 5-dihydro-1H-pyrazol-3-yl) phenoxy) substituted cobalt, iron and manganese phthalocyanines: Synthesis and electrochemical analysis. *Inorg. Chim. Acta* **2017**, *466*, 86–92. [CrossRef]
19. Hanabusa, K.; Shirai, H. Catalytic functions and application of metallophthalocyanine polymers. *ChemInform* **1993**, *24*. Available online: https://onlinelibrary.wiley.com/action/showCitFormats?doi=10.1002%2Fchin.199343294 (accessed on 1 April 2023). [CrossRef]
20. Bouvet, M. Phthalocyanine-based field-effect transistors as gas sensors. *Anal. Bioanal. Chem.* **2006**, *384*, 366–373. [CrossRef]
21. Yıldız, B.; Baygu, Y.; Kara, İ.; Dal, H.; Gök, Y. The synthesis, characterization and computational investigation of new metalloporphyrazine containing 15-membered S4 donor macrocyclic moieties. *Tetrahedron* **2016**, *72*, 6972–6981. [CrossRef]
22. Bayrak, R.; Ataşen, S.K.; Yılmaz, I.; Yalçın, İ.; Erman, M.; Ünver, Y.; Değirmencioğlu, İ. Synthesis and Spectro-Electrochemical Properties of New Metallophthalocyanines Having High Electron Transfer Capability. *J. Mol. Struct.* **2021**, *1231*, 129677. [CrossRef]
23. Günay, İ.; Orman, E.B.; Altındal, A.; Salih, B.; Özer, M.; Özkaya, A.R. Novel tetrakis 4-(hydroxymethyl)-2, 6-dimethoxyphenoxyl substituted metallophthalocyanines: Synthesis, electrochemical redox, electrocatalytic oxygen reducing, and volatile organic compounds sensing and adsorption properties. *Dye. Pigment.* **2018**, *154*, 172–187. [CrossRef]

24. Rana, M.K.; Sinha, M.; Panda, S. Gas sensing behavior of metal-phthalocyanines: Effects of electronic structure on sensitivity. *Chem. Phys.* **2018**, *513*, 23–34. [CrossRef]
25. Magna, G.; Nardis, S.; Di Natale, C.; Perdigon, V.M.; Torres, T.; Paolesse, R. The Skeleton Counts! A study of the porphyrinoid structure's influence on sensing properties. *J. Porphyr. Phthalocyanines* **2023**, *27*, 655–660. [CrossRef]
26. Şenoğlu, S.; Özer, M.; Dumludağ, F.; Acar, N.; Salih, B.; Bekaroğlu, Ö. Synthesis, characterization, DFT study, conductivity and effects of humidity on CO_2 sensing properties of the novel tetrakis-[2-(dibenzylamino) ethoxyl] substituted metallophthalocyanines. *Sens. Actuators B Chem.* **2020**, *310*, 127860. [CrossRef]
27. Su, H.C.; Tran, T.T.; Bosze, W.; Myung, N.V. Chemiresistive sensor arrays for detection of air pollutants based on carbon nanotubes functionalized with porphyrin and phthalocyanine derivatives. *Sens. Actuators Rep.* **2020**, *2*, 100011. [CrossRef]
28. Sarıoğulları, H.; Sengul, I.F.; Gürek, A.G. Comparative study on sensing and optical properties of carbazole linked novel zinc (II) and cobalt (II) phthalocyanines. *Polyhedron* **2022**, *227*, 116139. [CrossRef]
29. Young, J.G.; Onyebuagu, W. Synthesis and characterization of di-disubstituted phthalocyanines. *J. Org. Chem.* **1990**, *55*, 2155–2159. [CrossRef]
30. Magna, G.; Belugina, R.; Mandoj, F.; Catini, A.; Legin, A.V.; Paolesse, R.; Di Natale, C. Experimental determination of the mass sensitivity of quartz microbalances coated by an optical dye. *Sens. Actuators B Chem.* **2020**, *320*, 1373. [CrossRef]

Disclaimer/Publisher's Note: The statements, opinions and data contained in all publications are solely those of the individual author(s) and contributor(s) and not of MDPI and/or the editor(s). MDPI and/or the editor(s) disclaim responsibility for any injury to people or property resulting from any ideas, methods, instructions or products referred to in the content.

Article

Synthesis of New Amino-Functionalized Porphyrins: Preliminary Study of Their Organophotocatalytic Activity

Pol Torres [1], Marian Guillén [1], Marc Escribà [1], Joaquim Crusats [1,2] and Albert Moyano [1,*]

[1] Section of Organic Chemistry, Department of Inorganic and Organic Chemistry, Faculty of Chemistry, University of Barcelona, C. de Martí i Franquès 1-11, 08028 Barcelona, Spain
[2] Institute of Cosmos Science, C. de Martí i Franquès 1-11, 08028 Barcelona, Spain
* Correspondence: amoyano@ub.edu; Tel.: +34-93-4021245

Abstract: The design, synthesis, and initial study of amino-functionalized porphyrins as a new class of bifunctional catalysts for asymmetric organophotocatalysis is described. Two new types of amine–porphyrin hybrids derived from 5,10,15,20-tetraphenylporphyrin (TPPH$_2$), in which a cyclic secondary amine moiety is covalently linked either to a β-pyrrolic position (Type A) or to the *p*-position of one of the *meso* phenyl groups (Type B), were prepared by condensation, reductive amination, or amidation reactions from the suitable porphyrins (either formyl or methanamine derivatives) with readily available chiral amines. A preliminary study of the possible use of Type A amine–porphyrin hybrids as asymmetric, bifunctional organophotocatalysts was performed using the chiral, imidazolidinone-catalyzed Diels–Alder cycloaddition between cyclopentadiene 28 and *trans*-cinnamaldehyde 29 as a benchmark reaction. The yield and the stereochemical outcome of this process, obtained under purely organocatalytic conditions, under dual organophocatalysis, and under bifunctional organophotocatalysis, were compared.

Keywords: asymmetric catalysis; Diels–Alder reaction; imidazolidin-4-ones; organocatalysis; photocatalysis; porphyrins; reductive amination

Citation: Torres, P.; Guillén, M.; Escribà, M.; Crusats, J.; Moyano, A. Synthesis of New Amino-Functionalized Porphyrins: Preliminary Study of Their Organophotocatalytic Activity. *Molecules* **2023**, *28*, 1997. https://doi.org/10.3390/molecules28041997

Academic Editors: Carlos J. P. Monteiro, M. Amparo F. Faustino and Carlos Serpa

Received: 25 January 2023
Revised: 13 February 2023
Accepted: 16 February 2023
Published: 20 February 2023

Copyright: © 2023 by the authors. Licensee MDPI, Basel, Switzerland. This article is an open access article distributed under the terms and conditions of the Creative Commons Attribution (CC BY) license (https:// creativecommons.org/licenses/by/ 4.0/).

1. Introduction

The demand for the efficient production of enantiomerically pure chiral compounds has experienced an impressive growth in the past fifty years or so, driven particularly (but not exclusively) by the pharmaceutical industry. With its ability to afford structurally diverse sets of chiral molecules with high atom economies, asymmetric catalysis has played a central role in this challenge [1,2]. As a result of efforts in this area in the last two decades, asymmetric organocatalysis has become a well-established synthetic tool, together with enantioselective transition-metal catalysis and enzyme catalysis [3,4]. In approximately the same period, we have witnessed the development of visible-light-driven photocatalysis as a powerful synthetic method for performing radical transformations in a sustainable way, triggered either by triplet–triplet energy transfer (photosensitization) [5] or by single-electron transfer (photoredox catalysis) [6,7] between a photoexcited catalyst and the substrate. However, the interplay between photocatalysis and asymmetric catalysis is challenging due to the difficulty of controlling the stereochemistry of reaction steps that involve highly energetic radical intermediates. In the wake of the pioneering work of MacMillan, in a dual-catalysis approach involving the cooperative use of both a chiral organocatalyst and a photoredox catalyst [8,9], several successful, asymmetric photocatalytic transformations have been accomplished in the past years [10–13].

On the other hand, in an attempt to emulate the efficiency and the selectivity of enzymes, there is a growing interest in the development of bifunctional catalysis, in which a single molecule combines different catalytic activation modes [14,15]. Although it is conceptually more appealing, this approach also presents significant challenges. In the area of asymmetric organophotocatalysis, the pioneering work of Bach's group [16] paved the way to further developments by Melchiorre [17] and Alemán [18].

While the best-known molecular photocatalysts are inorganic or organometallic coordination complexes of Ir and Ru, the continued interest in visible light photoredox catalysis has led to the increasing use of organic dyes as cheaper, less toxic, and more environmentally friendly photocatalysts [7,19]. Among these, porphyrins and their metalated derivatives have been extensively used as photosensitizers for singlet oxygen generation via energy transfer; however, they have been studied far less as photoredox catalysts [20]. In particular, the application of porphyrins as photoredox catalysts for carbon–carbon bond formation has only been recently explored by the Gryko group, both for the α-alkylation of aldehydes with diazo acetates [21] and for the photoarylation of π-excedent heterocycles with aryldiazonium salts [22]. More recently, de Oliveira reported on the utilization of fluorinated porphyrins as photoredox catalysts in the arylation of enol acetates [23]. In this paper, we wish to disclose in the design, synthesis, and initial study of amino-functionalized porphyrins as a new class of bifunctional catalysts for asymmetric organophotocatalysis.

2. Results and Discussion

2.1. Synthesis of Novel Amino-Functionalized Porphyrins

In the context of our research program on novel catalytic applications of porphyrins [24–26] we have disclosed that the organocatalytic activity of both achiral [27] and of chiral [28] cyclic secondary amines linked to a 4-sulfonatophenylporphyrin scaffold can be efficiently regulated by the pH of the medium, which controls the aggregation state of the amphiphilic porphyrin moiety (Figure 1).

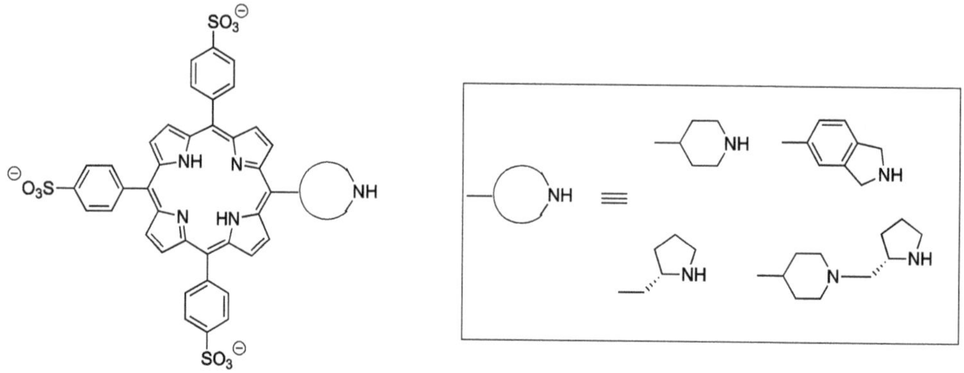

Figure 1. Amphiphilic Secondary Amine–porphyrin hybrids for pH-switchable aqueous organocatalysis.

These amine–porphyrin hybrids were prepared through a mixed-porphyrin synthesis [29] from pyrrole, benzaldehyde, and a suitable N-Boc-protected amino aldehyde, in which, after oxidation with p-chloranil, a careful chromatographic purification is necessary to obtain the desired mono-functionalized porphyrins from the statistical porphyrin mixture in low yields (4–15%). In order to overcome this drawback, we decided to investigate the synthesis of two new types of amine–porphyrin hybrids derived from 5,10,15,20-tetraphenylporphyrin (TPPH$_2$), in which the cyclic secondary amine moiety is linked either to a β-pyrrolic position (Type A) or to the p-position of one of the *meso* phenyl groups (Type B). As depicted in Figure 2, these two types of amine-functionalized porphyrins could be accessed from the known 2-formyl-5,10,15,20-tetraphenylporphyrin **1** [30] and from 5-(4-formylphenyl)-10,15,20-triphenylporphyrin **2**, [31] through condensation or reductive amination reactions with the suitable amines, taking advantage of the synthetic versatility of the formyl group.

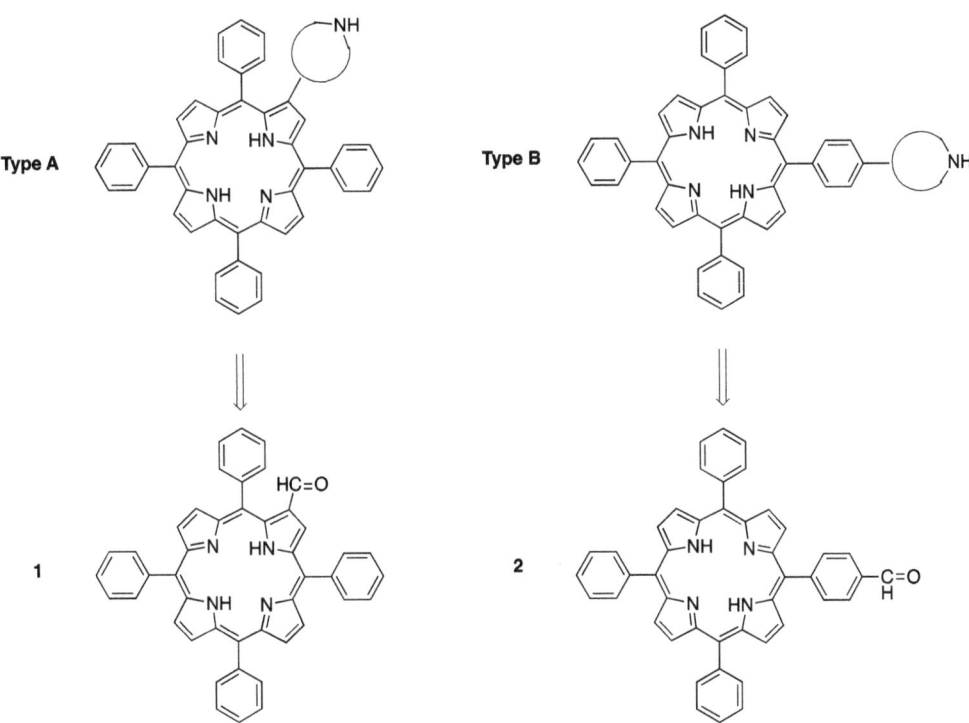

Figure 2. Secondary amine-TPPH$_2$ hybrids for bifunctional organophotocatalysis.

2-Formyl-tetraphenylporphyrin **1** and its metal complexes have been used in several instances as convenient starting materials for the synthesis of *meso*-tetraphenylporphyrins directly linked with an heterocyclic moiety at the β-position, either through 1,3-dipolar cycloadditions [32–35] or by lanthanum(III) triflate-catalyzed condensation reactions [36–39]. It is worth noting, however, that **1** had not been employed for the preparation of chiral imidazolidinone–porphyrin hybrids. The first step in the path towards amine–porphyrin hybrids of Type A was therefore the synthesis of **1**, involving the metalation and subsequent formylation of TPPH$_2$. The methodology used in this step was described by Richeter et al. in 2002 [40]. For the metalation of the porphyrin core, copper(II) acetate was added to a refluxing TPPH$_2$ solution in dichloromethane:methanol. After filtration, a purple solid was obtained with a 97% yield. Subsequently, the Vilsmeier–Haack formylation of the metalated porphyrin (TPPCu) was carried out, furnishing the copper(II) 2-formylporphyrinate complex **3** in a quantitative yield after hydrolysis with aqueous sodium acetate. In order to obtain the non-metalated porphyrin-2-carbaldehyde **1**, the crude Vilsmeier–Haack formylation mixture was treated with concentrated sulfuric acid before hydrolysis, according to the procedure of Bonfantini et al. [30] (Scheme 1).

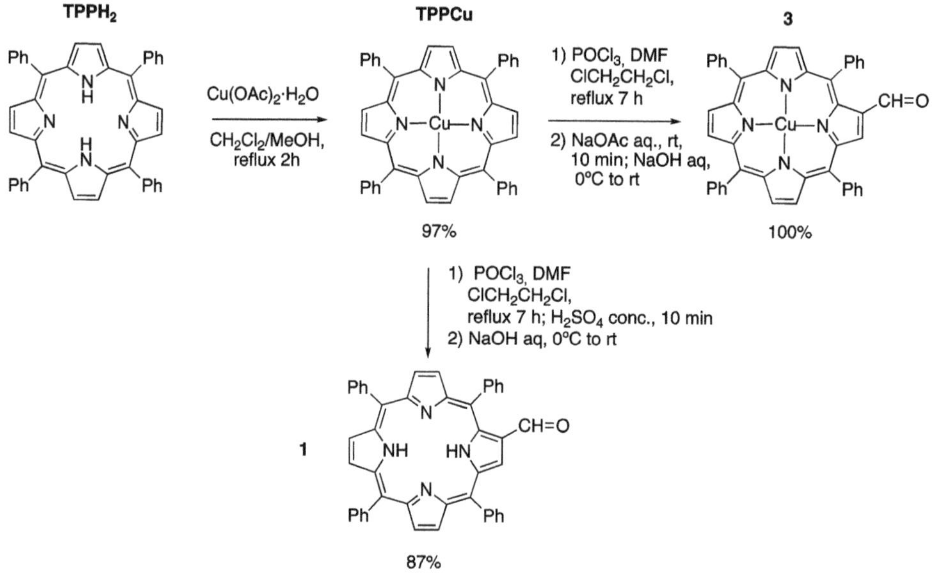

Scheme 1. Preparation of 2-formyl-5,10,15,20-tetraphenylporphyrin **1** and its Cu(II) complex **3**.

With compounds **1** and **3** in our hands, we set out to prepare the corresponding MacMillan imidazolidinones through the condensation of the formyl group with the *N*-methyl amides of the desired α-amino acids, according to the protocols described by Samulis and Tomkinson in 2011, which make use of ytterbium(III) triflate (YbTf$_3$) as a Lewis acid catalyst [41]. To that end, we obtained the *N*-methyl amides **4a**, **4b**, and **4c**, derived from L-alanine, L-phenylalanine, and L-*tert*-leucine, respectively, by a two step-route that implies the treatment of the methyl ester hydrochloride (either commercially available or easily obtained by the esterification of the amino acid in a methanol solution in the presence of thionyl chloride) [42] with an excess of methylamine, followed by neutralization with NaOH in ethanol or with aqueous ammonia [41,43]. However, all our attempts to obtain imidazolidinones derived from **1** by direct condensation with *N*-methyl amides were unsuccessful; even when a solution of **1** and of **4a** (1.1 equivalents) in chloroform was heated at reflux in the presence of YbTf$_3$ for 72 h, a mixture of starting aldehyde and the imine was obtained with no trace of the desired imidazolidinone **6a**. It is worth noting that under these conditions, Alemán and co-workers were able to achieve the formation of several thioxanthone-derived imidazolidinones from the corresponding aldehydes [18].

We reasoned that this failure could be due to the complexation of Yb(III) by the porphyrin core. In order to test this hypothesis, we submitted the Cu(II) complex **3** to the same reaction conditions. We were pleased to find that after 24 h, a 26% yield of the metalated imidazolidinone **5a** was obtained upon chromatographic purification on silica gel. A dichloromethane solution of **5a** was stirred for 4 min with concentrated H$_2$SO$_4$ (1 mL/mmol) at room temperature and poured over aqueous, concentrated NaOH at 0 °C. After extraction with chloroform, chromatographic purification afforded (with a 48% yield) the desired imidazolidinone **6a** as a single diastereomer, according to ^{1}H NMR (Scheme 2; see Figure S4 in the SM). On the precedent of the previously observed stereochemical preferences of 2-substituted imidazolidin-4-ones, [18,41] we tentatively assigned a *trans* relative stereochemistry to **6a**.

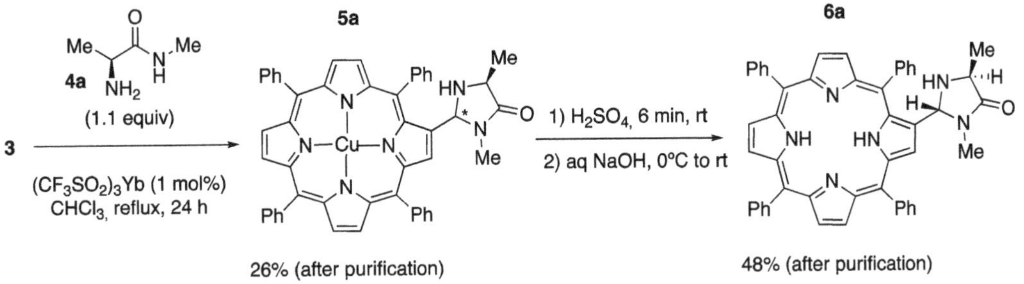

Scheme 2. Synthesis of 2-((2R,4S)-1,4-dimethyl-5-oxoimidazolidin-2-yl)-5,10,15,20-tetraphenylporphyrin **6a**. The symbol (*) indicates a stereogenic center of undetermined configuration.

On the other hand, when **3** was reacted with the methyl amide **4b**, derived from L-phenylalanine, and the crude product obtained (91% yield) was submitted to chromatographic purification, the Cu(II)-imidazolidinone **5b** was obtained with a 50% yield (as an unknown mixture of diastereomers, two spots by TLC). The demetalation of **5b** with concentrated sulfuric acid, followed by neutralization and filtration over a silica gel pad with dichloromethane:methanol 98:2, afforded the demetalated imidazolidinone as a mixture of the diastereomers **6b/6b'** with a 78% global yield. Careful chromatographic purification allowed us to isolate the pure diastereomers **6b** and **6b'** with an 18% and 26% yield, respectively (Scheme 3). As is explained in more detail in the Supplementary Materials, on the basis of their ^{1}H-NMR spectra, a relative *cis* stereochemistry was assigned to the minor isomer **6b** and a *trans* stereochemistry was assigned to the major isomer **6b'** (See Figures S1, S2 and S3 in the Supplementary Information).

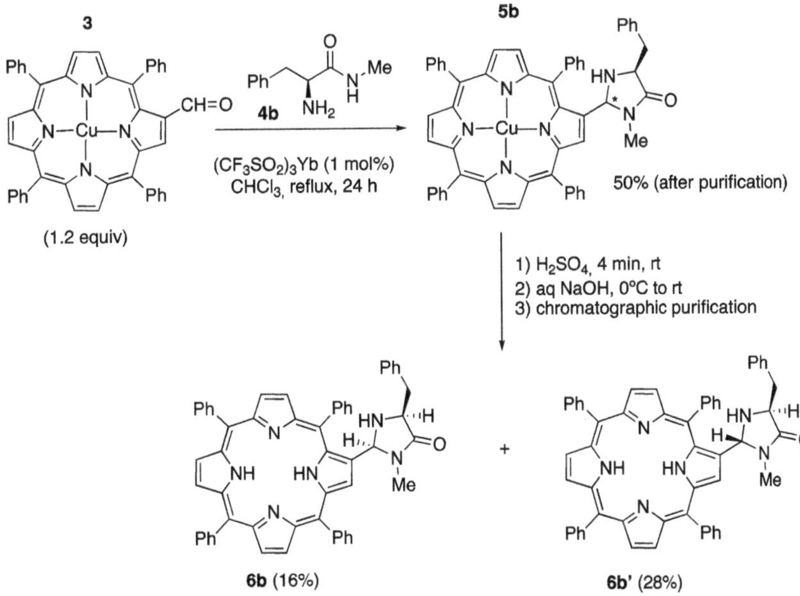

Scheme 3. Synthesis of imidazolidinone-porphyrin hybrids **6b** and **6b'**. The symbol (*) indicates a stereogenic center of undetermined configuration.

Finally, the condensation reaction between the Cu(II) 2-formylporphyrinate **3** and the *tert*-L-leucine-derived amide **4c** (1.3 equiv) was again carried out in refluxing chloroform and YbTf$_3$ as the catalyst. After 24 h, the crude product, obtained upon the evaporation of the solvent, was purified by column chromatography on silica gel to afford the compound **5c** (apparently as a diastereomer mixture). The demetalation was carried out with concentrated H$_2$SO$_4$ (4 min at rt), and the subsequent neutralization was carried out with NaOH. The crude product obtained was then purified by column chromatography, affording compound **6c** with an overall yield of 23% as a single diastereomer (Scheme 4; see Figure S9 in the Supplementary Materials). The relative stereochemistry of **6c** could not be determined by ^{1}H-^{1}H NOESY or by ^{1}H-^{1}H ROESY NMR. However, as in the case of the compounds **6a** and **6b′**, taking again into account the known preference for the formation of *trans*-imidazolidinones, [18,41] we assigned a *trans* relationship between the two substituents at C2 and C5 in the imidazolidinone ring.

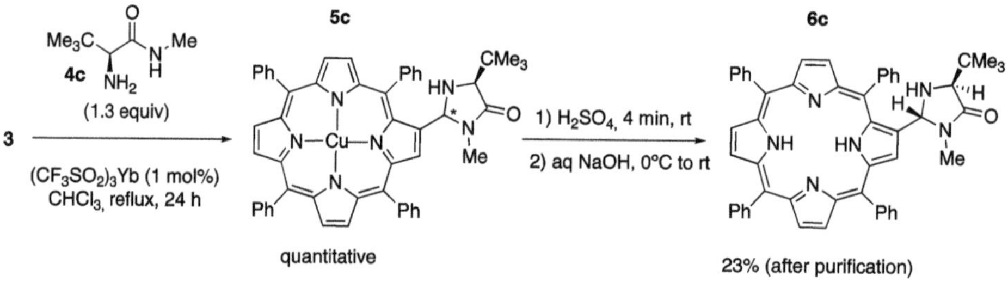

Scheme 4. Synthesis of 2-((2*R*,4*S*)-4-(*tert*-butyl)-1-methyl-5-oxoimidazolidin-2-yl)-5,10,15,20-tetraphenylporphyrin **6c**.

In summary, starting with the Cu(II) complex **3**, we were able to obtain three new tetraphenylporphyrin–imidazolidinone hybrids of Type I (**6a**, **6b**, and **6c**) through Ytterbium(III) catalyzed condensation with the suitable *N*-methyl amides **4a**, **4b**, and **4c**. Two hybrids (**6a** and **6c**) were obtained as single stereoisomers. In the case of **6b**, in which two stereoisomers were isolated, we could safely assign a *trans*-relative stereochemistry to the major one. The critical step in this sequence was the demetalation, which was accompanied by the hydrolysis of the imidazolidinone ring.

The synthesis of 5-(4-formylphenyl)-10,15,20-triphenylporphyrin **2**, the porphyrin building block for the construction of bifunctional photocatalysts of Type B, began with the preparation of 5-(4-(methoxycarbonyl)phenyl)-10,15,20-triphenylporphyrin **8** through a mixed porphyrin synthesis by the reaction of one equivalent of methyl-4-formylbenzoate **7** with three equivalents of benzaldehyde and four equivalents of pyrrole, in refluxing nitrobenzene in the presence of acetic acid [44]. After chromatographic purification, the resulting monofunctionalized porphyrin, which was obtained with a remarkably high yield (29%, corresponding to 66% of the maximum statistical yield), was reduced with lithium aluminum hydride to produce the corresponding carbinol **9**, [45] whose subsequent oxidation with pyridinium chlorochromate (PCC) afforded the target aldehyde **2** in an excellent yield. Finally, metalation with Cu(II) acetate furnished the corresponding complex **10** (Scheme 5) [31].

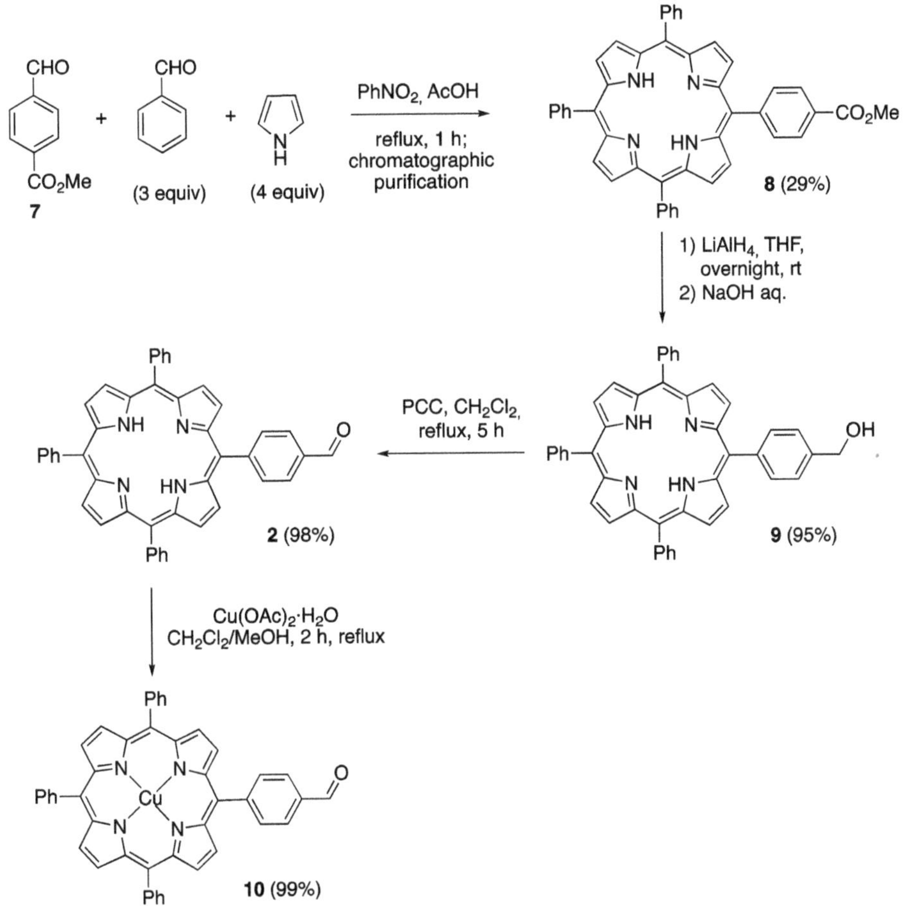

Scheme 5. Synthesis of 5-(4-formylphenyl)-10,15,20-triphenylporphyrin **2** and its Cu(II) derivative **10**.

The attempted condensation between **2** and the *N*-methyl amide **4a** (YbTf$_3$, CHCl$_3$, sealed tube, 24 h reflux) resulted, as in the case of **1**, in the formation of a complex mixture in which the imine **11** could be detected by NMR (singlet at 8.62 ppm in the ^{1}H NMR spectrum, CH=N; signal at 192.4 ppm in the ^{13}C NMR spectrum, CH=N). When the metalated aldehyde **10** was submitted to the same reaction conditions, the metalated imidazolidinone **12** was obtained with a 46% yield (after filtration through a silica gel pad). The treatment of this compound with concentrated sulfuric acid (3 min, rt) led to a complex reaction crude from which **13** was obtained in a low yield (27%) as a non-separable mixture of diastereomers, which could not be adequately characterized (Scheme 6).

In view of these difficulties, we decided to change our strategy and prepare a porphyrin–secondary amine hybrid of Type B through the reductive amination of **2** with the L-proline-derived primary amine ((*S*)-*N-tert*-butyloxycarbonyl(2-aminomethyl)pyrrolidine, **18**). This compound was prepared according to the route shown in Scheme 7, based on transformations previously described in the chemical literature [46]. The synthesis started with the reduction of L-proline to (*S*)-(2-pyrrolidin)methanol **14** with lithium aluminum hydride in a THF solution, followed by the subsequent *N*-Boc protection of the amine. The hydroxy group in **15** was activated via tosylate **16**, displaced with sodium azide to afford **17**, and finally reduced with triphenylphosphine in aqueous THF to the desired amine **18**.

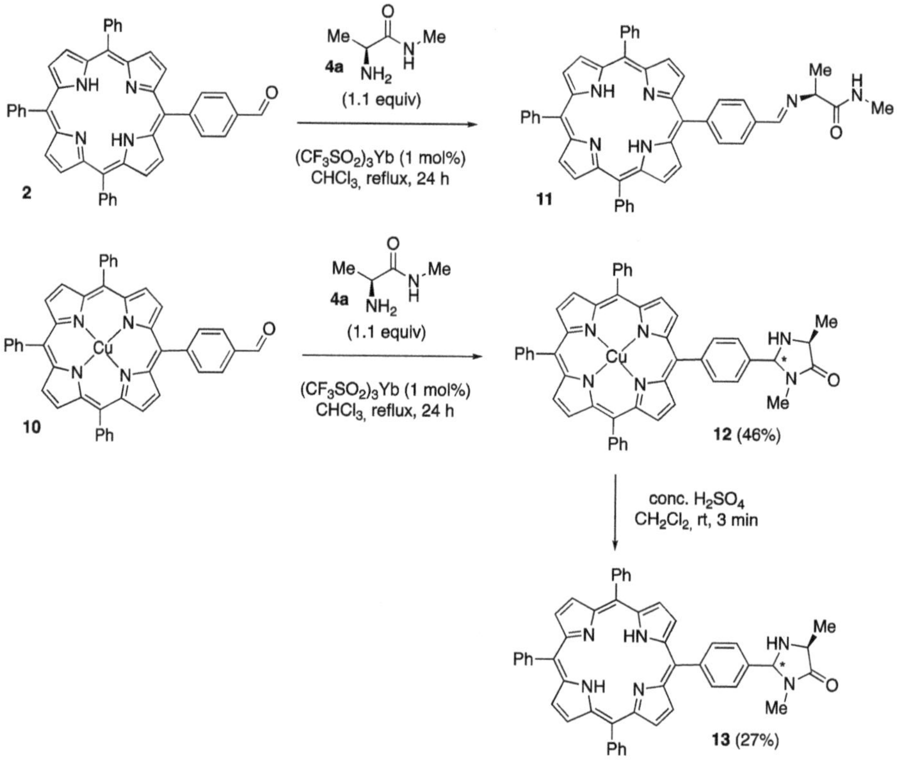

Scheme 6. Attempted synthesis of imidazolidinones derived from **2** and **10**.

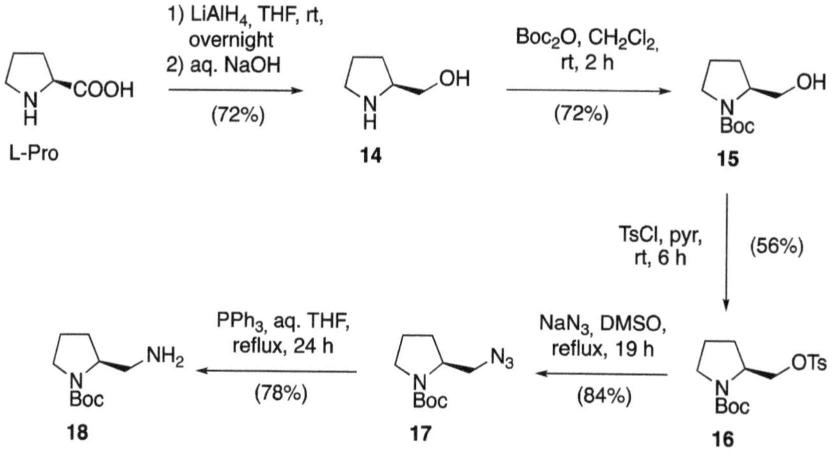

Scheme 7. Synthesis of (S)-N-tert-butyloxycarbonyl (2-aminomethyl)pyrrolidine **18**.

To obtain the desired bifunctional catalyst by the reductive amination of *meso*-tetra phenylporphyrin-carbaldehyde **2** with the primary amine **18**, we started with the conditions described by Liu et al. [46] for a related transformation (MeOH/THF, NaBH$_4$, reflux). However, we only observed the complete reduction of the aldehyde **2** to alcohol **9**. The same result was obtained when we used sodium cyanoborohydride in THF, or when we heated a chloroform solution of **2** and **50** in the presence of NaBH$_4$/YbTf$_3$.

At this point, we reasoned that the problem resided in the non-formation of the imine, and we decided to take advantage of the observed formation of the imine **11**, described above in Scheme 6. Thus, **2** was reacted with 2 equiv of the *N*-Boc amine **18** in CHCl$_3$ in the presence of a catalytic amount (2.4 mol%) of YbTf$_3$. After heating at reflux for 24 h, TLC monitoring showed that the formation of the imine intermediate was complete. After cooling to rt, 2 equiv of sodium cyanoborohydride was added in one portion and stirring was continued for 1 h. After chromatographic purification, we were pleased to find that the desired compound, **19**, could be isolated with a 67% yield. Finally, the essentially quantitative cleavage of the *N*-Boc group was achieved by treatment with an excess of trifluoroacetic acid in a chloroform solution at rt, affording the target porphyrine–pyrrolidine hybrid **20** (Scheme 8). It is worth noting that the more commonly used dichloromethane was not convenient for this purpose due to the low solubility of **19**.

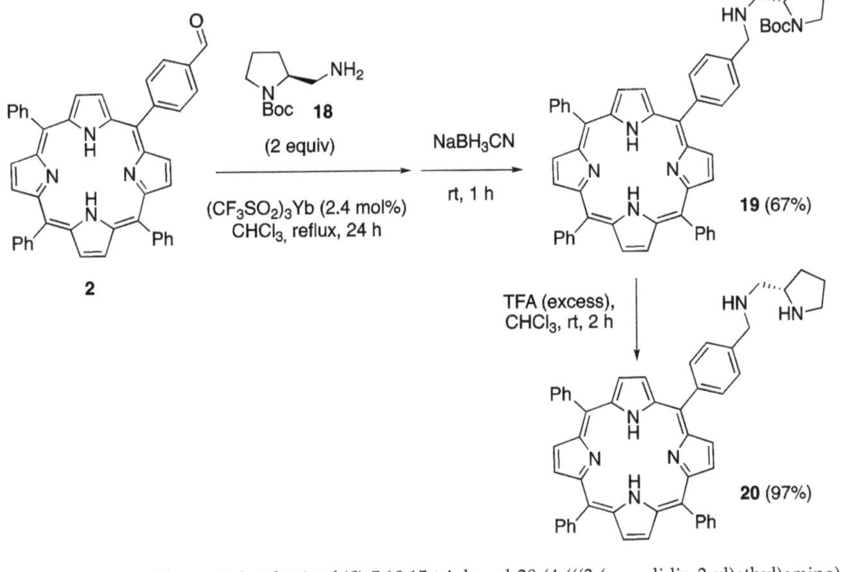

Scheme 8. Synthesis of (*S*)-5,10,15-triphenyl-20-(4-(((2-(pyrrolidin-2-yl)ethyl)amino)methyl)phenyl) porphyrin **20**.

When **2** was substituted by the Cu(II) complex **10**, reductive amination in the same conditions afforded the Cu(II) complex **21** with a good yield (51%, Scheme 9).

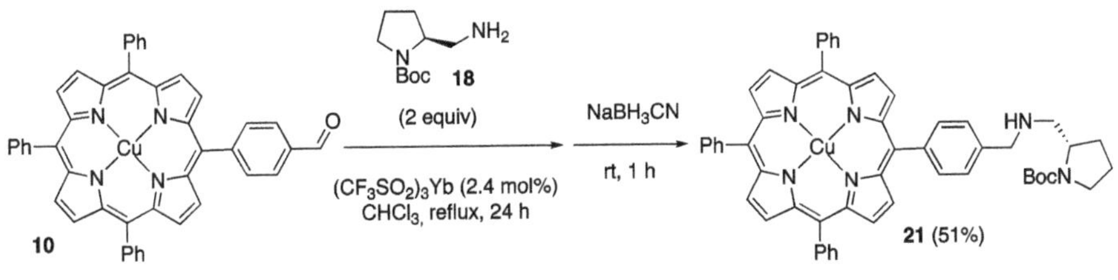

Scheme 9. Reductive amination of **10** with the pyrrolidine derivative **18**.

In order to obtain another example of an amine-functionalized porphyrin hybrid of Type B based on the formation of an amide bond, we decided to prepare the *meso*-

tetraphenylporphyrin amine derivative **24**, a compound that was first described by Bryden and Boyle [47]. To begin, the functionalized porphyrin **23**, having a nitrile group in the 4-position of one of the phenyl groups, was obtained by the mixed condensation of *p*-cyanobenzaldehyde **22**, benzaldehyde, and pyrrole in propionic acid under reflux. After two successive chromatographic purifications, **23** was obtained with a 21% yield (48% of the statistical yield). Next, the reduction of the cyano group was performed using lithium aluminum hydride, affording the desired porphyrin **24** with a 71% yield. Easily available *N*-Boc-L-proline **25** [48] was transformed to a mixed anhydride with ethyl chloroformate and triethylamine in THF at −15 °C, and a THF solution of **24** (0.83 equiv) was added dropwise. After warming to rt, the amide **26** was obtained in an almost quantitative yield following chromatographic purification. The cleavage of the *N*-Boc group by trifluoroacetic acid in chloroform provided the final porphyrin–amine hybrid **27** with a 72% yield (Scheme 10).

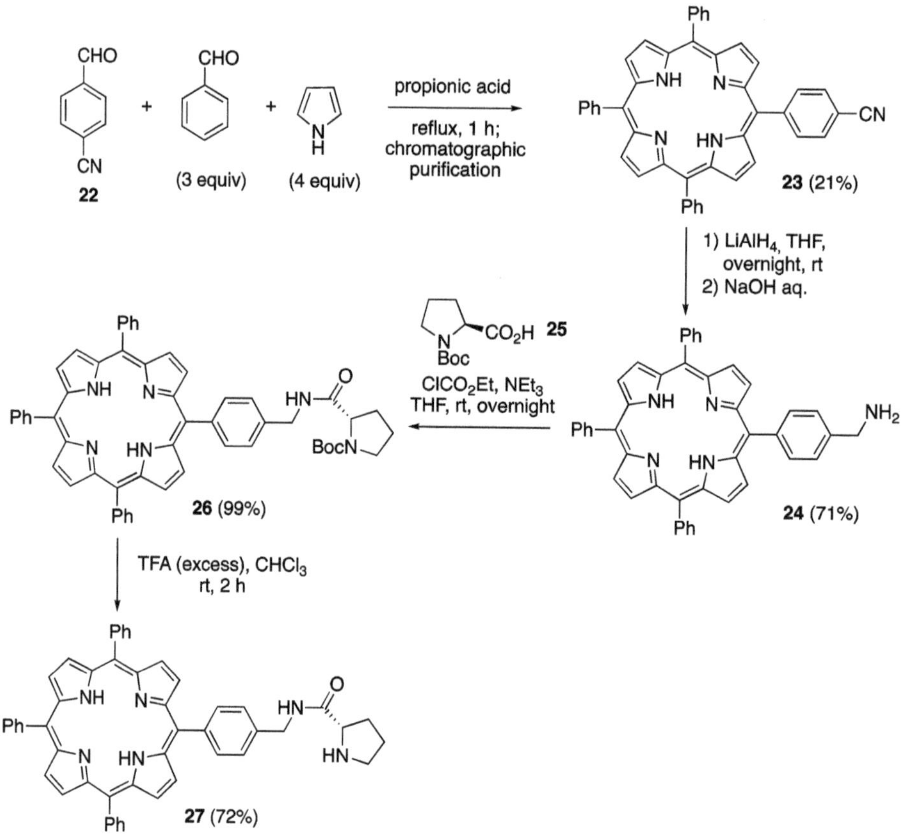

Scheme 10. Synthesis of 5,10,15-triphenyl-20-((*S*)-4-((pyrrolidine-2-carboxamido)methyl) phenyl) porphyrin **27**.

2.2. Study of the Catalytic Activity of Amine–Porphyrin Hybrids of Type A in the Organocatalyzed Diels–Alder Reaction

As a preliminary study of the possible use of amine–porphyrin hybrids as asymmetric bifunctional organophotocatalysts, we selected the chiral imidazolidinone-catalyzed Diels–Alder reaction between cyclopentadiene **28** and *trans*-cinnamaldehyde **29** as a prototypical, iminium-ion-mediated enantioselective process [49]. This reaction gives rise to a mixture of *endo*- (**30**) and *exo*-adducts (**31**) that, under optimized conditions (imidazolidinone **32** [50] as the chiral organocatalyst, HCl as the acid cocatalyst, 95:5 MeOH:H$_2$O), are obtained in a highly enantioenriched form (Scheme 11).

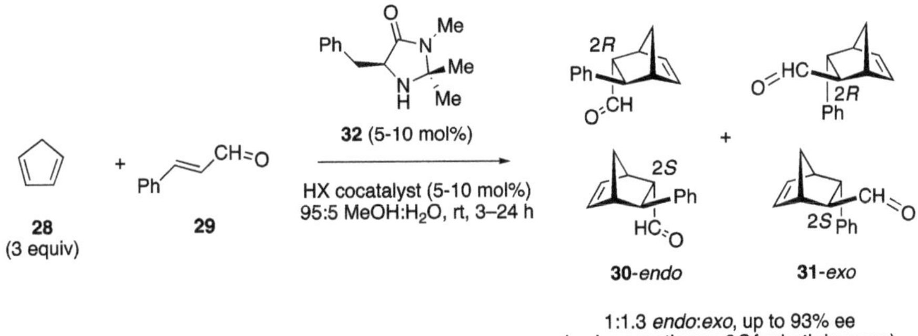

Scheme 11. Chiral imidazolidinone-catalyzed, asymmetric Diels–Alder cycloaddition.

To begin, we wanted to ascertain if the introduction of the bulky 5,10,15,20-tetra phenylporphyrin-2-yl substituent at the 2-position of the imidazolidin-4-one ring could negatively affect its organocatalytic activity (by inhibiting the formation of the cyclic iminium intermediate due to the increase in steric hindrance, for example). To that end, and taking as a benchmark reaction the catalysis by imidazolidinone **32** and *p*-toluenesulfonic acid in 95:5 MeOH-H$_2$O (Entry 1 in Table 1), we analyzed the results obtained with our Type A imidazolidinone–porphyrin hybrids **5** and **6**. The formation of the Diels–Alder adducts was monitored by TLC and ^{1}H NMR. After a partial chromatographic purification of the reaction crude (to remove unreacted **29**), the *endo:exo* ratio of the Diels–Alder adducts **30/31** was determined by the relative areas of the aldehydic protons in the ^{1}H NMR spectrum (9.60 and 9.92 ppm in CDCl$_3$, respectively; see Figure S14 in the Supplementary Materials). The adduct mixture was reduced to the corresponding alcohols by the procedure of Hayashi and co-workers (NaBH$_4$, MeOH, rt, 24 h) [51], and the enantiomeric composition was determined by chiral HPLC (see Figure S15 in the Supplementary Materials) [25].

Table 1. Asymmetric organocatalysis of the Diels–Alder cycloaddition by amine–porphyrin hybrids of Type A [a].

Entry	Organocatalyst	Solvent	% Yield [b] (dr) [c]	% ee (30/31) [d]
1	32	MeOH/H$_2$O [e]	48 (53:47)	71/81
2	5a	MeOH/H$_2$O [e]	17 (33:67)	24/17
3	5a	Toluene	30 (75:25)	4/1
4	5b	MeOH/H$_2$O [e]	36 (29:71)	37/43
5	6a	MeOH/H$_2$O [e]	0 (–)	–
6	6b	Toluene	9 (75:25)	3/1
7	6b'	Toluene	27 (68:32)	3/49
8	6c	Toluene	0 (–)	–

[a] Reaction conditions: cinnamaldehyde **29** (0.5 mmol, 1 equiv), cyclopentadiene **28** (1.5 mmol, 3 equiv), oerganocatalyst (5 mol%), *p*-TsOH (5 mol%), solvent (1 mL), rt, 72 h. [b] Isolated yield of adducts (**30** + **31**) after chromatographic purification. [c] *Endo:exo* ratio, determined by ^{1}H NMR (400 MHz). [d] Determined by chiral HPLC analysis of the mixture of alcohols obtained upon reduction with NaBH$_4$/MeOH. The major enantiomers had an absolute (2*S*) configuration in all instances. [e] 95:5 MeOH/H$_2$O.

When the metalated imidazolidinone **5a** (derived from L-alanine, see Scheme 2) was used as a catalyst, we observed that it was only partially soluble in 95:5 MeOH:H$_2$O, and that the isolated yield of the adducts was only 17% after 72 h (Entry 2 in Table 1). The *endo/exo* ratio and the enantiomeric purities were also very different from those observed for the reference reaction with imidazolidinone **32**. When the reaction was performed in toluene, in which the porphyrine complex **5a** was more soluble, the yield increased to 30% (entry 3). In this case, the *endo* adduct **30** was the major product; however, both adducts were obtained in essentially racemic form. With the metalated porphyrin **5b**, which was derived from L-phenylalanine (Scheme 3), the reaction could be performed in aqueous MeOH with a global yield of 36% (Entry 4 in Table 1). In this case, the *exo* adduct **31** was the main product, which was obtained with a moderate ee (43%). The *endo* isomer **30** showed a similar enantiomeric purity (37% ee). However, as **5b** was obtained as a *cis/trans* diastereomer mixture of unknown composition, these results are difficult to rationalize. We then redirected our attention to the non-metalated porphyrins **6**. The bifunctional porphyrin **6a** was not soluble in aqueous MeOH, and only traces of product were observed after 72 h (Entry 5 in Table 1). We then evaluated the organocatalytic activity of the two *cis* and *trans* isomers of **6b** in toluene. The *cis* isomer **6b** (Entry 6) gave a poor yield (9%), with a stereoselectivity (3:1 *endo:exo*, essentially racemic products) very similar to that observed for **5a** in the same solvent (Entry 3). On the other hand, the *trans* isomer **6b'** (Entry 7 in Table 1) provided a higher yield (27%) with a similar diastereoselectivity (*ca.* 70:30 *endo:exo*). While the *endo* adduct was obtained with a very low ee (3% ee), the *exo* was formed with a nearly 50% ee. The *tert*-leucine-derived porphyrin–imidazolidinone hybrid **6c** (Scheme 4) did not catalyze the reaction, apparently due to steric hindrance (Entry 8). In summary, while the introduction of the 5,10,15,20-tetraphenylporphyrin-2-yl moiety had a significant impact on both the yield and in the stereochemical outcome of the Diels–Alder cycloaddition, our imidazolidinone–porphyrin hybrids of Type A still present a significant organocatalytic activity in most cases (except for the *tert*-leucine-derived compound **6c**).

In order to evaluate the impact of photosensitization on the imidazolidinone-catalyzed Diels–Alder cycloaddition of Scheme 11, we decided to perform the reaction under visible light irradiation using both 5,10,15,20-tetraphenylporphyrin (TPPH$_2$) and 5,10,15,20-tetrakis(4-sufonatophenyl)porphyrin (as the sodium salt Na$_4$TPPH$_2$S$_4$ or in zwitterionic form (H$_3$O)$_2$TPPH$_4$S$_4$) [25] as photocatalysts. Additionally, (*S*)-5-benzyl-2,2,3-trimethyl imidazolidin-4-one **32**, (*S*)-2,2,3,5-tetramethylimidazolidin-4-one **33** [50], and (2*S*,5*S*)-5-benzyl-2-*tert*-butyl-3-methylimidazolidin-4-one **34** [43] were used as organocatalysts (Figure 3). The results of these dual organophotocatalyzed Diels–Alder reactions are summarized in Table 2.

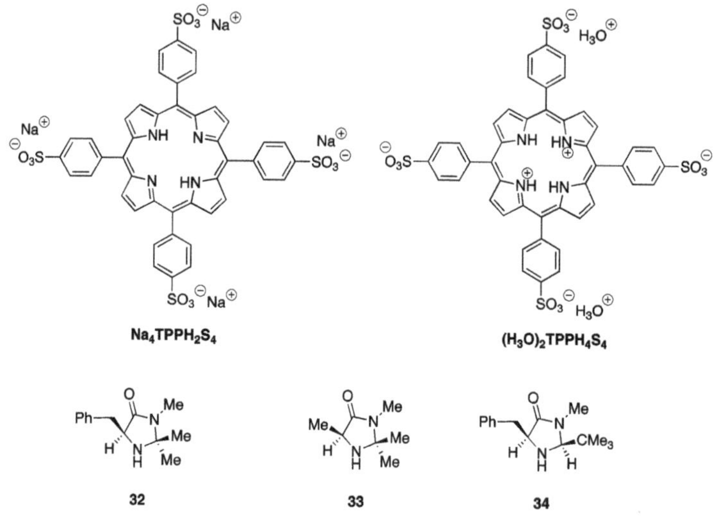

Figure 3. Porphyrines and chiral imidazolidinones used in the dual organophotocatalyzed asymmetric Diels–Alder cycloaddition. See Scheme 1 for the structure of **TPPH$_2$**.

Table 2. Dual organophotocatalyzed, asymmetric Diels–Alder cycloaddition [a].

Entry	Organocatalyst	Photocatalyst	Solvent	% Yield [b] (dr) [c]	% ee (30/31) [d]
1	32	TPPH$_2$	MeOH	99 (53:47)	69/76
2	32	TPPH$_2$	Toluene	0 (–)	–
3	32	(H$_3$O)$_2$TPPH$_4$S$_4$	MeOH	96 (55:45)	60/72
4 [e]	32	Na$_4$TPPH$_2$S$_4$	MeOH/H$_2$O [f]	34 (56:44)	37/24
5	33	TPPH$_2$	MeOH	90 (52:48)	61/67
6	34	TPPH$_2$	Toluene	0 (–)	–
7	34	TPPH$_2$	MeOH	0 (–)	–

[a] Reaction conditions: cinnamaldehyde 29 (0.5 mmol, 1 equiv), cyclopentadiene 28 (1.5 mmol, 3 equiv), organocatalyst (5 mol%), p-TsOH (5 mol%), photocatalyst (5 mol%), solvent (1 mL), irradiation with white light, rt, 72 h. [b] Isolated yield of adducts (30 + 31) after chromatographic purification. [c] Endo:exo ratio, determined by ^{1}H NMR (400 MHz). [d] Determined by chiral HPLC analysis of the mixture of alcohols obtained upon reduction with NaBH$_4$/MeOH. The major enantiomers had an absolute (2S) configuration in all instances. [e] Without p-TsOH. [f] 95:5 MeOH/H$_2$O.

When the reaction was performed in an MeOH solution with imidazolidinone 32 as the organocatalyst (i.e., under the conditions of the benchmark reaction, Scheme 11 and Entry 1 of Table 1) but under white light irradiation in the presence of *meso*-tetraphenylporphyrin (5 mol%), the stereochemical outcome (dr, %ee for both adducts) of the reaction was essentially the same as in the purely thermal reaction (Entry 1 of Table 2). However, we were pleased to find that the mixture of adducts 30/31 was obtained in an essentially quantitative yield after chromatographic purification. It is important to note that this was not due to external heating of the solution (see Experimental Section). A drastic change was observed when toluene was used as a solvent (Entry 2); in this case, no reaction was observed, probably due to the very low solubility of the p-toluenesulfonate salt of 32. When the zwitterionic form of 5,10,15,20-tetrakis(4-sulfonatophenyl)porphyrin ((H$_3$O)$_2$TPPH$_4$S$_4$) was used as a photocatalyst in MeOH solution, the results were also very similar to those obtained in Entry 1 with TPPH$_2$, except that in this case, a small decrease in the enantiomeric purities were observed (Entry 3 of Table 2). The use of the basic form of the same photocatalyst in aqueous methanol (Na$_4$TPPH$_2$S$_4$, Entry 4 in Table 2) resulted in a much lower yield and diminished enantiomeric purities. This likely reflects the important role played by the acid co-catalyst as, in this case, the reaction was performed in the absence of p-toluenesulfonic acid. The replacement of the 5-benzyl group with a methyl one (imidazolidinone 33, Entry 5 in Table 2) maintained both the yield and the *endo:exo* adduct ratio with respect to Entry 1; a small decrease in the enantiomeric purity took place, in accordance with the role attributed by MacMillan to the benzyl imidazolidinone substituent in determining the stereochemical outcome of the cycloaddition [4,49]. When the *tert*-butyl-substituted imidazolidinone 34 was used as the organocatalyst, no reaction was observed either in toluene (entry 6) or in methanol (entry 7), probably due to steric hindrance, which impeded the formation of the iminium ion intermediate. In summary, we found that the irradiation of the reaction mixture with visible light in the presence of a porphyrin photosensitizer did not appreciably change the stereochemical course of the Diels–Alder cycloaddition (as expected) but was accompanied by a great increase in the yield.

In light of these results, we finally performed the Diels–Alder cycloaddition between cyclopentadiene 28 and *trans*-cinnamaldehyde 29 under visible light irradiation, using the imidazolidinone–porphyrin hybrids of Type I 6b, 6b', and 6c as bifunctional organophotocatalysts in toluene, and in the presence of p-toluenesulfonic acid as a co-catalyst.

As revealed in Table 3, the resultant reactions, catalyzed by the imidazolidinone–porphyrin hybrids 6b (*cis*, entry 1) and 6b' (*trans*, entry 2), were similar to those obtained under purely thermal conditions: low yields and poor enantiocontrol (compared with Entries 6 and 7 in Table 1). However, it is important to note that in both instances a very high, unprecedented diastereoselectivity in favor of the formation of the *endo* adduct 30 (*ca.* 9:1) was observed. For the bifunctional catalyst 6c, the cycloaddition did not take place (Entry 3 in Table 3), in line with the results obtained under purely thermal conditions (Entry 8 in Table 1). This was likely due to the steric hindrance of the imidazolidinone moiety.

Table 3. Bifunctional organophotocatalysis of the Diels–Alder cycloaddition by amine–porphyrin hybrids of Type A [a].

Entry	Organophotocatalyst	Solvent	% Yield [b] (dr) [c]	% ee (30/31) [d]
1	6b	Toluene	12 (88:12)	1/4
2	6b'	Toluene	21 (85:15)	2/2
3	6c	Toluene	0 (–)	–

[a] Reaction conditions: cinnamaldehyde **29** (0.5 mmol, 1 equiv), cyclopentadiene **28** (1.5 mmol, 3 equiv), bifunctional organophotocatalyst (5 mol%), p-TsOH (5 mol%), toluene (1 mL), irradiation with white light, rt, 72 h. [b] Isolated yield of adducts (**30 + 31**) after chromatographic purification. [c] Endo:exo ratio, determined by ^{1}H NMR (400 MHz). [d] Determined by chiral HPLC analysis of the mixture of alcohols obtained upon reduction with NaBH$_4$/MeOH. The major enantiomers had an absolute (2S) configuration in all instances.

These important differences in the stereochemical outcome of the Diels–Alder cycloaddition under visible light irradiation, using a dual catalytic system (Table 2) on one hand and a bifunctional catalyst (Table 3) on the other, suggest that, in the last case, a competition between thermal and photochemical reaction paths may take place. Under dual catalysis conditions, in which only an increase in the yield is observed, photosensitization, followed by internal conversion, may facilitate non-radiative energy transfer to key intermediates in the organocatalytic cycle. For the bifunctional catalysis, however, the drastic change observed in the stereoselectivity of the process (increased *endo*-selectivity and practically total loss of the enantiocontrol) suggests that the reaction may follow an alternative mechanism, involving the photoexcitation of the porphyrin moiety in the unsaturated iminium ion, followed by an intramolecular single-electron transfer via oxidative quenching to provide a cationic diradical that can trigger a radical cyclization process with the diene (Scheme 12).

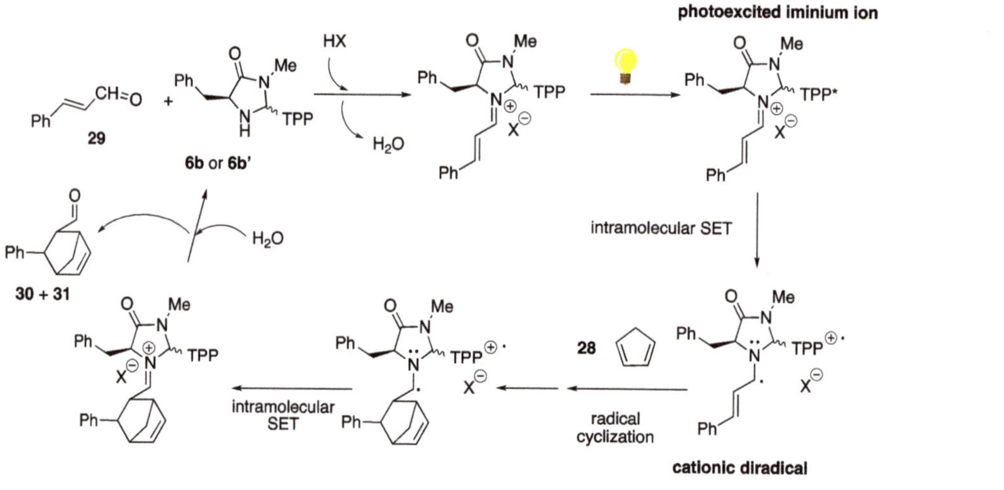

Scheme 12. Speculative reaction pathway for the bifunctional organophotocatalysis of the Diels–Alder cycloaddition (TPP = 5,10,15,20-tetraphenylporphyrin-2-yl).

In conclusion, we synthesized the first examples of a novel class of amine–porphyrin hybrids. In these hybrids, as potentially active bifunctional organophotocatalysts, a chiral cyclic moiety is covalently attached to a *meso*-tetraphenylporphyrin core. In these compounds, the cyclic secondary amine moiety can be linked either to a β-pyrrolic position (Type A) or to the *p*-position of one of the *meso* phenyl groups (Type B). These two types of amine-functionalized porphyrins have been accessed from the known 2-formyl-5,10,15,20-tetraphenylporphyrin **1** (compounds **6a**, **6b**, and **6c** and the corresponding metalated Cu(II) derivatives **5a–c**) and from 5-(4-formylphenyl)-10,15,20-triphenylporphyrin **2** (compound **20**) through condensation or reductive amination reactions with the suitable chiral amines. An additional example of a Type B bifunctional catalyst, 5,10,15-triphenyl-20-((S)-4-((pyrrolidine-2-carboxamido)methyl)phenyl)porphyrin **27**, was obtained by the formation of an amide between 5-(4-(aminomethyl)phenyl)-10,15,20-triphenylporphyrin **24** and N-Boc-L-proline **25**. A preliminary study of the organophotocatalytic activity of the Type A amine–porphyrin hybrids, performed using the chiral imidazolidinone-promoted Diels–Alder cycloaddition between cinnamaldehyde and cyclopentadiene, demonstrated that (a) the introduction of the 5,10,15,20-tetraphenylporphyrin-2-yl substituent at the 2-position of the imidazolidin-4-one ring did not suppress the organocatalytic activity the amine, except in the case of **6c**, which has the very bulky *tert*-butyl substituent at the 5-position, (b) the irradiation of the reaction mixture with visible light in the presence of a porphyrin photosensitizer (dual organophotocatalysis) did not appreciably change the stereochemical course of the chiral imidazolidinone-catalyzed Diels–Alder cycloaddition, but was accompanied by a great increase in the yield, and (c) the bifunctional catalysis of the Diels–Alder cycloaddition under visible light irradiation resulted in a dramatic change in the stereoselectivity of the process (increased *endo*-selectivity and a practically total loss of enantiocontrol); this suggests that, in this case, the reaction may follow an alternative mechanism, involving the photoexcitation of the porphyrin moiety in the bifunctional catalyst. Further assessment of the organophotocatalytic activity of these amine-functionalized porphyrins is being performed in our laboratory.

3. Materials and Methods

3.1. General Methods

Commercially available reagents, catalysts, and solvents were used as received from the supplier. Dichloromethane for porphyrin synthesis was distilled from CaH_2 prior to use, and THF was dried by distillation from $LiAlH_4$. Deuterated solvents were supplied by Merck Life Science. For normal phase HPLC chromatography, HPLC grade solvents (hexane and isopropyl alcohol) were used directly without any purification beyond that already applied by the supplier (VWR).

Thin-layer chromatography was carried out on silica gel plates Merck 60 F_{254}, and compounds were visualized by irradiation with UV light and/or and chemical developers ($KMnO_4$, *p*-anisaldehyde, and phosphomolybdic acid). Chromatographic purifications were performed under pressurized air in a column with silica gel Merck 60 (particle size: 0.040–0.063 mm, Merck Life Science S.L.U., Spain) as stationary phase and solvent mixtures (hexane, ethyl acetate, dichloromethane, and methanol) as eluents.

^1H (400 MHz) NMR spectra were recorded with a Varian Mercury 400 spectrometer (Agilent Technologies, Santa Clara, Cal, USA). Chemical shifts (δ) are provided in ppm relative to the peak of tetramethylsilane (δ = 0.00 ppm), and coupling constants (J) are provided in Hz. The spectra were recorded at room temperature. Data are reported as follows: s, singlet; d, doublet; t, triplet; q, quartet; m, multiplet; br, broad signal. IR spectra were obtained with a Nicolet 6700 FTIR instrument (Thermo Fisher Scientific, Waltham, MA, USA), using ATR techniques. UV–vis spectra were recorded on a double-beam Cary 500-scan spectrophotometer (Varian); cuvettes (quartz QS Suprasil, Hellma, Hellma GmbH & Co. KG, Mülheim, Germany) cm were used for measuring the absorption spectra. The porphyrin solutions in water were carefully degassed by gentle bubbling a nitrogen gas stream prior to the spectrophotometric measurement.

The chiral HPLC analyses of the Diels–Alder reaction products were performed on a Shimadzu instrument (Shimadzu Europa GmbH, Essen, Germany) containing a LC-20-AD solvent delivery unit, a DGU-20AS degasser unit, and a SPD-M20A UV/VIS Photodiode Array detector with a chiral stationary phase (250 mm × 4.6 mm Phenomenex® i-cellulose-5 column; Phenomenex España, S.I., C. de Valgrande 8, Alcobendas, 28018, Madrid, Spain). All solvents were of HPLC grade and were carefully degassed prior to use. The sample was injected at time 0.

Meso-tetraphenylporphyrin $TPPH_2$, [52] 5,10,15,20-tetrakis(4-sufonatophenyl)porphyrin tetrasodium salt $Na_4TPPH_2S_4$, [25] amides **4a**, [41,53] **4b**, [41] and **4c**, [43] N-Boc-(S)-2-

(aminomethyl)pyrrolidine **18**, [46] *N*-Boc-L-proline **25**, [48] and imidazolidinones **32**, [50] **33**, [50] and **34**, [43] were prepared according to previously described procedures. See the Supplementary Materials for more details.

3.2. Synthetic Procedures and Product Characterization

3.2.1. Synthesis of Porphyrin-Derived Imidazolidinones

Copper(II) meso-tetraphenylporphyrin (TPPCu). In a 250 mL round-bottomed flask, equipped with magnetic stirring and a Dimroth reflux condenser, 5,10,15,20-tetraphenyl porphyrin TPPH$_2$ (450 mg, 0.7 mmol) was dissolved in DCM (60 mL). Once all the solid was dissolved, MeOH (20 mL) and Cu(OAc)$_2$·H$_2$O (243 mg, 1.2 mmol) were added, and the reaction mixture was heated up to reflux and stirred for 2 h until all the starting material was consumed (one spot in TLC, Hexane/DCM (1/1)). The solvent was then removed by distillation under reduced pressure, and the residue was redissolved in the minimum amount of DCM and filtered through a short plug of silica gel, using DCM as an eluent. After filtration, the solvents were evaporated under reduced pressure to afford the desired compound as a purple solid (432 mg, 97% yield), which was not characterized [40].

2-Formyl-5,10,15,20-tetraphenylporphyrin (**1**). Dry *N*,*N*-dimethylformamide (11.8 mL, 152.04 mmol) was placed in a 500 mL round-bottomed flask under argon and cooled in an ice bath. Phosphorus oxychloride (9.3 mL, 99.12 mmol) was added slowly, forming a viscous, golden mixture containing the Vilsmeier–Haack reagent. In another 500 mL round-bottomed flask, copper(II) meso-tetraphenyl porphyrin TPPCu (1.05 g, 1.55 mmol) was stirred with dry 1,2-dichloroethane (105 mL) under argon and cooled in ice. The porphyrin solution was added to the Vilsmeier–Haack reagent, and the argon source was replaced with a drying tube before warming the mixture to room temperature. It was then heated at reflux for 7 h. The reaction was cooled overnight before adding concentrated sulfuric acid (19.2 mL, 360.8 mmol) to the vigorously stirred mixture. After stirring for 10 min, the green two-phase mixture was poured into ice-cold aq. sodium hydroxide (30.2 g in 1.05 L) in a separating funnel. Chloroform (0.59 L) was then added. The mixture was shaken until no green color remained or reappeared. The bottom layer was separated and washed twice with saturated aqueous NaHCO$_3$ (2 × 0.42 L). The organic layer was then dried with anhydrous MgSO$_4$ and the solvent was evaporated in vacuum. The solid was filtered through a silica pad with DCM as eluent to produce 0.863 g (87% yield) of the desired product **1** [30].

Purple solid. ^{1}**H NMR** (400 MHz, CDCl$_3$) δ 9.41 (s, 1H, CHO), 9.24 (s, 1H), 8.93–8.85 (m, 4H), 8.78–8.77 (m, 2H), 8.28–8.15 (m, 8H), 7.85–7.75 (m, 12H), −2.54 (s, 2H) ppm.

Copper (II) 2-formyl-5,10,15,20-tetraphenylporphyrinate (3). In a 500 mL round-bottomed flask equipped with magnetic stirring and a Liebig reflux condenser, under an Ar atmosphere, dry *N*,*N*-dimethylformamide (11.1 mL, 154 mmol) was added and cooled down in an ice-water bath. Then, phosphorous oxychloride (8.5 mL, 91.8 mmol) was added slowly, showing the formation of the Vilsmeier–Haack complex as a viscous, golden mixture. In another 500 mL round-bottomed flask, equipped with magnetic stirring, copper(II) meso-tetraphenyl porphyrin **TPPCu** (1.37 g, 2.0 mmol) was added and purged with Ar. At this point, dry 1,2-dichloroethane (210 mL) was added, and the resulting suspension was stirred in an ice-water bath until no solid remained. The porphyrin solution was then added via cannula to the Vilsmeier–Haack complex, and the resulting mixture was heated up to reflux for 7 h and stirred overnight at room temperature. The reaction mixture was poured into an aqueous solution of NaOAc (105 g in 500 mL) and stirred for 10 min. After this time, the green two-phase mixture was transferred to a separatory funnel and extracted with an aqueous solution of NaOH (36 g in 1.25 L) until no green color was observed. the aqueous phase was then extracted with CHCl$_3$ (3 × 200 mL), and the combined organic layers were washed with an aqueous saturated solution of NaHCO$_3$ (2 × 500 mL), dried over anhydrous MgSO$_4$, and evaporated under reduced pressure. The residue was redissolved in the minimum amount of DCM and filtered through a short plug of silica gel, using DCM as eluent to afford the desired product **3** (1.60 g, quantitative yield) as a purple solid [40].

Red-purple solid. **UV-vis** [toluene, λ_{max} (ε), 3.75×10^{-6} M]: 430 (181,400), 551 (9650), 591 (5700) nm.

Copper(II) 2-((2R/2S,4S)-1,4-dimethyl-5-oxoimidazolidin-2-yl)-5,10,15,20-tetra phenylporphyrinate (5a). A solution of the metalated porphyrincarbaldehyde 3 (0.50 g, 0.71 mmol), amide 4a (0,098 g, 0.96 mmol), and Yb(SO$_3$CF$_3$)$_3$ (5 mg, 0.008 mmol) in CHCl$_3$ (7 mL) in a sealed tube was heated to reflux for 24 h. After cooling, the reaction mixture was evaporated in vacuo. It was then purified via flash column chromatography in silica gel, using DCM:MeOH (98:2) as an eluent, to afford 0.147 g (26% yield) of the desired compound 5a as an unknown mixture of diastereomers.

Purple solid. **HRMS (ESI)** m/z calculated for C$_{49}$H$_{37}$CuN$_6$O ([M+H]$^+$), 788.2319; found 788.2317. **UV-vis** [toluene, λ_{max} (ε), 3.75×10^{-6} M]: 420 (801,300), 543 (42,700), 577 (850) nm.

2-((2R,4S)-1,4-Dimethyl-5-oxoimidazolidin-2-yl)-5,10,15,20-tetraphenylporphyrin (6a). In a 5 mL round-bottomed flask equipped with magnetic stirring, 98% sulfuric acid (0.38 mL) was added to a vigorously stirred solution of porphyrin 5a (25 mg, 0.316 mmol) in dry DCM (2 mL). After stirring for 4 min at rt, the green two-phase mixture was poured into ice-cold aqueous sodium hydroxide (0.576 g in 20 mL) in a separating funnel. CHCl$_3$ (11.2 mL) was then added. The mixture was shaken until no green color remained or reappeared in the organic phase. The bottom layer was separated and washed twice with saturated aqueous NaHCO$_3$ (2 × 8 mL). The organic layer was then dried over anhydrous MgSO$_4$, and the solvent was evaporated in vacuo. The resulting solid was filtered through a silica gel pad, using DCM:MeOH (97:3) as an eluent, to produce 11 mg (48% yield) of the desired product 6a as a single diastereomer to which a trans relative stereochemistry was tentatively assigned.

Purple solid. **^{1}H NMR** (400 MHz, CDCl$_3$) δ 8.85–8.75 (m, 3H), 8.69 (d, J = 4.7 Hz, 1H), 8.62 (s, 1H), 8.4 (d, J = 4.6 Hz, 1H), 8.25–8.15 (m, 8H), 8.09 (d, J = 7.4 Hz, 1H), 7.85–7.7 (m, 12H), 5.33 (s, 1H), 3.71 (q, J = 7.0 Hz, 1H), 2.78 (s, 3H), 1.24 (d, J = 7.0 Hz, 3H), −2,71 (s, 2H). **HRMS (ESI)** m/z calculated for C$_{49}$H$_{39}$N$_6$O ([M+H]$^+$), 727.3180; found 727.3179. **UV-vis** [toluene, λ_{max} (ε), 3.75×10^{-6} M]: 422 (292,600), 518 (15,300), 551 (7000), 595 (5300), 651 (4200) nm.

Copper(II) 2-(2R/2S,4S)-4-benzyl-1-methyl-5-oxoimidazolidin-2-yl)-5,10,15,20-tetraphenyl porphyrinate (5b). Freshly prepared freshly prepared L-phenylalanine methylamide 4b (198 mg, 1.11 mmol) was introduced in a 25 mL round-bottomed flask equipped with magnetic stirring and a Dimroth reflux condenser. A solution of copper(II) 2-formyl-5,10,15,20-tetraphenyl porphyrin 3 (960 mg, 1.36 mmol) in CHCl$_3$ (12 mL) and Yb(SO$_3$CF$_3$)$_3$ (6.0 mg, 0.010 mmol, 1 mol%) were then added sequentially. The stirred mixture was heated up to reflux for 24 h. After cooling to room temperature, the solvent was evaporated under reduced pressure and purified via flash column chromatography through Et$_3$N-pretreated silica gel (2.5% v/v Net$_3$). A mixture of DCM/MeOH 0.5% v/v was used an eluent. The fast-running red band, which contained unreacted starting material, was not collected. Compound 5b (purple solid, 481 mg, 50% yield) was obtained as a diastereomeric mixture (two spots in TLC, DCM/MeOH 0.5%). It was not further characterized.

2-((2S,4S)-4-Benzyl-1-methyl-5-oxoimidazolidin-2-yl)-5,10,15,20-tetraphenylporphyrin (6b-cis) and 2-((2R,4S)-4-benzyl-1-methyl-5-oxoimidazolidin-2-yl)-5,10,15,20-tetraphenyl porphyrin (6b'-trans). In a 50 mL round-bottomed flask equipped with magnetic stirring, 312 mg (0.36 mmol) of metalated porphyrin 5b was dissolved in DCM (20 mL). Concentrated H$_2$SO$_4$ (5 mL, 98%) was added, and the resulting green mixture was vigorously stirred for 4 min. The two-phase mixture was then poured over an ice-cold aqueous NaOH solution (9 g in 300 mL), transferred to a separatory funnel, and shaken until no green color was observed. The aqueous phase was extracted with DCM (3 × 100 mL) and the combined organic layers were washed with an aqueous saturated solution of NaHCO$_3$ (2 × 100 mL) and dried over anhydrous MgSO$_4$. The solvent was evaporated under reduced pressure to afford the crude product, which was purified via flash column chromatography through Et$_3$N-pretreated silica gel (2.5% v/v NEt$_3$), using DCM/MeOH 2% as an eluent. In this way, the pure *cis* (purple solid, 82 mg, 18% yield) and *trans* (purple solid, 116 mg, 26% yield) isomers of 6b could be isolated.

6b-cis. ^{1}**H-NMR** (CDCl$_3$, 400 MHz): δ = 8.85–8.80 (m, 3H), 8.77 (d, J = 4.8 Hz, 2H), 8.65 (d, J = 4.8 Hz, 1H), 8.61 (s, 1H), 8.26 (d, J = 7.4 Hz, 1H), 8.18 (d, J = 6.4 Hz, 4H), 8.16–8.12 (m, 2H), 8.10 (d, J = 7.4 Hz, 1H), 7.81–7.69 (m, 11H), 7.68 (dd, J = 7.6 Hz, J' = 1.3 Hz, 1H), 7.33–7.19 (m, 5H), 5.34 (s, 1H), 3.93 (dd, J = 9.1 Hz, J' = 3.6 Hz, 1H), 3.08 (dd, J = 14.0 Hz, J' = 3.6 Hz, 1H), 2.78 (dd, J = 14.0 Hz, J' = 9.1 Hz, 1H), 2.74 (s, 3H), 2.16 (br, 1H), −2.73 (br, 2H) ppm. ^{13}**C-NMR** (CDCl$_3$, 100 MHz): δ = 174.6, 142.2, 142.1, 142.0, 141.9, 141.82, 142.76, 138.2, 134.82/134.80 (2C), 134.73/134.72/134.72/134.71 (4C), 134.68/134.65/134.66 (3C), 134.6, 134.57, 133.60, 129.61/129.60 (2C), 128.9, 128.7, 128.6, 128.4, 128.04, 127.96/127.95 (2C), 126.95–126.93 (5C), 126.90/126.89/126.88 (3C), 126.8, 126.72/126.71 (2C), 120.7, 120.6, 120.5, 119.4 77.4, 77.3, 77.0, 76.6, 71.5, 37.2, 29.9 ppm. **HRMS (ESI)**: m/z calculated for C$_{55}$H$_{43}$N$_6$O [M+H]$^+$, 803.3493; found 803.3468. **UV-vis** [λ$_{max}$ (ε), c = 2.810 × 10^{-6} M, DCM]: 421 (178,000), 451 (173,300), 517 (202,000), 552 (80,800), 593 (63,250), 650 (50,900) nm.

6b'-trans. ^{1}**H-NMR** (CDCl$_3$, 400 MHz): δ = 8.84–8.79 (m, 3H), 8.77 (d, J = 4.7 Hz, 2H), 8.67 (d, J = 4.6 Hz, 1H), 8.38 (s, 1H), 8.32 (d, J = 7.5 Hz, 1H), 8.30–8.03 (m, 5H), 8.01 (d, J = 7.5 Hz, 2H), 7.86–7.70 (m, 10H), 7.67 (q, J = 7.6 Hz, 2H), 7.12–6.99 (m, 5H), 5.08 (s, 1H), 3.60 (t, J = 5.6 Hz, 1H), 3.11 (dd, J = 5.4 Hz, J' = 1.7 Hz, 2H), 2.73 (s, 3H), 1.89 (br, 1H), −2.73 (br, 2H) ppm. ^{13}**C-NMR** (CDCl$_3$, 100 MHz): δ = 174.1, 142.5, 142.2, 142.1, 141.94, 141.88, 139.5, 139.3, 134.9, 134.82/134.80 (2C), 134.72–134.70 (4C), 134.68/134.67 (2C), 134.63, 134.58, 134.5, 133.0, 132.5, 130.0, 129.2, 128.8, 128.7, 128.5, 128.3, 128.08, 128.05, 128.0, 127.1, 126.98–126.95 (4C), 126.96/126.95/126.94 (3C), 126.92/126.90 (2C), 126.86, 120.7, 118.5, 77.4, 77.3, 77.0, 76.7, 40.9, 29.8 ppm. **HRMS (ESI)**: m/z calculated for C$_{55}$H$_{43}$N$_6$O [M+H]$^+$, 803.3493; found 803.3458. **UV-vis** [λ$_{max}$ (ε), c = 1.338 × 10^{-6} M, DCM]: 422 (131,750), 519 (171,250), 553 (79,000), 595 (63,200), 655 (72,000) nm.

Copper(II) 2-((2R/2S,4S)-4-(tert-butyl)-1-methyl-5-oxoimidazolidin-2-yl)-5,10,15,20-tetraphenyl porphyrinate (5c). In a 25 mL round-bottomed flask equipped with magnetic stirring and a Dimroth reflux condenser, freshly prepared L-tert-leucine methylamide **4c** (85 mg, 0.59 mmol) was introduced. Then, a solution of copper(II) 2-formyl-5,10,15,20-tetraphenylporphyrinate **3** (318 mg, 0.45 mmol) in CHCl$_3$ (4.4 mL) and Yb(SO$_3$CF$_3$)$_3$ (3.0 mg, 0.005 mmol, 1 mol%) were added sequentially. The stirred mixture was heated up to reflux for 24 h. After cooling to room temperature, the solvent was evaporated under reduced pressure and the residue was purified using flash column chromatography through an NEt$_3$-pretreated silica gel (2.5% v/v NEt$_3$). A mixture of DCM/MeOH 0.5% v/v was used as an eluent. The fast-running red band, which contained unreacted starting material, was not collected. Compound **5c** (purple solid, 385 mg, quantitative yield) was obtained as a diastereomeric mixture (two spots in TLC, DCM/MeOH 0.5% v/v) and was not further characterized.

2-((2R,4S)-4-(tert-Butyl)-1-methyl-5-oxoimidazolidin-2-yl)-5,10,15,20-tetraphenylporphyrin (6c). In a 10 mL round-bottomed flask equipped with magnetic stirring, 385 mg of metalated porphyrin **5c** (0.46 mmol) was dissolved in 5 mL of DCM. Then, 6.5 mL (125.1 mmol) of concentrated sulfuric acid was added to the solution, which was vigorously stirred for 4 min. Then, the green two-phase mixture was poured into ice-cold aqueous NaOH (9 g in 300 mL) in a separatory funnel. The mixture was shaken until no green color remained or reappeared. The bottom layer was extracted with DCM (3 × 0.1 L) and washed twice with saturated aqueous NaHCO$_3$ (2 × 0.1 L). The organic layer was then dried over anhydrous MgSO$_4$, and the solvent was evaporated under reduced pressure. Finally, the obtained crude product was purified via flash column chromatography through Et$_3$N-pretreated silica gel (97.5:2.5 v/v), using DCM:MeOH (97:3) as an eluent. Finally, 85 mg (23% yield) of porphyrin **6c** was obtained as a single diastereomer to which a trans relative stereochemistry was tentatively assigned.

^{1}**H-NMR** (CDCl$_3$, 400 MHz): δ = 8.84–8.80 (m, 3H), 8.77 (d, J = 4.8 Hz, 2H), 8.69 (s, 1H), 8.62 (d, J = 4.9 Hz, 1H), 8.34 (d, J = 6.8 Hz, 1H), 8.24–8.10 (m, 7H), 7.83–7.38 (m, 12H), 5.47 (s, 1H), 3.41 (s, 1H), 2.69 (s, 3H), 2.24 (br, 1H), 0.95 (s, 9H), −2.69 (br, 2H$_1$) ppm. **HRMS**

(ESI): m/z calculated for $C_{52}H_{45}N_6O$ [M+H]$^+$, 769.3649; found 769.3618. **UV-vis** [λ_{max} (ε), c = 2.948 × 10^{-6} M, DCM]: 420 (284,500), 518 (14,100), 552 (5800), 593 (4300), 650 (3200) nm.

5-(4-(Methoxycarbonyl)phenyl)-10,15,20-triphenylporphyrin (8). In a 500 mL round-bottomed flask equipped with magnetic stirring and a Liebig reflux condenser, a mixture of glacial AcOH (200 mL) and nitrobenzene (150 mL) was heated up to reflux. Methyl 4-formylbenzoate **7** (1.44 g, 8.8 mmol) was then added in one portion. When all the solid was dissolved, benzaldehyde (2.25 mL, 21.9 mmol) and freshly distilled pyrrole (2.00 mL, 29.4 mmol) were added, and the mixture was stirred at reflux for 1 h. At this point, the solvents were distilled in vacuo and the remaining purple paste was purified via flash column chromatography, using hexane/DCM (7/3) as an eluent, to afford the desired porphyrin **8** (781 mg, 29% yield) as a purple solid [44].

^{1}H-NMR (CDCl$_3$, 400 MHz): δ = 8.87–8.78 (m, 8H), 8.42 (d, J = 8.9 Hz, 2H), 8.29 (d, J = 8.9 Hz, 2H), 8.22–8.20 (d, J = 7.4Hz, 6H), 7.79–7.72 (m, 9H), 4.10 (s, 3H), −2.77 (br, 2H) ppm.

5-(4-(Hydroxymethyl)phenyl)-10,15,20-triphenylporphyrin (9). In a 500 mL round-bottomed flask equipped with magnetic stirring, LiAlH$_4$ (296 mg, 7.8 mmol) was added, and the system was purged with Ar. Then, dry THF (6.1 mL) was added and a solution of 5-(4-(methoxycarbonyl)phenyl)-10,15,20-triphenylporphyrin **8** (1.32 g, 1.9 mmol) in dry THF (155 mL) was subsequently added via syringe. The reaction mixture was stirred at room temperature for 1 h. At this point, the reaction was quenched with 1% aqueous HCl (211 mL), and the resulting green solution was transferred to a separatory funnel and extracted with DCM until the aqueous phase became colorless. The combined organic layers were treated with a 32% w/w aqueous solution of NH$_3$, noting the changing of color from green to purple, and the aqueous phase was extracted with DCM (3 × 50 mL). The combined organic layers were dried over NaSO$_4$ and concentrated under reduced pressure to affect a purple solid, which was purified via flash column chromatography, using hexane/DCM (1/1) as an eluent. The desired porphyrin **9** (1.19 g, 95% yield) was obtained as a purple solid [45].

^{1}H-NMR (CDCl$_3$, 400 MHz): δ = 8.85 (s, 8H), 8.82 (d, J = 8.0 Hz, 8H), 7.78–7.73 (m, 11H), 5.30 (s, 2H), −2.77 (br, 2H) ppm.

4-(10,15,20-Triphenylporphyrin-5-yl)benzaldehyde (2). In a 500 mL round-bottomed flask equipped with magnetic stirring, 5-(4-(hydroxymethyl)phenyl)-10,15,20-triphenylporphyrin **9** (395 mg, 0.60 mmol) was dissolved in freshly distilled DCM (255 mL). Pyridinium chlorochromate (395 mg, 1.8 mmol) was then added, noting a change of color from purple to dark green, and the reaction mixture was stirred for 1 h at room temperature. At this point, silica gel (60 mL) was added to the reaction crude and the solvent was evaporated under reduced pressure. The resulting dark-grey solid was purified via flash column chromatography, using DCM as eluent, to afford the desired product **2** (390 mg, quantitative yield) as a purple solid [31].

^{1}H-NMR (CDCl$_3$, 400 MHz): δ = 10.39 (s, 1H), 8.88 (AB system (part A), J = 4.8 Hz, 2H), 8.85 (s, 4H), 8.79 (AB system (part B), J = 4.8 Hz, 2H) 8.42 (d, J = 7.8 Hz, 2H), 8.29 (d, J = 7.8 Hz, 2H), 8.22 (d, J = 6.4 Hz, 6H), 7.82–7.73 (m, 9H), −2.77 (br, 2H) ppm.

Copper(II) 5-(4-formylphenyl)-10,15,20-triphenylporphyrinate (10). In a 50 mL round-bottomed flask equipped with magnetic stirring and a Dimroth reflux condenser, 5-(4-formylphenyl)-10,15,20-triphenylporphyrin **2** (140 mg, 0.22 mmol) was dissolved in DCM (11 mL). MeOH (4 mL) and Cu(OAc)$_2$·H$_2$O (79 mg, 0.39 mmol) were then added, and the resulting solution was heated up to reflux for 2 h. After this time, the solvents were evaporated under reduced pressure; the residue was redissolved in the minimum amount of DCM and filtered through a short plug of silica gel, using DCM as eluent, to afford the desired product **10** (151 mg, quantitative yield) as a purple solid [31].

UV-vis [DCM, λ_{max} (ε), c = 4.59 × 10^{-5} M]: 412 (479900), 537 (15800) nm.

Copper(II) 5-(4-((2R/2S,4S)-1,4-dimethyl-5-oxoimidazolidin-2-yl)phenyl)-5,10,15-triphenyl porphyrinate (12). L-alanine methylamide **4a** (30 mg, 0.22 mmol) was introduced to a 25 mL round-bottomed flask equipped with magnetic stirring and a Dimroth reflux condenser. Then, a solution of copper(II) 5-(4-formylphenyl)-10,15,20-triphenylporphyrinate **10** (81 mg, 0.11 mmol) in CHCl$_3$ (6 mL) and Yb(SO$_3$CF$_3$)$_3$ (0.62 mg, 1 mol%) were added sequentially. The stirred mixture was heated up to reflux for 24 h. After cooling to room temperature, the solvent was evaporated under reduced pressure and the residue was purified via flash column chromatography through Et$_3$N-pretreated silica gel (2.5% v/v NEt$_3$), using mixtures of DCM/MeOH (from 1% v/v to 5% v/v) as eluents. The fast-running red band, which contained unreacted starting material, was not collected. Compound **12** (40 mg, 46% yield) was obtained as a diastereomeric mixture (two spots in TLC, DCM/MeOH 0.5%) that was not further characterized.

5-(4-((2R/2S,4S)-1,4-Dimethyl-5-oxoimidazolidin-2-yl)phenyl)-5,10,15-triphenylporphyrin (13). In a 25 mL round-bottomed flask equipped with magnetic stirring, 40 mg (0.05 mmol) of copper(II) porphyrinate **12** was dissolved in DCM (5 mL). Concentrated H$_2$SO$_4$ (98%, 0.7 mL) was added, and the resulting green mixture was vigorously stirred for 3 min. The two-phase mixture was then poured over a cold aqueous NaOH solution (1 g in 340 mL), transferred to a separatory funnel, and shaken until no green color was observed in the organic layer. The aqueous phase was extracted with DCM (3 × 50 mL), and the combined organic layers were washed with an aqueous saturated solution of NaHCO$_3$ (2 × 100 mL) and dried over Na$_2$SO$_4$. The solvent was then evaporated under reduced pressure. Finally, the obtained crude product was purified via flash column chromatography through Et$_3$N-pretreated silica gel (2.5% NEt$_3$ v/v), using DCM/MeOH 2% as an eluent, affording a complex diastereomeric mixture that could not be separated by column chromatography. The purple solid (10 mg, 27% yield) obtained, corresponding to compound **13**, was not characterized.

(S)-5-(4-((((1-(tert-Butoxycarbonyl)pyrrolidin-2-yl)methyl)amino)methyl)phenyl)-10,15,20-triphenylporphyrin (19). In a 25 mL round-bottomed flask equipped with magnetic stirring and a Dimroth reflux condenser, N-Boc-(S)-2-aminomethylpyrrolidine **18** (72 mg, 0.36 mmol) was introduced. Then, a solution of 5-(4-formylphenyl)-10,15,20-triphenylporphyrin **2** (115 mg, 0.18 mmol) in CHCl$_3$ (8 mL) and Yb(SO$_3$CF$_3$)$_3$ (2.8 mg, 0.0045 mmol, 1.2 mol%) were added sequentially. The stirred mixture was heated up to reflux for 24 h. At that point, TLC monitoring showed the complete formation of the imine intermediate. After cooling to rt, NaBH$_3$CN (22.6 mg, 0.36 mmol) was added in one portion. The reaction mixture was then stirred for 1 h at rt and quenched with H$_2$O (15 mL). The aqueous phase was then separated and extracted with DCM (3 × 10 mL), and the combined organic layers were dried over Na$_2$SO$_4$ and concentrated under reduced pressure. The resulting purple solid was purified via flash column chromatography, using a mixture of DCM/MeOH (from 1% to 5%) as an eluent, affording the desired product **19** (100 mg, 67% yield) as a purple solid.

^{1}H-NMR (CDCl$_3$, 400 MHz): δ = 8.84 (s, 8H), 8.21 (d, J = 7.7 Hz, 8H), 7.80–7.70 (m, 11H), 4.34–4.17 (m, 2H), 3.51–3.38 (m, 2H), 3.10–3.03 (m, 1H), 2.20 (m, 2H), 1.95–1.86 (m, 3H,), 1.51 (s, 11H), −2.77 (br, 2H) ppm. **HRMS (ESI)**: m/z calculated for C$_{55}$H$_{50}$N$_6$O$_2$ [M+H]$^+$, 827.3995.; found 827.4057.

(S)-5,10,15-Triphenyl-20-(4-(((2-(pyrrolidin-2-yl)ethyl)amino)methyl)phenyl)porphyrin (20). In a 25 mL round-bottomed flask equipped with magnetic stirring, (S)-5-(4-((((1-(tert-butoxycarbonyl)pyrrolidin-2-yl)methyl)amino)methyl)phenyl)-10,15,20-triphenylporphyrin **19** (100 mg, 0.12 mmol) was dissolved in CHCl$_3$ (6 mL). Then, trifluoroacetic acid (6 mL) was added and the resulting green solution was stirred for 2 h at rt. At this point, the reaction was concentrated under reduced pressure, redissolved in CHCl$_3$ (10 mL), and an aqueous solution of NaOH (3.45 g in 100 mL) was added, noting a change of color from green to purple. The aqueous phase was separated and extracted with DCM (10 mL portions) until colorless. The combined organic phases were dried over Na$_2$SO$_4$ and concentrated under reduced pressure, affording a purple solid that was purified via flash column chromatography through silica gel, using a mixture of DCM/MeOH (from 0.5 to 10%) to produce the desired porphyrin **20** (84 mg, 97% yield) as a purple solid.

^{1}**H-NMR** (CDCl$_3$, 400 MHz): δ = 8.82 (s, 8H), 8.20–8.16 (m, 8H), 7.76–7.69 (m, 11H), 4.41–4.27 (m, 2H), 3.49–3.34 (m, 3H), 2.32 (m, 1H), 2.17–2.07 (m, 2H), 1.82 (br, 1H), 1.31–1.25 (m, 2H), 0.88 (m, 2H) ppm. **HRMS (ESI):** m/z calculated for C$_{50}$H$_{42}$N$_6$ [M+H]$^+$, 727.3471; found 727.3549. **UV-vis** [λ$_{max}$ (ε), 1.382 × 10^{-6} M, DCM]: 418 (533,700), 515 (49,200), 549 (22,200), 590 (5,300), 647 (5,050) nm.

Copper(II) (S)-5-(4-((((1-(*tert*-butoxycarbonyl)pyrrolidin-2-yl)methyl)amino)methyl) phenyl)-10,15,20-triphenylporphyrinate (**21**). This compound was obtained using the same procedure described above for the non-metalated porphyrin **19** but began with copper(II) 5-(4-formylphenyl)-10,15,20-triphenylporphyrinate **10** (73 mg, 0.10 mmol). Complex **21** (46 mg, 51% yield) was obtained as a red-colored solid that was not further characterized. The identity of **21** was confirmed through treatment with concentrated sulfuric to provide a compound identical to **20**.

5-(4-Cyanophenyl)-10,15,20-triphenylporphyrin (**23**). In a 500 mL round-bottomed flask equipped with magnetic stirring and a Liebig reflux condenser, propionic acid (350 mL) and 4-cyanobenzaldehyde **22** (1.93 g, 14.7 mmol) were added sequentially. Once all the aldehyde was dissolved, the mixture was heated up to reflux and benzaldehyde (4.5 mL, 44.1 mmol) and freshly distilled pyrrole (4.1 mL, 58.8 mmol) were added. The black reaction mixture was stirred for 1 h under reflux and was protected from light. The reaction mixture was then cooled down to rt and the solvent was removed via vacuum distillation. Next, the obtained purple paste was purified by flash column chromatography through silica gel, using DCM as an eluent. In this way, the less-polar **TPPH$_2$** could be separated from another fraction containing the mixture of substituted porphyrins. After the elimination of the solvent, the crude mixture was purified again by flash column chromatography in silica gel, using DCM/Hexane (4/1) as an eluent, affording the desired product **23** (868 mg, 21% yield) as a metallic purple solid [47].

^{1}**H-NMR** (CDCl$_3$, 400 MHz): δ = 8.89–8.72 (m, 8H), 8.33 (d, J = 7.9 Hz, 2H), 8.25–8.18 (m, 6H), 8.06 (d, J = 8.2 Hz, 2H), 7.80–7.70 (m, 9H), −2.77 (br, 2H) ppm.

5-(4-(Aminomethyl)phenyl)-10,15,20-triphenylporphyrin (**24**). In a 500 mL round-bottomed flask equipped with magnetic stirring, LiAlH$_4$ (310 mg, 8.2 mmol) was introduced, and the system was purged with Ar. Then, dry THF (6.1 mL) was added and, subsequently, a solution of 5-(4-cyanophenyl)-10,15,20-triphenylporphyrin **23** (868 mg, 1.4 mmol) in dry THF (155 mL) was added via syringe. The reaction mixture was stirred at room temperature for 1 h. At this point, the reaction was quenched with aqueous 1% HCl (211 mL), and the resulting green solution was transferred to a separatory funnel and extracted with DCM until colorless. The combined organic layers were treated with a 32% (w/w) aqueous solution of NH$_3$, which brought about a change of color from green to purple. The aqueous phase then was separated and extracted with DCM (3 × 50 mL). The combined organic layers were dried over Na$_2$SO$_4$ and concentrated under reduced pressure to afford a purple solid. The purple solid was purified via flash column chromatography through silica gel, using a mixture of DCM/MeOH (from 2% to 5%) as an eluent. The desired porphyrin **24** (618 mg, 71% yield) was obtained as a purple solid [47].

^{1}**H-NMR** (CDCl$_3$, 400 MHz): δ = 8.83 (m, 8H), 8.26 (br, 2H) 8.25–8.10 (m, 8H), 7.85–7.65 (m, 11H), 4.25 (s, 2H), −2.67 (br, 2H) ppm.

(S)-5-(4-((1-(*tert*-Butoxycarbonyl)pyrrolidine-2-carboxamido)methyl)phenyl)-10,15, 20-triphenyl porphyrin (**26**). In a 25 mL round-bottomed flask equipped with magnetic stirring, a solution of N-Boc-L-Proline **25** (251 mg, 1.2 mmol) and NEt$_3$ (270 mL, 2.1 mmol) in dry THF (2.1 mL) was cooled down to −15 °C. A solution of ethyl chloroformate (110 mL, 125 mg, 1.15 mmol) in dry THF (1.6 mL) was then added, and the mixture was stirred for 30 min at −15 °C. Next, a solution of 5-(4-(aminomethyl)phenyl)-10,15,20-triphenylporphyrin **24** (618 mg, 0.96 mmol) in dry THF (5.1 mL) was added dropwise, and the reaction mixture was stirred overnight at room temperature. At this point, DCM (15 mL) was added. The organic phase was then washed with NaHCO$_3$ (15 mL) and brine (15 mL), dried over Na$_2$SO$_4$, and concentrated under reduced pressure, affording a purple solid that was purified by flash column chromatography through silica gel, using a mixture of DCM/MeOH (from 0.5% to 2%) as an eluent. The desired porphyrin **26** (806 mg, quantitative yield) was obtained as a purple solid.

^{1}**H-NMR** (CDCl$_3$, 400 MHz): δ = 8.84 (s, 8H), 8.22–8.16 (m, 8H), 7.76–7.63 (m, 11H), 4.93 (dd, J = 10.2 Hz, J' = 6.4 Hz, 1H), 4.48 (m, 1H), 3.55 (m, 2H), 2.54–2.28 (m, 1H), 2.06–1.99 (m, 3H), 1.49 (s, 9H), 0.88 (m, 2H), −2.78 (br, 2H) ppm. **HRMS (ESI):** m/z calculated for C$_{55}$H$_{48}$N$_6$O$_3$ [M+H]$^+$, 841.3861; found 841.3900.

(S)-5,10,15-Triphenyl-20-(4-((pyrrolidine-2-carboxamido)methyl)phenyl)porphyrin (27). In a 250 mL round-bottomed flask equipped with magnetic stirring, (S)-5-(4-((1-(tert-butoxycarbonyl)pyrrolidine-2-carboxamido)methyl)phenyl)-10,15,20-triphenylporphyrin **26** (806 mg, 0.96 mmol) was dissolved in CHCl$_3$ (51 mL). Then, trifluoroacetic acid (51 mL) was added, and the resulting green solution was stirred for 2 h at rt. At this point, the reaction was concentrated under reduced pressure, and the residue was redissolved in CHCl$_3$ (50 mL). An aqueous solution of NaOH (7.639 g in 200 mL) was then added, noting a change of color from green to purple. The aqueous phase was then extracted with DCM until colorless, dried over Na$_2$SO$_4$, and concentrated under reduced pressure, affording a purple solid that was purified via flash column chromatography through silica gel, using a mixture of DCM/MeOH (from 0.5% to 5%), to afford the desired porphyrin **27** (541 mg, 72% yield) as a purple solid.

^{1}**H-NMR** (CDCl$_3$, 400 MHz): δ = 9.09 (br, 1H), 8.82 (s, 8H), 8.21–8.14 (m, 8H), 7.74–7.65 (m, 11H), 4.97 (dd, J = 10.2 Hz, J' = 6.4 Hz, 1H), 4.82–4.76 (m, 1H), 3.34–3.23 (m, 2H), 2.60 (br, 1H), 2.05–2.00 (m, 3H), 1.70–1.40 (m, 2H), −2.79 (br, 2H). **HRMS (ESI):** m/z calculated for C$_{50}$H$_{40}$N$_6$O$_2$ [M+H]$^+$, 741.3264; found 741.3332. **UV-Vis** [λ$_{max}$ (ε), c = 3.892 × 10^{-6} M, DCM]: 418 (562,400), 515 (49,200), 550 (22,300), 590 (7000), 646 (5800) nm.

3.2.2. Imidazolidinone-Catalyzed Diels–Alder Reactions

General procedure for the imidazolidinone-catalyzed Diels–Alder cycloaddition of cyclopentadiene **28** and (E)-cinnamaldehyde **29** under thermal conditions. The corresponding organocatalyst (0.025 mmol, 5 mol%), p-toluenesulfonic acid monohydrate (5 mg, 0.026 mmol, 5 mol%), and 1 mL of solvent (toluene or MeOH) were placed in a 3 mL vial equipped with magnetic stirring. The resulting mixture was stirred for 10 min. At this point, (E)-cinnamaldehyde **29** (63 µL, 66 mg, 0.50 mmol) and, after 30 min, freshly distilled cyclopentadiene **28** (125 µL, 98 mg, 1.48 mmol) were added, and the reaction mixture was stirred for 72 h at room temperature. The mixture was then poured over DCM (10 mL) and washed with an aqueous saturated solution of NaHCO$_3$ (2 × 30 mL) until no green color remained or reappeared. The organic layer was dried over anhydrous MgSO$_4$ and concentrated under reduced pressure. The ratio of the Diels–Alder adducts **30** (endo) and **31** (exo) was determined via ^{1}H-NMR analysis of the crude reaction mixture (Figure S14 in the Supplementary Materials).

When using MeOH as solvent, the Diels–Alder adducts were obtained in their acetal form, which was not purified any further. Subsequently, the deprotection was performed. The crude reaction mixture was diluted with DCM (2 mL) and TFA (1 equivalent), and H$_2$O (1 equivalent) was added. After stirring for 2 h at rt, the mixture was diluted with DCM and washed with an aqueous saturated solution of NaHCO$_3$ and brine. The organic layer was dried over MgSO$_4$ and concentrated under reduced pressure to obtain the aldehyde adducts **30** and **31**. When using toluene as solvent, the Diels–Alder adducts were purified via flash column chromatography on silica gel, using hexane/DCM (1/1) as an eluent.

General procedure for the imidazolidinone-catalyzed Diels–Alder cycloaddition of cyclopentadiene **28** and (E)-cinnamaldehyde **29** under photochemical conditions. In a 3 mL vial equipped with magnetic stirring, (E)-cinnamaldehyde **29** (63 µL, 66 mg, 0.50 mmol), freshly distilled cyclopentadiene **28** (125 µL, 98 mg, 1.48 mmol), p-toluenesulfonic acid monohydrate (5 mg, 0.026 mmol, 5 mol%), and the corresponding catalyst[a] (0.025 mmol, 5 mol%) were dissolved in 1 mL of previously degassed solvent. The vial was capped with a septum, a high vacuum was applied for 10 min, and the vial was backfilled with Ar. The process was repeated three times in order to remove all the dissolved gasses. At this point, the vial was sealed with Parafilm®, placed in the photoreactor (Figure 4), and stirred under irradiation for 72 h. The whole assembly was cooled by an air stream from a fan situated above the beaker with the aid of a clamp. The mixture was then diluted with DCM (10 mL) and washed with an aqueous saturated solution of NaHCO$_3$ (2 × 30 mL) until no green color remained or reappeared. The organic layer was dried over MgSO$_4$ and concentrated under reduced pressure. The ratio of the Diels–Alder adducts was determined via ^{1}H-NMR analysis of the crude reaction mixture (Figure S14 in the Supplementary Materials).

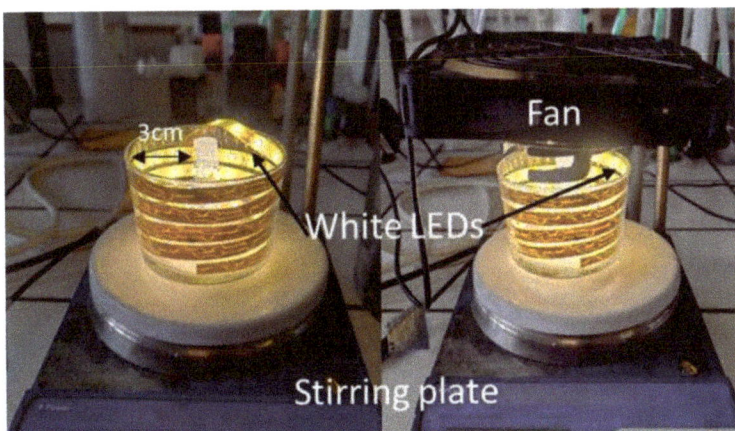

Figure 4. Experimental setting for the photocatalytic reactions. LED coil (white light, 18 W, 1080 lumens), stirring plate. Left: Without fan. Right: With fan.

^{a} Employing a bifunctional catalytic system: 0.025 mmol of catalyst (5 mol%); employing a dual catalytic system: 0.025 mmol of **Na$_4$TPPH$_2$S$_4$** or of the zwitterionic **(H$_3$O)$_2$TPPH$_4$S$_4$** as the photocatalyst (5 mol%) and 0.025 of imidazolidinone (**32**, **33** or **34**) as the organocatalyst (5 mol%).

(1R*,2S*,3S*,4S*)-3-Phenylbicyclo [2.2.1]hept-5-ene-2-carbaldehyde (30, endo). Yellowish oil. ^{1}**H-NMR** (CDCl$_3$, 400 MHz): δ = 9.60 (d, J = 2.2 Hz, 1H), 7.36–7.12 (m, 5H), 6.42 (dd, J = 5.5 Hz, J' = 3.4 Hz, 1H), 6.18 (dd, J = 5.5 Hz, J' = 2.7 Hz, 1H), 3.35–3.32 (m, 1H), 3.15–3.10 (m, 1H), 3.09 (d, J = 4.7 Hz, 1H), 3.00–2.95 (m, 1H), 1.81 (d, J = 9.2 Hz, 1H), 1.62 (d, J = 9.2 Hz, 1H) ppm. **HPLC** (Phenomenex i-cellulose 5 column; hexane/IPA 0.8%; flow rate 1 mL/min; 210 nm) t$_R$ = 44.3 min (2R,3R), 51.6 min (2S,3S). After conversion to the corresponding alcohol with excess of NaBH$_4$ in MeOH.

(1S*,2S*,3S*,4R*)-3-Phenylbicyclo [2.2.1]hept-5-ene-2-carbaldehyde (31, exo). Yellowish oil. ^{1}**H-NMR** (CDCl$_3$, 400 MHz): δ = 9.92 (d, J = 2 Hz, 1H), 7.36–7.12 (m, 5H), 6.34 (dd, J = 5.8 Hz, J' = 3.4 Hz, 1H), 6.08 (dd, J = 5.4 Hz, J' = 2.8 Hz, 1H), 3.73 (t, J = 3.9 Hz, 1H), 3.25–3.20 (m, 1H), 3.09 (d, J = 4.7 Hz, 1H), 3.00–2.95 (m, 1H), 2.59 (d, J = 4.9 Hz, 1H), 1.42 (d, J = 9.2 Hz, 1H) ppm. **HPLC** (Phenomenex i-cellulose 5 column; hexane/IPA 0.8%; flow rate 1 mL/min; 210 nm) t$_R$ = 37.1 min (2R,3R), 48.5 (2S,3S). After conversion to the corresponding alcohol with an excess of NaBH$_4$ in MeOH (See Figure S15 in the Supplementary Materials).

Supplementary Materials: The following supporting information can be downloaded at https://www.mdpi.com/article/10.3390/molecules28041997/s1. Figures S1–S3: Stereochemical assignation of imidazolidinone–porphyrin hybrids **6b** and **6b'**; Experimental procedures for the preparation of compounds TPPH$_2$, Na$_4$TPPH$_2$S$_4$, (H$_3$O)$_2$TPPH$_4$S$_4$, **4a–c**, **18**, **25**, **32**, **33**, and **34**; Figures S4–S13: NMR spectra of new compounds; Figures S14 and S15: determination of the stereochemical composition of the products of the Diels–Alder cycloaddition of cyclopentadiene (**28**) and (E)-cinnamaldehyde (**29**).

Author Contributions: Conceptualization and methodology, A.M. and J.C.; investigation, P.T., M.G. and M.E.; writing—original draft preparation, A.M.; writing—review and editing, A.M., P.T. and J.C.; supervision, A.M.; project administration, A.M.; funding acquisition, A.M. All authors have read and agreed to the published version of the manuscript.

Funding: This research was funded by FEDER, Ministerio de Ciencia e Innovación (MICINN), Agencia Estatal de Investigación, grant number PID2020-116846GB-C21.

Institutional Review Board Statement: Not applicable.

Informed Consent Statement: Not applicable.

Data Availability Statement: No new data were created in addition to those reported here and in the Supplementary Materials.

Acknowledgments: The administrative and technical support of the CCTiUB is gratefully acknowledged. We thank L.A. Zeoly for the MMFF calculations of compounds **6b** and **6b'**.

Conflicts of Interest: The authors declare no conflict of interest. The funder had no role in the design of the study; in the collection, analyses, or interpretation of data; in the writing of the manuscript, or in the decision to publish the results.

Sample Availability: Samples of the compounds **6a**, **6b**, **6b'**, **6c**, **20**, and **27** are available from the authors.

References

1. Della Sala, G.; Schettini, R. (Eds.) *New Trends in Asymmetric Catalysis*; MDPI: Basel, Switzerland, 2021.
2. Akiyama, T.; Ojima, I. (Eds.) *Catalytic Asymmetric Synthesis*, 4th ed.; John Wiley & Sons: Hoboken, NJ, USA, 2022.
3. Dalko, P.I. (Ed.) *Comprehensive Enantioselective Organ Catalysis*; Wiley-VCH: Weinheim, Germany, 2013; Volumes 1–3.
4. Rios Torres, R. (Ed.) *Stereoselective Organocatalysis. Bond Formation Methodologies and Activation Modes*; John Wiley & Sons: Hoboken, NJ, USA, 2013.
5. Strieth-Kalthoff, F.; James, M.J.; Teders, M.; Pitzer, L.; Glorius, F. Energy transfer catalysis mediated by visible light: Principles, applications, directions. *Chem. Soc. Rev.* **2018**, *47*, 7190–7202. [CrossRef] [PubMed]
6. Prier, C.K.; Rankic, D.A.; MacMillan, D.W.C. Visible Light Photoredox Catalysis with Transition Metal Complexes: Applications in Organic Synthesis. *Chem. Rev.* **2013**, *113*, 5322–5363. [CrossRef] [PubMed]
7. Romero, N.A.; Nicewicz, D.A. Organic Photoredox Catalysis. *Chem. Rev.* **2013**, *113*, 10075–10166. [CrossRef] [PubMed]
8. Nicewicz, D.A.; MacMillan, D.W.C. Merging Photoredox Catalysis with Organocatalysis: The Direct Asymmetric Alkylation of Aldehydes. *Science* **2008**, *322*, 77–80. [CrossRef] [PubMed]
9. Nagib, D.A.; Scott, M.E.; MacMillan, D.W.C. Enantioselective α-Trifluoromethylation of Aldehydes vis Photoredox Organocatalysis. *J. Am. Chem. Soc.* **2009**, *131*, 10875–10877. [CrossRef]
10. Hopkinson, M.N.; Sahoo, B.; Li, J.-L.; Glorius, F. Dual Catalysis Sees the Light: Combining Photoredox with Organo-, Acid, and Transition-Metal Catalysis. *Chem.—Eur. J.* **2014**, *20*, 3874–3886. [CrossRef]
11. Skubi, K.L.; Blum, T.R.; Yoon, T.P. Dual Catalysis Strategies in Photochemical Synthesis. *Chem. Rev.* **2016**, *116*, 10035–10074. [CrossRef]
12. Zou, Y.-Q.; Hörmann, F.M.; Bach, T. Iminium and Enamine Catalysis in Enantioselective Photochemical Reactions. *Chem. Soc. Rev.* **2018**, *47*, 278–290. [CrossRef]
13. Yin, Y.; Zhao, X.; Qiao, B.; Jiang, Z. Cooperative photoredox and chiral hydrogen-bonding catalysis. *Org. Chem. Front.* **2020**, *7*, 1283–1296. [CrossRef]
14. Shibasaki, M.; Kanai, M.; Matsunaga, S.; Kumagai, N. Recent Progress in Asymmetric Bifunctional Catalysis Using Multimetallic Systems. *Acc. Chem. Res.* **2009**, *42*, 1117–1127. [CrossRef]
15. Paull, D.H.; Abraham, C.J.; Scerba, M.T.; Alden-Danforth, E.; Lectka, T. Bifunctional Asymmetric Catalysis: Cooperative Lewis Acid/base Systems. *Acc. Chem. Res.* **2008**, *41*, 655–663. [CrossRef]
16. Bauer, A.; Westkamper, F.; Grimme, S.; Bach, T. Catalytic enantioselective reactions driven by photoinduced electron transfer. *Nature* **2005**, *436*, 1139–1140. [CrossRef]
17. Arceo, E.; Jurberg, I.D.; Álvarez-Fernández, A.; Melchiorre, P. Photochemical activity of a key donor acceptor complex that can drive stereoselective catalytic α-alkylations of aldehydes. *Nat. Chem.* **2013**, *5*, 750–756. [CrossRef]
18. Rigotti, T.; Casado-Sánchez, A.; Cabrera, S.; Pérez-Ruiz, R.; Liras, M.; de la Peña O'Shea, V.A.; Alemán, J. A Bifunctional Photoaminocatalyst for the Alkylation of Aldehydes: Design, Analysis, and Mechanistic Studies. *ACS Catal.* **2018**, *8*, 5928–5940. [CrossRef]
19. Wu, Y.; Kim, D.; Teets, T.S. Photophysical Properties and Redox Potentials of Photosensitizers for Organic Photoredox Transformations. *Synlett* **2022**, *33*, 1154–1179.

20. Costa e Silva, R.; da Silva, L.O.; de Andrade, A.B.; Brocksom, T.J.; de Oliveira, K.T. Recent applications of porphyrins as photocatalysts in organic synthesis: Batch and continuous flow approaches. *Beilstein J. Org. Chem.* **2020**, *16*, 917–955. [CrossRef]
21. Rybika-Jasinska, K.; Shan, W.; Zawada, K.; Kadish, K.M.; Gryko, D. Porphyrins as Photoredox Catalysts: Experimental and Theoretical Studies. *J. Am. Chem. Soc.* **2016**, *138*, 15451–15458. [CrossRef]
22. Rybika-Jasinska, K.; König, B.; Gryko, D. Porphyrin-Catalyzed Photochemical C–H Arylation of Heteroarenes. *Eur. J. Org. Chem.* **2017**, *2017*, 2104–2107. [CrossRef]
23. De Souza, A.A.N.; Silva, N.S.; Müller, A.V.; Polo, A.S.; Brocksom, T.J.; de Oliveira, K.T. Porphyrins as Photoredox Catalysts in Csp^2–H Arylations: Batch and Continuous Flow Approaches. *J. Org. Chem.* **2018**, *83*, 10577–15086. [CrossRef]
24. Arlegui, A.; El-Hachemi, Z.; Crusats, J.; Moyano, A. 5-Phenyl-10,15,20-Tris(4-sulfonatophenyl)porphyrin: Synthesis, Catalysis, and Structural Studies. *Molecules* **2018**, *23*, 3363. [CrossRef]
25. Arlegui, A.; Soler, B.; Galindo, A.; Orteaga, O.; Canillas, A.; Ribó, J.M.; El-Hachemi, Z.; Crusats, J.; Moyano, A. Spontaneous mirror-symmetry breaking coupled to top-bottom chirality transfer: From porphyrin self-assembly to scalemic Diels–Alder adducts. *Chem. Commun.* **2019**, *55*, 12219–12222. [CrossRef]
26. Crusats, J.; Moyano, A. Absolute Asymmetric Catalysis, from Concept to Experiment: A Narrative. *Synlett* **2021**, *32*, 2013–2035. [CrossRef]
27. Arlegui, A.; Torres, P.; Cuesta, V.; Crusats, J.; Moyano, A. A pH-Switchable Aqueous Organocatalysis with Amphiphilic Secondary Amine–Porphyrin Hybrids. *Eur. J. Org. Chem.* **2020**, *2020*, 4399–4407. [CrossRef]
28. Arlegui, A.; Torres, P.; Cuesta, V.; Crusats, J.; Moyano, A. Chiral Amphiphilic Secondary Amine-Porphyrin Hybrids for Aqueous Organocatalysis. *Molecules* **2020**, *25*, 3420. [CrossRef] [PubMed]
29. Lindsey, J.S. Synthesis of meso-Substituted Porphyrins. In *The Porphyrin Handbook*; Kadish, K.M., Smith, K.M., Guilard, R., Eds.; Academic Press: San Diego, CA, USA, 2000; Volume 1, pp. 45–118.
30. Bonfantini, E.E.; Burrell, A.K.; Campbell, W.M.; Crossley, M.J.; Gosper, J.J.; Harding, M.M.; Officer, D.L.; Reid, D.C.W. Efficient synthesis of free-base 2-formyl-5,10,15,20-tetraarylporphyrins, their reduction and conversion to [(porphyrin-2-yl)methyl]phosphonium salts. *J. Porphyr. Phthalocyanines* **2002**, *6*, 708–719. [CrossRef]
31. Liu, G.; Khlobystov, A.N.; Charalambidis, G.; Coutsolelos, A.G.; Briggs, G.A.D.; Porfyrakis, K. N@C60-Porphyrin: A Dyad of Two Radical Centers. *J. Am. Chem. Soc.* **2012**, *134*, 1938–1941. [CrossRef]
32. Drovetskaya, T.; Reed, C.A.; Boyd, P. A Fullerene Porphyrin Conjugate. *Tetrahedron Lett.* **1995**, *36*, 7971–7974. [CrossRef]
33. Silva, A.M.G.; Tomé, A.C.; Neves, M.G.P.M.S.; Silva, A.M.S.; Cavaleiro, J.A.S. Synthesis of New β-Substituted meso-Tetraphenylporphyrins via 1,3-Dipolar Cycloaddition Reactions. *J. Org. Chem.* **2002**, *67*, 726–732. [CrossRef]
34. Liu, X.-G.; Feng, Y.-Q.; Tan, C.-J.; Chen, H.-L. Cycloaddition of Porphyrins to Isatin: Synthesis of Novel Spiro Porphyrin Derivatives. *Synth. Commun.* **2006**, *36*, 2655–2659. [CrossRef]
35. Ladeira, B.M.F.; Dias, C.J.; Gomes, A.T.P.C.; Tomé, A.C.; Neves, M.G.P.M.S.; Moura, N.M.M.; Almeida, A.; Faustino, M.A.F. Cationic Pyrrolidine/Pyrroline-Substituted Porphyrins as Efficient Photosensitizers against *E. coli*. *Molecules* **2021**, *26*, 464. [CrossRef]
36. Moura, N.M.M.; Faustino, M.A.F.; Neves, M.G.P.M.S.; Almeida Paz, F.A.; Silva, A.M.S.; Tomé, A.C.; Cavaleiro, J.A.S. A new synthetic approach to benzoporphyrins and Kröhnke-type porphyrin-2-ylpyridines. *Chem. Commun.* **2012**, *48*, 6142–6144. [CrossRef] [PubMed]
37. Sharma, S.; Nath, M. Novel 5-benzazolyl-10,15,20-triphenylporphyrins and β,meso-benzoxazolyl- bridged porphyrin dyads: Synthesis, characterization and photophysical properties. *Dyes Pigments* **2012**, *92*, 1241–1249. [CrossRef]
38. Moura, N.M.M.; Ramos, C.I.V.; Linhares, I.; Santos, M.S.; Faustino, M.A.F.; Almeida, A.; Cavaleiro, J.A.S.; Amado, F.M.L.; Lodeiro, C.; Neves, M.G.P.M.S. Synthesis, characterization and biological evaluation of cationic porphyrin–terpyridine derivatives. *RSC Adv.* **2016**, *6*, 110674. [CrossRef]
39. Moreira, X.; Santos, P.; Faustino, M.A.F.; Raposo, M.M.M.; Costa, S.P.G.; Moura, N.M.M.; Gomes, A.T.P.C.; Almeida, A.; Neves, M.G.P.M.S. An insight into the synthesis of cationic porphyrin-imidazole derivatives and their photodynamic inactivation efficiency against *Escherichia coli*. *Dyes Pigments* **2020**, *178*, 108330. [CrossRef]
40. Richeter, S.; Jeandon, C.; Gisselbrecht, J.-P.; Ruppert, R.; Callot, H. Synthesis and Optical and Electrochemical Properties of Porphyrin Dimers Linked by Metal Ions. *J. Am. Chem. Soc.* **2002**, *124*, 6168–6179. [CrossRef]
41. Samulis, L.; Tomkinson, N.C.O. Preparation of the MacMillan Imidazolidinones. *Tetrahedron* **2011**, *67*, 4263–4267. [CrossRef]
42. Li, R.; Yao, S.; Cui, J.; Zhu, Y.; Ge, Z.; Cheng, T. Tripetide Boronic Acid or Boronic Ester, Preparative Method and Use Thereof. U.S. Patent 2012/0135921 A1, 31 May 2012.
43. Welin, E.R.; Warkentin, A.A.; Conrad, J.C.; MacMillan, D.W.C. Enantioselective α-Alkylation of Aldehydes by Photoredox Organocatalysis: Rapid Access to Pharmacophore Fragments from β-Cyanoaldehydes. *Angew. Chem. Int. Ed.* **2015**, *54*, 9668–9672. [CrossRef]
44. Tomé, J.P.C.; Neves, M.G.P.M.S.; Tomé, A.C.; Cavaleiro, J.A.S.; Mendonça, A.F.; Pegado, I.N.; Duarte, R.; Valdeira, M.L. Synthesis of glycoporphyrin derivatives and their antiviral activity against herpes simplex virus types 1 and 2. *Bioorg. Med. Chem.* **2005**, *13*, 3878–3888. [CrossRef]
45. Tang, X.; Feng, S.; Wang, Y.; Yang, F.; Zheng, Z.; Zhao, J.; Wu, Y.; Yin, H.; Liu, G.; Meng, Q. Bifunctional metal-free photo-organocatalysts for enantioselective aerobic oxidation of β−dicarbonyl compounds. *Tetrahedron* **2018**, *74*, 3624–3633. [CrossRef]

46. Liu, J.; Li, P.; Zhang, Y.; Ren, K.; Wang, L.; Wang, G. Recyclable Merrifield Resin-Supported Organocatalysts Containing Pyrrolidine Unit through A^3-Coupling Reaction Linkage for Asymmetric Michael Addition. *Chirality* **2010**, *22*, 432–441. [CrossRef]
47. Bryden, F.; Boyle, R.W. A Mild, Facile, One-Pot Synthesis of Zinc Azido Porphyrins as Substrates for Use in Click Chemistry. *Synlett* **2013**, *24*, 1978–1982.
48. Huy, P.; Neudörfl, J.-M.; Schmalz, H.-G. A Practical Synthesis of *Trans*-3-Substituted Proline Derivatives through 1,4-Addition. *Org. Lett.* **2011**, *13*, 216–219. [CrossRef]
49. Ahrendt, K.A.; Borths, C.J.; MacMillan, D.W.C. New Strategies for Organic Catalysis: The First Highly Enantioselective Organocatalytic Diels–Alder Reaction. *J. Am. Chem. Soc.* **2000**, *122*, 4243–4244. [CrossRef]
50. Holland, M.C.; Metternich, J.B.; Daniliuc, C.; Schweizer, W.B.; Gilmour, R. Aromatic Interactions in Organocatalyst Design: Augmenting Selectivity reversal in Iminium Ion Activation. *Chem. Eur. J.* **2015**, *6*, 10031–10038. [CrossRef]
51. Hayashi, Y.; Samana, S.; Gotoh, H.; Ishikawa, H. Asymmetric Diels–Alder Reactions of α,β-Unsaturated Aldehydes Catalyzed by a Diarylprolinol Silyl ether Salt in the Presence of Water. *Angew. Chem. Int. Ed.* **2008**, *47*, 11616–11617. [CrossRef]
52. Adler, A.D.; Longo, F.R.; Finarelli, J.D.; Goldmacher, J.; Assour, J.; Korsaloff, L.A. A Simplified Synthesis for *meso*-Tetraphenylporphin. *J. Org. Chem.* **1966**, *32*, 476. [CrossRef]
53. López, M.C.; Royal, G.; Philouze, C.; Chavant, P.Y.; Blandin, V. Imidazolidinone nitroxides as catalysts in the aerobic oxidation of alcohols, en route to atroposelective oxidative desymmetrization. *Eur. J. Org. Chem.* **2014**, *2014*, 4884–4896. [CrossRef]

Disclaimer/Publisher's Note: The statements, opinions and data contained in all publications are solely those of the individual author(s) and contributor(s) and not of MDPI and/or the editor(s). MDPI and/or the editor(s) disclaim responsibility for any injury to people or property resulting from any ideas, methods, instructions or products referred to in the content.

Article

Green Aromatic Epoxidation with an Iron Porphyrin Catalyst for One-Pot Functionalization of Renewable Xylene, Quinoline, and Acridine

Gabriela A. Corrêa, Susana L. H. Rebelo * and Baltazar de Castro

LAQV/REQUIMTE, Departamento de Química e Bioquímica, Faculdade de Ciências, Universidade do Porto, Rua do Campo Alegre s/n, 4169-007 Porto, Portugal
* Correspondence: susana.rebelo@fc.up.pt

Abstract: Sustainable functionalization of renewable aromatics is a key step to supply our present needs for specialty chemicals and pursuing the transition to a circular, fossil-free economy. In the present work, three typically stable aromatic compounds, representative of products abundantly obtainable from biomass or recycling processes, were functionalized in one-pot oxidation reactions at room temperature, using H_2O_2 as a green oxidant and ethanol as a green solvent in the presence of a highly electron withdrawing iron porphyrin catalyst. The results show unusual initial epoxidation of the aromatic ring by the green catalytic system. The epoxides were isolated or evolved through rearrangement, ring opening by nucleophiles, and oxidation. Acridine was oxidized to mono- and di-oxides in the peripheral ring: 1:2-epoxy-1,2-dihydroacridine and *anti*-1:2,3:4-diepoxy-1,2,3,4-tetrahydroacridine, with TON of 285. *o*-Xylene was oxidized to 4-hydroxy-3,4-dimethylcyclohexa-2,5-dienone, an attractive building block for synthesis, and 3,4-dimethylphenol as an intermediate, with TON of 237. Quinoline was directly functionalized to 4-quinolone or 3-substituted-4-quinolones (3-ethoxy-4-quinolone or 3-hydroxy-4-quinolone) and corresponding hydroxy-tautomers, with TON of 61.

Keywords: green chemistry; C-H functionalization; epoxidation; iron porphyrin; oxidation catalysis; renewable aromatics

Citation: Corrêa, G.A.; Rebelo, S.L.H.; de Castro, B. Green Aromatic Epoxidation with an Iron Porphyrin Catalyst for One-Pot Functionalization of Renewable Xylene, Quinoline, and Acridine. *Molecules* **2023**, *28*, 3940. https://doi.org/10.3390/molecules28093940

Academic Editor: Wim Dehaen

Received: 16 March 2023
Revised: 22 April 2023
Accepted: 5 May 2023
Published: 7 May 2023

Copyright: © 2023 by the authors. Licensee MDPI, Basel, Switzerland. This article is an open access article distributed under the terms and conditions of the Creative Commons Attribution (CC BY) license (https://creativecommons.org/licenses/by/4.0/).

1. Introduction

Biomimetic systems allow to reproduce enzyme activity while avoiding expensive enzyme extractions from natural sources and the expensive procedures of bio- and enzymatic catalysis. Iron and manganese porphyrins have shown the ability to mimic the remarkable activity of oxygenase enzymes, such as cytochrome P450, leading to efficient catalytic systems for sustainable oxidation of aromatic substrates, in mild conditions, and with novel reactivity patterns [1,2]. The porphyrin structure, the microenvironment, such as the solvent, co-catalyst, or catalyst support, have shown to play a key role on biomimetic efficiency [3,4].

A remarkable reaction of P450 during metabolism is the epoxidation of polycyclic aromatic compounds (PACs) in peripheral positions [5]. This reaction is not common in chemical systems, where PACs are mostly oxidized on the *meso*-rings to afford phenols, quinones, and analogues [6]. These are also observed using catalytic systems based on polyoxometalates [7], metallophthalocyanines [8], and metalloporphyrins [9].

In recent years, biomimetic aromatic epoxidations have been disclosed using Mn (2,6-dichlorophenyl)porphyrins (MnP) as catalysts (Scheme 1A–C), using non-green conditions, and epoxides of naphthalene, anthracene [10], tetracene [11], and acridine [12] have been obtained (Scheme 1A).

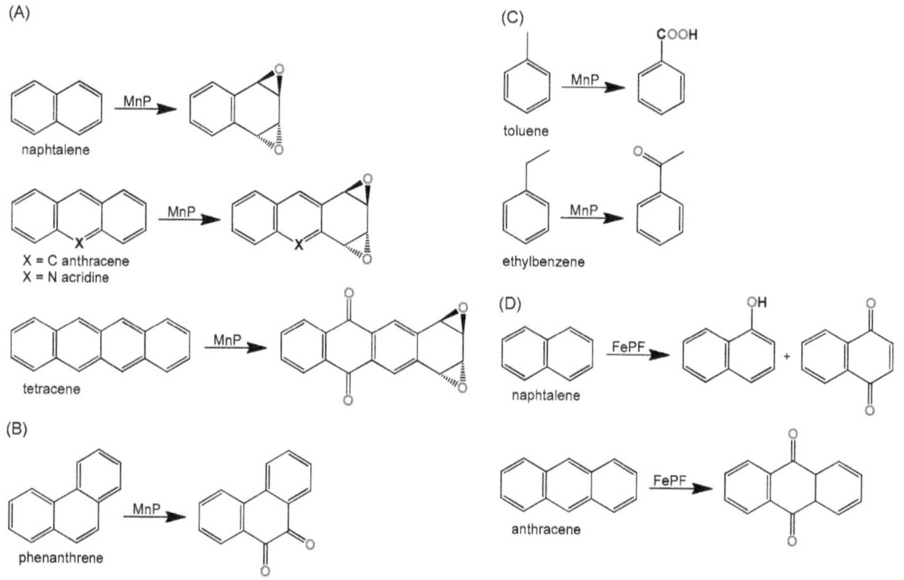

Scheme 1. Main direct oxidations of aromatic compounds by biomimetic catalysis with (**A–C**) Mn porphyrin (MnP) and (**D**) Fe porphyrin (FePF).

In some cases, epoxide formation was considered an intermediate step and products resulting from rearrangement of the epoxide ring were obtained, e.g., the o-diketone obtained in phenanthrene oxidation (Scheme 1B) [10,13]. It should be noted that alkylbenzenes oxidation in the presence of MnP catalytic systems afforded selectively alkyl group oxidation, e.g., toluene was oxidized mainly to benzoic acid and ethylbenzene to acetophenone (Scheme 1C) [14].

Interest in producing valued aromatic compounds from renewable sources has grown enormously in recent years, aiming to implement a circular economy and decrease dependence on fossil-based materials as feedstock in the fine chemicals industry [15–17].

Promising techniques for the production of platform green aromatics are the catalytic pyrolysis of biomass or wastes [18,19] and gas-phase Diels–Alder condensation of furan derivatives [20], among others. Optimization studies have been directed towards the increased production of the BTX (benzene, toluene, and xylenes) fraction, where xylenes are of major interest, for use as solvents and as intermediates for synthesis [21].

The chemical oxidation of o-xylene is described in the literature using harsh conditions, resulting in methyl groups functionalization [22], or degradation/removal from the environment [23]. Still, in biological systems, o-xylene is selectively oxidized in the aromatic ring by a diiron monooxygenase [24].

Quinoline and quinolone scaffolds are present in a vast number of natural compounds and pharmacologically active substances, comprising a significant segment of the pharmaceutical market [25]. Much has been achieved in developing greener syntheses of quinoline and its derivatives. In effect, quinoline can be obtained from biomass derivatives, such as glycerol or levulinic acid by reactions with aniline [26,27].

The direct oxidation of quinoline has been studied to obtain its degradation [28] or site-selective oxidation using enzymatic catalysis [29].

Acridine derivatives are also an important class of bioactive compounds with antibacterial and antimalarial activity and these have been studied as therapeutic agents for cancer and Alzheimer's disease [12]. Acridine can be obtained from non-fossil sources by catalytic pyrolysis of amino acids [30]. Previous studies on biomimetic oxidation of acridine with

Mn porphyrins led to direct and unprecedent epoxidation of the peripheral aromatic rings, disclosing the possibility of new functionalization routes (Scheme 1A) [12]. However, it would be desirable to obtain greener conditions, namely the substitution of acetonitrile as the solvent and improve product selectivity.

A green metalloporphyrin system for catalytic oxidation was described, using hydrogen peroxide as a green oxidant, producing water as the only byproduct, a highly electron withdrawing iron porphyrin [Fe(TPFPP)Cl] (FePF; Scheme 2), and ethanol as a green solvent, without any other additives or co-catalyst [FePF@H_2O_2_EtOH]. Moreover, improved methodologies for metalloporphyrin synthesis in eco-sustainable conditions have been reported [31]. This system has been effective in the epoxidation of alkenes and aromatic ring hydroxylation, but direct epoxidation of the aromatic ring has not been observed (Scheme 1D) [32,33].

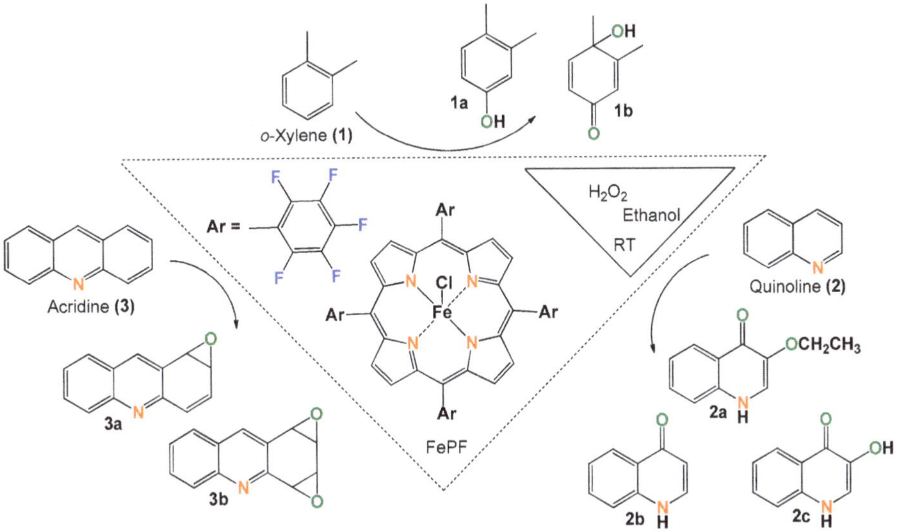

Scheme 2. Biomimetic oxidation of renewable aromatics.

The different catalytic activity of the MnPs and FePF has been ascribed to the formation of different active species in the catalytic cycle [3,14]. With the FePF, a hydroperoxyl species [PFe(III)-OOH] has been ascribed as the active oxidant, while an oxo-species is considered the active oxidant in the catalytic cycle of Mn porphyrins [PMn(V)=O] [32].

The present work describes the application of the [FePF@H_2O_2_EtOH] catalytic system, at room temperature (RT), in the oxidative valorization of the renewable aromatic compounds (Scheme 2).

2. Results and Discussion

The oxidation of o-xylene (**1**), quinoline (**2**), and acridine (**3**) was carried out by progressive addition of H_2O_2 at a rate of 0.6 mmol·h^{-1} in ethanol and at room temperature (RT), using the fluorinated iron porphyrin [Fe(TPFPP)Cl] (FePF) as catalyst. Control reactions performed in the absence of catalyst showed no substrate conversion during the catalytic reaction time.

2.1. o-Xylene (**1**)

The catalytic oxidation of o-xylene afforded two products, the 3,4-dimethylphenol (**1a**) and 4-hydroxy-3,4-dimethylcyclohexa-2,5-dienone (**1b**). The substrate conversion and product selectivity were monitored by GC-FID and the results are summarized in Table 1.

Using catalyst loadings of 0.3 and 0.6 mol %, the xylene conversion was 30% and 80%, respectively. These values are relevant in the context of C-H bond functionalization, where catalyst loadings between 2.5 and 15 mol % are commonly used [34]. The selectivity for the main product **1b** is 86% and is independent of catalyst loading. The kinetic plot of the reaction described in entry 2 (Figure 1) shows a nearly constant yield of **1a** during the reaction time, which indicates it as an intermediate of the final product **1b**. The maximum conversion was reached after 2.5 h of reaction with a turnover number (TON) of 237.

Figure 1. Kinetic curve of *o*-xylene catalytic oxidation reaction.

Table 1. Green oxidation of *o*-xylene by Fe porphyrin catalysis in ethanol at room temperature [a].

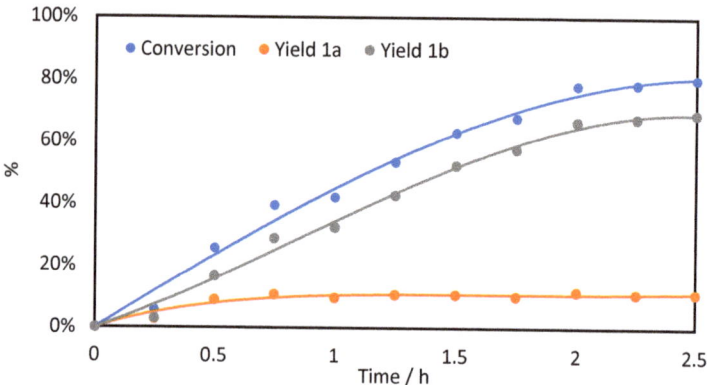

Entry	[FeP] (mol%)	Time (min) / H$_2$O$_2$ (eq.) [b]	Conversion (%) [c]	Selectivity (%) [c] 1a	Selectivity (%) [c] 1b	TON [d]
1	0.3	90 / 3 eq.	30	14	86	178
2	0.6	150 / 5 eq.	80	14	86	237

[a] Reaction conditions: *o*-xylene (0,3 mmol), [Fe(TPFPP)Cl] (1–2 mg), ethanol (2 mL), H$_2$O$_2$ (5 mol eq.), at RT for 2.5 h; [b] H$_2$O$_2$ added at 2 mol equivalents/h; [c] Conversion and selectivity measured by GC-FID analysis; [d] Turnover number (TON), two catalytic cycles were considered for product **1b** [35].

The MS spectrum obtained by GC-MS(EI) of the reaction mixtures are reported in Supplementary Material (SM). Compound **1a** shows [M$^{+\bullet}$] m/z 122 and loss of CO and CH$_3$ fragments as main peaks (Figure S1), matching 3,4-dimethylphenol [36]. The MS spectrum of **1b** shows a di-oxygenated product with [M$^{+\bullet}$] m/z 138 (Figure S2). Compound **1b** was isolated by fractionation of the reaction mixture using preparative thin layer chromatography (TLC) on silica gel and was fully characterized by ^{1}H, ^{13}C and 2D-NMR techniques.

The ^{1}H NMR spectrum shows the two methyl groups at 1.46 and 2.09 ppm (Figure 2). The latter peak is a doublet with ^4J = 1.5 Hz, due to a four-bonds coupling with H-2. The three signals in the alkene region corroborate the 4-hydroxycyclodienone structure. The H-2 signal is a quintet due long-range coupling with H-6 and CH$_3$(C-3). A double doublet is ascribed to H-6 with ^3J = 9.9 Hz and ^4J = 2.0 Hz from coupling with adjacent H-5 and at four-bonds with H-2, respectively. These assignments are corroborated by the HMBC (^{1}H^APT) spectrum (Figure 3) and by APT and HSQC spectra in SM (Figures S3 and S4).

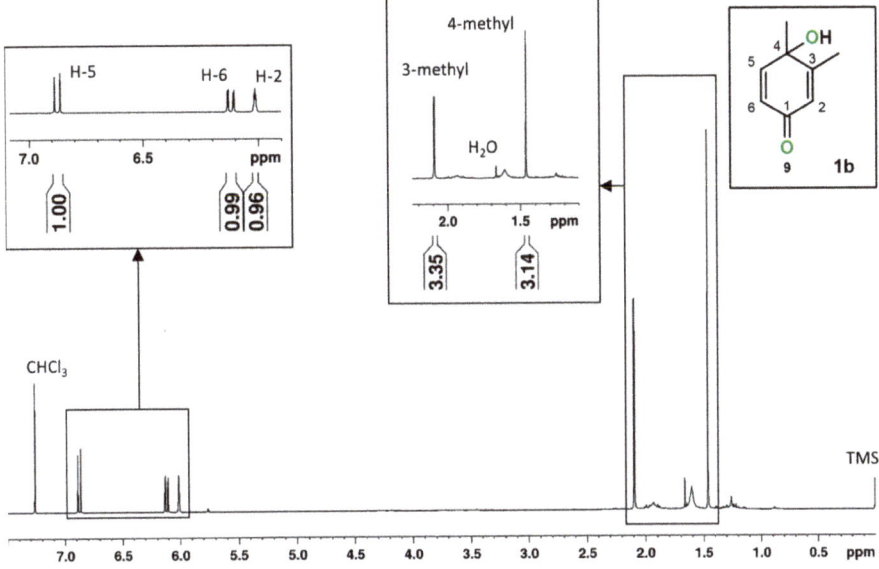

Figure 2. ^{1}H NMR spectrum for product **1b** in CDCl$_3$ using tetramethylsilane (TMS) as reference.

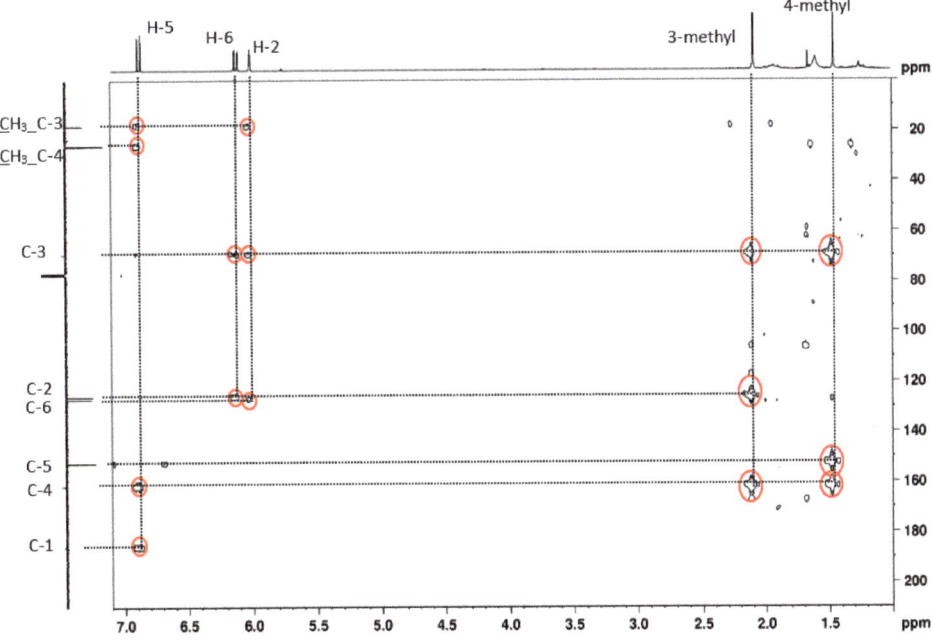

Figure 3. HMBC spectrum (^{1}H^APT) for product **1b** in CDCl$_3$. Red circles mark key correlation signals.

To our knowledge, this compound has not been previously characterized or isolated and is an attractive building block for synthesis, e.g., as a dienophile in cycloaddition reactions or as a structural analogue of ring C of tetracycline family antibiotics [11]. The different chemoselectivity relative to the previously reported Mn porphyrin catalytic systems, which promote selective alkyl group oxidation [14], highlights this reaction as a new and completely green pathway for the selective functionalization of the aromatic ring of alkylbenzenes.

2.2. Quinoline (2)

The direct one-pot oxidation of quinoline afforded the quinolone products 3-ethoxy-4-quinolone (**2a**), 4-quinolone (**2b**) and 3-hydroxy-4-quinolone (**2c**). The latter two products were observed also as hydroxyquinoline tautomeric compounds: 4-hydroxyquinoline (**2b***) and 3,4-dihydroxyquinoline (**2c***). The results are collected in Table 2, where for simplicity, selectivity and yield for products **2b** and **2c** corresponds to joint values observed for tautomer compounds (**2b** and **2b***) or (**2c** and **2c***).

Table 2. Green oxidation of quinoline by Fe porphyrin catalysis in ethanol at room temperature [a].

Entry	[FeP] (mol%)	Conversion (%)	Selectivity (Yield) (%)				TON [h]
			2a	2b [d]	2c [d]	Other	
1	0.9	57 [e,f]	100 (57) [g]	-	-	-	61
2	1.3	53 [f,g]	-	46 (24) [g]	22 (12) [g]	27 (14) [g]	-
3	1.9	70 [e]	-	-	-	-	38

[a] Reaction conditions: quinoline (0.3 mmol), [Fe(TPFPP)Cl] (3 mg), ethanol (2 mL), H_2O_2 (8 mol eq.), at RT for 4 h; [b] H_2O_2 added at 2 mol equivalents/h; [c] During the work-up, the evaporation of reaction mixture was performed in the rotavapor at 60 °C (I) or at RT (II); [d] For simplicity, this value corresponds to the joint selectivity of the two tautomer compounds observed; [e] Measured by GC-FID analysis; [f] Reaction mixture separated by TLC; [g] Measured by ^{1}H NMR spectrum of the final reaction mixture; [h] Turnover number (TON).

Quinoline consumption during the reaction was monitored by GC-FID, but reaction products were not observable by this technique. The reaction mixtures were fractionated by preparative TLC and all the collected fractions were analyzed by NMR and HR-ESI-MS^2 to obtain products identification/characterization. Subsequently, product selectivity and substrate conversion were obtained by ^{1}H NMR analysis of the final reaction mixtures in DMSO-d_6. Similar values of substrate conversion were observed by both techniques for identical reaction conditions (Table 2, entries 1 and 2). Using catalyst loadings of 0.6 and 1.9 mol%, the substrate conversion was 57% and 70%, respectively, after 4 h of reaction time and upon addition of 8 mol equivalents of H_2O_2 (Table 2, entries 1 and 3).

The presence of a substituent at 3-position (ethoxy or hydroxy) was dependent on the work-up conditions, namely the temperature of solvent evaporation. Solvent evaporation at 60 °C in the rotary evaporator in Path I and at RT in Path II.

Upon Path I, the fractionation of the reaction mixture by preparative TLC, afforded the 3-ethoxyquinolin-4-(1H)-one (3-ethoxy-4-quinolone, **2a**), which was isolated as the single reaction product (Table 2, entry 1).

The ^{1}H NMR spectrum of **2a** (Figure 4) shows the selective functionalization on the pyridyl ring, as the four signals in the aromatic region show a multiplicity and coupling pattern typical of a non-functionalized aromatic ring (COSY spectrum in SM, Figure S5). The two doublets at δ 4.01 ppm and 4.73 ppm (broad), coupling with each other (*J* = 4.8 Hz), are ascribed to H-2 and H-1(NH), respectively. The low δ values observed for H-2 and C-2 (HSQC spectrum in Figure 5) are expected for 3-substituted-4-quinolones [37], carrying an electron donor substituent at position 3. High electron density on C-2/H-2 is justified by the presence of mesomerism in compound **2a** (Figure 4, upper insert), with a significant contribution of two zwitterionic resonance hybrids to describe its structure. This is confirmed by the multiplet signal at δ 3.91–4.00 ppm, ascribed to ethoxy -CH$_2$ group. The contribution of the two zwitterionic structures leads to a hindrance in Ar-OEt bond rotation, resulting in distinct chemical environment on the -CH$_2$ protons [38,39]. The hydroxyl tautomer of compound **2a**, 3-ethoxy-4-hydroxyquinoline, was not observed.

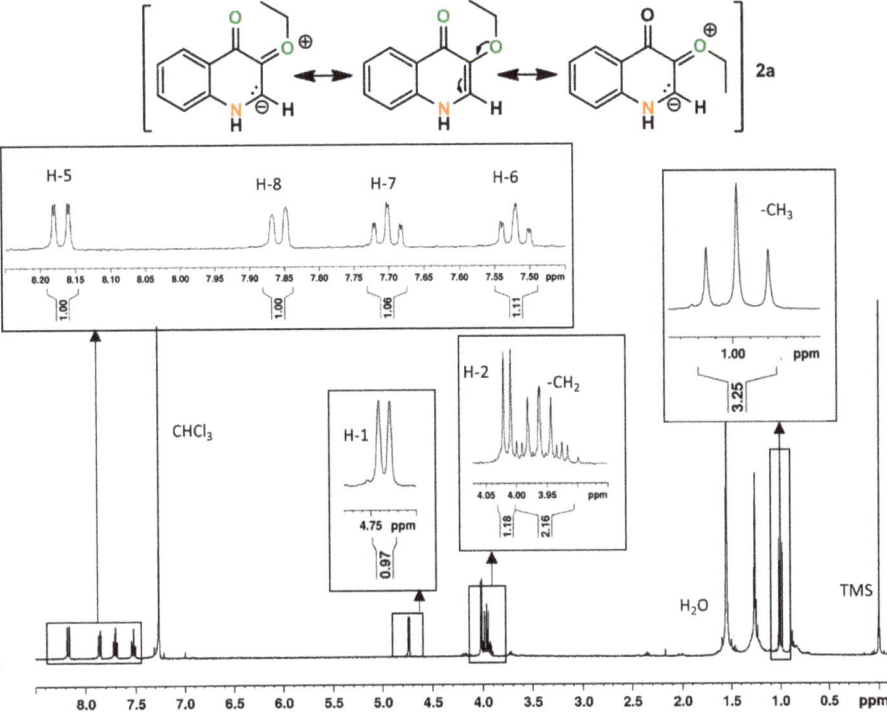

Figure 4. ^{1}H NMR spectrum of product **2a** in CDCl$_3$ using tetramethylsilane (TMS) as reference.

We found no previous references in the literature for compound **2a**, and this methodology may be an effective and green way to produce new 3-substituted quinolone derivatives, by direct functionalization of quinoline.

Upon Path II (solvent evaporation at RT, 17–22 °C), the fractionation of the reaction mixture by preparative TLC, afforded quinolin-4(1H)-one (4-quinolone, **2b**) and 3-hydroxyquinolin-4(1H)-one (3-hydroxy-4-quinolone, **2c**) and the corresponding tautomers 4-hydroxyquinoline (**2b***) and 3,4-dihydroxyquinoline (**2c***). The selectivity was 46% and 22% for the mixtures of tautomers (**2b** + **2b***) and (**2c** + **2c***), respectively (Table 2, entry 2).

Compounds **2b** and **2b***, **2c** and **2c*** were identified by HR-ESI-MS^2 (SM, Figures S7 and S8). The [M + H]$^+$ ions in the MS spectra were m/z 146.060 and m/z 162.055 for compounds **2b/2b*** and **2c/2c***, respectively. NMR studies in DMSO-d6 (^{1}H, APT, COSY, and HSQC; SM, Figures S9 and S10) confirmed the identification of these compounds.

The ^{1}H NMR spectrum of the total reaction, after evaporation at RT (Path II, Table 2, entry 2), was obtained in DMSO-d6 (SM, Figures S11 and S12). The area of chosen non-overlapping peaks from quinoline and reaction products were used for quantification of product selectivity, product yield and substrate conversion. The 4-hydroxy-tautomers were observed as the major products. It should be noted that the ferric center of [Fe(TPFPP)Cl] has a markedly acidic character [40] and may confers acidity to the reaction media, favoring the presence of hydroxyl-tautomers **2b*** and **2c***.

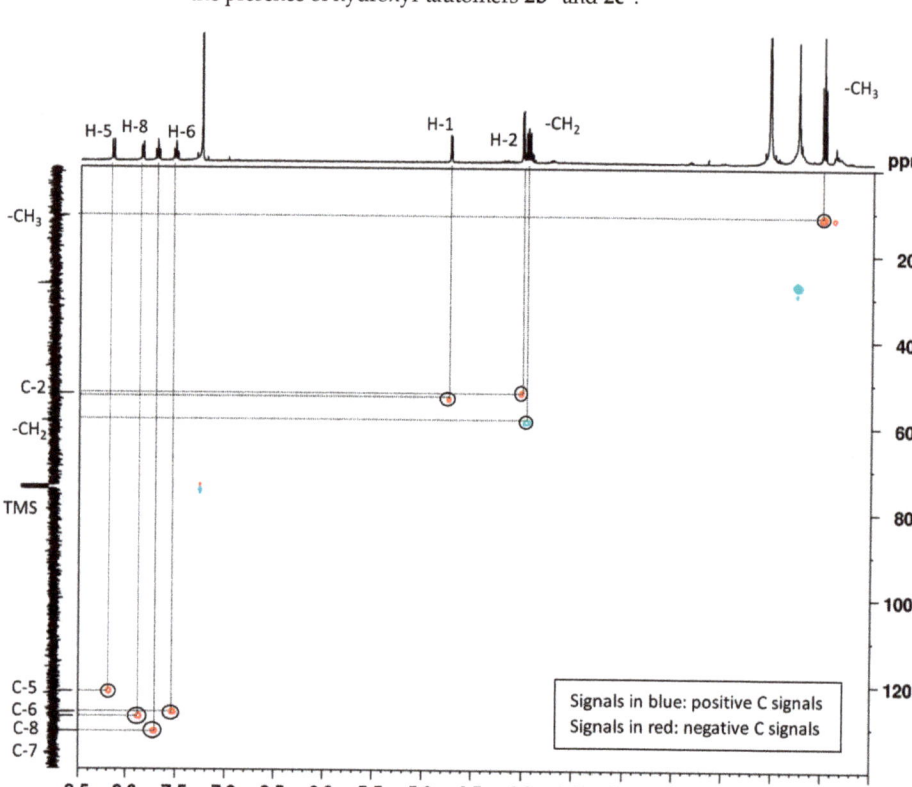

Figure 5. HSQC (^{1}H^APT) spectrum from product **2a** in CDCl$_3$. Positive signals are in blue and negative signals are in red.

2.3. Acridine (3)

Acridine oxidation yielded 1:2-epoxyacridine (**3a**) and *anti*-1:2,3:4-diepoxyacridine (**3b**). NMR studies of the reaction mixture after chromatographic separation of the catalyst allowed compounds' identification by comparison with previously described data [12]. Better selectivity was obtained for compound **3a** (90%, Table 1, entry 1) than in those studies using the MnP catalytic system (70%, Scheme 1) [12].

As acridine has an N atom in the structure, which confers basicity to the substrate that might influence the oxidation reaction, one reaction was carried out with addition of HNO$_3$. The results are presented in Table 3. It is observed that the pH does not lead to significant changes in the conversion of acridine, resulting only in a small increase in the yield of the monoepoxide (**3a**).

According to the ^{1}H spectrum of the reaction mixture, shown in Figure 6, there is a total of 18 protons, which indicates the presence of a mixture of the two products, **3a** and **3b**. The only two singlets present in the spectrum, at δ 8.38 and 8.20 ppm, correspond to H-9 of both compounds and their areas were used to quantify products selectivity. The ^{1}H, APT, and COSY NMR spectra are reported in SM (Figures S13 and S14).

Table 3. Green oxidation of acridine by Fe porphyrin catalysis in ethanol at room temperature [a].

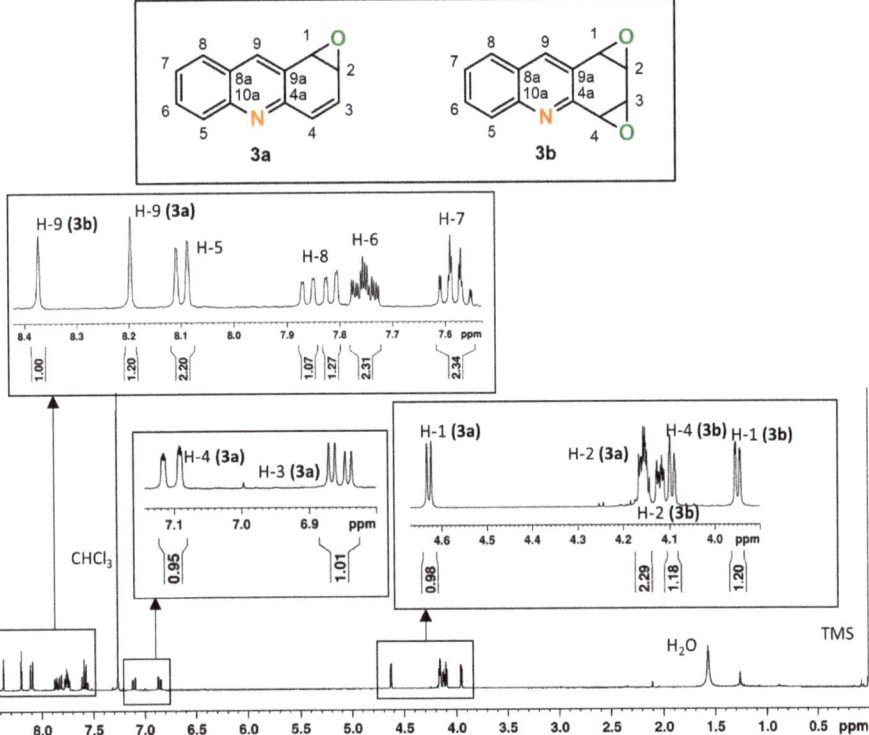

Entry	[FeP] (mol%)	Additive	Conversion (%) [c]	Selectivity (%) [c] 3a	Selectivity (%) [c] 3b	TON [d]
1	0.6	None	94	10	90	285
2	0.6	HNO$_3$ (pH 4.6)	89	16	84	261

[a] Reaction conditions: 2 mg of catalyst, 2 mL of solvent, 0.3 mmol of substrate, 5 eq. H$_2$O$_2$, 2h30 of reaction time; [b] H$_2$O$_2$ added at 2 mol equivalents/h; [c] Conversion and selectivity measured by ^{1}H RMN; [d] Turnover number (TON), two catalytic cycles were considered for product **1b**.

Figure 6. ^{1}H NMR spectrum of products **3a** and **3b** from fraction (A2) in CDCl$_3$ using tetramethylsilane (TMS) as reference.

2.4. Catalyst Stability

The reactions were followed by UV–vis. At the beginning of the reaction, the Fe porphyrin Soret band, at 410 nm, is observed and its intensity decreases as the reaction proceeds. This indicates the concomitant oxidation of the porphyrin macrocycle (SM, Figures S15 and S16). The cessation of substrate conversion relates to the complete disappearance of the Soret band after the TON maximum of 237, 61, and 285 for o-xylene, quinoline, and acridine, respectively.

2.5. Considerations on the Mechanism

Catalytic performance of the [Fe(TPFPP)Cl] (FePF) might be associated with the typical acidic character of iron porphyrins [40], which is intensified by the strong electron-withdrawing porphyrin ligand due to extensive fluorination. The iron-hydroperoxy-species [PFe(III)-OOH] formed by coordination of hydrogen peroxide to the metal center and subsequent deprotonation, has been considered the active oxidant in the catalytic cycle. In the absence of a co-catalyst, it is not expected that this species evolve into an oxo species [32].

Metallo-hydroperoxy species have been described as the active oxidant in epoxidation reactions or in the generation hydroxyl radicals [3]. Previous studies showed that the Fe(III) porphyrin is effective in alkene epoxidation and aromatic hydroxylation and an EPR spin trap study confirmed the absence of free hydroxyl radicals in these conditions [3]. The [PFe-OOH] is a strong oxidant, which can be further activated in the presence of a protic solvent by hydrogen bond formation with the hydroperoxide group [41]. This leads to an enhanced δ^+ at the distal oxygen (Scheme 3, central species).

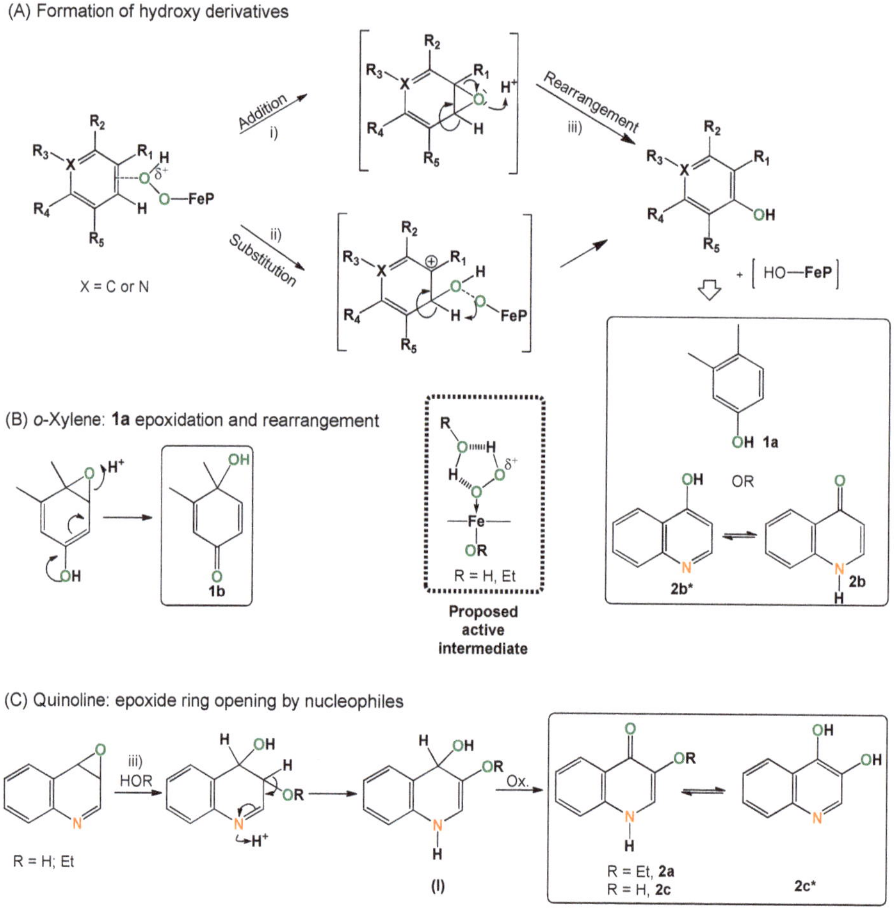

Scheme 3. Mechanistic proposals.

In the present work, it was observed that, unlike Mn porphyrins in aprotic solvent (acetonitrile) and with a co-catalyst, the present [FePF@H_2O_2_EtOH] system performs the selective aromatic oxidation of *o*-xylene, the oxidation of the aza-ring of quinoline and the epoxidation of the peripheral ring of acridine.

In previous studies [31], it was pointed out that the action of this catalyst in the formation of naphthoquinone by naphthalene oxidation (Scheme 1D) can be explained by the formation of naphtol via an electrophilic substitution mechanism (Scheme 3A, path ii). The electrophilic attack of [Fe-OO(δ^+)H] on the aromatic ring π-system, with formation of a carbocation intermediate and recovery of aromaticity by deprotonation of the adjacent proton.

However, the results of the present work suggest a direct epoxidation of the aromatic π-system (Scheme 3A, path i) by an addition reaction, and the hydroxylated derivatives formed result from subsequent rearrangement of the epoxide ring in acidic medium (path iii).

The presence of path (i) is confirmed by: (a) isolation of acridine epoxides; (b) formation of the quinoline derivatives **2a** (3-ethoxy-4-quinolone) and **2c** (3-hydroxy-4-quinolone) (Scheme 3C), either resulting by epoxide ring opening through nucleophilic attack of EtOH (60 °C) or H_2O (RT); (c) regioselectivity of hydroxylation in the formation of the final xylene product **1b** (Scheme 3B), suggesting epoxidation of the aromatic ring and not electrophilic substitution (path ii), as in the latter case, the derivatives should result from the formation of a tertiary carbocation intermediate, namely, 1,3-dihydroxy-4,5-dimethylbenzene or 3,4-dimethyl-*o*-benzoquinone.

The oxidation of intermediate **I** might occur also non-catalytically in the oxidizing reaction media, similarly to the formation of benzoquinones from hydroquinones previously observed [14].

3. Materials and Methods

3.1. Materials

The chloro [5,10,15,20-tetraquis(pentafluorophenyl)porphyrinate] iron (III) [Fe(TPFPP)Cl] (FePF) was prepared by a literature procedure, in environmentally compatible conditions, using microwave heating [31]. Quinoline (98%), acridine (97%), and H_2O_2 30% *w/w* were purchased from Sigma-Aldrich (St. Louis, MO, USA). *o*-Xylene (98%) *n*-hexane, and ethyl acetate were acquired from Fisher Scientific (Waltham, MA, USA). Ethanol was from AppliChem (Gatersleben, Germany). All solvents were p.a. grade. Nitric acid (65%) was from PanReac (Barcelona, Spain). The chromatographic purifications were carried out using silica gel 60 F254 from Merck (Darmstadt, Germany).

3.2. Instrumentation

The GC-FID analyses were performed using a Varian 3900 chromatograph (Palo Alto, CA, USA), using nitrogen as carrier, and GC-MS analyses were performed in a Thermo Scientific Trace 1300, coupled to a Thermo Scientific ISQ Single quadropole MS apparatus (Waltham, MA, USA), using helium as the carrier gas. In both cases, DB-5-type-fused silica Supelco (Sigma-Aldrich) capillary columns were used (30 m, 0.25 mm i.d.; 0.25 μm film thickness) and the temperature program was: 70 °C (1 min), 20 °C min^{-1} to 200 °C (5 min). The injector temperature was set at 200 °C and the detector temperature was set at 250 °C.

UV–vis absorption spectra were recorded at room temperature on a Genesys 10s Thermo Scientific Spectrophotometer in the region 300–800 nm.

High-resolution electrospray ionization mass spectra (HR-ESI-MS) were obtained using an LTQOrbitrap XL mass spectrometer (Thermo Scientific). Evaporated samples were dissolved in acetonitrile while reaction mixtures were directly injected and infused into the electrospray ion source at 10 μL·min^{-1}. The spectrometer was operated in the positive ionization mode with the capillary voltage set to +3.1 kV, sheath gas flow to 6, and the temperature of the ion transfer capillary to 275 °C.

NMR spectra (1D and 2D) were recorded on Bruker Avance instruments operating at a frequency of 400 MHz for ^1H experiments and 100 MHz for ^{13}C experiments, with sample temperatures of 22 °C and using $CDCl_3$ or DMSO-d_6 as solvent (Euroisotop, Cambridge, UK).

NMR and MS analyses were performed at CEMUP (Centro de Materiais da Universidade do Porto).

3.3. Catalytic Oxidation Reactions

The catalytic experiments were performed using the following general procedure: The substrate (0.3 mmol), the catalyst [Fe(TPFPP)Cl] (FePF), from 0.3 mol % (1 mg, 1 μmol) to 1.9 mol% (5 mg, 5 μmol), as indicated in Tables 1–3, were dissolved in 2 mL of ethanol and stirred at RT (17–22 °C) protected from light. H_2O_2 (aq.) 30% w/w was progressively added to the reaction mixture, by addition of aliquots of 0.5 mol equivalents relatively to the substrate every 15 min. The reactions were terminated when the substrate conversion did not change despite the addition of H_2O_2. When specified, the acridine reaction media was acidified by addition of HNO_3 until pH~4.6.

At the end of reactions, the solvent was evaporated at 60 °C in the rotary evaporator (Work-up-1, used for quinoline) or at RT (Work-Up 2, used for all substrates) and the reaction mixtures were separated by TLC, using mixtures of ethyl acetate: *n*-hexane as eluent: (40:60% v/v) for xylene (**1**) and (50:50% v/v) for quinoline (**2**) and acridine (**3**). Compounds were revealed on TLC plates using a UV lamp and were removed from silica with the same eluent used for chromatography.

For ^{1}H NMR analyses of the total reaction mixtures, the final reaction was passed through a small plug of silica-gel and eluted with DMSO-d_6.

Conversion (%) = [n (sum of products)/n of substrate]; Selectivity (%) = [n of product P/n (sum of products)]; Yield (%) = [n of product P/n of substrate].

3.4. Spectroscopic Data of Products

4-hydroxy-3,4-dimethylcyclohexa-2,5-dienone (**1b**) ^{1}H NMR (CDCl$_3$, 400 MHz) δ 1.46 (3H, s, CH$_3$_C-4), 2.09 (3H, d, J = 1.5 Hz, CH$_3$_C-3), 6.01 (1H, quintet, J = 1.5 Hz, H-2), 6.12 (1H, dd, J = 9.9, 2.0 Hz, H-6), 6.88 (1H, d, J = 9.9 Hz, H-5); ^{13}C NMR (CDCl$_3$, 400 MHz) δ 18.0 (CH$_3$_C-3), 26.0 (CH$_3$_C-4), 69.2 (C, C-4), 125.9 (CH, C-2), 127.0 (CH, C-6), 152.5 (CH, C-5), 161.7 (C, C-3), 185.8 (C, C-1); EIMS m/z (relative abundance %) 138 [M]$^{+\bullet}$ (43), 123 (100), 110 (54), 95 (77).

3-ethoxy-4-quinolone (**2a**) ^{1}H NMR (CDCl$_3$, 400 MHz) δ 1.00 (3H, t, J = 7.1 Hz, -CH$_3$), 3.96 (2H, m, -CH$_2$), 4.01 (1H, d, J = 4.8 Hz, H-2), 4.73 (1H, d-broad, J = 4.8 Hz, H-1), 7.52 (1H, t, J = 8.2 Hz, H-6), 7.70 (1H, t, J = 8.0 Hz, H-7), 7.85 (1H, d, J = 8.0 Hz, H-8), 8.18 (1H, dd, J = 8.2, 1.3 Hz, H-5); ^{13}C NMR (CDCl$_3$, 400 MHz) δ 14.1 (-CH$_3$), 54.6 (CH, C-2), 60.8 (-CH$_2$), 124.6 (CH, C-5), 129.3 (CH, C-6), 130.1 (CH, C-8), 133.6 (CH, C-7).

4-quinolone (**2b**) ^{1}H NMR (DMSO-d_6, 400 MHz) δ 6.51 (1H, d, J = 9.0 Hz, H-3), 7.19 (1H, t, J = 7.8 Hz, H-6), 7.32 (1H, d, J = 7.9 Hz, H-8), 7.50 (1H, t, H-7), 7.66 (1H, d, J = 7.8 Hz, H-5), 7.91 (1H, d, J = 9.0 Hz, H-2); ^{13}C NMR (DMSO-d_6, 400 MHz) δ 115.6 (CH, C-8), 122.2 (CH, C-3), 122.3 (CH, C-6), 128.4 (CH, C-5), 130.9 (CH, C-7), 140.8 (CH, C-2); HRESIMS m/z (relative abundance %) 162.055 [M + H]$^+$, (100), 144.045 (7), 134.060 (7), 116.050 (10).

4-hydroxyquinoline (**2b***) ^{1}H NMR (DMSO-d_6, 400 MHz) δ 7.50 (1H, d, J = 7.3 Hz, H-3), 7.76 (1H, t, J = 7.5 Hz, H-6), 7.84 (1H, t, J = 8.4 Hz, H-7), 7.99 (1H, d, J = 7.3 Hz, H-2), 8.11 (1H, d, J = 7.5 Hz, H-5), 8.55 (1H, d, J = 8.4 Hz, H-8); ^{13}C NMR (DMSO-d_6, 400 MHz) δ 119.3 (CH, C-8), 122.4 (CH, C-3), 126.1 (CH, C-2), 129.1 (CH, C-5), 129.3 (CH, C-6), 130.9 (CH, C-7), 119.6 (C, C-4a), 141.2 (C, C-8a), 162.6 (C, C-4); HRESIMS m/z (relative abundance %) 146.060 [M + H]$^+$, (100), 129.057 (2), 128.050 (3).

4. Conclusions

Inert aromatic compounds have been selectively functionalized by catalytic oxidation in mild and green conditions, using H_2O_2 as green oxidant, ethanol as green solvent, in the absence of other additives, at room temperature and using an electron withdrawing iron porphyrin catalyst, obtainable in eco-sustainable conditions, and used in a low loading of <2 mol%. The results support the occurrence of an initial direct epoxidation of the aromatic ring, leading to an unusual selectivity for *o*-xylene oxidation, as it occurred exclusively on the aromatic ring and not on the methyl groups, as previously observed. Moreover, the functionalization of acridine on the peripheral ring, instead of on the *meso* 9-position was unusual. The new methodology can be very attractive for the preparation of new 3-substituted quinolone derivatives, which have high potential for biological activity. The oxidations resulted in loss of aromaticity in products or in one of the aromatic rings. Two new compounds with attractive application potential were isolated and characterized.

The results point to the future relevance of aromatic epoxidation reactions, still largely unexplored in organic synthesis. This is mainly of importance in the valorization of aromatic products resulting from recycling processes based on pyrolysis of biomass and waste. Further developments of the catalytic system can be pursued developing more easily obtainable highly electron withdrawing iron catalysts.

Supplementary Materials: The following supporting information can be downloaded at: https://www.mdpi.com/article/10.3390/molecules28093940/s1, Figures S1 and S2: MS spectrum (EI) of products **1a** and **1b**; Figures S3 and S4: NMR spectra (APT and HSQC) of product **1b**; Figures S5 and S6: NMR spectra (COSY and APT) of product **2a**; Figures S7 and S8: HR-MS2 (ESI) spectra of products **2b** and **2c**; Figures S9 and S10: 2D-NMR spectra (COSY and HSQC) of a fraction containing compounds **2b**, **2b*** and **2c***; Figures S11 and S12: NMR spectra (1H and COSY) of a quinoline total reaction mixture; Figures S13 and S14: NMR spectra (COSY and APT) of a fraction containing compounds **3a** and **3b**; Figures S15 and S16: UV-vis of quinoline and acridine reaction mixtures.

Author Contributions: Conceptualization, S.L.H.R.; methodology, G.A.C. and S.L.H.R.; validation, G.A.C., S.L.H.R. and B.d.C.; investigation, G.A.C.; resources, S.L.H.R.; writing—original draft preparation, G.A.C. and S.L.H.R.; writing—review and editing, S.L.H.R. and B.d.C.; supervision, S.L.H.R. and B.d.C.; funding acquisition, B.d.C. All authors have read and agreed to the published version of the manuscript.

Funding: This work received financial support from PT national funds (FCT/MCTES, Fundação para a Ciência e Tecnologia and Ministério da Ciência, Tecnologia e Ensino Superior) through the projects UIDB/50006/2020 and UIDP/50006/2020.

Institutional Review Board Statement: Not applicable.

Informed Consent Statement: Not applicable.

Data Availability Statement: Not applicable.

Acknowledgments: The work was supported through the projects UIDB/50006/2020 and UIDP/50006/2020, funded by FCT/MCTES through national funds. S.L.H.R. thanks FCT (Fundação para a Ciência e Tecnologia) for funding through program DL 57/2016—Norma transitória (Ref. REQUIMTE/EEC2018/30). The authors thank M. Graça P.M.S. Neves, from the University of Aveiro, for her contribution to compound **2a** NMR spectra interpretation.

Conflicts of Interest: The authors declare no conflict of interest.

References

1. Huang, X.; Groves, J.T. Oxygen Activation and Radical Transformations in Heme Proteins and Metalloporphyrins. *Chem. Rev.* **2018**, *118*, 2491–2553. [CrossRef] [PubMed]
2. Calvete, M.J.F.; Piñeiro, M.; Dias, L.D.; Pereira, M.M. Hydrogen Peroxide and Metalloporphyrins in Oxidation Catalysis: Old Dogs with Some New Tricks. *ChemCatChem* **2018**, *10*, 3615–3635. [CrossRef]
3. Rebelo, S.L.H.; Moniz, T.; Medforth, C.J.; de Castro, B.; Rangel, M. EPR spin trapping studies of H_2O_2 activation in metaloporphyrin catalyzed oxygenation reactions: Insights on the biomimetic mechanism. *Mol. Catal.* **2019**, *475*, 110500. [CrossRef]
4. Lipińska, M.E.; Rebelo, S.L.H.; Pereira, M.F.R.; Figueiredo, J.L.; Freire, C. Photoactive Zn(II)Porphyrin–multi-walled carbon nanotubes nanohybrids through covalent β-linkages. *Mater. Chem. Phys.* **2013**, *143*, 296–304. [CrossRef]
5. Shimada, T.; Fujii-Kuriyama, Y. Metabolic activation of polycyclic aromatic hydrocarbons to carcinogens by cytochromes P450 1A1 and 1B1. *Cancer Sci.* **2004**, *95*, 1–6. [CrossRef]
6. Haynes, J.P.; Miller, K.E.; Majestic, B.J. Investigation into Photoinduced Auto-Oxidation of Polycyclic Aromatic Hydrocarbons Resulting in Brown Carbon Production. *Environ. Sci. Technol.* **2019**, *53*, 682–691. [CrossRef] [PubMed]
7. Estrada, A.C.; Simões, M.M.Q.; Santos, I.C.M.S.; Neves, M.G.P.M.S.; Cavaleiro, J.A.S.; Cavaleiro, A.M.V. Oxidation of Polycyclic Aromatic Hydrocarbons with Hydrogen Peroxide in the Presence of Transition Metal Mono-Substituted Keggin-Type Polyoxometalates. *ChemCatChem* **2011**, *3*, 771–779. [CrossRef]

8. Sorokin, A.; Meunier, B. Oxidation of Polycyclic Aromatic Hydrocarbons Catalyzed by Iron Tetrasulfophthalocyanine FePcS: Inverse Isotope Effects and Oxygen Labeling Studies. *Eur. J. Inorg. Chem.* **1998**, *1998*, 1269–1281. [CrossRef]
9. Giri, N.G.; Chauhan, S.M.S. Oxidation of polycyclic aromatic hydrocarbons with hydrogen peroxide catalyzed by Iron(III)porphyrins. *Catal. Commun.* **2009**, *10*, 383–387. [CrossRef]
10. Rebelo, S.L.H.; Simões, M.M.Q.; Neves, M.G.P.M.S.; Silva, A.M.S.; Cavaleiro, J.A.S. An efficient approach for aromatic epoxidation using hydrogen peroxide and Mn(iii) porphyrins. *Chem. Commun.* **2004**, *5*, 608–609. [CrossRef]
11. Costa, P.; Linhares, M.; Rebelo, S.L.H.; Neves, M.G.P.M.S.; Freire, C. Direct access to polycyclic peripheral diepoxy-meso-quinone derivatives from acene catalytic oxidation. *RSC Adv.* **2013**, *3*, 5350–5353. [CrossRef]
12. Linhares, M.; Rebelo, S.L.; Biernacki, K.; Magalhães, A.L.; Freire, C. Biomimetic one-pot route to acridine epoxides. *J. Org. Chem.* **2015**, *80*, 281–289. [CrossRef] [PubMed]
13. Rebelo, S.L.H.; Pires, S.M.G.; Simões, M.M.Q.; de Castro, B.; Neves, M.G.P.M.S.; Medforth, C.J. Biomimetic Oxidation of Benzofurans with Hydrogen Peroxide Catalyzed by Mn(III) Porphyrins. *Catalysts* **2020**, *10*, 62. [CrossRef]
14. Rebelo, S.L.H.; Simões, M.M.Q.; Neves, M.G.P.M.S.; Cavaleiro, J.A.S. Oxidation of alkylaromatics with hydrogen peroxide catalysed by manganese(III) porphyrins in the presence of ammonium acetate. *J. Mol. Catal. A Chem.* **2003**, *201*, 9–22. [CrossRef]
15. Dutta, S.; Bhat, N.S.; Anchan, H.N. Nanocatalysis for Renewable Aromatics. In *Heterogeneous Nanocatalysis for Energy and Environmental Sustainability*; John Wiley & Sons, Inc.: Hoboken, NJ, USA, 2022; pp. 61–90.
16. Wollensack, L.; Budzinski, K.; Backmann, J. Defossilization of pharmaceutical manufacturing. *Curr. Opin. Green Sustain. Chem.* **2022**, *33*, 100586. [CrossRef]
17. Niziolek, A.M.; Onel, O.; Guzman, Y.A.; Floudas, C.A. Biomass-Based Production of Benzene, Toluene, and Xylenes via Methanol: Process Synthesis and Deterministic Global Optimization. *Energy Fuels* **2016**, *30*, 4970–4998. [CrossRef]
18. Wang, S.; Li, Z.; Yi, W.; Fu, P.; Zhang, A.; Bai, X. Renewable aromatic hydrocarbons production from catalytic pyrolysis of lignin with Al-SBA-15 and HZSM-5: Synergistic effect and coke behaviour. *Renew. Energy* **2021**, *163*, 1673–1681. [CrossRef]
19. Genuino, H.C.; Muizebelt, I.; Heeres, A.; Schenk, N.J.; Winkelman, J.G.M.; Heeres, H.J. An improved catalytic pyrolysis concept for renewable aromatics from biomass involving a recycling strategy for co-produced polycyclic aromatic hydrocarbons. *Green Chem.* **2019**, *21*, 3802–3806. [CrossRef]
20. Gancedo, J.; Faba, L.; Ordóñez, S. From Biomass to Green Aromatics: Direct Upgrading of Furfural–Ethanol Mixtures. *ACS Sustain. Chem. Eng.* **2022**, *10*, 7752–7758. [CrossRef]
21. Zeng, Y.; Wang, Y.; Liu, Y.; Dai, L.; Wu, Q.; Xia, M.; Zhang, S.; Ke, L.; Zou, R.; Ruan, R. Microwave catalytic co-pyrolysis of waste cooking oil and low-density polyethylene to produce monocyclic aromatic hydrocarbons: Effect of different catalysts and pyrolysis parameters. *Sci. Total Environ.* **2022**, *809*, 152182. [CrossRef]
22. Wellmann, A.; Grazia, L.; Bermejo-Deval, R.; Sanchez-Sanchez, M.; Lercher, J.A. Effect of promoters on o-xylene oxidation pathways reveals nature of selective sites on TiO_2 supported vanadia. *J. Catal.* **2022**, *408*, 330–338. [CrossRef]
23. Mei, J.; Shen, Y.; Wang, Q.; Shen, Y.; Li, W.; Zhao, J.; Chen, J.; Zhang, S. Roles of Oxygen Species in Low-Temperature Catalytic o-Xylene Oxidation on MOF-Derived Bouquetlike CeO_2. *ACS Appl. Mater. Interfaces* **2022**, *14*, 35694–35703. [CrossRef] [PubMed]
24. Zhou, T.-P.; Deng, W.-H.; Wu, Y.; Liao, R.-Z. QM/MM Calculations Suggest Concerted O–O Bond Cleavage and Substrate Oxidation by Nonheme Diiron Toluene/o-Xylene Monooxygenase. *Chem. Asian J.* **2022**, *17*, e202200490. [CrossRef]
25. Batista, V.F.; Pinto, D.C.G.A.; Silva, A.M.S. Synthesis of Quinolines: A Green Perspective. *ACS Sustain. Chem. Eng.* **2016**, *4*, 4064–4078. [CrossRef]
26. Nasseri, M.A.; Zakerinasab, B.; Kamayestani, S. Proficient Procedure for Preparation of Quinoline Derivatives Catalyzed by $NbCl_5$ in Glycerol as Green Solvent. *J. Appl. Chem.* **2015**, *2015*, 743094. [CrossRef]
27. Ortiz-Cervantes, C.; Flores-Alamo, M.; García, J.J. Synthesis of pyrrolidones and quinolines from the known biomass feedstock levulinic acid and amines. *Tetrahedron Lett.* **2016**, *57*, 766–771. [CrossRef]
28. Jiao, Z.; Zhang, X.; Gong, H.; He, D.; Yin, H.; Liu, Y.; Gao, X. CuO-doped Ce for catalytic wet peroxide oxidation degradation of quinoline wastewater under wide pH conditions. *J. Ind. Eng. Chem.* **2022**, *105*, 49–57. [CrossRef]
29. Wang, Z.; Zhao, L.; Mou, X.; Chen, Y. Enzymatic approaches to site-selective oxidation of quinoline and derivatives. *Org. Biomol. Chem.* **2022**, *20*, 2580–2600. [CrossRef]
30. Sharma, R.K.; Chan, W.G.; Hajaligol, M.R. Product compositions from pyrolysis of some aliphatic α-amino acids. *J. Anal. Appl. Pyrolysis* **2006**, *75*, 69–81. [CrossRef]
31. Rebelo, S.L.; Silva, A.M.; Medforth, C.J.; Freire, C. Iron(III) Fluorinated Porphyrins: Greener Chemistry from Synthesis to Oxidative Catalysis Reactions. *Molecules* **2016**, *21*, 481. [CrossRef]
32. Rebelo, S.L.H.; Pereira, M.M.; Simões, M.M.Q.; Neves, M.G.P.M.S.; Cavaleiro, J.A.S. Mechanistic studies on metalloporphyrin epoxidation reactions with hydrogen peroxide: Evidence for two active oxidative species. *J. Catal.* **2005**, *234*, 76–87. [CrossRef]
33. Rebelo, S.L.H.; Pires, S.M.G.; Simões, M.M.Q.; Medforth, C.J.; Cavaleiro, J.A.S.; Neves, M.G.P.M.S. A Green and Versatile Route to Highly Functionalized Benzofuran Derivatives Using Biomimetic Oxygenation. *ChemistrySelect* **2018**, *3*, 1392–1403. [CrossRef]
34. Dalton, T.; Faber, T.; Glorius, F. C–H Activation: Toward Sustainability and Applications. *ACS Cent. Sci.* **2021**, *7*, 245–261. [CrossRef] [PubMed]
35. Kozuch, S.; Martin, J.M.L. "Turning Over" Definitions in Catalytic Cycles. *ACS Catal.* **2012**, *2*, 2787–2794. [CrossRef]
36. Kurganova, E.A.; Frolov, A.S.; Koshel', G.N.; Nesterova, T.N.; Shakun, V.A.; Mazurin, O.A. A Hydroperoxide Method for 3,4-Xylenol Synthesis. *Pet. Chem.* **2018**, *58*, 451–456. [CrossRef]

37. Zalibera, Ľ.; Milata, V.; Ilavský, D. ^{1}H and ^{13}C NMR spectra of 3-substituted 4-quinolones. *Magn. Reson. Chem.* **1998**, *36*, 681–684. [CrossRef]
38. Bergman, J.J.; Chandler, W.D. A Study of the Barriers to Rotation in Some Highly Substituted Diphenyl Ethers. *Can. J. Chem.* **1972**, *50*, 353–363. [CrossRef]
39. Eddahmi, M.; Moura, N.M.M.; Bouissane, L.; Gamouh, A.; Faustino, M.A.F.; Cavaleiro, J.A.S.; Paz, F.A.A.; Mendes, R.F.; Lodeiro, C.; Santos, S.M.; et al. New nitroindazolylacetonitriles: Efficient synthetic access via vicarious nucleophilic substitution and tautomeric switching mediated by anions. *New J. Chem.* **2019**, *43*, 14355–14367. [CrossRef]
40. Martins, M.B.M.S.; Corrêa, G.A.; Moniz, T.; Medforth, C.J.; de Castro, B.; Rebelo, S.L.H. Nanostructured binuclear Fe(III) and Mn(III) porphyrin materials: Tuning the mimics of catalase and peroxidase activity. *J. Catal.* **2023**, *419*, 125–136. [CrossRef]
41. Rocha, M.; Rebelo, S.L.H.; Freire, C. Enantioselective arene epoxidation under mild conditions by Jacobsen catalyst: The role of protic solvent and co-catalyst in the activation of hydrogen peroxide. *Appl. Catal. A Gen.* **2013**, *460–461*, 116–123. [CrossRef]

Disclaimer/Publisher's Note: The statements, opinions and data contained in all publications are solely those of the individual author(s) and contributor(s) and not of MDPI and/or the editor(s). MDPI and/or the editor(s) disclaim responsibility for any injury to people or property resulting from any ideas, methods, instructions or products referred to in the content.

Article

Photodynamic Inactivation of Microorganisms Using Semisynthetic Chlorophyll *a* Derivatives as Photosensitizers

Marciana Pierina Uliana [1,2,3,*], Andréia da Cruz Rodrigues [3], Bruno Andrade Ono [1], Sebastião Pratavieira [1], Kleber Thiago de Oliveira [2] and Cristina Kurachi [1]

[1] Instituto de Física de São Carlos, Universidade de São Paulo, São Carlos, São Paulo CEP 13560-970, Brazil
[2] Departamento de Química, Universidade Federal de São Carlos, Rodovia Washington Luís, km 235-SP-310, São Carlos, São Paulo CEP 13565-905, Brazil
[3] Universidade Federal da Integração Latino-Americana, Foz do Iguaçu CEP 85866-000, Brazil
* Correspondence: marciana.machado@unila.edu.br

Abstract: In this study, we describe the semisynthesis of cost-effective photosensitizers (PSs) derived from chlorophyll a containing different substituents and using previously described methods from the literature. We compared their structures when used in photodynamic inactivation (PDI) against *Staphylococcus aureus*, *Escherichia coli*, and *Candida albicans* under different conditions. The PSs containing carboxylic acids and butyl groups were highly effective against *S. aureus* and *C. albicans* following our PDI protocol. Overall, our results indicate that these nature-inspired PSs are a promising alternative to selectively inactivate microorganisms using PDI.

Keywords: *Spirulina maxima*; chlorophyll *a* derivatives; photosensitizers; semisynthesis; photodynamic inactivation

1. Introduction

Photodynamic reactions have been demonstrated to be an efficient alternative for the treatment of cancer [1–5], psoriasis [6], herpes [7], dermatological treatment [8], periodontics [9], canine otitis [10], control of cariogenic bacteria [11,12], and pneumonia [13]. In particular, antimicrobial photodynamic therapy (aPDT) or photodynamic inactivation (PDI) represents an interesting alternative for microbiological control because this technique is minimally toxic, noninvasive, minimizes the use of antibiotics, and due to its action over a broad spectrum of biomolecules, the risk of resistance is unlikely [14–18].

PDI has been used in the treatment of diseases caused by various microorganisms, including Gram-positive and Gram-negative bacteria and fungi [19], such as the inactivation of *Candida albicans*, which causes diseases in patients with low immunity [20,21]. It has also been used in the inactivation of *Aedes aegypti* mosquito larvae, a vector of the dengue, Zika, and chikungunya arboviruses [22–26]. Furthermore, PDI has gained attention in the inactivation of viruses, which have also shown resistance to drugs, such as an alternative to antiviral treatments against human papillomavirus and hepatitis B virus [27]. Recent studies have also shown the use of PDI for the treatment of severe acute respiratory syndrome caused by the new coronavirus (COVID-19) [28]. As PDI can inactivate DNA- or RNA-based viruses, these studies suggest considerable potential for use in virus photoinactivation in the future [29].

In general, photodynamic reactions require the presence of a photosensitizer, which is activated by light at a specific wavelength, allowing the production of reactive oxygen species (ROS). The main ROS are singlet oxygen, superoxide anions, hydroxyl radicals, and hydrogen peroxide. The quantum yields of each of these ROS depend on the PS and the conditions of the medium. After the formation of ROS, they interact with the target cells, causing death [30–34].

The most commonly used photosensitizers are methylene blue [35], porphyrins [36], chlorins [16,37], curcuminoids [38,39], and bacteriochlorins [40]. However, natural product derivatives, such as chlorin-e6 (chl-e6) obtained from chlorophyll a, are relatively low-cost photosensitizers and present many advantages in terms of pharmacokinetics, as they are easily eliminated from the body [11,41].

Chlorophyll a is formally a chlorin derivative, with four nitrogen atoms surrounding a central magnesium atom, along with numerous attached side chains and a hydrocarbon chain. Chlorins are excellent photosensitizers, and several synthetic chlorin analogues, such as m-tetrahydroxyphenylchlorin and mono-L aspartyl chlorin e6, have been used. Some substances, such as porphyrins, chlorins, and bacteriochlorins, stand out for their application in PDI: they present selected photophysical characteristics, allowing structural modifications to promote better solubility and amphiphilicity and improve their properties in several treatments [42]. Natural products are sources of inspiration for the development of several drugs [43]. The natural chlorophyll pigments from the cyanobacterium *Spirulina maxima* are abundant and easy to obtain. According to our protocol, [44] once dried, *S. maxima* is treated with methanol with 5% sulfuric acid, it gives rise to methyl pheophorbide *a*, and, after treatment with different amines or molecular oxygen, generates methyl pheophorbide *a* derivatives, including purpurin-18.

In this study, we propose the diverse semisynthesis of chlorophyll *a* derivatives, using simple reactions to evaluate these photosensitizers against microorganisms. All the chemical modifications performed with chlorophyll a were also aimed at conferring amphiphilicity to the PS, as this strategy has succeeded well in many approaches for PDT or PDI studies [45–48].

2. Results and Discussion

2.1. Semisynthesis of Photosensitizers

All the compounds were prepared using previously described methods from the literature. Initially, methyl-pheophorbide *a* (**1**) was obtained as previously described by our research group [44] (Scheme 1). Subsequently, primary aliphatic amines (butylamine, hexylamine, and octylamine) were reacted with **1** to give compounds **2–4**, all with absorption bands around 660 nm [49–51].

The purpurin-18 methyl ester (**5**) was also obtained from the oxidation of methyl pheophorbide *a* (**1**) [52,53], which allowed the addition of different aliphatic amines (butylamine, hexylamine, and octylamine) into the anhydride ring, resulting in products **6–8**. Compound **5** exhibited a characteristic absorption at 700 nm, which became approximately 660 nm when the anhydride ring was opened by the insertion of the amines (products **6–8**) [54–56].

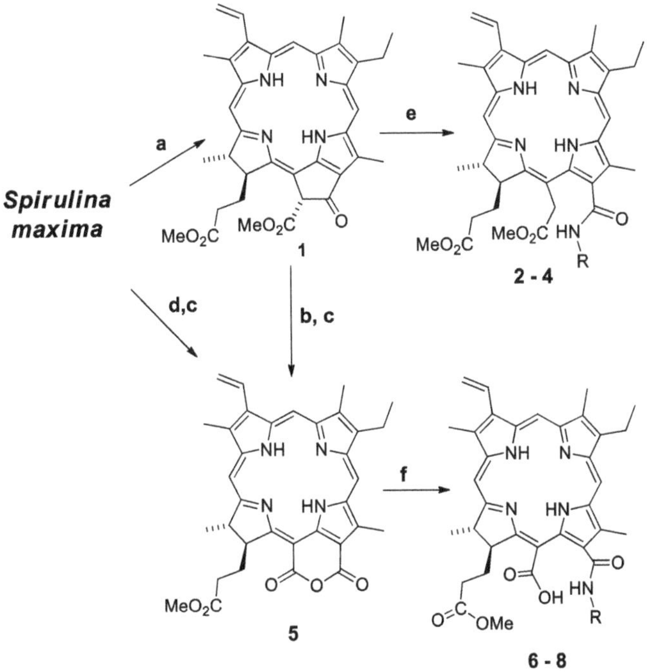

Scheme 1. Semisynthesis of chlorophyll *a* derivatives.

Conditions:
a) MeOH, H$_2$SO$_4$, 48h, **1** (0.8%)
b) i) O$_2$, NaOH, acetone, 3h, ii) citric acid, **5** (60%);
c) CH$_2$N$_2$, CH$_2$Cl$_2$, 10 min, 78%
d) i) O$_2$, NaOH, acetone, ii) citric acid 0.4%;
e) amine, CH$_2$Cl$_2$, 20 min, **2** (72%), **3** (95%), **4** (78%)
f) amine, CH$_2$Cl$_2$, 2h, **6** (90%), **7** (86%), **8** (80%)

Substituents
R= Butyl (**2, 6**), Hexyl (**3, 7**) or Octyl (**4, 8**)

The obtained PS **2–4** and **6–8**, were characterized using NMR, UV-vis, and high-resolution mass spectrometry (HRMS Q-TOFF) and presented absorption bands in the red region. See more details on the characterizations in Supplementary Materials, in which we present our compound band wavelengths and additional literature data (Table S1). Overall, these photosensitizers were semisynthesized at a low cost because we performed small structural modifications in the natural chlorophyll a, having the methyl-pheophorbide *a* (**1**) as a direct and versatile molecular template. With these modifications, we obtained six chlorin derivatives, **2–4** and **6–8**, all with absorption bands near 660 nm, different substituents, and amphiphilicity, which are desirable for use in PDT (Figures 1 and 2) [57]. The photostability of all compounds was also checked, showing that the photobleaching was not measurable even after 10 min of irradiation (see Supplementary Materials, Figure S18).

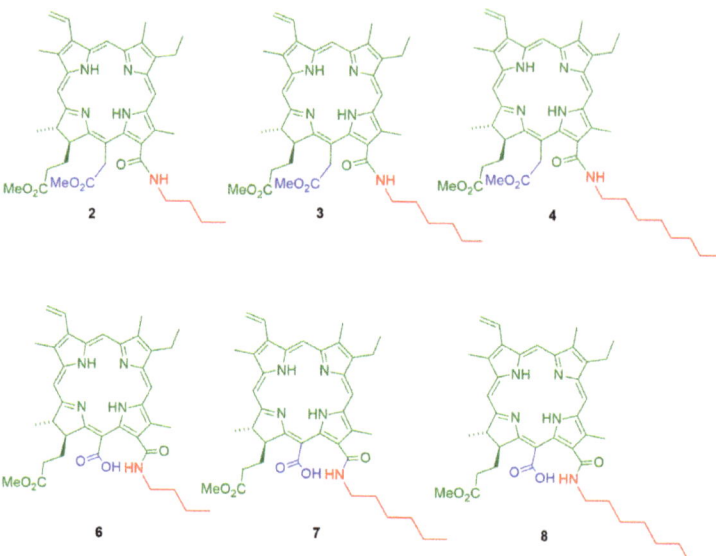

Figure 1. Chlorophyll *a* derivatives **2–4** and **6–8**.

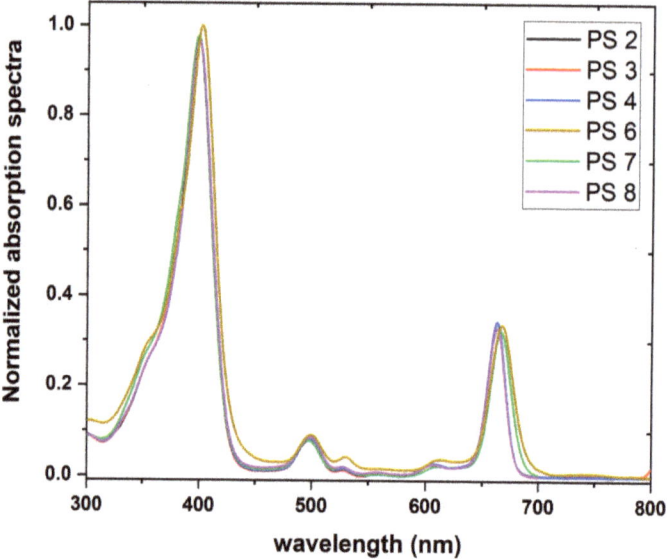

Figure 2. UV-vis absorption spectra of chlorophyll *a* derivatives **2–4** and **6–8** in ethyl acetate.

2.2. Photodynamic Inactivation

The compounds **2–4** and **6–8** were evaluated against three microorganisms—*S. aureus*, *C. albicans*, and *E. coli*—for the inactivation of these microorganisms. First, we investigated the dark toxicity of the chlorophyll derivatives. The microorganisms were incubated for 20 min in the dark with the respective photosensitizer (10 µM) and their mortality was evaluated after 24 h. No mortality was observed after 24 h with only the irradiation of the microorganisms without photosensitizers (30 Jcm^{-2}). The inactivation study was then performed with each microorganism. All photosensitizers **2–4** and **6–8** were used in the evaluation of *S. aureus* (Gram-positive bacteria), and different concentrations of photosensitizer (1 µM and 10 µM) and light fluences (15 Jcm^{-2} and 30 Jcm^{-2}) were utilized. The results obtained from these initial photoinactivation studies (Figure 3) show that with the increase in the carbon chain (from four to eight C atoms), the photosensitizers presented a decrease in the inactivation levels for the derivatives **2–4** and **6–8**.

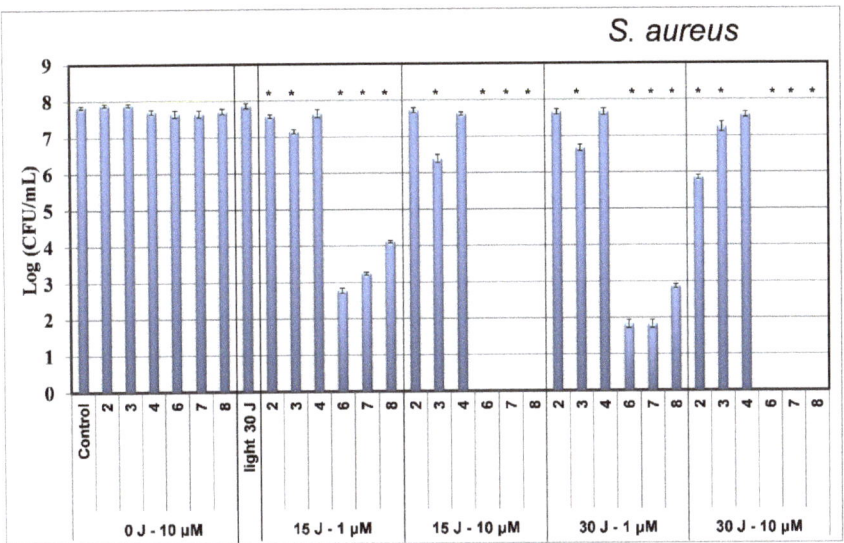

Figure 3. *Staphyloccocus aureus* counts (log CFU/mL) for 1 µM and 10 µM of photosensitizers **2–4** and **6–8** with the light doses of 0 Jcm^{-2} (control), 15 Jcm^{-2} and 30 Jcm^{-2}. Results are presented as mean ± SE. All the experiments were statistically evaluated (Tukey's test) and are in agreement with significance level of at least $p < 0.05$. Experiments pointed out with (*) represent $p < 0.05$ when compared to the control.

In addition, we observed that the photoinactivation of both *S. aureus* (Figure 3) and *C. albicans* (Figure 4) was influenced by light fluence and photosensitizer concentration, with increased photoinactivation at high concentrations and fluences.

Evaluating the photoinactivation of *S. aureus* (Figure 3), we observed that at 1 µM and 15 Jcm^{-2}, the methyl pheophorbide derivatives **2–4** did not present relevant inactivation, whereas the purpurin-18 derivatives **6–8** allowed significant inactivation, with a reduction of 3 log using PS **8**, 3.5 log with PS **7**, and 4 log with PS **6**. Maintaining the same concentration of photosensitizer and increasing the dose of light from 15 Jcm^{-2} to 30 Jcm^{-2} resulted in the inactivation of microorganisms by derivatives **2–4**; however, the photoinactivation promoted by PS **8** was approximately 4 log, and that by derivatives **6** and **7** was approximately 5 log.

It is possible to observe in Figure 3 that at 10 µM, both light doses (15 Jcm^{-2} and 30 Jcm^{-2}) were not very effective in the photoinactivation of *S. aureus* with photosensitizers **2–4**. In contrast, purpurin-18 derivatives **6–8** completely inhibited the growth of these microorganisms, proving that these derivatives are much more effective than those derived from methyl pheophorbide **2–4**.

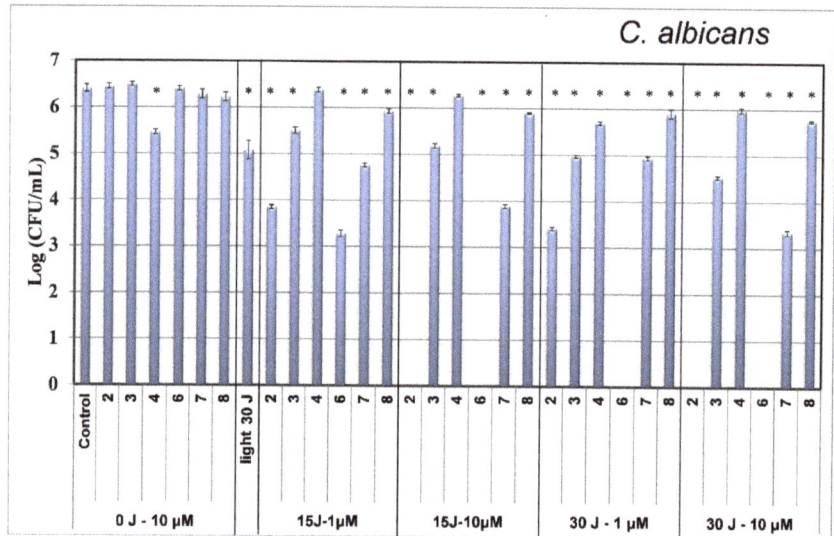

Figure 4. *Candida albicans* counts (log CFU/mL) for 1 µM and 10 µM of photosensitizers **2–4** and **6–8** with the light doses of 0 Jcm^{-2} (control), 15 Jcm^{-2} and 30 Jcm^{-2}. Results are presented as mean ± SE. All the experiments were statistically evaluated (Tukey's test) and are in agreement with significance level of at least $p < 0.05$. Experiments pointed out with (*) represent $p < 0.05$ when compared to the control.

These results suggest that the purpurin-18 derivatives **6–8**, due to the presence of a carboxylic acid group in the molecules may facilitate their microorganism uptake, whereas the derivatives of methyl pheophorbide **2–4** have an ester group with lower uptake.

The results presented in Figure 4 for *C. albicans* were similar to those obtained for *S. aureus*, with purpurin-18 derivatives **6–8** presenting better results in terms of photoactivation than methyl pheophorbide *a* derivatives **2–4**. However, PS **2** with the ester and butyl group also showed promising inactivation of *C. albicans*.

Overall, the results were similar to those for *S. aureus* and the higher the carbon chain present in the photosensitizers, the more hydrophobic the PS is used, and lower microorganism uptake was observed. As a consequence, we observed lower photoinactivation.

Photosensitizers **2–4** and **6–8** were also tested against *E. coli*, (Gram-negative bacteria), at 500 µM and light dose of 45 Jcm^{-2}. No photoinactivation by the photosensitizers **2–4** and **6–8** was observed, even using high doses of light, which was completely expected as Gram-negative microorganisms are preferentially inactivated by cationic photosensitizers [58,59]. Compared with our previous results using chlorin e6 (**chl-e6**) [44], we observed that *S. aureus* was completely inactivated at **chl-e6** concentrations of 1.6 µM and 16 µM when using a light dose of 30 Jcm^{-2}, which is very similar to the results obtained with derivatives **6** and **7**. However, considering *C. albicans* when compared with **chl-e6**, we found total inactivation at concentrations equivalent to 33 µM (20 µg mL^{-1}) and 50 µM (30 µg mL^{-1}) using a light dose of 30 Jcm^{-2}. We obtained the best results with derivative **6**, which completely inactivated *C. albicans* at a concentration of 10 µM (15 Jcm^{-2}) and at concentrations of 1 µM and 10 µM with 30 Jcm^{-2}. Chlorin e6 and **2, 3, 4, 6, 7**, and **8** did not show efficacy against *E. coli* [44].

Comparing among photosensitizers, when methylene blue was evaluated against the microorganism *S. aureus* at a concentration of 50 µM and a light dose of 9 J at 660 nm [60], a reduction of approximately 1.5 log CFU was obtained, whereas the photosensitizers **6, 7**, and **8** used in this study completely inactivated the *S. aureus* microorganism at a concentration of 10 µM and a light dose of 15 Jcm^{-2}, demonstrating it was more effective using these semisynthetic PS. When using 50 μM and a light dose of 9 J to perform photodynamic inactivation with methylene blue for *C. albicans*, a reduction of approximately 1 log CFU was obtained, whereas in this study, PS **7** at a concentration of 10 μM and a light dose of 15 Jcm^{-2} allowed a reduction of approximately 2 log CFU. Using the PS **3**, we obtained a reduction of approximately 1 log CFU, and PS **2** and **6** completely inactivated the microorganism.

Indocyanine green (ICG) [61], was also used as a photosensitizer for the photoinactivation of *S. aureus*. At a concentration of 25 μg mL^{-1} (32 μM) ICG and exposure to 411 Jcm^{-2} with near-infrared (NIR) light (808 nm laser), a significant reduction in the viable count was achieved (5.56 log10); complete inactivation of the microorganism occurred at a concentration of 200 μg mL^{-1} (258 μM) and 411 Jcm^{-2}. In another study using ICG [62], PDI against *S. aureus* with an ICG concentration of 8 μg mL^{-1} (10 μM) at an energy dose of 84 Jcm^{-2} resulted in 100% inactivation of this microorganism. In the current study, PS **6**, **7**, and **8** also showed complete inactivation of *S. aureus* at a concentration of 10 μM using 15 Jcm^{-2} and 30 Jcm^{-2} (660 nm).

In an in vitro study, the inactivation of *C. albicans* was evaluated with ICG at 1 μg mL^{-1} using a light dose of 228 J/cm^{-2} (810 nm); a satisfactory result showing a reduction of 1.2 log was obtained, similar to the results with nystatin. Compared to the control, the elimination of *C. albicans* increased by 92% when treated with ICG (1 mg/mL) with infrared (IR) laser irradiation (810 nm, 55 J/cm^{-2}).

3. Materials and Methods

Nuclear magnetic resonance (NMR) analyses were performed on a Bruker Avance 400 spectrometer at 400.15 MHz (^{1}H) and 100.13 MHz (^{13}C). Tetramethylsilane was used as an internal reference.

High-resolution mass spectrometry (HRMS) was performed using ESI-TOF (Waters Xevo G2-S Qtof). Ultraviolet-visible spectrophotometry analyses were performed using a Lambda 25 spectrometer (PerkinElmer, Waltham, MA, USA).

Spirulina maxima powder was purchased from Pharma Nostra (Rio de Janeiro, Brazil), and other reagents were purchased from Sigma-Aldrich (St. Louis, MO, USA).

3.1. Semisynthesis of Photosensitizers

*Isolation of methyl pheophorbide a (**1**) from Spirulina maxima (Mepheo a):* 300 g of the *Spirulina maxima* was treated with a 1.5 L of the 5% methanolic solution of H$_2$SO$_4$ for 24 h at room temperature. This mixture was filtered and washed with methanol (900 mL) and ethyl acetate (900 mL), and the organic phases were evaporated under reduced pressure. After that, 150 g of crushed ice was added to the crude residue. The residue was neutralized with solid NaHCO$_3$ and placed on a silica gel plug. The chlorophyll derivatives were retained on the plug, and the residual proteins and peptides (pale yellow in color) were eluted with water. The chlorophyll derivatives were then eluted with ethyl acetate (900 mL) and washed with water (3 × 400 mL). The organic phase was separated, dried over Na$_2$SO$_4$, and the solvent evaporated under reduced pressure. The methyl pheophorbide *a* was purified by silica gel flash chromatography using as eluent toluene:ethyl acetate (9:1), yielding the methyl pheophorbide *a* (**1**) (2.4 g, 3.9 mmol, 0.8% yield from natural dried *Spirulina maxima*) [44,63,64].

UV-vis (CH$_2$Cl$_2$): λmax (nm): 666, 609, 534, 505, 409 [65].
^{1}H-NMR (CDCl$_3$, 400 MHz) δ: 9.51 (s, 1H, H-10); 9.37 (s, 1H, H-5); 8.56 (s, 1H, H-20); 7.97 (dd, 1H, J = 11.6, J = 17.8 Hz, H-3^1); 6.25 (s, 1H, H-13^2); 6.24 (dd, 1H, J = 1.5 and J = 17.8 Hz, H-3^{2B}); 6.16 (dd, 1H, J = 1.5 and J = 17.8 Hz, H-3^{2a}); 4.48–4.43 (m, 1H, H-18^1); 4.21–4.19 (m, 1H, H-17); 3.88 (s, 3H, H-13^4); 3.68 (s, 3H, H-12^1); 3.60 (q, 2H, J = 7.7 Hz, H-8^1); 3.57 (s, 3H, H-17^4); 3.40 (s, 3H, H-2^1); 3.22 (s, 3H, H-7^1); 2.68–2.48 (m, 2H, H-17^1); 2.36–2.20 (m, 2H, H-17^2); 1.81 (d, 3H, J = 7.3 Hz, H-18^2); 1.69 (t, 3H, J = 7.6 Hz, H-8^2); 0.54 (br. s, 1H, H-21); −1.63 (br. s, 1H, H-23).

^{13}C-NMR (CDCl$_3$, 100 MHz) δ: 192.1, 174.7, 172.5, 169.4, 162.9, 155.9, 151.1, 150.1, 145.3, 142.3, 137.9, 136.8, 136.4, 136.3, 132.0, 129.1, 129.0, 127.9, 122.9, 104.6, 104.5, 97.5, 93.4, 65.9, 65.1, 63.3, 52.2, 51.1, 50.3, 30.8, 30.0, 23.2, 19.4, 17.4, 12.1, 11.3

13-(Butylcarbamoyl)-chlorin e6 15,17-dimethyl ester (**2**). To a solution of 30 mg (0.049 mmol) of **1** in 3 mL of dry tetrahydrofuran, 0.5 mL of butylamine was added and the reaction was stirred for 20 min at room temperature. After that, the solvent was evaporated, the mixture was diluted with dichloromethane and washed with HCl solution (1%). The resulting solution was dried over anhydrous sodium sulfate, and the solvent evaporated under reduced pressure. Compound 2 was isolated by chromatography over silica gel using toluene:ethyl acetate (9:1) as eluent (22 mg, 72% yield) [50,51].

UV-vis (CH$_2$Cl$_2$) λmax (nm): 662, 607, 528, 498, 399 [66].

^{1}H-NMR (CDCl$_3$, 400 MHz) δ: 9.69 (s, 1H); 9.64 (s, 1H); 8.80 (s, 1H); 8.08 (dd, 1H, J = 17.8 and 11.6 Hz); 6.36–6.40 (m, 1H, CONH); 6.33 (dd, 1H, J = 16.7, 1.4 Hz); 6.15 (dd, 1H, J = 11.6, 1.4 Hz); 5.54 and 5.26 (d, 1H each, J = 18.9 Hz); 4.46 (q, 1H, J = 7.2 Hz); 4.35 (d, 1H, J = 9.5 Hz); 3.91–3.76 (m, 4H); 3.81 (s, 3H); 3.60 (s, 3H); 3.56 (s, 3H); 3.49 (s, 3H); 3.32 (s, 3H); 2.56–2.09 (m, 4H); 1.82–1.69 (m, 10H); 1.06 (t, 3H, J = 7.3 Hz); −1.61 (br.s, NH); −1.83 (br.s, NH).

HRMS (ESI): *m/z* calculated [M + H]$^+$ for C$_{40}$H$_{50}$O$_5$N$_5^+$ = 680.38065; found 680.38379.

13-(Hexylcarbamoyl)chlorin e6 15,17-dimethyl ester (**3**). To a solution of 40 mg (0.066 mmol) of **1** in 3 mL of dry tetrahydrofuran, 0.5 mL of hexylamine was added and the reaction was stirred for 20 min at room temperature. After that, the solvent was evaporated, the mixture was diluted with dichloromethane and washed with HCl solution (1%). The resulting solution was dried over anhydrous sodium sulfate, and the solvent evaporated under reduced pressure. Compound 3 was isolated by chromatography over silica gel using toluene:ethyl acetate (9:1) as eluent (45 mg, 95% yield) [50,51].

UV-vis (CH$_2$Cl$_2$): λmax (nm): 663, 605, 527, 497, 398 [66].

^{1}H-NMR (CDCl$_3$, 400 MHz) δ: 9.70 (s, 1H), 9.64 (s, 1H); 8.80 (s, 1H); 8.10 (dd,1H, J = 11.5 and 17.8 Hz), 6.36–6.39 (m, CONH), 6.35 (dd, 1H, J = 1.4 and 17.8 Hz), 6.13 (dd, 1H, J = 1.4 and 11.5 Hz), 5.52 (d, 1H, J = 18.9 Hz), 5.25 (d, 1H, J = 19.0 Hz), 4.46 (q, 1H, J = 7.3 Hz), 4.35 (d, 1H, J = 11.2 Hz), 3.90–3.68 (m, 4H), 3.80 (s, 3H), 3.60 (s, 3H), 3.57 (s, 3H), 3.49 (s, 3H), 3.32 (s, 3H), 2.56–2.07 (m, 4H), 1.83–1.37 (m, 16H), 0.94 (t, 1H, J = 7.2 Hz), −1.63 (NH), −1.83 (NH).

HRMS (ESI): m/z calculated [M + H]$^+$ for C$_{42}$H$_{54}$O$_5$N$_5^+$ = 708.41195; found 708.41516.

13-(Octylcarbamoyl)chlorin e6 15,17-dimethyl ester (**4**). To a solution of 30 mg (0.049 mmol) of **1** in 3 mL of dry tetrahydrofuran, 0.5 mL of octylamine was added and the reaction was stirred for 20 min at room temperature. After that, the solvent was evaporated, the mixture was diluted with dichloromethane and washed with HCl solution (1%). The resulting solution was dried over anhydrous sodium sulfate, and the solvent evaporated under reduced pressure. Compound 4 was isolated by chromatography over silica gel using toluene:ethyl acetate (9:1) as eluent (28 mg, 78% yield) [51].

UV-vis (CH$_2$Cl$_2$): λmax (nm): 663, 609, 527, 497, 399 [66].

^{1}H-NMR (CDCl$_3$, 400 MHz) δ: 9.70 (s, 1H); 9.64 (s, 1H); 8.80 (s, 1H); 8.09 (dd, 1H, J = 17.8, 11.5 Hz); 6.36–6.39 (m, 1H, CONH); 6.36 (dd, 1H, J = 18.2, 1.4 Hz); 6.14 (dd, 1H, J = 11.5, 1.4 Hz), 5.54 and 2.25 (d, 1H each, J = 18.9 Hz); 4.47 (q, 1H, J = 7.1 Hz); 4.36 (d. 1H, J = 7.6 Hz); 3.90–3.77 (m, 4H); 3.80 (s, 3H); 3.60 (s, 3H); 3.57 (s, 3H); 3.49 (s, 3H); 3.32 (s, 3H); 2.52–2.09 (m, 4H); 1.83–1.69 (m, 10 H); 1.46–1.33 (m, 8H); 0.90 (t, 3H, J = 7.1 Hz); −1.62 (br, s, NH); −1.83 (br, s, NH).

HRMS (ESI): *m/z* calculated [M + H]$^+$ for C$_{44}$H$_{58}$O$_5$N$_5^+$ = 736.44325; found 736.44147.

Purpurin-18 Methyl Ester (**5**) Method 1: Methyl pheophorbide *a* **1** (151 mg, 0.25 mmol) was dissolved in 500 mL of diethyl ether and 5 mL of pyridine. After, a potassium hydroxide solution in 1-propanol (2 g of KOH was dissolved in 10 mL of 1-propanol) was added into the first solution and oxygen was bubbled into the resulting reaction mixture for 1 h. The reaction mixture was extracted with water (500 mL). The aqueous layer was collected and the pH adjusted to 2–4 using cold H_2SO_4 solution (25%). The aqueous layer was extracted with CH_2Cl_2 and the solvent evaporated to give a purple residue. The product was purified by chromatography over silica gel using hexane:ethyl acetate 3:1 as eluent. After that, the carboxylic acid precursor was obtained in 55% yield (0.078 g, 0.119 mmol). This product was further reacted with a diazomethane solution in dichloromethane to produce purpurin-18 methyl ester (**5**) for 10 min at 0 °C. The residue was crystallized with dichloromethane/hexane, thus obtaining **5** in 60% yield (63.0 mg, 0.109 mmol) as purple red crystals [51]. Method 2: Pigments were extracted twice from *S. maxima* dried powder (10 g) with acetone (4 × 100 mL) under magnetic stirring at 60 °C (4 × 30 min). The dark green extract was filtered off and the filtrate (ca. 400 mL) was reduced to 200 mL by partial evaporation under reduced pressure. NaOH (40 mL, 6 M) was added to the previous extract (200 mL); the mixture was vigorously stirred and oxygen was bubbled during 3 h. The solution was then acidified with concentrated HCl. The oxidized extract was evaporated to dryness. Carotenoids and part of xanthophylls were removed by extraction with petroleum ether (2 × 100 mL). The resulting residue was purified by flash chromatography (eluent CH_2Cl_2:MeOH 8:2) obtaining 40.0 mg (0.070 mmol, 0.4% yield) of the purpurin-18 carboxylic acid. The product was esterified with diazomethane obtaining 32.0 mg (0.055 mmol, 78% yield) of the **5** [53,67].

UV-vis (CH_2Cl_2): λmax (nm): 699, 642, 546, 508, 478, 410.
^{1}-H NMR (400 MHz, $CDCl_3$, ppm): 9.57 (s, 1H, H-10); 9.36 (s, 1H, H-5); 8.56 (s, 1H, H-20); 7.89 (dd, 1H, J = 17.8, 11.6 Hz, H-3^1); 6.27 (dd, 1H, J = 17.8 and J = 9.4, 2.5 Hz, H-3^{2B}); 6.20 (dd, 1H, J = 11.6 and J = 9.4, 2.5 Hz, H-3^{2a}); 4.39 (q, 1H, J = 7.3 Hz, H-18^1); 3.77 (s, 3H, H-12^1); 3.64 (q, 2H, J = 7.9 Hz, H-8^1); 3.60 (s, 3H, H-17^4); 3.35 (s, 3H, H-2^1); 3.15 (s, 3H, H-7^1); 2.78–2.70 (m, 2H H-17^1); 2.49–2.43 (m, 2H, H-17^2); 1.74 (d, 3H, J = 7.3 Hz, H-18^2); 1.65 (t, 3H, J = 7.6 Hz, H-8^2); 0.21 (br s, 1H, H-21); −0.08 (br s, 1H, H-23).
^{13}C-NMR ($CDCl_3$, 100 MHz) δ: 178.6, 173.5, 169.8, 165.4, 163.4, 158.4, 154.2, 146.9, 145.1, 141.9, 140.0, 137.9, 135.5, 133.5, 132.8, 132.8, 131.4, 128.0, 125.6, 107.4, 102.5, 97.9, 94.5, 55.5, 51.7, 49.6, 32.4, 31.4, 20.4, 19.6, 16.9, 12.9, 12.1, 11.2

Chlorin-p, 6-N-Butylamide-7-methyl Ester (**6**). To a solution of **5** (40.0 mg, 0.069 mmol) in dichloromethane (5 mL), butylamine (0.5 mL) was added and the reaction mixture was stirred at room temperature in the dark under nitrogen for 2 h. After that, the UV-vis analysis and TLC showed the absence of starting material. The product was purified by crystallization with dichloromethane-hexane to give the product **6** (48.0 mg, 0.070 mmol, 90% yield) [54–56].

UV-vis (CH_2Cl_2): λmax (nm): 662, 606, 526, 498, 398.
^{1}H-NMR ($CDCl_3$, 400 MHz) δ: 9.67(s, 1H), 9.62 (s, 1H), 8.74 (s, 1H), 8.12 (dd, J = 18, 11 Hz 1H), 6.96–7.05 (m, NH), 6.37 (d, J = 18 Hz, 1H), 6.15 (d, J = 11 Hz, 1H), 5.06 (d, J = 7.7 Hz,1H), 4.33–4.37 (m, 1H) 3.74–3.78 (m, 4H), 3.57 (s, 3H), 3.46 (s, 3H), 3.31 (s, 3H), 3.23 (s, 3H), 2.49–2.27 (m, 2H), 1.70 (d, 3H, J = 7.2Hz), 1.68 (t, 3H, J = 7.5Hz), 0.79–1.10 (m, 4H), 0.95 (t, 3H, J = 7.2Hz), 0.63(t, 3H, J = 7.2Hz), −1.66 (s, NH) and −1.94 (s, NH).
HRMS (ESI): m/z calculated $[M + H]^+$ for $C_{38}H_{46}O_5N_5^+$ = 652.34935; found 652.35046.

Chlorin-p, 6-N-hexylamide-7-methyl Ester (**7**). To a solution of **5** (40 mg, 0.069 mmol) in dichloromethane (0.3 mL), hexylamine (0.2 mL) was added and the reaction mixture was stirred at room temperature in the dark under nitrogen for 2 h. After that, UV-vis analysis and TLC showed the absence of starting material. The product was purified by crystallization with dichloromethane-hexane to give the product **7** (42.0 mg, 0.062 mmol, 86% yield) [54–56].

UV-vis (CH_2Cl_2): λmax (nm): 666, 608, 528, 497, 398.
^{1}H-NMR ($CDCl_3$, 400 MHz) δ: 9.66 (s, 1H), 9.55 (s, 1H), 8.70 (s, 1H), 7.98 (dd, 1H, J = 11.5 and 17.6 Hz), 6.33–6.37 (m, NH), 6.28 (d, 1H, J = 16.8Hz), 6.10 (d, 1H, J = 12.8Hz), 4.93 (d, 1H, J = 7.7Hz), 4.40 (q, 1H, J = 7.0Hz), 4.13 (s, 3H), 3.82–3.60 (m, 4H), 3.56 (s, 3H), 3.40 (s, 3H), 3.22 (s, 3H), 2.70–2.12 (m, 4H), 1.77 (d, 3H, J = 7.2Hz), 1.64 (t, 3H, J = 7.5Hz), 0.92–0.77 (m, 9H), 0.67 (t, 3H, J = 7.2Hz), −1.32 (s, NH), −1.54 (s, NH).

HRMS (ESI): m/z calculated [M + H]$^+$ for $C_{40}H_{50}O_5N_5^+$ = 680.38065; found 680.38251.

Chlorin-p, 6-N-octylamide-7-methyl Ester (8). To a solution of **5** (22.0 mg, 0.038 mmol) in dichloromethane (1.0 mL), octylamine (0.2 mL) was added. The reaction mixture was stirred at room temperature in the dark under nitrogen for 2 h. After that, UV-vis analysis and TLC showed the absence of starting material. The product was purified by crystallization with dichloromethane-hexane to give the product **8** (21.0 mg, 0.030 mmol, 80% yield) [54–56].

UV-vis (CH$_2$Cl$_2$): λmax (nm): 663, 605, 528, 499, 399.
^{1}H-NMR (CD$_3$)$_2$CO, 400 MHz) δ: 9.71 (s, 1H), 9.41 (s, 1H), 9.04 (s, 1H), 8.17 (dd, 1H), 7.78–7.82 (m, NH), 6.37 (d, 1H), 6.15 (d, 1H), 5.16 (d, 1H), 4.54 (q, 1H), 3.67 (s, 3H), 3.60–3.64 (m, 4H), 3.48 (s, 3H), 3.40 (s, 3H), 3.25 (s, 3H), 2.70–2.12 (m, 6H), 1.78 (d, 3H), 1.63 (t, 3H, J = 7.5Hz), 0.92–0.77 (m, 9H), 0.85 (t, 3H, J = 7.2Hz), −1.46 (s, NH), −1.82 (s, NH).
HRMS (ESI): m/z calculated [M + H]$^+$ for $C_{42}H_{54}O_5N_5^+$ = 708.41195; found 708.41364.

3.2. Photodynamic Inactivation

Staphylococcus aureus (American Type Culture Collection, ATCC 25923) and *Escherichia coli* (ATCC 25922) were grown in brain and heart infusion media. *Candida albicans* (ATCC 10231) was grown in Sabouraud dextrose broth. For experimental purposes, the microorganism concentration was adjusted to 10^7–10^8 cells/mL in sterile distilled water, and 500 µL of each microorganism culture was added to 24-well plates with photosensitizers **2–4** and **6–8**. Solutions were prepared by diluting the photosensitizer powder (1 mg for *S. aureus* and *C. albicans* and 2 mg for *E. coli*) in 100 µL of dimethyl sulfoxide (DMSO) (to dissolve the PS) and 900 µL of sterile water; the initial concentration of DMSO in the stock solutions was 10%. After dilution of these stock solutions to final concentrations of 1 µM and 10 µM for *S. aureus* and *C. albicans*, respectively, and of 500 µM for *E. coli*, the final concentration of DMSO was less than 2% in all the solutions. After preparing the solutions, they were protected from light. The 24-well plates were kept in the dark at 37 °C for 20 min.

A homemade LED-based device with emission centered at 660 nm was used to irradiate the culture plates. The 24-well plates were irradiated at 30 mWcm^{-2} using this device for 8, 16, and 25 min, resulting in fluences of 15, 30, and 45 Jcm^{-2}, respectively, which were used in the PDI against microorganisms. The fluence levels used were 15 Jcm^{-2} and 30 Jcm^{-2} for *S. aureus* and *C. albicans*, and 45 Jcm^{-2} for *E. coli*. After irradiation, 10-fold serial dilutions were performed and cells were cultured in agar plates. The colony-forming units (CFUs) were determined 24 h after initiation of the experimental procedure. All experiments were performed in triplicate.

In the control group (no treatment), 24-well plates were maintained at room temperature for 32 min. Using the same incubation time, the dark toxicity of the photosensitizers was evaluated in 24-well plates covered with aluminum foil to avoid light exposure. Phototoxicity was determined by irradiation at 30 Jcm^{-2} for *S. aureus* and *C. albicans*, and 45 Jcm^{-2} for *E. coli*.

Survival fractions (SFs) were expressed as ratios of CFUs of treated groups to the control group. The SF at 0 J/cm^2 provides a measure of the dark toxicity of chlorins.

3.3. Statistical Analysis

All the results reported in Figures 3 and 4 were statistically analyzed using the RStudio software (R version 4.1.1 (10 August 2021), R Core Team (2021), R: A language and environment for statistical computing; R Foundation for Statistical Computing, Vienna, Austria (https://www.R-project.org, accessed on 10 August 2021)) using a significance level of at least $p < 0.05$ and a confidence level of approximately 95%. The data were analyzed and approved by the normality test. Comparisons between the experimental groups were verified by Tukey's test.

4. Conclusions

The photosensitizers **2–4** and **6–7** derived from chlorophyll a were successfully semisynthesized. The characterization of the compounds is in accordance with data described in the literature and the main photophysical data are compiled and organized in Table S1 (SI). Overall, methyl pheophorbide *a* or purpurin-18 derivatives with different side chains (from the butyl to octyl groups) were prepared and studied. Subsequently, we investigated the photoinactivation of *S. aureus* and observed that the methyl pheophorbide derivatives **2–4** did not show great inactivation, whereas the purpurin-18 derivatives **6–8** allowed significant PDT inactivation. For *C. albicans*, the purpurin-18 derivative with the butyl group showed relevant inactivation, and the methyl pheophorbide with the butyl group also exhibited PDI. Photosensitizers **2–4** and **6–8** were also tested against the Gram-negative bacterium *E. coli*; however, no significant photoinactivation was observed. In general, the higher the carbon chain present in the photosensitizers, the more hydrophobic the compounds, with consequently lower photoinactivation efficacy. These results suggest that the use of chlorin derivatives with lower hydrophobic properties can be more effective for the photoinactivation of such microorganisms as *S. aureus* and *C. albicans*.

Supplementary Materials: The following supporting information can be downloaded at: https://www.mdpi.com/article/10.3390/molecules27185769/s1, Figure S1: ^{1}H-NMR (CDCl$_3$) spectrum of methyl pheophorbide *a* (1); Figure S2: ^{13}C-NMR (CDCl$_3$) spectrum of methyl pheophorbide *a* (1); Figure S3: ^{1}H-NMR (CDCl$_3$) spectrum of 13-(Butylcarbamoyl)chlorin *e*6 15,17-dimethyl ester (2); Figure S4: HRMS (ESI) of 13-(Butylcarbamoyl)chlorin *e*6 15,17-dimethyl ester (2); Figure S5: ^{1}H-NMR (CDCl$_3$) spectrum of 13-(Hexylcarbamoyl)chlorin *e*6 15,17-dimethyl ester (3); Figure S6: HRMS (ESI) of 13-(Hexylcarbamoyl)chlorin *e*6 15,17-dimethyl ester (3); Figure S7: ^{1}H-NMR (CDCl$_3$) spectrum of 13-(Octylcarbamoyl)chlorin *e*6 15,17-dimethyl ester (4); Figure S8: HRMS (ESI) of 13-(Octylcarbamoyl)chlorin *e*6 15,17-dimethyl ester (4); Figure S9: ^{1}H-NMR (CDCl$_3$) spectrum of Purpurin-18 Methyl Ester (5); Figure S10: ^{13}C-NMR (CDCl$_3$) spectrum of Purpurin-18 Methyl Ester (5); Figure S11: ^{1}H-NMR (CDCl$_3$) spectrum of Chlorin-p, 6-N-Butylamide-7-methyl Ester (6); Figure S12: HRMS (ESI) of Chlorin-p, 6-N-Butylamide-7-methyl Ester (6); Figure S13: ^{1}H-NMR (CDCl$_3$) spectrum of Chlorin-p, 6-N-hexylamide-7-methyl Ester (7); Figure S14: HRMS (ESI) of Chlorin-p, 6-N-hexylamide-7-methyl Ester (7); Figure S15: ^{1}H-NMR (CDCl$_3$) spectrum of Chlorin-p, 6-N-octylamide-7-methyl Ester (8); Figure S16: HRMS (ESI) of Chlorin-p, 6-N-octylamide-7-methyl Ester (8); Figure S17: Homemade engineered Biotable model for PDI studies (660 nm). Figure S18: Photodegradation experiment at 660 nm at 63.7 mWcm-2 using ethyl acetate as solvent. Table S1: Wavelengths data and Quantum yield of Singlet Oxygen $^1O_2$ data of the literature for some of the compounds.

Author Contributions: Formal analysis, M.P.U., A.d.C.R. and B.A.O.; Methodology, M.P.U.; Writing—original draft, M.P.U., A.d.C.R., B.A.O. and S.P.; Writing—review and editing, M.P.U., K.T.d.O. and C.K. All authors have read and agreed to the published version of the manuscript.

Funding: The authors thank FAPESP (São Paulo Research Foundation) grant numbers: 2013/06532-4, 2011/19720-8, 2011/13993-2, 2008/57858-9, 2009/54035-4 (EMU) and 2013/07276-1 (CEPOF); 2019/27176-8 (KTO), 2020/06874-6 (KTO); PRPPG-UNILA (80/2019/PRPPG, 104/2020/PRPPG and 105/2020/PRPPG); CNPq (465360/2014-9 and 306919/2019-2) and CAPES for financial support and fellowships.

Institutional Review Board Statement: Not applicable.

Informed Consent Statement: Not applicable.

Data Availability Statement: Not applicable.

Acknowledgments: Thanks are also due to Waters laboratory in Brazil for ESI-TOF measurements, and P. T. Weis and V. A. Kuana for the technical appointment.

Conflicts of Interest: The authors declare no conflict of interest.

References

1. Agostinis, P.; Berg, K.; Cengel, K.A.; Foster, T.H.; Girotti, A.W.; Gollnick, S.O.; Hahn, S.M.; Hamblin, M.R.; Juzeniene, A.; Kessel, D.; et al. Photodynamic Therapy of Cancer: An Update. *CA-A Cancer J. Clin.* **2011**, *61*, 250–281. [CrossRef] [PubMed]
2. Betrouni, N.; Boukris, S.; Benzaghou, F. Vascular targeted photodynamic therapy with TOOKAD (R) Soluble (WST11) in localized prostate cancer: Efficiency of automatic pre-treatment planning. *Lasers Med. Sci.* **2017**, *32*, 1301–1307. [CrossRef] [PubMed]

3. Mazor, O.; Brandis, A.; Plaks, V.; Neumark, E.; Rosenbach-Belkin, V.; Salomon, Y.; Scherz, A. WST11, a novel water-soluble bacteriochlorophyll derivative; Cellular uptake, pharmacokinetics, biodistribution and vascular-targeted photodynamic activity using melanoma tumors as a model. *Photochem. Photobiol.* **2005**, *81*, 342–351. [CrossRef] [PubMed]
4. Mfouo-Tynga, I.S.; Dias, L.D.; Inada, N.M.; Kurachi, C. Features of third generation photosensitizers used in anticancer photodynamic therapy: Review. *Photodiagn. Photodyn. Ther.* **2021**, *34*, 102091. [CrossRef] [PubMed]
5. Alzeibak, R.; Mishchenko, T.A.; Shilyagina, N.Y.; Balalaeva, I.V.; Vedunova, M.V.; Krysko, D.V. Targeting immunogenic cancer cell death by photodynamic therapy: Past, present and future. *J. Immunother. Cancer* **2021**, *9*, e001926. [CrossRef]
6. Zhang, P.; Wu, M.X. A clinical review of phototherapy for psoriasis. *Lasers Med. Sci.* **2018**, *33*, 173–180. [CrossRef]
7. Marotti, J.; Aranha, A.C.C.; Eduardo, C.D.; Ribeiro, M.S. Photodynamic Therapy Can Be Effective as a Treatment for Herpes Simplex Labialis. *Photomed. Laser Surg.* **2009**, *27*, 357–363. [CrossRef]
8. Issa, M.C.A.; Manela-Azulay, M. Photodynamic therapy: A review of the literature and image documentation. *An. Bras. Dermatol.* **2010**, *85*, 501–511. [CrossRef]
9. De Carvalho, G.G.; Sanchez-Puetate, J.C.; Donatoni, M.C.; Huacho, P.M.M.; Rastelli, A.N.D.; de Oliveira, K.T.; Spolidorio, D.M.P.; Zandim-Barcelos, D.L. Photodynamic inactivation using a chlorin-based photosensitizer with blue or red-light irradiation against single-species biofilms related to periodontitis. *Photodiagn. Photodyn. Ther.* **2020**, *31*, 101916. [CrossRef]
10. Seeger, M.G.; Ries, A.S.; Gressler, L.T.; Botton, S.A.; Iglesias, B.A.; Cargnelutti, J.F. In vitro antimicrobial photodynamic therapy using tetra-cationic porphyrins against multidrug-resistant bacteria isolated from canine otitis. *Photodiagn. Photodyn. Ther.* **2020**, *32*, 101982. [CrossRef]
11. Hirose, M.; Yoshida, Y.; Horii, K.; Hasegawa, Y.; Shibuya, Y. Efficacy of antimicrobial photodynamic therapy with Rose Bengal and blue light against cariogenic bacteria. *Arch. Oral Biol.* **2021**, *122*, 105024. [CrossRef] [PubMed]
12. Reis, A.C.M.; Regis, W.F.M.; Rodrigues, L.K.A. Scientific evidence in antimicrobial photodynamic therapy: An alternative approach for reducing cariogenic bacteria. *Photodiagn. Photodyn. Ther.* **2019**, *26*, 179–189. [CrossRef] [PubMed]
13. Geralde, M.C.; Leite, I.S.; Inada, N.M.; Salina, A.C.G.; Medeiros, A.I.; Kuebler, W.M.; Kurachi, C.; Bagnato, V.S. Pneumonia treatment by photodynamic therapy with extracorporeal illumination—An experimental model. *Physiol. Rep.* **2017**, *5*, e13190. [CrossRef] [PubMed]
14. Babu, B.; Sindelo, A.; Mack, J.; Nyokong, T. Thien-2-yl substituted chlorins as photosensitizers for photodynamic therapy and photodynamic antimicrobial chemotherapy. *Dyes Pigm.* **2021**, *185*, 108886. [CrossRef]
15. Kate, C.; Blanco, N.M.I.; Carbinatto, F.M.; Bagnato, V.S. Antimicrobial Efficacy of Curcumin Formulations by Photodynamic Therapy. *J. Pharm. Pharmacol.* **2017**, *5*, 506–511.
16. Amos-Tautua, B.M.; Songca, S.P.; Oluwafemi, O.S. Application of Porphyrins in Antibacterial Photodynamic Therapy. *Molecules* **2019**, *24*, 2456. [CrossRef] [PubMed]
17. Sampaio, L.S.; de Annunzio, S.R.; de Freitas, L.M.; Dantas, L.O.; de Boni, L.; Donatoni, M.C.; de Oliveira, K.T.; Fontana, C.R. Influence of light intensity and irradiation mode on methylene blue, chlorin-e6 and curcumin-mediated photodynamic therapy against Enterococcus faecalis. *Photodiagn. Photodyn. Ther.* **2020**, *31*, 101925. [CrossRef]
18. Mesquita, M.Q.; Dias, C.J.; Neves, M.; Almeida, A.; Faustino, M.A.F. Revisiting Current Photoactive Materials for Antimicrobial Photodynamic Therapy. *Molecules* **2018**, *23*, 2424. [CrossRef] [PubMed]
19. da Silva, A.P.; Uliana, M.P.; Guimaraes, F.E.G.; de Oliveira, K.T.; Blanco, K.C.; Bagnato, V.S.; Inada, N.M. Investigation on the in vitro anti-Trichophyton activity of photosensitizers. *Photochem. Photobiol. Sci.* **2022**, *21*, 1185–1192. [CrossRef]
20. Habermeyer, B.; Guilard, R. Some activities of PorphyChem illustrated by the applications of porphyrinoids in PDT, PIT and PDI. *Photochem. Photobiol. Sci.* **2018**, *17*, 1675–1690. [CrossRef]
21. Ziganshyna, S.; Guttenberger, A.; Lippmann, N.; Schulz, S.; Bercker, S.; Kahnt, A.; Rffer, T.; Voigt, A.; Gerlach, K.; Werdehausen, R. Tetrahydroporphyrin-tetratosylate (THPTS)-based photodynamic inactivation of critical multidrug-resistant bacteria in vitro. *Int. J. Antimicrob. Agents* **2020**, *55*, 105976. [CrossRef]
22. De Souza, L.M.; Inada, N.M.; Venturini, F.P.; Carmona-Vargas, C.C.; Pratavieira, S.; de Oliveira, K.T.; Kurachi, C.; Bagnato, V.S. Photolarvicidal effect of curcuminoids from Curcuma longa Linn. against Aedes aegypti larvae. *J. Asia-Pac. Entomol.* **2019**, *22*, 151–158. [CrossRef]
23. Oriel, S.; Nitzan, Y. Photoinactivation of Candida albicans by Its Own Endogenous Porphyrins. *Curr. Microbiol.* **2010**, *60*, 117–123. [CrossRef] [PubMed]
24. Huang, L.; Dai, T.; Hamblin, M.R. Antimicrobial photodynamic inactivation and photodynamic therapy for infections. *Methods Mol. Biol.* **2010**, *635*, 155–173. [CrossRef] [PubMed]
25. Gois, M.M.; Kurachi, C.; Santana, E.J.B.; Mima, E.G.O.; Spolidorio, D.M.P.; Pelino, J.E.P.; Bagnato, V.S. Susceptibility of Staphylococcus aureus to porphyrin-mediated photodynamic antimicrobial chemotherapy: An in vitro study. *Lasers Med. Sci.* **2010**, *25*, 391–395. [CrossRef] [PubMed]
26. De Souza, L.M.; Inada, N.M.; Pratavieira, S.; Corbi, J.J.; Kurachi, C.; Bagnato, V.S. Efficacy of Photogem®(Hematoporphyrin Derivative) as a Photoactivatable Larvicide against Aedes aegypti (Diptera: Culicidae) Larvae. *J. Life Sci.* **2017**, *11*, 74–81. [CrossRef]
27. Wiehe, A.; O'Brien, J.M.; Senge, M.O. Trends and targets in antiviral phototherapy. *Photochem. Photobiol. Sci.* **2019**, *18*, 2565–2612. [CrossRef]

28. Conrado, P.C.V.; Sakita, K.M.; Arita, G.S.; Galinari, C.B.; Goncalves, R.S.; Lopes, L.D.G.; Lonardoni, M.V.C.; Teixeira, J.J.V.; Bonfim-Mendonca, P.S.; Kioshima, E.S. A systematic review of photodynamic therapy as an antiviral treatment: Potential guidance for dealing with SARS-CoV-2. *Photodiagn. Photodyn. Ther.* **2021**, *34*, 102221. [CrossRef]
29. Svyatchenko, V.A.; Nikonov, S.D.; Mayorov, A.P.; Gelfond, M.L.; Loktev, V.B. Antiviral photodynamic therapy: Inactivation and inhibition of SARS-CoV-2 in vitro using methylene blue and Radachlorin. *Photodiagn. Photodyn Ther.* **2021**, *33*, 102112. [CrossRef]
30. Sternberg, E.D.; Dolphin, D.; Bruckner, C. Porphyrin-based photosensitizers for use in photodynamic therapy. *Tetrahedron* **1998**, *54*, 4151–4202. [CrossRef]
31. Ben Amor, T.; Jori, G. Sunlight-activated insecticides: Historical background and mechanisms of phototoxic activity. *Insect Biochem. Mol. Biol.* **2000**, *30*, 915–925. [CrossRef]
32. Nesi-Reis, V.; Lera-Nonose, D.; Oyama, J.; Silva-Lalucci, M.P.P.; Demarchi, I.G.; Aristides, S.M.A.; Teixeira, J.J.V.; Silveira, T.G.V.; Lonardoni, M.V.C. Contribution of photodynamic therapy in wound healing: A systematic review. *Photodiagn. Photodyn. Ther.* **2018**, *21*, 294–305. [CrossRef] [PubMed]
33. Scoditti, S.; Chiodo, F.; Mazzone, G.; Richeter, S.; Sicilia, E. Porphyrins and Metalloporphyrins Combined with N-heterocyclic Carbene (NHC) Gold (I) Complexes for Photodynamic Therapy Application. What Is the Weight of the Heavy Atom Effect? *Molecules* **2022**, *27*, 4046. [CrossRef] [PubMed]
34. Ma, Q.L.; Sun, X.; Wang, W.L.; Yang, D.L.; Yang, C.J.; Shen, Q.; Shao, J.J. Diketopyrrolopyrrole-derived organic small molecular dyes for tumor phototheranostics. *Chin. Chem. Lett.* **2022**, *33*, 1681–1692. [CrossRef]
35. Sampaio, F.J.P.; de Oliveira, S.; Crugeira, P.J.L.; Monteiro, J.S.C.; Fagnani, S.; Pepe, I.M.; de Almeida, P.F.; Pinheiro, A.L.B. aPDT using nanoconcentration of 1,9-dimethylmethylene blue associated to red light is eficacious in killing Enterococcus faecalis ATCC 29212 in vitro. *J. Photochem. Photobiol. B* **2019**, *200*, 11654. [CrossRef]
36. Sobotta, L.; Skupin-Mrugalska, P.; Piskorz, J.; Mielcarek, J. Porphyrinoid photosensitizers mediated photodynamic inactivation against bacteria. *Eur. J. Med. Chem.* **2019**, *175*, 72–106. [CrossRef]
37. Yook, K.; Kim, J.; Song, W. In vitro study on the effects of photodynamic inactivation using methyl pheophorbide a, PhotoMed, PhotoCure, and 660 nm diode laser on *Candida albicans*. *Photodiagn. Photodyn. Ther.* **2022**, *38*, 102871. [CrossRef]
38. Cieplik, F.; Deng, D.M.; Crielaard, W.; Buchalla, W.; Hellwig, E.; Al-Ahmad, A.; Maisch, T. Antimicrobial photodynamic therapy—What we know and what we don't. *Crit. Rev. Microbiol.* **2018**, *44*, 571–589. [CrossRef]
39. Santezi, C.; Reina, B.D.; Dovigo, L.N. Curcumin-mediated Photodynamic Therapy for the treatment of oral infections—A review. *Photodiagn. Photodyn. Ther.* **2018**, *21*, 409–415. [CrossRef]
40. Pratavieira, S.; Uliana, M.P.; Lopes, N.S.D.; Donatoni, M.C.; Linares, D.R.; Anibal, F.D.; de Oliveira, K.T.; Kurachi, C.; de Souza, C.W.O. Photodynamic therapy with a new bacteriochlorin derivative: Characterization and in vitro studies. *Photodiagn. Photodyn. Ther.* **2021**, *34*, 102251. [CrossRef]
41. Kharkwal, G.B.; Sharma, S.K.; Huang, Y.Y.; Dai, T.H.; Hamblin, M.R. Photodynamic Therapy for Infections: Clinical Applications. *Lasers Surg. Med.* **2011**, *43*, 755–767. [CrossRef] [PubMed]
42. Simplicio, F.I.; Maionchi, F.; Hioka, N. Photodynamic therapy: Pharmacological aspects, applications and news from medications development. *Quim. Nova* **2002**, *25*, 801–807. [CrossRef]
43. Kubrak, T.P.; Kołodziej, P.; Sawicki, J.; Mazur, A.; Koziorowska, K.; Aebisher, D. Some Natural Photosensitizers and Their Medicinal Properties for Use in Photodynamic Therapy. *Molecules* **2022**, *27*, 1192. [CrossRef]
44. Uliana, M.P.; Pires, L.; Pratavieira, S.; Brocksom, T.J.; de Oliveira, K.T.; Bagnato, V.S.; Kurachi, C. Photobiological characteristics of chlorophyll a derivatives as microbial PDT agents. *Photochem. Photobiol. Sci.* **2014**, *13*, 1137–1145. [CrossRef] [PubMed]
45. De Oliveira, K.T.; Silva, A.M.S.; Tomé, A.C.; Neves, M.G.P.M.S.; Neri, C.R.; Garcia, V.S.; Serra, O.A.; Iamamoto, Y.; Cavaleiro, J.A.S. Synthesis of new amphiphilic chlorin derivatives from protoporphyrin-IX dimethyl ester. *Tetrahedron* **2008**, *64*, 8709–8715. [CrossRef]
46. Petrilli, R.; Praça, F.S.G.; Carollo, A.R.H.; Medina, W.S.G.; de Oliveira, K.T.; Fantini, M.C.A.; Neves, M.G.P.M.S.; Cavaleiro, J.A.S.; Serra, O.A.; Yassuko Iamamoto, Y.; et al. Nanoparticles of Lyotropic Liquid Crystals: A Novel Strategy for the Topical Delivery of a Chlorin Derivative for Photodynamic Therapy of Skin Cancer. *Curr. Nanosci.* **2013**, *9*, 434–444. [CrossRef]
47. Dos Santos, F.A.B.; Uchoa, A.F.; Baptista, M.S.; Iamamoto, Y.; Serra, O.A.; Brocksom, T.J.; de Oliveira, K.T. Synthesis of functionalized chlorins sterically-prevented from self-aggregation. *Dyes Pigm.* **2013**, *99*, 402–411. [CrossRef]
48. Uchoa, A.F.; de Oliveira, K.T.; Baptista, M.S.; Bortoluzzi, A.J.; Iamamoto, Y.; Serra, O.A. Chlorin Photosensitizers Sterically Designed To Prevent Self-Aggregation. *J. Org. Chem.* **2011**, *76*, 8824–8832. [CrossRef]
49. Gushchina, O.I.; Larkina, E.A.; Nikolskaya, T.A.; Mironov, A.F. Synthesis of amide derivatives of chlorin e(6) and investigation of their biological activity. *J. Photochem. Photobiol. B* **2015**, *153*, 76–81. [CrossRef]
50. Belykh, D.V.; Pushkareva, E.I. Amidation of the Ester Group of Methylpheoforbide a with Sterically Nonhindered Primary and Secondary Aliphatic Amines. *Russ. J. Gen. Chem.* **2011**, *81*, 1216–1221. [CrossRef]
51. Belykh, D.V.; Kopylov, E.A.; Gruzdev, I.V.; Kuchin, A.V. Opening of the extra ring in pheophorbide a methyl ester by the action of amines as a one-step method for introduction of additional fragments at the periphery of chlorin macroring. *Russ. J. Org. Chem.* **2010**, *46*, 577–585. [CrossRef]
52. Louda, J.W.; Li, J.; Liu, L.; Winfree, M.N.; Baker, E.W. Chlorophyll-a degradation during cellular senescence and death. *Org. Geochem.* **1998**, *29*, 1233–1251. [CrossRef]

53. Drogat, N.; Barriere, M.; Granet, R.; Sol, V.; Krausz, P. High yield preparation of purpurin-18 from Spirulina maxima. *Dyes Pigm.* **2011**, *88*, 125–127. [CrossRef]
54. Lee, S.J.H.; Jagerovic, N.; Smith, K.M. Use of the chlorophyll derivative, purpurin-18, for syntheses of sensitizers for use in photodynamic therapy. *J. Chem. Soc.-Perkin Trans. 1* **1993**, *19*, 2369–2377. [CrossRef]
55. Zheng, G.; Potter, W.R.; Camacho, S.H.; Missert, J.R.; Wang, G.S.; Bellnier, D.A.; Henderson, B.W.; Rodgers, M.A.J.; Dougherty, T.J.; Pandey, R.K. Synthesis, photophysical properties, tumor uptake, and preliminary in vivo photosensitizing efficacy of a homologous series of 3-(1′-alkyloxy) ethyl-3-devinaylpurpurin-18-N-alkylimides with variable lipophilicity. *J. Med. Chem.* **2001**, *44*, 1540–1559. [CrossRef] [PubMed]
56. Kozyrev, A.N.; Zheng, G.; Zhu, C.F.; Dougherty, T.J.; Smith, K.M.; Pandey, R.K. Syntheses of stable bacteriochlorophyll-a derivatives as potential photosensitizers for photodynamic therapy. *Tetrahedron Lett.* **1996**, *37*, 6431–6434. [CrossRef]
57. Ash, C.; Dubec, M.; Donne, K.; Bashford, T. Effect of wavelength and beam width on penetration in light-tissue interaction using computational methods. *Lasers Med. Sci.* **2017**, *32*, 1909–1918. [CrossRef]
58. Moreira, X.; Santos, P.; Faustino, M.A.F.; Raposo, M.M.M.; Costa, S.P.G.; Moura, N.M.M.; Gomes, A.; Almeida, A.; Neves, M. An insight into the synthesis of cationic porphyrin-imidazole derivatives and their photodynamic inactivation efficiency against *Escherichia coli*. *Dyes Pigm.* **2020**, *178*, 108330. [CrossRef]
59. Ladeira, B.M.F.; Dias, C.J.; Gomes, A.T.P.C.; Tomé, A.C.; Neves, M.G.P.M.S.; Moura, N.M.; Almeida, A.; Faustino, M.A.F. Cationic Pyrrolidine/Pyrroline-Substituted Porphyrins as Efficient Photosensitizers against *E. coli*. *Molecules* **2021**, *26*, 464. [CrossRef]
60. De Oliveira, B.P.; Lins, C.C.d.S.A.; Diniz, F.A.; Melo, L.L.; de Castro, C.M.M.B. In Vitro antimicrobial photoinactivation with methylene blue in different microorganisms. *Braz. J. Oral Sci.* **2014**, *13*, 53–57. [CrossRef]
61. Omar, G.S.; Wilson, M.; Nair, S.P. Lethal photosensitization of wound-associated microbes using indocyanine green and near-infrared light. *BMC Microbiol.* **2008**, *8*, 10. [CrossRef] [PubMed]
62. Topaloglu, N.; Gulsoy, M.; Yuksel, S. Antimicrobial Photodynamic Therapy of Resistant Bacterial Strains by Indocyanine Green and 809-nm Diode Laser. *Photomed. Laser Surg.* **2013**, *31*, 155–162. [CrossRef]
63. Hargus, J.A.; Fronczek, F.R.; Graca, M.; Vicente, H.; Smith, K.M. Mono-(L)-aspartylchlorin-e(6). *Photochem. Photobiol.* **2007**, *83*, 1006–1015. [CrossRef]
64. Ma, L.; Dolphin, D. Nucleophilic reaction of 1,8- diazabicyclo[5.4.0]undec-7-ene and 1,5- diazabicyclo[4.3.0]non-5-ene with methyl pheophorbide a. Unexpected products. *Tetrahedron* **1996**, *52*, 849–860. [CrossRef]
65. Kustov, A.V.; Belykh, D.V.; Startseva, O.M.; Kruchin, S.O.; Venediktov, E.A.; Berezin, D.B. New Sensitizers Developed on a Methylpheophorbide a Platform for Photodynamic Therapy: Synthesis, Singlet Oxygen Generation and Modeling of Passive Membrane Transport. *Pharm. Anal. Acta* **2016**, *7*, 480. [CrossRef]
66. Belykh, D.V.; Tarabukinaa, I.S.; Gruzdevb, I.V.; Kodessc, M.I.; Kutchin, A.V. Aminomethylation of chlorophyll a derivatives using bis(N,N-dimethylamino)methane. *J. Porphyr. Phthalocyanines* **2009**, *13*, 949–956. [CrossRef]
67. Redmond, R.W.; Gamlin, J.N. A Compilation of Singlet Oxigen Yields from Biologically Relevant Molecules. *Photochem. Photobiol.* **1999**, *70*, 391–475. [CrossRef]

Article

Sulfonamide Porphyrins as Potent Photosensitizers against Multidrug-Resistant *Staphylococcus aureus* (MRSA): The Role of Co-Adjuvants

Sofia N. Sarabando [1], Cristina J. Dias [1], Cátia Vieira [2], Maria Bartolomeu [2], Maria G. P. M. S. Neves [1], Adelaide Almeida [2], Carlos J. P. Monteiro [1,*] and Maria Amparo F. Faustino [1,*]

[1] LAQV-Requimte and Department of Chemistry, University of Aveiro, 3810-193 Aveiro, Portugal
[2] CESAM, Department of Biology, University of Aveiro, 3810-193 Aveiro, Portugal
* Correspondence: cmonteiro@ua.pt (C.J.P.M.); faustino@ua.pt (M.A.F.F.)

Abstract: Sulfonamides are a conventional class of antibiotics that are well-suited to combat infections. However, their overuse leads to antimicrobial resistance. Porphyrins and analogs have demonstrated excellent photosensitizing properties and have been used as antimicrobial agents to photoinactivate microorganisms, including multiresistant *Staphylococcus aureus* (MRSA) strains. It is well recognized that the combination of different therapeutic agents might improve the biological outcome. In this present work, a novel *meso*-arylporphyrin and its Zn(II) complex functionalized with sulfonamide groups were synthesized and characterized and the antibacterial activity towards MRSA with and without the presence of the adjuvant KI was evaluated. For comparison, the studies were also extended to the corresponding sulfonated porphyrin TPP(SO$_3$H)$_4$. Photodynamic studies revealed that all porphyrin derivatives were effective in photoinactivating MRSA (>99.9% of reduction) at a concentration of 5.0 μM upon white light radiation with an irradiance of 25 mW cm^{-2} and a total light dose of 15 J cm^{-2}. The combination of the porphyrin photosensitizers with the co-adjuvant KI during the photodynamic treatment proved to be very promising allowing a significant reduction in the treatment time and photosensitizer concentration by six times and at least five times, respectively. The combined effect observed for TPP(SO$_2$NHEt)$_4$ and ZnTPP(SO$_2$NHEt)$_4$ with KI seems to be due to the formation of reactive iodine radicals. In the photodynamic studies with TPP(SO$_3$H)$_4$ plus KI, the cooperative action was mainly due to the formation of free iodine (I$_2$).

Keywords: porphyrins; sulfonamides; MRSA; photodynamic therapy; antimicrobial resistance; photosensitizer; singlet oxygen; potassium iodide; gram-positive bacteria; *Staphylococcus aureus*

1. Introduction

Sulfonamides or sulfa drugs were the first effective synthetic chemical entities used systematically against a broad spectrum of bacteria [1,2]. They act as competitive inhibitors of dihydropteroate synthase, mimetizing *p*-aminobenzoic acid and therefore inhibiting the synthesis of tetrahydrofolic acid, which is essential for the formation of nucleic acids precursors in bacteria [3]. Since their discovery in 1935 [4,5], many sulfonamides have been developed, but their importance has declined due to an increase in microbial resistance. However, recent strategies have focused on the development of novel treatment options and alternative antimicrobial therapies. Among them, antimicrobial Photodynamic Treatment (aPDT) has been described as a promising alternative to conventional antibiotics as it leads to bacterial cell death without the development of microbial-resistant strains [6,7]. This technique requires the administration of a photosensitizer (PS), that, after being photoactivated with visible light, interacts with dioxygen (O$_2$) producing reactive oxygen species (ROS) via two different mechanisms. Singlet oxygen (^{1}O$_2$), the main ROS responsible for microbial death, is produced by mechanism Type II, while hydroxyl radicals (•OH), superoxide anion radical (O$_2$•$^-$), and H$_2$O$_2$ are produced by mechanism Type I [8].

These ROS can act in the external cell wall of bacterial cells leading to their inactivation. This therapeutic modality is being recognized as an effective method for inactivating a broad spectrum of microorganisms, including multidrug-resistant Gram-(+) and Gram-(−) bacterial strains [9–20].

Porphyrins are the most common PS used in aPDT due to their high efficacy against a large spectrum of microorganisms [21–23]. Furthermore, porphyrins are easily modified allowing the insertion of groups to improve their photophysical properties and affinity to microbial cells. Research efforts have been focused on the development of cationic porphyrin derivatives that clearly demonstrates aPDT inactivation efficacy [24–28]. Their biocide effect arises from the possibility of the interaction of these compounds with the bacterial cell membrane [29–34]. In contrast, the development of anionic porphyrins, regardless of their efficiency to produce ROS, does not receive great attention because of their low interaction with the bacterial membrane [35–37]. However, negatively charged functional groups induce an inherent solubility of the porphyrins in aqueous media, which would greatly benefit their biological application [38,39]. In addition to negative and positive PS, neutral PS has also been studied and it was found that amphiphilicity can promote the enhancement of cell membrane interaction, playing an important role in the therapeutic effect [8]. The functionalization of organic compounds with sulfonamides is a versatile method to modulate amphiphilicity [5,40]. For instance, it was reported that the insertion of sulfonamide groups at the periphery of phthalocyanines can enhance their efficacy against bacteria [41], even suggesting that the presence of these groups may increase the interaction of the PS-sulfonamide conjugates towards target components of the bacterial cells' outer structures. Porphyrins and metalloporphyrins containing sulfonamides and positively charged peripheral N-alkyl pyridinium groups have also been prepared and considered for biological applications such as DNA binding and intercalation [42,43]. Furthermore, synthetic strategies for preparing libraries of porphyrin-sulfonamides as photosensitizing agents have been disclosed in the literature [40,44–47]. However, as far as we know, their biological potential as aPDT agents is as yet poorly evaluated.

In 2017, Michael Hamblin and co-workers [48] unveiled the potentiation of aPDT (up to 6 logs of extra killing) by the addition of salts (KI, KBr, NaSCN, NaN$_3$, NaNO$_2$). The most powerful and versatile salt was potassium iodide (KI). This salt can enhance the photodynamic efficiency of neutral and cationic porphyrins resulting in a higher inactivation rate when compared to the use of PS alone. This synergic effect is caused by the initial production of peroxyiodide (HOOI$_2^-$) as a result of the interaction of KI with $^1O_2$ produced by the PS. Then, the peroxyiodide can undergo further degradation into cytotoxic species like free iodine/tri-iodine ion (I$_2$/I$_3^-$) and hydrogen peroxide (H$_2$O$_2$) or iodine radicals (I$_2^{-\bullet}$) [18,22,49]. Several photodynamic studies were performed since then to improve antimicrobial outcomes [16,50–53]. In this context, Hamblin and co-workers selected the tetracationic 5,10,15,20-tetrakis(1-methylpyridinium-4-yl)porphyrin tetratosylate (**TMPyP**) and the tetra-anionic 5,10,15,20-tetrakis(4-sulfonatophenyl)porphyrin dihydrochloride [**TPP(SO$_3$H)$_4$.2HCl**] in order to obtain knowledge about porphyrin net charges and the effect of KI as a co-adjuvant in bacteria targeting [54]. **TPP(SO$_3$H)$_4$·2HCl** (200 nM) under blue light at 415 nm at an irradiance of 50 mW cm^{-2} and a total light dose of 10 J cm^{-2} in the presence of KI (100 mM) was more effective than **TMPyP** in eradicating the Gram-(+) bacterium, methicillin-resistant *Staphylococcus aureus* (MRSA) and the fungal yeast *Candida albicans*. Since **TPP(SO$_3$H)$_4$** was used in the dihydrochloride salt form, the authors explained the effectiveness of **TPP(SO$_3$H)$_4$.2HCl** by the existence of a degree of cationic character when in the presence of bacteria [54].

Considering the potential of porphyrins as PSs and sulfonamides as bacteriostatic compounds, we decided to take advantage of both chemical entities, by preparing porphyrins containing sulfonamides for bacterial photoinactivation. The capability of the porphyrin core to complex with a large number of metal cations, giving metalloporphyrins, strongly influences their electronic and photophysical properties by internal heavy-atom effects [55], enhancing the intersystem crossing to the triplet state which is the precursor to the formation of ROS and $^1O_2$ [56]. Thus, the present work describes the synthetic access and photodynamic action of the free-base 5,10,15,20-tetrakis[4-(N-ethylsulfamoyl)phenyl]porphyrin (**TPP(SO$_2$NHEt)$_4$**) and of its zinc(II) complex, **ZnTPP(SO$_2$NHEt)$_4$** towards the methicillin-resistant *Staphylococcus aureus* (MRSA), selected as a bacterial model of Gram-(+). The biological studies performed in the presence of these PSs bearing four sulfonamide substituents at the *para* position of each phenyl ring were carried out with and without the adjuvant KI. For comparison, the 5,10,15,20-tetrakis(4-sulfophenyl)porphyrin, **TPP(SO$_3$H)$_4$** just bearing sulfonic acid groups was also considered. An investigation into the photophysical/photochemical features of the molecules under study was also carried out in order to corroborate their photodynamic action.

2. Results

2.1. Synthesis and Structural Characterization of Meso-Aryl Porphyrins Functionalized with Sulfonamides and Sulfonic Acids

The required porphyrins functionalized with sulfonamides and sulfonic acid substituents were obtained from 5,10,15,20-tetraphenylporphyrin (TPP) prepared according to the literature (Scheme 1) [57,58]. To synthesize the required amphiphilic porphyrin sulfonamide derivative, the functionalization of **TPP** into the corresponding chlorosulfonated derivatives was achieved by mixing **TPP** with an excess of chlorosulfonic acid at room temperature over one hour [44–46]. After reaction completion and work-up, the **TPP(SO$_2$Cl)$_4$** derivative was obtained in a quantitative yield, without chromatographic purification.

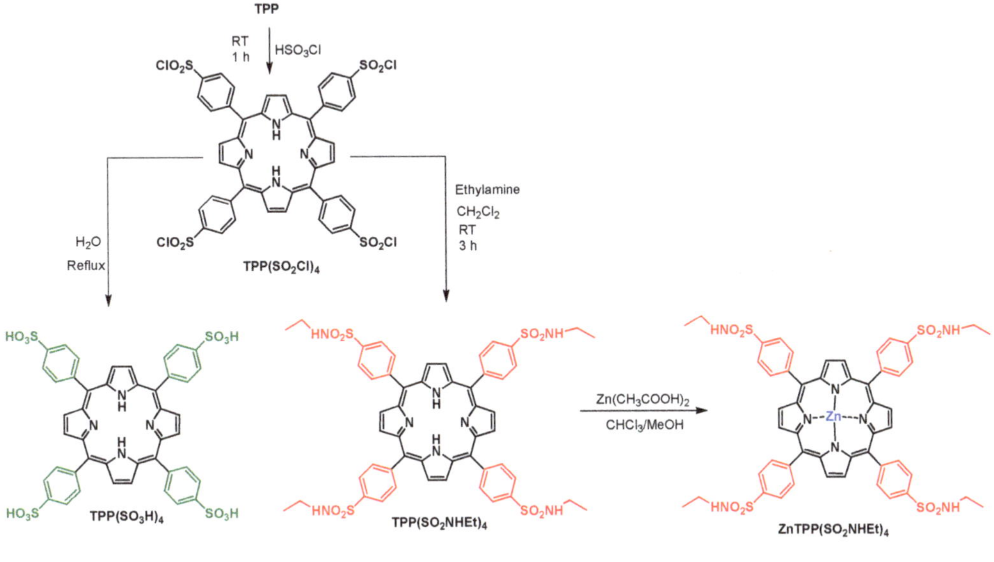

Scheme 1. Synthetic route for the preparation of the **TPP(SO$_3$H)$_4$**, **TPP(SO$_2$NHEt)$_4$**, and **ZnTPP(SO$_2$NHEt)$_4$** used as PS.

The nucleophilic step was performed at room temperature for 3 h between the **TPP(SO$_2$Cl)$_4$** dissolved in dichloromethane and a large excess of ethylamine to afford the corresponding sulfonamide derivative. After work-up, and silica gel chromatography, the amphiphilic sulfonamide derivative 5,10,15,20-tetrakis[4-(N-ethylsulfamoyl)phenyl]porphyrin (**TPP(SO$_2$NHEt)$_4$**) was isolated with 88% yield. The compound structure was confirmed by ^{1}H NMR (Figure S1, Supplementary Material) and high-resolution mass spectrometry (HRMS) (Figure S2, Supplementary Material). With the aim of modulating the photophysical and photochemical features, the complexation of the free base sulfonamide, **TPP(SO$_2$NHEt)$_4$** with zinc(II) acetate was carried out in chloroform/methanol [59,60]. After confirming the success of the metalation process, by UV-Vis and TLC, the organic layer was washed with distilled water, giving 5,10,15,20-tetrakis[4-(*N*-ethylsulfamoyl)phenyl]porphyrinatezinc(II) (**ZnTPP(SO$_2$NHEt)$_4$**) quantitatively. The structure of the complex was confirmed by UV-Vis analysis, ^{1}H NMR (Figure S3, Supplementary Material), and HRMS (Figure S5, Supplementary Material).

The **TPP(SO$_2$Cl)$_4$** was also used to prepare the **TPP(SO$_3$H)$_4$** quantitatively, suspending the chlorosulfonated derivative in water under reflux for 12 h (see Scheme 1) [61]. The ^{1}H NMR confirmed the chemical structure of **TPP(SO$_3$H)$_4$** which is in accordance with the literature (Figure S6, Supplementary Material) [44].

2.2. Photophysical Characterization

2.2.1. UV-Vis Absorption Properties

The absorption spectra of all prepared compounds show the typical profile of *meso*-tetraarylporphyrins (Figure 1), which are dominated by a Soret band (B band) around 420 nm and the Q bands varying between four and two depending on the macrocycle symmetry (Table 1). The free-base derivatives **TPP(SO$_3$H)$_4$** and **TPP(SO$_2$NHEt)$_4$** present four Q bands with relative intensities of "etio" type while the complex **ZnTPP(SO$_2$NHEt)$_4$** displays the expected two Q bands between 560–600 nm (Table 1).

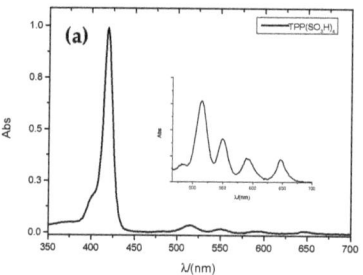

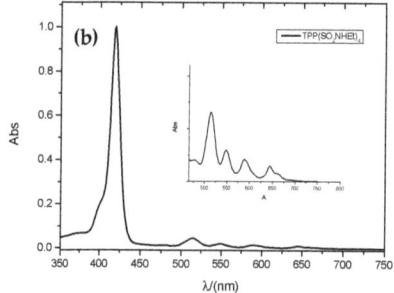

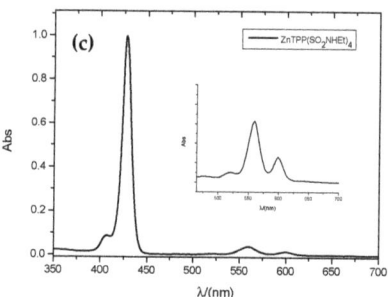

Figure 1. Normalized UV-Vis absorption spectra and respective Q-band magnifications of porphyrins (a) **TPP(SO$_3$H)$_4$**, (b) **TPP(SO$_2$NHEt)$_4$**, and (c) **ZnTPP(SO$_2$NHEt)$_4$** in DMF.

The molar absorption coefficients of the prepared PS were determined from the Beer–Lambert Law using DMF solutions and are presented in Table 1.

Table 1. Maximum of absorption and molar absorption coefficients of **TPP(SO₃H)₄**, **TPP(SO₂NHEt)₄**, and **ZnTPP(SO₂NHEt)₄** and **TPP** used as reference and maximum of fluorescence emission and fluorescence quantum yield in DMF.

Photosensitizer	Absorption λ_{max}/nm (ε /M^{-1} cm^{-1})					Fluorescence λ_{max}/nm		Φ_F [a]
	B (0-0)	Q_x (1-0)	Q_x (0-0)	Q_y (1-0)	Q_y (0-0)	Q (0-0)	Q (0-1)	
TPP [62]	420 (5.3 × 10⁵)	516 (2.5 × 10⁴)	550 (1.1 × 10⁴)	592 (7.4 × 10³)	648 (6.7 × 10³)	651	716	0.11
TPP(SO₃H)₄	420 (3.1 × 10⁵)	515 (1.3 × 10⁴)	550 (6.6 × 10³)	591 (3.9 × 10³)	647 (3.8 × 10³)	652	716	0.13
TPP(SO₂NHEt)₄	420 (3.5 × 10⁵)	514 (1.6 × 10⁴)	549 (7.3 × 10³)	589 (5.1 × 10³)	645 (3.6 × 10³)	654	719	0.14
ZnTPP(SO₂NHEt)₄	427 (4.9 × 10⁵)	561 (2.1 × 10⁴)	600 (9.2 × 10³)	—	—	607	659	0.04

[a] Fluorescence quantum yields (Φ_F) were measured using **TPP** as a reference in DMF, ϕ_F = 0.11 [63].

2.2.2. Fluorescence Spectroscopy

The emissive features of the prepared molecules were evaluated and Figure 2 depicts the recorded emissive spectra of all derivatives including **TPP** in DMF as a reference. All spectra show two emission bands: a band at a shorter wavelength attributed to the Q (0,0) transition, and the adjacent band attributed to the Q (0,1) transition (Table 1). Looking at the relative intensity of the bands, it is shown that the intensity of the first band is always higher than the second one whose relative intensity increases when **TPP(SO₂NHEt)₄** is complexed with zinc(II). The free-base derivatives **TPP(SO₃H)₄** and **TPP(SO₂NHEt)₄** showed similar fluorescence quantum yields (Φ_F ~0.13) to **TPP** (Φ_F = 0.11). However, the zinc(II) complex **ZnTPP(SO₂NHEt)₄** displays a lower fluorescence quantum yield (Φ_F = 0.04).

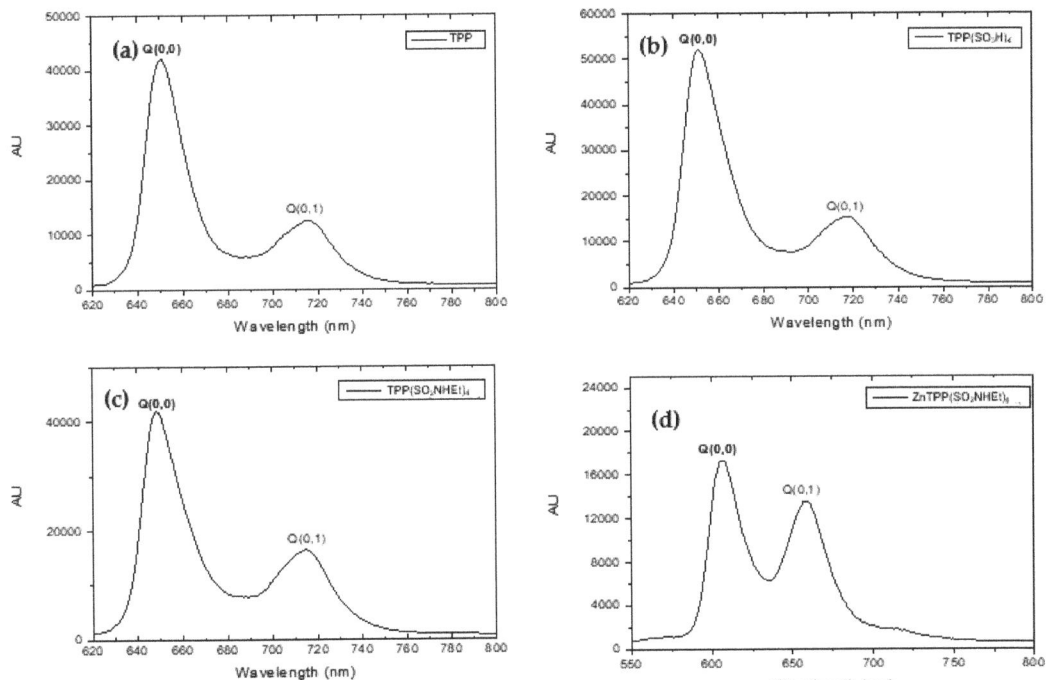

Figure 2. Fluorescence emission spectra and assignment of fluorescence emission bands: (**a**) **TPP**, (**b**) **TPP(SO₃H)₄**, (**c**) **TPP(SO₂NHEt)₄**, and (**d**) **ZnTPP(SO₂NHEt)₄** in DMF at 20 °C.

2.2.3. Singlet Oxygen Generation

The generation of $^1O_2$ was assessed by using a well-established method [64]. The photo-oxidation of 9,10-dimethylanthracene (DMA) was followed by UV-Vis spectroscopy in the presence of each porphyrin derivative upon irradiation at 420 ± 5 nm, in DMF. The photodecay produced by the generation of $^1O_2$ follows a pseudo-first-order kinetic and Figure 3 shows the results as $\ln(A_0/A)f(t)$.

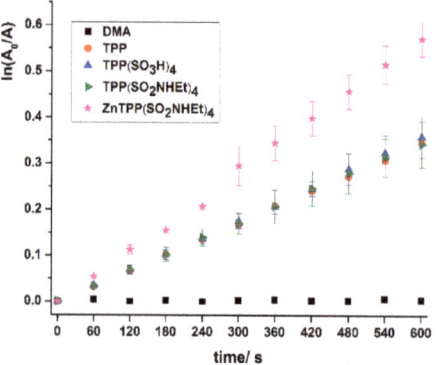

Figure 3. Photo-oxidation of DMA, in DMF, photosensitized by **TPP** (used as reference), **TPP(SO$_3$H)$_4$**, **TPP(SO$_2$NHEt)$_4$**, and **ZnTPP(SO$_2$NHEt)$_4$** during 600 s at 420 nm ± 5 nm. The symbols may be overlapped. DMA (black squares) is a solution of DMA irradiated without the addition of any PS in order to assure that no photodegradation occurs.

2.2.4. Detection of Iodine Formation

The ability of **TPP(SO$_2$NHEt)$_4$**, **ZnTPP(SO$_2$NHEt)$_4$)**, and **TPP(SO$_3$H)$_4$** to generate iodine (I$_2$) was analyzed, and the assays were performed by monitoring its absorbance at 340 nm, upon white light irradiation of each porphyrin derivative (25 mW cm^{-2}) at a concentration of 5.0 µM in the presence of KI at a concentration of 100 mM (Figure S7, Supplementary Material). The results in Figure S7 show that the highest amount of I$_2$ was observed in the presence of **TPP(SO$_3$H)$_4$**, which attained the maximum after ca. 15 min of irradiation. Although in lower amounts, the free-base **TPP(SO$_2$NHEt)$_4$** was also able to generate I$_2$, which slowly increased in the course of the experiment. The formation of I$_2$ by the complex **ZnTPP(SO$_2$NHEt)$_4$** was negligible.

2.3. Photodynamic Inactivation of MRSA

In this study, the photodynamic efficiency of **TPP(SO$_3$H)$_4$**, **TPP(SO$_2$NHEt)$_4$**, and **ZnTPP(SO$_2$NHEt)$_4$** was evaluated against MRSA, selected as a Gram-(+) bacterial model. We also evaluated the combined effect of each PS with KI, a well-known co-adjuvant of aPDT [48], allowing in some cases to reduce the treatment time and the concentration of PS applied.

Photodynamic assays were performed in the presence of the PSs alone at 1.0 µM and 5.0 µM [and also at 0.1 µM and 0.5 µM for **TPP(SO$_3$H)$_4$**] and in combination with KI at 100 mM, for 120 min of white light irradiation (400–700 nm) at an irradiance of 25 mW cm^{-2}. The results obtained for the PSs **TPP(SO$_3$H)$_4$**, **TPP(SO$_2$NHEt)$_4$**, and **ZnTPP(SO$_2$NHEt)$_4$** are represented in Figures 4–6, respectively.

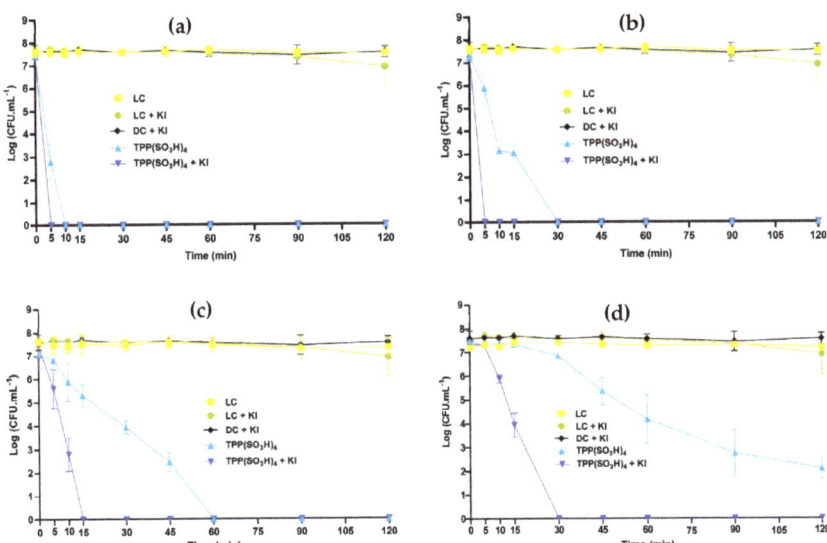

Figure 4. Photodynamic inactivation of MRSA in the presence of **TPP(SO₃H)₄** at 5.0 µM (**a**), 1.0 µM (**b**), 0.5 µM (**c**), and 0.1 µM (**d**) alone or in combination with KI at 100 mM, along 120 min of irradiation (white light at an irradiance of 25 mW cm^{-2}). Each value represents the mean ± standard deviation of three independent assays, with three replicates each.

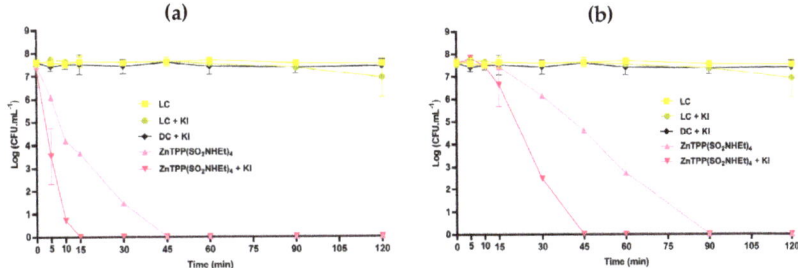

Figure 5. Photodynamic inactivation of MRSA in the presence of **ZnTPP(SO₂NHEt)₄** at 5.0 µM (**a**) and 1.0 µM (**b**), in the presence and absence of KI at 100 mM, during 120 min of irradiation (white light at an irradiance of 25 mW cm^{-2}). Each value represents the mean ± standard deviation of three independent assays, with three replicates each.

The results obtained indicate that the porphyrinic PSs were efficient in reducing the viability of MRSA and that the inactivation profiles were PS- and concentration-dependent (Figures 4–6). In the case of **TPP(SO₃H)₄** at 5.0 µM (Figure 4a), a decrease in MRSA viability was observed until the detection limit of the method (~7.5 Log reduction of CFU mL^{-1}) after 10 min (light dose of 15 J cm^{-2}) of photodynamic treatment (ANOVA, $p < 0.0001$). The same effect was achieved after 30 min of treatment (light dose of 45 J cm^{-2}) at a concentration of 1.0 µM of **TPP(SO₃H)₄** (Figure 4b). The combination of **TPP(SO₃H)₄** at 5.0 and 1.0 µM with KI at 100 mM improved the photodynamic effect of this PS and in both cases allowed for a reduction in the treatment time to 5 min (light dose of 7.5 J cm^{-2}) (minimal time point considered) (ANOVA, $p < 0.0001$). These results indicate that the **TPP(SO₃H)₄** + KI combination is promising and allows not only a reduction in the treatment time by about 6 times but also a reduction in the PS concentration to at least 1.0 µM. The outstanding efficiency of **TPP(SO₃H)₄**, towards MRSA strain at 5.0 and 1.0 µM and the short time required for reducing the viability of MRSA until the method detection limit, led us to investigate the inactivation profile at lower PS concentrations (0.1 µM and 0.5 µM). When a concentration of 0.5 µM of **TPP(SO$_3$H)$_4$** was used (Figure 4c), a decrease in MRSA viability was observed until the detection limit of the method (~7.5 Log reduction of CFU mL^{-1}) after 60 min (light dose of 90 J cm^{-2}) of photodynamic treatment (ANOVA, $p < 0.0001$). Moreover, when 0.1 µM of PS was employed, the reduction viability of MRSA was not so effective: ~5.1 Log reduction of CFU mL^{-1} (99.999%), after 120 min of irradiation (light dose of 180 J cm^{-2}). Still, the combination of **TPP(SO$_3$H)$_4$** at 0.5 and 0.1 µM with KI at 100 mM (Figure 4c,d) improved the photodynamic effect of this PS, and in both cases allowed a reduction in the treatment time to 15 min and 30 min (light doses of 22.5 and 45 J cm^{-2}), respectively, to reach the method detection limit (ANOVA, $p < 0.0001$).

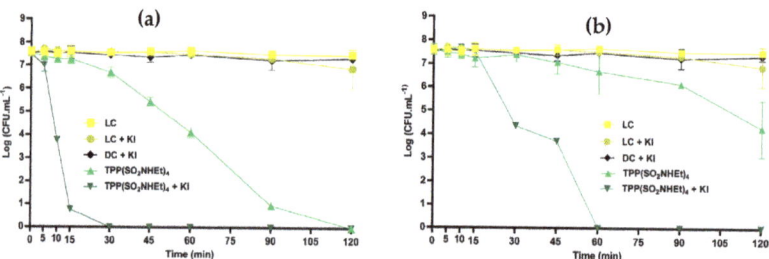

Figure 6. Photodynamic inactivation of MRSA in the presence of PS **TPP(SO$_2$NHEt)$_4$** at 5.0 µM (**a**) and 1.0 µM (**b**), in the presence and absence of KI at 100 mM, during 120 min of irradiation (white light at an irradiance of 25 mW cm^{-2}). Each value represents the mean ± standard deviation of three independent assays, with three replicates each.

ZnTPP(SO$_2$NHEt)$_4$ still proved to be a promising compound for inactivating MRSA (Figure 5). At 5.0 and 1.0 µM, this PS led to a total reduction of bacterial viability until the method detection limit (~7.5 log CFU mL^{-1} reduction) after 45 min and 90 min of treatment (light doses of 67.5 and 135 J cm^{-2}), respectively (ANOVA, $p < 0.0001$) (Figure 5a,b). In the presence of KI, the effect of PS was also improved, allowing the treatment time to be reduced to 15 and 45 min (light doses of 22.5 and 67.5 J cm^{-2}), at a PS concentration of 5.0 and 1.0 µM, respectively (ANOVA, $p < 0.0001$). These results indicate that the use of KI as a co-adjuvant not only allowed a reduction in the treatment by about 3 times (from 45 to 15 min) but also reduces by five times the concentration of PS required to reach the method detection limit (from 5.0 to 1.0 µM).

The **TPP(SO$_2$NHEt)$_4$** showed a lower photodynamic activity rate (ANOVA, $p < 0.0001$) (Figure 6); at 5.0 µM, the reduction in bacterial viability until the detection limit of the method (~7.5 Log of CFU mL^{-1} of reduction, $p < 0.0001$) was attained after 120 min of treatment (light dose of 180 J cm^{-2}) (Figure 6a). At 1.0 µM and under the same irradiation conditions a lower decrease in MRSA, by about ~3.4 Log CFU mL^{-1} (ANOVA, $p < 0.0001$), was observed (Figure 6b). The combination of **TPP(SO$_2$NHEt)$_4$** + KI resulted in faster bacterial inactivation rates until the detection limit of the method: at 5.0 µM the time required of white light irradiation was 30 min while at 1.0 µM was 60 min (light doses of 45 and 90 J cm^{-2}) (Figure 6a,b) (ANOVA, $p < 0.0001$). It is noteworthy that the combination of **TPP(SO$_2$NHEt)$_4$** (1.0 µM) + KI was more effective in inactivating MRSA than the PS alone at 5.0 µM. These results indicate that KI was effective in improving the effect of **TPP(SO$_2$NHEt)$_4$**, allowing a decrease in the treatment time by about four-fold, and also the PS concentration by five-fold.

For all assays, light controls (LC), KI controls (LC+KI), and dark controls performed for the tested PS (DC) showed no significant variation in the number of viable MRSA cells (ANOVA, $p > 0.05$). These results indicated that the bacterial viability was not affected by light and other test conditions, that KI is not toxic at the concentration tested throughout the treatment time, and the PSs did not induce cytotoxicity in the dark at the highest concentration tested (5.0 µM) (Figures 4–6).

3. Discussion

The synthetic route to obtain the porphyrin sulfonamides **TPP(SO₃NHEt)₄** and **ZnTPP(SO₃NHEt)₄** and the sulfonic acid derivative **TPP(SO₃H)₄** using 5,10,15,20-tetraphenylporphyrin (TPP) as starting material, was planned according to well-established procedures and comprise the following steps: (i) chlorosulfonation of **TPP** to the corresponding 5,10,15,20-tetrakis(p-chlorosulfophenyl)porphyrin derivative **TPP(SO₂Cl)₄**; (ii) reaction of this derivative with water or ethylamine to give the water-soluble sulfonic acid derivative **TPP(SO₃H)₄** or the amphiphilic sulfonamide porphyrin **TPP(SO₃NHEt)₄**, respectively, and (iii) complexation of the free base sulfonamide with the Zn(II) acetate to afford the metalloporphyrin complex **ZnTPP(SO₃NHEt)₄** (Scheme 1).

The introduction of sulfonate groups in aromatic compounds [65] is a reaction widely used in the functionalization of organic molecules [38]. However, one of the hurdles with these reactions is that very polar sulfonic groups are obtained, the purification of which becomes quite difficult. Therefore, the chlorosulfonation reaction is an alternative for the peripheral functionalization of porphyrins, as the chlorosulfonates formed are stable enough to further react in the presence of many nucleophiles such as water [44,61], amines [46], or alcohols [45,66] affording porphyrins with sulfonic acids, sulfonamides or sulfo esters moieties with high biological interest [41,45,46,67,68]. The TPP chlorosulfonation reaction was carried out and we proceed with our synthetic strategy of functionalization of porphyrins with the aim of obtaining PS with different physicochemical properties, such as different solubilities in physiological or aqueous media. Due to the high reactivity of chlorosulfonates with amines, and as we wanted to develop amphiphilic PS, this led us to prepare porphyrins functionalized with sulfonamide groups. We added an excess of ethylamine to **TPP(SO₂Cl)₄** dissolved in CH_2Cl_2, yielding **TPP(SO₂NHEt)₄**. To modulate the photophysical and photochemical features, it was decided to prepare porphyrin metal complexes since it is well established that the coordination of metals in tetrapyrrolic macrocycles leads to an increase in the production of $^1O_2$ and other ROS, owing to the heavy atom effect that potentiates formation of the triplet state ($^3PS^*$) and reduces the quantum yield of fluorescence (Φ_F) [69,70]. Following this, a complexation reaction of the sulfonamide **TPP(SO₂NHEt)₄** with zinc(II) salts was carried out, yielding purple crystals of **ZnTPP(SO₂NHEt)₄**. When **TPP(SO₂Cl)₄** was hydrolyzed with water, under reflux, the water soluble **TPP(SO₃H)₄**. was obtained quantitatively.

The UV-Vis absorption (Figure 1) as well as the fluorescence emission (Figure 2) features and the fluorescence quantum yield of the studied PS are presented in Table 1. It is shown that the sulfonic acids or sulfonamide groups at the periphery of **TPP** do not cause a significant change in the UV-Vis spectra when compared with the starting *meso*-phenylporphyrin, **TPP**. Moreover, the complexation of **TPP(SO₂NHEt)₄** with zinc(II) acetate, **ZnTPP(SO₂NHEt)₄** changes the free base symmetry from D_{2h} to D_{4h}, and, consequently, the four Q bands are converted in two for the metal complex [69]. Additionally, complexation with zinc(II) leads to a redshift (~7 nm) of the Soret band. This shift is typical due to an interaction of the metal ion d orbital with the porphyrin aromatic π system [69].

All fluorescence emission spectra of *meso*-substitutedporphyrins display similar profiles. The fluorescence quantum yields calculated for the free base porphyrins **TPP(SO₃H)₄** and **TPP(SO₂NHEt)₄** are very similar to the values found for **TPP**, which indicates that the introduction of sulfonic acids or sulfonamides onto the peripheral phenyl rings does not affect significantly the singlet state properties ($^1S^*$). Conversely, the presence of zinc(II) led to a significant decrease in the fluorescence quantum yield compared to **TPP** and **TPP(SO₂NHEt)₄**. It is well-established from the literature [69,71] that metalloporphyrins with closed metal shells, such as Zn(II), are less fluorescent than the corresponding free-base porphyrins, have shorter fluorescence lifetimes, and higher intersystem crossing to triplet states, promoted by the spin-orbit coupling mechanism (heavy atom effect) [55].

The success of the prepared porphyrin derivatives as PS depends strongly on their capability to generate ROS, such as $^1O_2$, the main oxidative species accountable to cause microorganisms inactivation [8,72]. The $^1O_2$ formation of the studied PS was determined by the photo-oxidation of 9,10-dimethylanthracene (DMA) [64]. $^1O_2$ origins the photo-oxidation of DMA, and its conversion to endoperoxide. In this sense, it is possible to track the decrease in DMA absorption in solution at 378 nm. As observed in Figure 3, all the evaluated PSs were able to produce $^1O_2$ and no photo-oxidation of DMA was observed in their absence. In this series, **ZnTPP(SO$_2$NHEt)$_4$** merits to be highlighted since is the most efficient compound to generate $^1O_2$. This observation is in accordance with the literature data, since it is expected that zinc(II) complexes are better $^1O_2$ producers due to the heavy atom effect, as stated above [69,70]. **TPP(SO$_3$H)$_4$** and **TPP(SO$_2$NHEt)$_4$** revealed similar $^1O_2$ generation as the reference TPP which is considered a good $^1O_2$ generator ($\Phi_\Delta = 0.65$ in DMF) [63]. Therefore, all molecules are considered promising for application as PS in aPDT.

aPDT relies upon the use of a PS molecule whose structure and substituent groups are highly important for the antimicrobial success of this approach [18,56,73,74]. A few literature studies reported the use of sulfonamide phthalocyanine [41] and *ortho*-halogenated sulfonamide porphyrin (containing a sulfonamide at the 3′ position of the phenyl ring) [68,75] as PS against bacteria. However, the aPDT effect on bacteria using 4′-sulfonamide groups on non-halogenated porphyrins is yet to be studied.

Co-adjuvant KI has been reported to potentiate the photodynamic efficiency of a broad spectrum of molecules [22,76]. The mechanism under the potentiation of photodynamic efficiency is related to the I_2 formation since $^1O_2$ reacts with KI producing peroxyiodide (HOOI$_2^-$). This species can be further decomposed into free iodine (I_2/I_3^-) and iodine radicals ($I_2^{•-}$), which are responsible to cause cell killing [22,48,76]. Thus, the formation of iodine by the prepared PS at 5.0 µM was evaluated in the presence of KI at 100 mM, under irradiation with white light (Figure S7). In the presence of KI the absorbance at 340 nm of all solutions increased as a result of I_2 formation, although **TPP(SO$_3$H)$_4$** remarkably stands out with a significative absorbance increase in the initial 15 min of irradiation being the most efficient PS to generate I_2, followed by **TPP(SO$_2$NHEt)$_4$** and the less efficient compound **ZnTPP(SO$_2$NHEt)$_4$** to generate I_2. Considering all the photophysical data, the *meso*-arylporphyrins functionalized with sulfonamide groups **TPP(SO$_2$NHEt)$_4$** and **ZnTPP(SO$_2$NHEt)$_4$** and the sulfonic acid derivative **TPP(SO$_3$H)$_4$** proceeded to biological evaluation due to it promising properties as PS. Thus, all three derivatives were assessed for their photodynamic action in combination with the adjuvant KI against a methicillin-resistant *Staphylococcus aureus* (MRSA) strain, selected as a bacterial model of Gram-(+) bacteria.

Our findings indicate that, at low concentrations, the **TPP(SO$_3$H)$_4$** (5.0, 1.0, 0.5, and 0.1 µM), **TPP(SO$_2$NHEt)$_4$** (5.0 and 1.0 µM), and **ZnTPP(SO$_2$NHEt)$_4$** (5.0 and 1.0 µM) were effective in the photoinactivation of MRSA strain, promoting a bacterial reduction always higher than 3.0 Log CFU mL^{-1} (> 99.9%, ANOVA, $p < 0.0001$) (Figures 4–6). Thus, according to the guideline of the American Society for Microbiology, the PSs synthesized in this study meet the necessary conditions to be considered bactericidal agents [77]. Comparing the photodynamic efficiency of all prepared PS, **TPP(SO$_3$H)$_4$** proved to be the most effective in reducing the viability of MRSA cells, followed by **ZnTPP(SO$_2$NHEt)$_4$** and finally by **TPP(SO$_2$NHEt)$_4$** (ANOVA, $p < 0.0001$). It was clear that the **ZnTPP(SO$_2$NHEt)$_4$** was more efficient for the inactivation of MRSA when compared with the parent freebase **TPP(SO$_2$NHEt)$_4$**. The different effectiveness rates exhibited by the tested PSs might be attributed to their physicochemical properties. It is well known that the aggregation behavior of a PS when in physiological media tends to significantly diminish the amount of $^1O_2$ produced, which leads to a reduction in the photoinactivation rate [78]. The **TPP(SO$_3$H)$_4$** is the most soluble PS in an aqueous solution and consequently less affected by aggregation phenomena, which can explain its greater effectiveness in the photoinactivation of MRSA cells. The two sulfonamide-porphyrin derivatives **TPP(SO$_2$NHEt)$_4$** and **ZnTPP(SO$_2$NHEt)$_4$** are amphiphilic compounds and when in aqueous solution (PBS) some aggregation phenomena might occur, which substantiates the decrease in inactivation rate when compared with **TPP(SO$_3$H)$_4$**. This phenomenon can explain why **ZnTPP(SO$_2$NHEt)$_4$** with the highest efficiency to generate $^1O_2$ presented a slower rate of inactivation towards the MRSA strain than **TPP(SO$_3$H)$_4$** (Figure 3). Even so, the photodynamic effect of all synthesized PS is even more remarkable when compared with those of the literature [68,75]. In 2017, Dabrowski [68] and co-authors studied the photodynamic efficacy of fluoro and chloro, *ortho*-halogenated porphyrins containing sulfonic acid groups and sulfonamides at the 3′ position: 5,10,15,20-tetrakis(2,6-difluoro-3-sulfophenyl)porphyrin (**F$_2$POH**), 5,10,15,20-tetrakis(2,6-dichloro-3-sulfophenyl)porphyrin (**Cl$_2$POH**), and 5,10,15,20-tetrakis[2,6-dichloro-3-(*N*-ethylsulfamoyl)phenyl]porphyrin (**Cl$_2$PEt**). These PSs were studied against Gram-(+) (*S. aureus*, *E. faecalis*), Gram-(−) bacteria (*E. coli*, *P. aeruginosa*, *S. marcescens*), and fungal yeast (*C. albicans*). The best photoinactivation results against *S. aureus* were found for both sulfonic acid derivatives (**F$_2$POH**, **Cl$_2$POH**) at a concentration of 20 μM and after 10 min irradiation (10 J cm^{-2}) reaching a 6 Log CFU mL^{-1} reduction. In addition, for the same strain and at a concentration of 20 μM the *ortho*-halogenated porphyrins, **Cl$_2$PEt** gave only a reduction of 99% (2 Log CFU mL^{-1} reduction) in the *S. aureus* survival after 10 min irradiation. Later on, the same group [75] reported a comparison study between neutral (**TPP**), positive (**TMPyP**), and negative (**TPPSO$_3$H**) non-halogenated *meso*-arylporphyrins, with a series of neutral (**ClTPP**, **Cl$_2$TPP**) or anionic (**Cl$_2$TPPS**) *ortho*-chlorophenylporphyrins. The photodynamic treatments with the **Cl$_2$TPPS** used as PS at a concentration of 20 μM led to the 5 Log of CFU mL^{-1} reduction of *S. aureus* after to have received a low light dose (5 J cm^{-2}). Moreover, high efficiency in the *S. aureus* photoinactivation was found when 100 mM of KI was used as a co-adjuvant, and the photodynamic inactivation was enhanced by 2–3 Log for *S. aureus*. When the authors increased the light dose to 40 J cm^{-2}, a complete destruction of Gram-(+) bacterium was observed.

In the present study, the porphyrinic PSs evaluated were more efficient against *S. aureus* since the bacterial viability reached the detection limit of the methodology (reduction of 7.5 Log of CFU mL^{-1}) with a lower concentration of PS (5.0 μM without co-adjuvant KI) and a total light dose ranging from 15 to 180 J cm^{-2} [**TPP(SO$_3$H)$_4$**: 15 J cm^{-2}; **ZnTPP(SO$_2$NHEt)$_4$**: 67.5 J cm^{-2}; **TPP(SO$_2$NHEt)$_4$**: 180 J cm^{-2}].

Although the synthesized PSs were effective in inactivating MRSA, their photodynamic effect was significantly improved in the presence of the co-adjuvant KI. Previous studies have reported that the bacterial death curve can give an indication of the iodine species responsible for the extra inactivation [22,79]. When the death curve assumes an abrupt bacterial inactivation profile, free I$_2$ is the main specie responsible for microbial inactivation. If there is a gradual increase in the bacterial death rate, there is a more pronounced contribution from short-lived reactive iodine species. In this way, the PS + KI inactivation profiles were evaluated to identify the potential species responsible for the additional bacterial killing. In the case of **TPP(SO$_3$H)$_4$** for all tested concentrations (0.1—5.0 μM) + KI, an abrupt decrease in the MRSA inactivation profile (reaching the detection limit of the method) was observed (5 min at 1.0 μM—30 min at 0.1 μM, Figure 4). These results can be explained by the preferential decomposition of peroxyiodide into free iodine (I$_2$), a species recognized for its high antimicrobial action after reaching a threshold concentration, justifying the rapid inactivation of MRSA and the high efficacy of this combination. This production of free iodine (I$_2$) was confirmed by the high and rapid detection of I$_2$ by spectrophotometry (Figure S7).

In the case of the **TPP(SO$_2$NHEt)$_4$** + KI and **ZnTPP(SO$_2$NHEt)$_4$** + KI combinations, a gradual decrease in the MRSA inactivation profile was observed, mainly evident at the lowest concentration of PS (1.0 μM) (Figures 5 and 6, respectively). These results lead us to assume that the mechanism of action is related to the preferential decomposition of peroxyiodide into iodine radicals which, due to their short diffusion distance, cause a gradual decrease in the inactivation profile. This fact was confirmed by the low detection of free I$_2$ by spectrophotometric analysis, which did not affect the high combined effect observed with **TPP(SO$_2$NHEt)$_4$** and **ZnTPP(SO$_2$NHEt)$_4$** in the presence of KI.

4. Materials and Methods

The most relevant spectroscopic and structural features of the prepared porphyrin precursors are described in detail in the Supplementary Material section. To the best of our knowledge, compounds **TPP(SO$_2$NHEt)$_4$** and **ZnTPP(SO$_2$NHEt)$_4$** were synthesized here for the first time and the synthesis and full structural characterization of these compounds are described below.

4.1. Photosensitizers Preparation and Characterization

5,10,15,20-tetrakis(4-chlorosulfonylphenyl)porphyrin, TPP(SO$_2$Cl)$_4$ was obtained following a literature method [46] through the aromatic electrophilic chlorosulfonation of **TPP**. This porphyrin was used directly (without further purification) in the subsequent reactions.

5,10,15,20-tetrakis(4-sulfophenyl)porphyrin, TPP(SO$_3$H)$_4$ was obtained by hydrolysis of TPP(SO$_2$Cl)$_4$ in water under reflux over 18 h [44] and the ^{1}H NMR is in agreement with the literature. ^{1}H NMR (300 MHz, DMSO-d$_6$), δ ppm: 8.86 (s, 8H, β-H); 8.19 (d, J = 8.1 Hz, 8H, Ar-H); 8.05 (d, J = 8.1 Hz, 8H, Ar-H); -2.98 (s, 2H, NH).

5,10,15,20-tetrakis[4-(N-ethylsulfamoyl)phenyl]porphyrin, TPP(SO$_2$NHEt)$_4$. To a round bottom flask with **TPP(SO$_2$Cl)$_4$** (100.9 mg, 0.1 mmol) dissolved in dichloromethane (50 mL) an excess of ethylamine (1.2 mmol, 0.1 mL) was added. The reaction mixture was stirred for 3 h at room temperature. The reaction was monitored by TLC using dichloromethane/ethyl acetate (7:3) as the eluent. Then, the organic layer was washed three times with 1M HCl solution, followed by a neutralization procedure with a saturated solution of sodium hydrogen carbonate and dried over anhydrous sodium sulfate. After solvent evaporation under reduced pressure, the solid residue was purified on a silica gel column chromatography using dichloromethane/ethyl acetate (7:3) as the eluent. The main compound **TPP(SO$_2$NHEt)$_4$** (92.6 mg) was isolated, yielding 88%. ^{1}H NMR (300 MHz, DMSO-d$_6$), δ ppm: 8.87 (s, 8H, β-H); 8.47 (d, 8H, J = 8.3 Hz, Ar-H); 8.25 (d, 8H, J = 8.3 Hz, Ar-H); 7.93 (t, 4H, J = 5.8 Hz, -NH-); 3.08–3.17 (m, 8H, -CH$_2$-), 1.16–1.24 (m, 12H, -CH$_3$), -2.95 (s, 2H, NH). HRMS-ESI(+): m/z calcd for C$_{52}$H$_{51}$N$_8$O$_8$S$_4$: 1043.2690 [M+H]$^+$; found 1043.2707.

5,10,15,20-tetrakis(4-(N-ethylsulfamoyl)phenyl)porphyrinate zinc(II), ZnTPP(SO$_2$NHEt)$_4$. In a round bottom flask, 20 mg of **TPP(SO$_2$NHEt)$_4$** (0.019 mmol) were dissolved in chloroform (10 mL). The reaction mixture was heated to 50 °C and then 35 mg of zinc(II) acetate (0.19 mmol), previously dissolved in methanol (5 mL), was added. The reaction remained under heating (50 °C) and magnetic stirring, being controlled by UV-Vis. In the end, the reaction was cooled to room temperature, the solvent was removed under a vacuum and the solid was dissolved in dichloromethane. The organic phase was washed with distilled water (three times) and dried over anhydrous sodium sulfate. After evaporation, the purple solid formed was dried under a vacuum. The compound **ZnTPP(SO$_2$NHEt)$_4$** (21 mg, 0.019 mmol) was obtained in quantitative yield. ^{1}H NMR (300 MHz, CDCl$_3$/CD$_3$OD), δ ppm: 8.83 (s, 8H, β -H); 8.38 (d, 8H, J = 8.4 Hz, Ar-H); 8.25 (d, 8H, J = 8.4 Hz, Ar-H); 3.30 (q, 8H, J = 7.3 Hz, -CH$_2$-), 1.32 (t, 12H, J = 7.3 Hz, -CH$_3$). ^{13}C (75 MHz, DMSO-d$_6$), δ ppm: 149.6, 147.6, 139.4, 134.9, 131.7, 125.0, 119.1, 40.2, 38.3, 15.0. HRMS-ESI(+): m/z calcd for C$_{52}$H$_{49}$N$_8$O$_8$S$_4$Zn: 1105.1842 [M+H]$^+$; found 1105.1817.

4.2. UV-Vis Absorption Spectroscopy

The UV-Vis spectra were acquired using a Shimadzu UV-2501PC spectrophotometer on 1×1 cm quartz optical cells in DMF as a solvent. The molar absorption coefficients were calculated through the Beer–Lambert law.

4.3. Fluorescence Quantum Yield Measurements

The fluorescence excitation and emission spectra were acquired at room temperature using a Horiba Spex Fluoromax 4 Plus Spectrofluorimeter with DMF as a solvent. The quantum fluorescence yield (Φ_F) of all PSs was determined using **TPP** as a reference in

DMF ($\Phi_F = 0.11$) [63]. Solutions of **TPP** in DMF and each PS were freshly prepared, and the emission spectra were measured upon excitation at 420 nm of solutions with an OD of 0.02 in quartz cells with four faces of 1 × 1 cm of optical path.

4.4. Singlet Oxygen Generation

Solutions of each compound (**TPP(SO$_3$H)$_4$, TPP(SO$_2$NHEt)$_4$, and ZnTPP(SO$_2$NHEt)$_4$**) and **TPP** in DMF (2.5 mL) were placed in a quartz cuvette and the absorbance adjust to ≈ 0.1 at 420 nm. Then, 9,10-dimethylanthracene (DMA) in DMF was added at a concentration of 30 µM. The solutions were irradiated at 420 ± 5 nm and the absorbance of each solution was monitored at 378 nm, every 60 s across 600 s, in a UV-2501PC SHIMADZU spectrophotometer and registered in a first-order kinetic plot.

4.5. Detection of Iodine Formation

In a 96-well microplate were prepared solutions of each PS (**TPP(SO$_3$H)$_4$, TPP(SO$_2$NHEt)$_4$, and ZnTPP(SO$_2$NHEt)$_4$**) at a concentration of 5.0 µM in PBS into which were added a solution of KI at 100 mM in PBS. All solutions were incubated under stirring in the dark for 15 min and then irradiated with white light (400–700 nm) at an irradiance of 25 mW cm^{-2}. The formation of iodine (I$_2$) during the experiment progression was monitored by measuring the absorbance at 340 nm at different pre-defined irradiation times in a Synergy™ HTX Multi-Mode Microplate Reader from BioTek Instruments.

4.6. Light Source

An LED system (LUMECO, 30 W, 2000 lm) was used as an artificial white light (400–700 nm) source in the iodine (I$_2$) formation assay as well as the photodynamic treatment assays. Before each assay, the LED system was placed above the samples at a distance that allows homogeneous irradiation of the samples at an irradiance of 25 mW cm^{-2}. The irradiance was measured and adjusted to 25 mW cm^{-2} with a FieldMaxII–TOP energy meter combined with a PowerSens PS19Q (Coherent).

4.7. Biological Assays

4.7.1. Photosensitizer and Potassium Iodide Stock Solutions

The stock solutions of **TPP(SO$_3$H)$_4$, TPP(SO$_2$NHEt)$_4$, and ZnTPP(SO$_2$NHEt)$_4$** were prepared at 500 µM in DMSO and kept in the dark. Prior to each assay, the PS stock solution was sonicated for 30 min at room temperature (ultrasonic bath, Nahita 0.6 L, 40 kHz).

Potassium iodide (KI) was purchased from Sigma-Aldrich (St. Louis, MO, USA) and KI solutions were prepared at 5 M in sterile PBS immediately before each assay.

4.7.2. Photoinactivation Conditions

For the biological assays, the methicillin-resistant *S. aureus* strain DSM 25693 was selected as a Gram-(+) bacterial model. This strain produces the staphylococcal enterotoxins A, C, H, G, and I. The bacterium was maintained in the laboratory on Tryptic Soy Agar medium (TSA, Merck) at 4 °C. Before each assay, three colonies were transferred to 30 mL of Tryptic Soy Broth medium (TSB, Merck) and then incubated at 37 °C for 18 h under stirring (120 rpm). Then, 300 µL of the previously grown bacterial suspension was transferred to 30 mL of new TSB medium and incubated under the conditions described above, to promote the bacterial growth until the stationary phase (approximately 10^9 CFU mL^{-1}).

4.7.3. Photodynamic Treatment Experiments

To evaluate the photodynamic inactivation efficiency of the synthesized PSs against MRSA, the PSs **TPP(SO$_3$H)$_4$, TPP(SO$_2$NHEt)$_4$, and ZnTPP(SO$_2$NHEt)$_4$** were tested at the concentrations of 5.0 and 1.0 µM, in combination with KI at 100 mM. The **TPP(SO$_3$H)$_4$** was also tested at concentrations of 0.5 and 0.1 µM. Photodynamic assays were performed for 120 min under irradiation with white light (400–700 nm) at an irradiance of 25 mW cm^{-2}.

The bacterial culture (approximately 10^9 CFU mL^{-1}) was 100-fold diluted in phosphate-buffered saline (PBS) and distributed in 12-well plates. An appropriate volume of each PS and PS + KI was added to obtain the desired PS concentrations. The total volume used per well was 4 mL. Simultaneously, the following controls were also performed: light control (LC), containing only the bacterial suspension and PBS, exposed to light; light control with KI (LC KI), containing the bacterial suspension in PBS and KI, exposed to light; dark control (DC) containing the bacterial suspension in PBS, KI and PS at the maximum tested PS concentration (5.0 µM) and protected from light with aluminium foil during the experiment time course.

The samples and controls were incubated in the dark for 15 min, under stirring (120 rpm) and at room temperature, to promote PS binding to bacterial cells. Afterward, samples and light controls were irradiated for 120 min under agitation, and simultaneously, dark controls were protected from light. To evaluate the effect of the treatments on the MSRA cells, aliquots (150 µL) of each sample and control were collected at pre-defined periods: 0 (after the dark incubation period) and 5, 10, 15, 30, 45, 60, 90, and 120 min of treatment. The collected aliquots were then serially diluted in PBS, drop-plated (10 µL) in triplicate on TSA, and incubated at 37 °C for 18–24 h. Later, colonies were counted at the most appropriate dilution and expressed as Colony Forming Units per mL (CFU mL^{-1}). For each condition tested, three independent assays, each assay in triplicate, were performed.

4.7.4. Statistical Analysis

Statistical analysis was performed using the GraphPad Prism 9 program. Normal distributions were analyzed using the Kolmogorov–Smirnov test and the homogeneity of variance was verified using the Brown–Forsythe test. Differences between the results were evaluated by 2-way ANOVA and by Tukey's multiple comparison tests. p values < 0.05 were considered significant. For each condition tested at least three independent assays were performed and each assay was in triplicate.

5. Conclusions

The sulfonamide derivatives **TPP(SO$_2$NHEt)$_4$** and **ZnTPP(SO$_2$NHEt)$_4$** were efficiently prepared by well-established synthetic procedures. Photodynamic studies revealed that **TPP(SO$_3$H)$_4$**, **TPP(SO$_2$NHEt)$_4$**, and **ZnTPP(SO$_2$NHEt)$_4$** were effective PSs in the photoinactivation of Gram-(+) MRSA. **TPP(SO$_3$H)$_4$** showed greater efficacy than the other tested PSs, followed by **ZnTPP(SO$_2$NHEt)$_4$** and **TPP(SO$_2$NHEt)$_4$**. These results seem to be related to the physicochemical properties of these compounds, which modulate the solubility in aqueous media, the $^1O_2$ production, and the production of iodine reactive species when combined with KI. Furthermore, the potential of these compounds is evidenced by the low PS concentrations required to photoinactivate a multiresistant *S. aureus* strain and by the low irradiance and light dose needed to achieve total inactivation of MRSA, being an advantage for clinical and environmental applications of these PSs. The co-adjuvant of KI improved the photodynamic effect of the three PSs, allowing a reduction in the treatment time by about 6 times but also reducing the PS concentration to at least 1.0 µM. The combined effect observed for the **TPP(SO$_3$H)$_4$** + KI is mainly due to the formation of free iodine, but for **TPP(SO$_2$NHEt)$_4$** and **ZnTPP(SO$_2$NHEt)$_4$** the effect of its combination with the KI co-adjuvant is mainly due to the formation of reactive iodine radicals.

Supplementary Materials: The following supporting information can be downloaded at: https://www.mdpi.com/article/10.3390/molecules28052067/s1, *Instrumentation and reagents* used to prepare the PSs (and respective precursors); Figure S1: ^{1}H NMR of **TPP(SO$_2$NHEt)$_4$** in DMSO-d$_6$; Figure S2: HRMS of **TPP(SO$_2$NHEt)$_4$**; Figure S3: ^{1}H NMR of **ZnTPP(SO$_2$NHEt)$_4$** in CDCl$_3$/CD$_3$OD (98:2); Figure S4: ^{13}C NMR of **ZnTPP(SO$_2$NHEt)$_4$** in CDCl$_3$/CD$_3$OD (98:2); Figure S5: HRMS of **ZnTPP(SO$_2$NHEt)$_4$**; Figure S6: ^{1}H NMR of **TPP(SO$_3$H)$_4$** in DMSO-d6; Figure S7: Iodine monitoring formation at 340 nm after irradiation with white light (25 mW cm^{-2}) across 120 min with the tested PS.

Author Contributions: Conceptualization, C.J.P.M. and M.A.F.F.; methodology and investigation, S.N.S., C.V., M.B. and C.J.D.; writing—original draft preparation, S.N.S., C.V., C.J.D., C.J.P.M. and M.A.F.F.; writing—review and editing, C.J.P.M., M.G.P.M.S.N., A.A. and M.A.F.F.; supervision, C.J.P.M., M.A.F.F. and A.A.; project administration and funding acquisition, M.A.F.F. and A.A. All authors have read and agreed to the published version of the manuscript.

Funding: This work received support from PT national funds (FCT/MCTES, Fundação para a Ciência e Tecnologia and Ministério da Ciência, Tecnologia e Ensino Superior) through the projects UIDB/50006/2020 and UIDP/50006/2020 (LAQV-REQUIMTE); UIDP/50017/2020 + UIDB/50017/2020 + LA/P/0094/2020 (CESAM) and the Project FCT PREVINE—FCT-PTDC/ASP-PES/29576/2017. C.V. (SFRH/BD/150358/2019) and C.J.D. (SFRH/BD/150676/2020) thank FCT for their Ph.D. grants.

Data Availability Statement: Not applicable.

Acknowledgments: The authors thank the University of Aveiro and FCT/MCT for the financial support provided to LAQV-REQUIMTE (UIDB/50006/2020 and UIDP/50006/2020), CESAM (UIDP/50017/2020 + UIDB/50017/2020 + LA/P/0094/2020) and to Project PREVINE—FCT-PTDC/ASP-PES/29576/2017, through national funds (OE) and where applicable co-financed by the FEDER-Operational Thematic Program for Competitiveness and Internationalization-COMPETE 2020, within the PT2020 Partnership Agreement. Thank are also due to the Portuguese NMR and Mass Networks.

Conflicts of Interest: The authors declare no conflict of interest.

References

1. da Cunha, B.R.; Fonseca, L.P.; Calado, C.R.C. Antibiotic Discovery: Where Have We Come from, Where Do We Go? *Antibiotics* **2019**, *8*, 45. [CrossRef] [PubMed]
2. Haller, J.S., Jr. The First Miracle Drugs: How the Sulfa Drugs Transformed Medicine. *J. Hist. Med. Allied Sci.* **2007**, *63*, 119–121. [CrossRef]
3. Sköld, O. Sulfonamide resistance: Mechanisms and trends. *Drug Resist. Updat.* **2000**, *3*, 155–160. [CrossRef] [PubMed]
4. Bentley, R. Different roads to discovery; Prontosil (hence sulfa drugs) and penicillin (hence beta-lactams). *J. Ind. Microbiol. Biotechnol.* **2009**, *36*, 775–786. [CrossRef]
5. Feng, M.H.; Tang, B.Q.; Liang, S.H.; Jiang, X.F. Sulfur Containing Scaffolds in Drugs: Synthesis and Application in Medicinal Chemistry. *Curr. Top. Med. Chem.* **2016**, *16*, 1200–1216. [CrossRef]
6. Almeida, A. Photodynamic Therapy in the Inactivation of Microorganisms. *Antibiotics* **2020**, *9*, 138. [CrossRef]
7. Almeida, A.; Cunha, A.; Faustino, M.A.F.; Tomé, A.C.; Neves, M.G.P.M.S. Chapter 5 Porphyrins as Antimicrobial Photosensitizing Agents. In *Photodynamic Inactivation of Microbial Pathogens: Medical and Environmental Applications*; The Royal Society of Chemistry: London, UK, 2011; Volume 11, pp. 83–160.
8. Dabrowski, J.M.; Arnaut, L.G. Photodynamic therapy (PDT) of cancer: From local to systemic treatment. *Photochem. Photobiol. Sci.* **2015**, *14*, 1765–1780. [CrossRef]
9. Gomes, M.; Bartolomeu, M.; Vieira, C.; Gomes, A.T.P.C.; Faustino, M.A.F.; Neves, M.G.P.M.S.; Almeida, A. Photoinactivation of Phage Phi6 as a SARS-CoV-2 Model in Wastewater: Evidence of Efficacy and Safety. *Microorganisms* **2022**, *10*, 659. [CrossRef]
10. Santos, I.; Gamelas, S.R.D.; Vieira, C.; Faustino, M.A.F.; Tomé, J.P.C.; Almeida, A.; Gomes, A.T.P.C.; Lourenço, L.M.O. Pyrazole-pyridinium porphyrins and chlorins as powerful photosensitizers for photoinactivation of planktonic and biofilm forms of E. coli. *Dyes Pigm.* **2021**, *193*, 109557. [CrossRef]
11. Lopes, M.M.; Bartolomeu, M.; Gomes, A.T.P.C.; Figueira, E.; Pinto, R.; Reis, L.; Balcão, V.M.; Faustino, M.A.F.; Neves, M.G.P.M.S.; Almeida, A. Antimicrobial Photodynamic Therapy in the Control of Pseudomonas syringae pv. actinidiae Transmission by Kiwifruit Pollen. *Microorganisms* **2020**, *8*, 1022. [CrossRef]
12. Ndemueda, A.; Pereira, I.; Faustino, M.A.F.; Cunha, Â. Photodynamic inactivation of the phytopathogenic bacterium Xanthomonas citri subsp. citri. *Lett. Appl. Microbiol.* **2020**, *71*, 420–427. [CrossRef] [PubMed]
13. do Prado-Silva, L.; Gomes, A.T.P.C.; Mesquita, M.Q.; Neri-Numa, I.A.; Pastore, G.M.; Neves, M.G.P.M.S.; Faustino, M.A.F.; Almeida, A.; Braga, G.Ú.L.; Sant'Ana, A.S. Antimicrobial photodynamic treatment as an alternative approach for Alicyclobacillus acidoterrestris inactivation. *Int. J. Food Microbiol.* **2020**, *333*, 108803. [CrossRef] [PubMed]

14. Braz, M.; Salvador, D.; Gomes, A.T.P.C.; Mesquita, M.Q.; Faustino, M.A.F.; Neves, M.G.P.M.S.; Almeida, A. Photodynamic inactivation of methicillin-resistant Staphylococcus aureus on skin using a porphyrinic formulation. *Photodiagnosis Photodyn. Ther.* **2020**, *30*, 101754. [CrossRef] [PubMed]
15. Sousa, V.; Gomes, A.T.P.C.; Freitas, A.; Faustino, M.A.F.; Neves, M.G.P.M.S.; Almeida, A. Photodynamic Inactivation of Candida albicans in Blood Plasma and Whole Blood. *Antibiotics* **2019**, *8*, 221. [CrossRef] [PubMed]
16. Santos, A.R.; Batista, A.F.P.; Gomes, A.T.P.C.; Neves, M.G.P.M.S.; Faustino, M.A.F.; Almeida, A.; Hioka, N.; Mikcha, J.M.G. The Remarkable Effect of Potassium Iodide in Eosin and Rose Bengal Photodynamic Action against Salmonella Typhimurium and Staphylococcus aureus. *Antibiotics* **2019**, *8*, 211. [CrossRef]
17. Beirão, S.; Fernandes, S.; Coelho, J.; Faustino, M.A.F.; Tomé, J.P.C.; Neves, M.G.P.M.S.; Tomé, A.C.; Almeida, A.; Cunha, A. Photodynamic Inactivation of Bacterial and Yeast Biofilms With a Cationic Porphyrin. *Photochem. Photobiol.* **2014**, *90*, 1387–1396. [CrossRef] [PubMed]
18. Vieira, C.; Santos, A.; Mesquita, M.Q.; Gomes, A.T.P.C.; Neves, M.G.P.M.S.; Faustino, M.A.F.; Almeida, A. Advances in aPDT based on the combination of a porphyrinic formulation with potassium iodide: Effectiveness on bacteria and fungi planktonic/biofilm forms and viruses. *J. Porphyr. Phthalocyanines.* **2019**, *23*, 534–545. [CrossRef]
19. Yin, R.; Agrawal, T.; Khan, U.; Gupta, G.K.; Rai, V.; Huang, Y.Y.; Hamblin, M.R. Antimicrobial photodynamic inactivation in nanomedicine: Small light strides against bad bugs. *Nanomedicine* **2015**, *10*, 2379–2404. [CrossRef]
20. Almeida, A.; Faustino, M.A.F.; Tome, J.P.C. Photodynamic inactivation of bacteria: Finding the effective targets. *Future Med. Chem.* **2015**, *7*, 1221–1224. [CrossRef]
21. Alves, E.; Faustino, M.A.F.; Neves, M.G.P.M.S.; Cunha, A.; Nadais, H.; Almeida, A. Potential applications of porphyrins in photodynamic inactivation beyond the medical scope. *J. Photochem. Photobiol. C-Photochem. Rev.* **2015**, *22*, 34–57. [CrossRef]
22. Vieira, C.; Gomes, A.T.P.C.; Mesquita, M.Q.; Moura, N.M.M.; Neves, M.G.P.M.S.; Faustino, M.A.F.; Almeida, A. An Insight Into the Potentiation Effect of Potassium Iodide on aPDT Efficacy. *Front. Microbiol.* **2018**, *9*, 2665. [CrossRef] [PubMed]
23. Mesquita, M.Q.; Dias, C.J.; Neves, M.G.P.M.S.; Almeida, A.; Faustino, M.A.F. Revisiting Current Photoactive Materials for Antimicrobial Photodynamic Therapy. *Molecules* **2018**, *23*, 2424. [CrossRef] [PubMed]
24. Marciel, L.; Mesquita, M.Q.; Ferreira, R.; Moreira, B.; Graca, M.; Neves, M.G.P.M.S.; Faustino, M.A.F.; Almeida, A. An efficient formulation based on cationic porphyrins to photoinactivate Staphylococcus aureus and Escherichia coli. *Future Med. Chem.* **2018**, *10*, 1821–1833. [CrossRef] [PubMed]
25. Simoes, C.; Gomes, M.C.; Neves, M.G.P.M.S.; Cunha, A.; Tome, J.P.C.; Tome, A.C.; Cavaleiro, J.A.S.; Almeida, A.; Faustino, M.A.F. Photodynamic inactivation of Escherichia coli with cationic meso-tetraarylporphyrins—The charge number and charge distribution effects. *Catal. Today* **2016**, *266*, 197–204. [CrossRef]
26. Li, L.; Wang, Y.; Huang, T.; He, X.; Zhang, K.; Kang, E.-T.; Xu, L. Cationic porphyrin-based nanoparticles for photodynamic inactivation and identification of bacteria strains. *Biomater. Sci.* **2022**, *10*, 3006–3016. [CrossRef]
27. Moura, N.M.M.; Ramos, C.I.V.; Linhares, I.; Santos, S.M.; Faustino, M.A.F.; Almeida, A.; Cavaleiro, J.A.S.; Amado, F.M.L.; Lodeiro, C.; Neves, M.G.P.M.S. Synthesis, characterization and biological evaluation of cationic porphyrin–terpyridine derivatives. *RSC Adv.* **2016**, *6*, 110674–110685. [CrossRef]
28. Castro, K.A.D.F.; Moura, N.M.M.; Simões, M.M.Q.; Cavaleiro, J.A.S.; Faustino, M.A.F.; Cunha, Â.; Almeida Paz, F.A.; Mendes, R.F.; Almeida, A.; Freire, C.S.R.; et al. Synthesis and characterization of photoactive porphyrin and poly(2-hydroxyethyl methacrylate) based materials with bactericidal properties. *Appl. Mater. Today.* **2019**, *16*, 332–341. [CrossRef]
29. Angell, N.G.; Lagorio, M.G.; San Roman, E.A.; Dicelio, L.E. Meso-substituted cationic porphyrins of biological interest. Photophysical and physicochemical properties in solution and bound to liposomes. *Photochem. Photobiol.* **2000**, *72*, 49–56. [CrossRef]
30. Merchat, M.; Bertolini, G.; Giacomini, P.; Villanueva, A.; Jori, G. Meso-substituted cationic porphyrins as efficient photosensitizers of gram-positive and gram-negative bacteria. *J. Photochem. Photobiol. B* **1996**, *32*, 153–157. [CrossRef]
31. Malik, Z.; Ladan, H.; Nitzan, Y. Photodynamic Inactivation of Gram-Negative Bacteria—Problems and Possible Solutions. *J. Photochem. Photobiol. B* **1992**, *14*, 262–266. [CrossRef]
32. Preuß, A.; Zeugner, L.; Hackbarth, S.; Faustino, M.A.F.; Neves, M.G.P.M.S.; Cavaleiro, J.A.S.; Roeder, B. Photoinactivation of Escherichia coli (SURE2) without intracellular uptake of the photosensitizer. *J. Appl. Microbiol.* **2013**, *114*, 36–43. [CrossRef]
33. Alves, E.; Santos, N.; Melo, T.; Maciel, E.; Dória, M.L.; Faustino, M.A.F.; Tomé, J.P.C.; Neves, M.G.P.M.S.; Cavaleiro, J.A.S.; Cunha, Â.; et al. Photodynamic oxidation of Escherichia coli membrane phospholipids: New insights based on lipidomics. *Rapid Commun. Mass Spectrom.* **2013**, *27*, 2717–2728. [CrossRef]
34. Alves, E.; Melo, T.; Simões, C.; Faustino, M.A.F.; Tomé, J.P.C.; Neves, M.G.P.M.S.; Cavaleiro, J.A.S.; Cunha, Â.; Gomes, N.C.M.; Domingues, P.; et al. Photodynamic oxidation of Staphylococcus warneri membrane phospholipids: New insights based on lipidomics. *Rapid Commun. Mass Spectrom.* **2013**, *27*, 1607–1618. [CrossRef]
35. Vejzovic, D.; Piller, P.; Cordfunke, R.A.; Drijfhout, J.W.; Eisenberg, T.; Lohner, K.; Malanovic, N. Where Electrostatics Matter: Bacterial Surface Neutralization and Membrane Disruption by Antimicrobial Peptides SAAP-148 and OP-145. *Biomolecules* **2022**, *12*, 1252. [CrossRef] [PubMed]
36. Sohlenkamp, C.; Geiger, O. Bacterial membrane lipids: Diversity in structures and pathways. *FEMS Microbiol. Rev.* **2015**, *40*, 133–159. [CrossRef] [PubMed]
37. George, S.; Hamblin, M.R.; Kishen, A. Uptake pathways of anionic and cationic photosensitizers into bacteria. *Photochem. Photobiol. Sci.* **2009**, *8*, 788–795. [CrossRef] [PubMed]

38. Luciano, M.; Brückner, C. Modifications of Porphyrins and Hydroporphyrins for Their Solubilization in Aqueous Media. *Molecules* **2017**, *22*, 980. [CrossRef] [PubMed]
39. Dabrowski, J.M.; Arnaut, L.G.; Pereira, M.M.; Monteiro, C.J.P.; Urbanska, K.; Simoes, S.; Stochel, G. New halogenated water-soluble chlorin and bacteriochlorin as photostable PDT sensitizers: Synthesis, spectroscopy, photophysics, and in vitro photosensitizing efficacy. *ChemMedChem* **2010**, *5*, 1770–1780. [CrossRef] [PubMed]
40. Pisarek, S.; Maximova, K.; Gryko, D. Strategies toward the synthesis of amphiphilic porphyrins. *Tetrahedron* **2014**, *70*, 6685–6715. [CrossRef]
41. da Silva, R.N.; Cunha, Â.; Tomé, A.C. Phthalocyanine–sulfonamide conjugates: Synthesis and photodynamic inactivation of Gram-negative and Gram-positive bacteria. *Eur. J. Med. Chem.* **2018**, *154*, 60–67. [CrossRef]
42. Manono, J.; Marzilli, P.A.; Fronczek, F.R.; Marzilli, L.G. New Porphyrins Bearing Pyridyl Peripheral Groups Linked by Secondary or Tertiary Sulfonamide Groups: Synthesis and Structural Characterization. *Inorg. Chem.* **2009**, *48*, 5626–5635. [CrossRef]
43. Manono, J.; Marzilli, P.A.; Marzilli, L.G. New Porphyrins Bearing Positively Charged Peripheral Groups Linked by a Sulfonamide Group to meso-Tetraphenylporphyrin: Interactions with Calf Thymus DNA. *Inorg. Chem.* **2009**, *48*, 5636–5647. [CrossRef] [PubMed]
44. Gonsalves, A.M.A.R.; Johnstone, R.A.W.; Pereira, M.M.; deSantAna, A.M.P.; Serra, A.C.; Sobral, A.; Stocks, P.A. New procedures for the synthesis and analysis of 5,10,15,20-tetrakis(sulphophenyl)porphyrins and derivatives through chlorosulphonation. *Heterocycles* **1996**, *43*, 829–838.
45. Sobral, A.; Eleouet, S.; Rousset, N.; Gonsalves, A.M.D.; Le Meur, O.; Bourre, L.; Patrice, T. New sulfonamide and sulfonic ester porphyrins as sensitizers for photodynamic therapy. *J. Porphyr. Phthalocyanines* **2002**, *6*, 456–462. [CrossRef]
46. Monteiro, C.J.P.; Pereira, M.M.; Pinto, S.M.A.; Simões, A.V.C.; Sá, G.F.F.; Arnaut, L.G.; Formosinho, S.J.; Simões, S.; Wyatt, M.F. Synthesis of amphiphilic sulfonamide halogenated porphyrins: MALDI-TOFMS characterization and evaluation of 1-octanol/water partition coefficients. *Tetrahedron* **2008**, *64*, 5132–5138. [CrossRef]
47. Vinagreiro, C.S.; Goncalves, N.P.F.; Calvete, M.J.F.; Schaberle, F.A.; Arnaut, L.G.; Pereira, M.M. Synthesis and characterization of biocompatible bimodal meso-sulfonamide-perfluorophenylporphyrins. *J. Fluor. Chem.* **2015**, *180*, 161–167. [CrossRef]
48. Hamblin, M.R. Potentiation of antimicrobial photodynamic inactivation by inorganic salts. *Expert Rev. Anti-Infect. Ther.* **2017**, *15*, 1059–1069. [CrossRef] [PubMed]
49. Huang, Y.-Y.; Wintner, A.; Seed, P.C.; Brauns, T.; Gelfand, J.A.; Hamblin, M.R. Antimicrobial photodynamic therapy mediated by methylene blue and potassium iodide to treat urinary tract infection in a female rat model. *Sci. Rep.* **2018**, *8*, 7257. [CrossRef]
50. Bartolomeu, M.; Oliveira, C.; Pereira, C.; Neves, M.G.P.M.S.; Faustino, M.A.F.; Almeida, A. Antimicrobial Photodynamic Approach in the Inactivation of Viruses in Wastewater: Influence of Alternative Adjuvants. *Antibiotics* **2021**, *10*, 767. [CrossRef]
51. Castro, K.A.D.F.; Brancini, G.T.P.; Costa, L.D.; Biazzotto, J.C.; Faustino, M.A.F.; Tomé, A.C.; Neves, M.G.P.M.S.; Almeida, A.; Hamblin, M.R.; da Silva, R.S.; et al. Efficient photodynamic inactivation of Candida albicans by porphyrin and potassium iodide co-encapsulation in micelles. *Photochem. Photobiol. Sci.* **2020**, *19*, 1063–1071. [CrossRef]
52. Calmeiro, J.M.D.; Gamelas, S.R.D.; Gomes, A.T.P.C.; Faustino, M.A.F.; Neves, M.G.P.M.S.; Almeida, A.; Tomé, J.P.C.; Lourenço, L.M.O. Versatile thiopyridyl/pyridinone porphyrins combined with potassium iodide and thiopyridinium/methoxypyridinium porphyrins on E. coli photoinactivation. *Dyes Pigm.* **2020**, *181*, 108476. [CrossRef]
53. Moreira, X.; Santos, P.; Faustino, M.A.F.; Raposo, M.M.M.; Costa, S.P.G.; Moura, N.M.M.; Gomes, A.T.P.C.; Almeida, A.; Neves, M.G.P.M.S. An insight into the synthesis of cationic porphyrin-imidazole derivatives and their photodynamic inactivation efficiency against *Escherichia coli*. *Dye. Pigment.* **2020**, *178*, 108330. [CrossRef]
54. Huang, L.; El-Hussein, A.; Xuan, W.; Hamblin, M.R. Potentiation by potassium iodide reveals that the anionic porphyrin TPPS4 is a surprisingly effective photosensitizer for antimicrobial photodynamic inactivation. *J. Photochem. Photobiol. B Biol.* **2018**, *178*, 277–286. [CrossRef]
55. Azenha, E.G.; Serra, A.C.; Pineiro, M.; Pereira, M.M.; de Melo, J.S.; Arnaut, L.G.; Formosinho, S.J.; Gonsalves, A. Heavy-atom effects on metalloporphyrins and polyhalogenated porphyrins. *Chem. Phys.* **2002**, *280*, 177–190. [CrossRef]
56. Mesquita, M.Q.; Dias, C.J.; Neves, M.G.P.M.S.; Almeida, A.; Faustino, M.A.F. The Role of Photoactive Materials Based on Tetrapyrrolic Macrocycles in Antimicrobial Photodynamic Therapy. In *Handbook of Porphyrin Science*; World Scientific: Singapore, 2022; Volume 46, pp. 201–277.
57. Gonsalves, A.M.A.R.; Varejão, J.M.T.B.; Pereira, M.M. Some new aspects related to the synthesis of mesosubstituted porphyrins. *J. Heterocycl. Chem.* **1991**, *28*, 635–640. [CrossRef]
58. Pereira, M.M.; Monteiro, C.J.P.; Peixoto, A.F. Meso-substituted porphyrin synthesis from monopyrrole: An overview. In *Targets in Heterocyclic Systems-Chemistry and Properties*; Attanasi, O.A., Spinelli, D., Eds.; Italian Society of Chemistry: Rome, Italy, 2009; Volume 12, pp. 258–278.
59. Adler, A.D.; Longo, F.R.; Kampas, F.; Kim, J. On preparation of metalloporphyrins. *J. Inorg. Nucl. Chem.* **1970**, *32*, 2443–2445. [CrossRef]
60. Smith, K.M. *Porphyrins and Metalloporphyrins*, 1st ed.; Smith, K.M., Ed.; Elsevier: Amsterdam, The Netherlands, 1975.
61. Monteiro, C.J.P.; Pereira, M.M.; Azenha, M.E.; Burrows, H.D.; Serpa, C.; Arnaut, L.G.; Tapia, M.J.; Sarakha, M.; Wong-Wah-Chung, P.; Navaratnam, S. A comparative study of water soluble 5,10,15,20-tetrakis(2,6-dichloro-3-sulfophenyl)porphyrin and its metal complexes as efficient sensitizers for photodegradation of phenols. *Photochem. Photobiol. Sci.* **2005**, *4*, 617–624. [CrossRef] [PubMed]

62. Kempa, M.; Kozub, P.; Kimball, J.; Rojkiewicz, M.; Kuś, P.; Gryczyński, Z.; Ratuszna, A. Physicochemical properties of potential porphyrin photosensitizers for photodynamic therapy. *Spectrochim. Acta A Mol. Biomol.* **2015**, *146*, 249–254. [CrossRef]
63. Ermilov, E.A.; Tannert, S.; Werncke, T.; Choi, M.T.M.; Ng, D.K.P.; Röder, B. Photoinduced electron and energy transfer in a new porphyrin–phthalocyanine triad. *Chem. Phys.* **2006**, *328*, 428–437. [CrossRef]
64. Gomes, A.; Fernandes, E.; Lima, J.L.F.C. Fluorescence probes used for detection of reactive oxygen species. *J. Biochem. Biophys. Methods* **2005**, *65*, 45–80. [CrossRef] [PubMed]
65. Cremlyn, R.J. *Chlorosulfonic Acid-A Versatile Reagent*; Royal Society of Chemistry: Cambridge, UK, 2002.
66. Simoes, A.V.C.; Adamowicz, A.; Dabrowski, J.M.; Calvete, M.J.F.; Abreu, A.R.; Stochel, G.; Arnaut, L.G.; Pereira, M.M. Amphiphilic meso(sulfonate ester fluoroaryl)porphyrins: Refining the substituents of porphyrin derivatives for phototherapy and diagnostics. *Tetrahedron* **2012**, *68*, 8767–8772. [CrossRef]
67. Domingues, M.R.M.; Santana-Marques, M.G.; Ferrrer-Correia, A.J.; Tome, A.C.; Neves, M.G.P.M.S.; Cavaleiro, J.A.S. Liquid secondary ion mass spectrometry and collision-induced dissociation mass spectrometry of sulfonamide derivatives of meso-tetraphenylporphyrin. *J. Porphyr. Phthalocyanines* **1999**, *3*, 172–179. [CrossRef]
68. Pucelik, B.; Paczynski, R.; Dubin, G.; Pereira, M.M.; Arnaut, L.G.; Dabrowski, J.M. Properties of halogenated and sulfonated porphyrins relevant for the selection of photosensitizers in anticancer and antimicrobial therapies. *PLoS ONE* **2017**, *12*, e0185984. [CrossRef] [PubMed]
69. Arnaut, L.G. Design of porphyrin-based photosensitizers for photodynamic therapy. In *Advances in Inorganic Chemistry*; VanEldik, R.S.G., Ed.; Academic Press: Cambridge, MA, USA, 2011; Volume 63, pp. 187–233.
70. Moura, N.M.M.; Faustino, M.A.F.; Neves, M.G.P.M.S.; Tomé, A.C.; Rakib, E.M.; Hannioui, A.; Mojahidi, S.; Hackbarth, S.; Röder, B.; Almeida Paz, F.A.; et al. Novel pyrazoline and pyrazole porphyrin derivatives: Synthesis and photophysical properties. *Tetrahedron* **2012**, *68*, 8181–8193. [CrossRef]
71. Taniguchi, M.; Lindsey, J.S.; Bocian, D.F.; Holten, D. Comprehensive review of photophysical parameters (ε, Φf, τs) of tetraphenyl-porphyrin (H2TPP) and zinc tetraphenylporphyrin (ZnTPP)—Critical benchmark molecules in photochemistry and photosynthesis. *J. Photochem. Photobiol. C Photochem. Rev.* **2021**, *46*, 100401. [CrossRef]
72. Mesquita, M.Q.; Dias, C.J.; Gamelas, S.; Fardilha, M.; Neves, M.G.P.M.S.; Faustino, M.A.F. An Insight on the role of photosensitizer nanocarriers for Photodynamic Therapy. *An. Acad. Bras. Ciênc.* **2018**, *90*, 1101–1130. [CrossRef] [PubMed]
73. Amos-Tautua, B.M.; Songca, S.P.; Oluwafemi, O.S. Application of Porphyrins in Antibacterial Photodynamic Therapy. *Molecules* **2019**, *24*, 2456. [CrossRef] [PubMed]
74. Bartolomeu, M.; Rocha, S.; Cunha, Â.; Neves, M.G.P.M.S.; Faustino, M.A.F.; Almeida, A. Effect of Photodynamic Therapy on the Virulence Factors of *Staphylococcus aureus*. *Front. Microbiol.* **2016**, *7*, 267. [CrossRef]
75. Sułek, A.; Pucelik, B.; Kobielusz, M.; Barzowska, A.; Dąbrowski, J.M. Photodynamic Inactivation of Bacteria with Porphyrin Derivatives: Effect of Charge, Lipophilicity, ROS Generation, and Cellular Uptake on Their Biological Activity In Vitro. *Int. J. Mol. Sci.* **2020**, *21*, 8716. [CrossRef]
76. Huang, L.; Szewczyk, G.; Sarna, T.; Hamblin, M.R. Potassium Iodide Potentiates Broad-Spectrum Antimicrobial Photodynamic Inactivation Using Photofrin. *ACS Infect. Dis.* **2017**, *3*, 320–328. [CrossRef]
77. Pankey, G.A.; Sabath, L.D. Clinical Relevance of Bacteriostatic versus Bactericidal Mechanisms of Action in the Treatment of Gram-Positive Bacterial Infections. *Clin. Infect. Dis.* **2004**, *38*, 864–870. [CrossRef]
78. Safar Sajadi, S.M.; Khoee, S. The simultaneous role of porphyrins' H- and J- aggregates and host–guest chemistry on the fabrication of reversible Dextran-PMMA polymersome. *Sci. Rep.* **2021**, *11*, 2832. [CrossRef]
79. Huang, L.; Bhayana, B.; Xuan, W.; Sanchez, R.P.; McCulloch, B.J.; Lalwani, S.; Hamblin, M.R. Comparison of two functionalized fullerenes for antimicrobial photodynamic inactivation: Potentiation by potassium iodide and photochemical mechanisms. *J. Photochem. Photobiol. B Biol.* **2018**, *186*, 197–206. [CrossRef]

Disclaimer/Publisher's Note: The statements, opinions and data contained in all publications are solely those of the individual author(s) and contributor(s) and not of MDPI and/or the editor(s). MDPI and/or the editor(s) disclaim responsibility for any injury to people or property resulting from any ideas, methods, instructions or products referred to in the content.

Article

Interaction of Some Asymmetrical Porphyrins with U937 Cell Membranes–In Vitro and In Silico Studies

Dragos Paul Mihai [1], Rica Boscencu [1,*], Gina Manda [2], Andreea Mihaela Burloiu [1,*], Georgiana Vasiliu [1], Ionela Victoria Neagoe [2], Radu Petre Socoteanu [3,*] and Dumitru Lupuliasa [1]

1. Faculty of Pharmacy, "Carol Davila" University of Medicine and Pharmacy, 6 Traian Vuia St., 020956 Bucharest, Romania
2. "Victor Babeș" National Institute of Pathology, 99-101 Splaiul Independentei, 050096 Bucharest, Romania
3. "Ilie Murgulescu" Institute of Physical Chemistry, Romanian Academy, 202 Splaiul Independentei, 060021 Bucharest, Romania
* Correspondence: rica.boscencu@umfcd.ro (R.B.); andreea-mihaela.burloiu@drd.umfcd (A.M.B.); psradu@yahoo.com (R.P.S.)

Abstract: The aim of the present study was to assess the effects exerted in vitro by three asymmetrical porphyrins (5-(2-hydroxyphenyl)-10,15,20-tris-(4-acetoxy-3-methoxyphenyl)porphyrin, 5-(2-hydroxyphenyl)-10,15,20-tris-(4-acetoxy-3-methoxyphenyl)porphyrinatozinc(II), and 5-(2-hydroxyphenyl)-10,15,20-tris-(4-acetoxy-3-methoxyphenyl)porphyrinatocopper(II)) on the transmembrane potential and the membrane anisotropy of U937 cell lines, using bis-(1,3-dibutylbarbituric acid)trimethine oxonol (DiBAC4(3)) and 1-(4-trimethylammoniumphenyl)-6-phenyl-1,3,5-hexatriene p-toluenesulfonate (TMA-DPH), respectively, as fluorescent probes for fluorescence spectrophotometry. The results indicate the hyperpolarizing effect of porphyrins in the concentration range of 0.5, 5, and 50 µM on the membrane of human U937 monocytic cells. Moreover, the tested porphyrins were shown to increase membrane anisotropy. Altogether, the results evidence the interaction of asymmetrical porphyrins with the membrane of U937 cells, with potential consequences on cellular homeostasis. Molecular docking simulations, and Molecular mechanics Poisson–Boltzmann surface area (MM/PBSA) free energy of binding calculations, supported the hypothesis that the investigated porphyrinic compounds could potentially bind to membrane proteins, with a critical role in regulating the transmembrane potential. Thus, both the free base porphyrins and the metalloporphyrins could bind to the SERCA2b (sarco/endoplasmic reticulum ATPase isoform 2b) calcium pump, while the metal complexes may specifically interact and modulate calcium-dependent (large conductance calcium-activated potassium channel, Slo1/KCa1.1), and ATP-sensitive (K_{ATP}), potassium channels. Further studies are required to investigate these interactions and their impact on cellular homeostasis and functionality.

Keywords: asymmetrical porphyrins; human U937 monocytic cells; transmembrane potential; membrane anisotropy; sarco/endoplasmic reticulum Ca^{2+}-ATPase (SERCA2b); Slo1; SUR2; molecular docking

Citation: Mihai, D.P.; Boscencu, R.; Manda, G.; Burloiu, A.M.; Vasiliu, G.; Neagoe, I.V.; Socoteanu, R.P.; Lupuliasa, D. Interaction of Some Asymmetrical Porphyrins with U937 Cell Membranes–In Vitro and In Silico Studies. *Molecules* 2023, 28, 1640. https://doi.org/10.3390/molecules28041640

Academic Editors: Carlos J. P. Monteiro, M. Amparo F. Faustino and Carlos Serpa

Received: 29 December 2022
Revised: 4 February 2023
Accepted: 7 February 2023
Published: 8 February 2023

Copyright: © 2023 by the authors. Licensee MDPI, Basel, Switzerland. This article is an open access article distributed under the terms and conditions of the Creative Commons Attribution (CC BY) license (https://creativecommons.org/licenses/by/4.0/).

1. Introduction

Porphyrins are tetrapyrrolic structures that have been extensively studied, especially for their applications as photosensitizers (PS) for photodynamic therapy (PDT) of malignant tumors [1,2].

The selectivity for tumor tissues manifested by porphyrinic photosensitizers, as well as the use of visible light for their activation, implying lower energies compared to those utilized in radiotherapy, are arguments that recommend PDT as an efficient alternative treatment to chemotherapy and radiotherapy, with a good potential for the management of some tumors. Another key advantage of PDT consists in the fact that the method can be applied as a unique treatment procedure as well as associated with other therapeutic approaches, such as chemotherapy or radiotherapy [3–5]. The therapeutic effect of porphyrins is determined by the ability of this structural type to accumulate preferentially in the malignant tissue and to generate, in the presence of light and molecular oxygen, reactive oxygen species (ROS), which are able to damage tumor cells [6,7]. Moreover, the structural and spectral profiles of porphyrins allows their use as non-invasive diagnosis devices in the detection of tumor cells [4,5]. One of the advantages of the simultaneous use of porphyrins in both therapeutic and diagnosis purposes, is the fact that once internalized in the tumor cell, PS becomes an important indicator in monitoring antitumor treatment, the fluorescence signal decreasing with the damage to the tumor cells [3].

Although a series of pharmaceutical forms containing a tetrapyrrole compound as the active substance (Photofrin®, Foscan®, Purlytin®, Radachlorin®) have been approved and clinically used, there are some disadvantages for each case, that limit their therapeutic effects [3–6]. These disadvantages are mainly related to an insufficient uptake by tumor cells, due to their structural profile having a low ratio between the hydrophilic and hydrophobic substituents. Cellular internalization of porphyrins depends not only on their hydrophobic/hydrophilic balance, but also on other factors such as the presence of a metallic ion in the core of the porphyrin ring, the distribution charge on the molecule, its aggregation state, etc. [2,4].

Studies of interactions with the cell membrane can be very useful in highlighting some strategies for designing, and optimizing the potential of, cellular internalization of novel theranostic agents [8]. The transmembrane potential and membrane fluidity are among the most important biophysical properties of the cell membrane, with important impacts on cellular homeostasis. Therefore, studies on its changes in the presence of molecules with therapeutic potential are of the utmost importance in biomedical research [9–11].

Membrane potential changes can be evaluated using *bis*-(1,3-dibutylbarbituric acid) trimethine oxonol (DiBAC4 (3), a fluorescent voltage-sensitive probe that binds to the intracellular or inner cell membrane proteins, leading to an increase of the fluorescent signal. Membrane depolarization generates an influx of the probe into the cell, with an increase of the emission intensity, while hyperpolarization of the membrane reduces the flow of the probe into the cell, and the fluorescent signal is reduced [12].

Membrane fluidity measurements use fluorescence determinations in polarized light of compounds belonging to the class of diphenylhexatriene, (e.g. 1-(4-trimethylammoniumphenyl)-6–phenyl-1,3,5-hexatriene p-toluenesulfonate (TMA-DPH)), as fluorescent probes.

Considering the fact that anisotropy is inversely correlated with membrane fluidity, the assessment of anisotropy proves to be a useful approach for obtaining information regarding the permeability of membranes for therapeutically active substances. By decreasing membrane fluidity, the membrane permeability for a drug is expected to decrease [12], but partial immobilization of drugs at the membrane level may also support their interaction with membrane proteins, hence reinforcing their therapeutic efficacy.

Taking into account that cellular internalization of photosensitizers is directly influenced by their structural profile and interactions with the cell membrane, our research has been focused on obtaining and characterizing some porphyrins with various degrees of hydrophobic/hydrophilic substitutions, that favor their cellular uptake [13–16].

In this paper we investigated three asymmetrical porphyrins: 5-(2-hydroxyphenyl)-10,15,20-tris-(4-acetoxy-3-methoxyphenyl)porphyrin, (TMAPOHo); 5-(2-hydroxyphenyl)-10,15,20-tris-(4-acetoxy-3-methoxyphenyl)porphyrinatozinc(II), (Zn(II)TMAPOHo); and 5-(2-hydroxyphenyl)-10,15,20–tris-(4-acetoxy-3-methoxyphenyl)porphyrinatocopper(II), (Cu(II)TMAPOHo) (Figure 1), from the point of view of their interactions with the cell membrane.

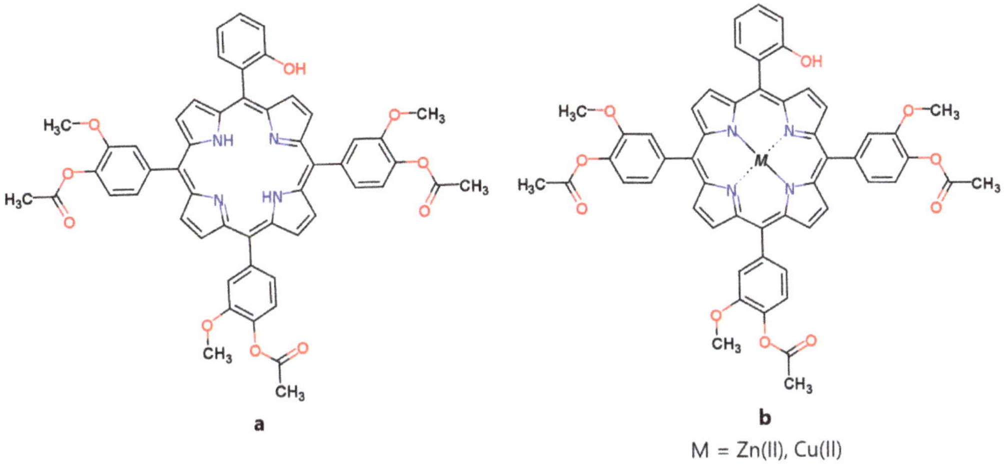

Figure 1. The molecular structures of the investigated porphyrins. (**a**) 5-(2-hydroxyphenyl)-10,15,20-tris-(4-acetoxy-3-methoxyphenyl) porphyrin, (**b**) 5-(2-hydroxyphenyl)-10,15,20-tris-(4-acetoxy-3-methoxyphenyl)porphyrinatozinc(II), 5-(2-hydroxyphenyl)-10,15,20–tris-(4-acetoxy-3-methoxyphenyl)porphyrinatocopper(II).

The experimental study was performed on the U937 human monocytic cell line deriving from histiocytic lymphoma. Moreover, we used in silico methods to predict the permeability across the cell membrane and the impact on membrane fluidity.

2. Results

2.1. Asymmetric Porphyrins Induce Cell Membrane Hyperpolarization and Anysotropy

The action of the asymmetric porphyrinic compounds on the transmembrane potential and membrane anisotropy of U937 cells was investigated in vitro. The rationale for using this cell line was the fact that the investigated porphyrinic compounds are designed for intravenous administration, and that phagocytic cells are among the first blood cells to capture and internalize photosensitizers [17]. Along with granulocytes, monocytes can be physiological carriers of photosensitizers into tumors, and this may increase the PDT efficacy [18,19]. Moreover, the accumulation of the photosensitizer in tumor-associated macrophages (TAM) is also expected to sustain PDT, by modulating the local anti-tumor immune response [20].

In order to confirm the penetration of the potential-sensitive probe DiBAC4(3) into the cell membrane, we assessed the fluorescence spectra of U937 cells with and without this probe. DiBAC4(3) produced an approx. 100 times increase of the fluorescent signal of the cells, indicating that the probe penetrated into the cell membrane. In the case of U937 cells incubated for 24 h with porphyrins, the fluorescent signal of DiBAC4(3) was lower than in the untreated control samples, indicating a membrane hyperpolarization effect for all the tested compounds and at all tested concentrations. The results suggest the modulation of the voltage-gated membrane channels by the porphyrinic compounds, with potential consequences on ion exchange and cellular homeostasis.

For the porphyrinic ligand TMAPOHo, the experimental results showed a hyperpolarizing effect on the U937 cell membrane of approx. 10% (Figure 2b), at all the tested concentrations (0.5 µM, 5 µM, and 50 µM).

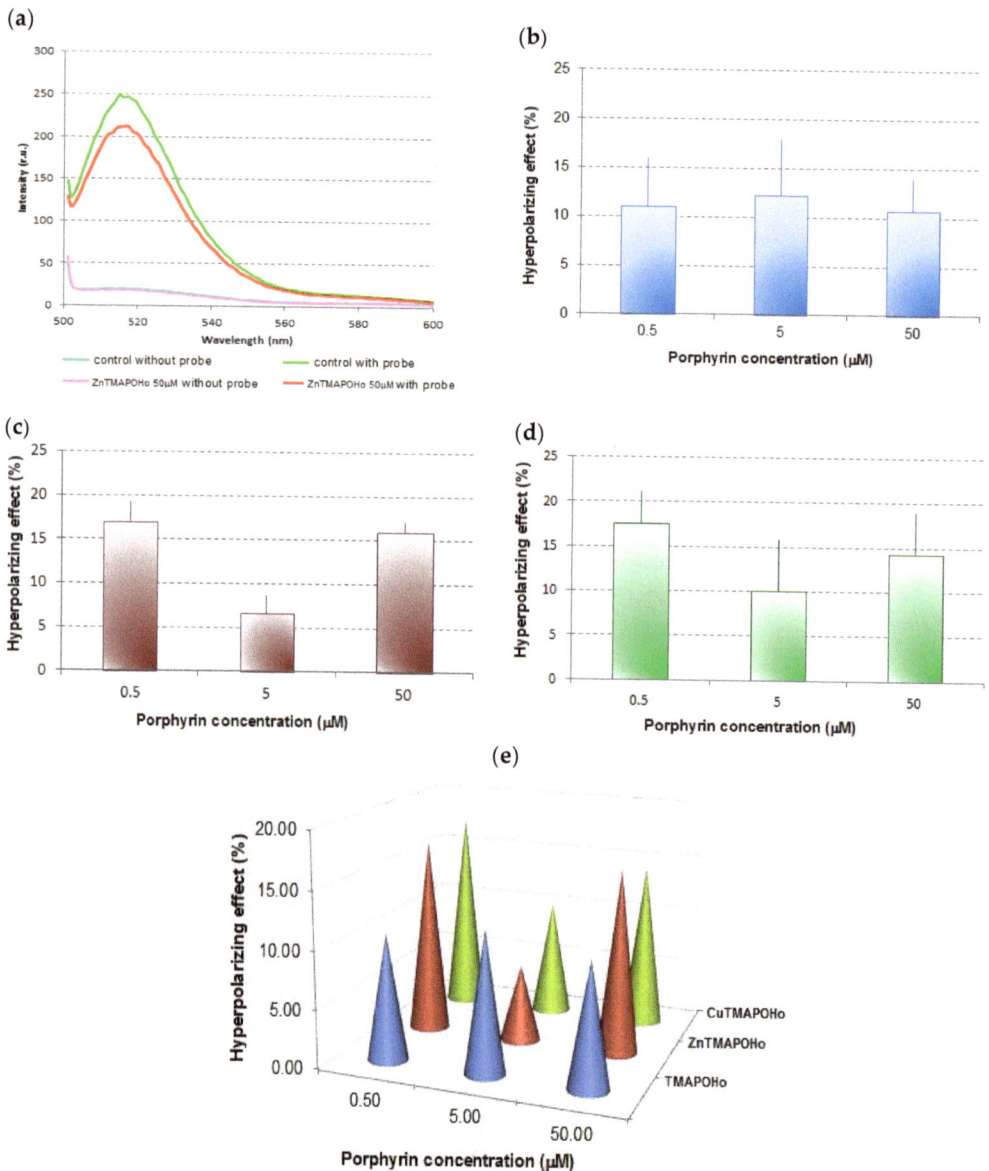

Figure 2. Representative effects of porphyrinic compounds on the transmembrane potential of U937 cells. (**a**)—Fluorescence spectra of U937 cells, with and without the potential-sensitive probe DiBAC$_4$(3), when Zn(II)TMAPOHo was tested (—— control without probe, —— control with probe, —— 50 µM ZnTMAPOHo without probe, —— 50 µM ZnTMAPOHo with probe); (**b**)—Hyperpolarizing effect for TMAPOHo; (**c**)—Hyperpolarizing effect for Zn(II)TMAPOHo; (**d**)—Hyperpolarizing effect for Cu(II)TMAPOHo; (**e**)—Comparative effect of porphyrinic compounds (24 h incubation time and 0.5 µM, 5 µM, 50 µM concentrations) on the transmembrane potential of U937 cells. In Figure 2b,c,d results are presented as mean hyperpolarizing effect (%) ± standard deviation (SD) for triplicate samples.

The hyperpolarizing effects induced by Zn(II)TMAPOHo on U937 cell membranes (Figure 2c) was more pronounced at low and high concentrations (5 µM and 50 µM), suggesting a potential biphasic action.

The Cu(II)TMAPOHo compound, having a low fluorescence compared to the free base porphyrin TMAPOHo, had a behavior at the membrane level similar to that of Zn(II)TMAPOHo (Figure 2d).

A relatively constant effect for the free base porphyrin TMAPOHo, regardless of the dose, is highlighted in Figure 2b, but, complex combinations of the assessed porphyrin with zinc and copper exerted a dose-dependent hyperpolarizing action, suggesting distinctive mechanisms of action (Figure 2e).

When using TMA-DPH, an increase in the intensity of the response signal was observed, indicating the internalization of this probe into the cell membrane (Figure 3a). The signals were different for the four relative positions of the polarizers (Figure 3b).

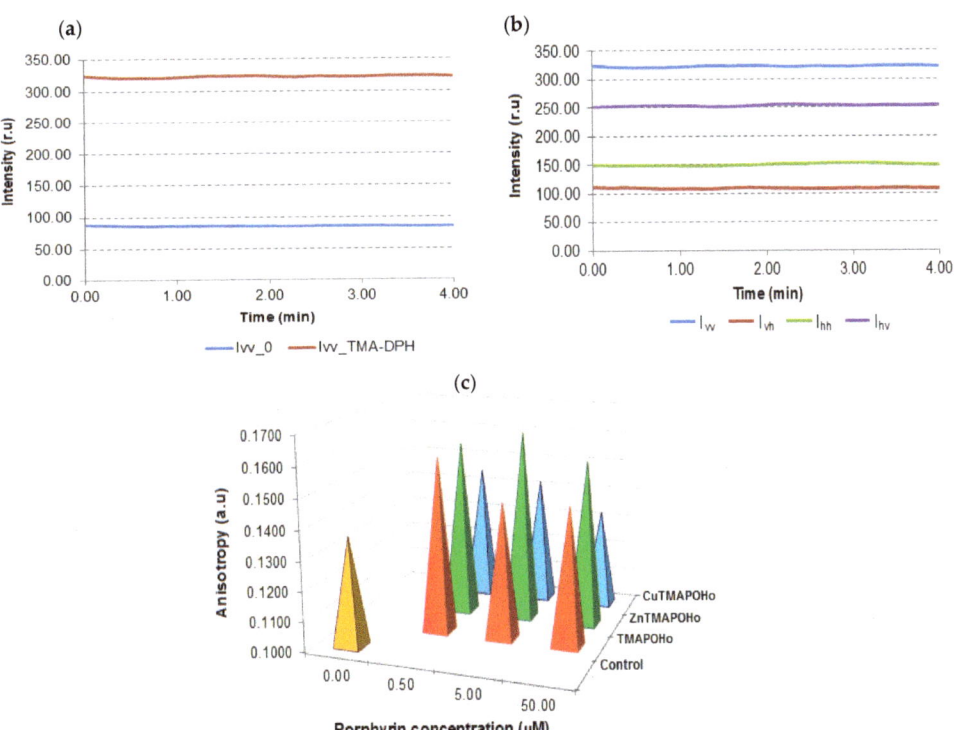

Figure 3. Investigation of the effects of porphyrinic compounds on membrane anisotropy. (a)—Intensity of signals obtained in Time Drive mode for U937 cells labelled with TMA-DPH vs. unlabeled cells; (b)—Intensity of response signals obtained in Time Drive mode for TMA-DPH-labelled cells; (c)—The effect of porphyrinic compounds (24 h incubation time and 0.5 µM, 5 µM, 50 µM concentrations) on the membrane anisotropy of U937 cells. Data of a representative experiments are presented.

The comparative analysis of the values of the TMA-DPH anisotropy in the cell membrane (Figure 3c), indicated an increased anisotropy in the case of porphyrin-treated cells as compared to the untreated controls.

2.2. Prediction of Permeability across the Cell Membrane

Previous studies revealed that some porphyrin derivatives can modulate ion channels by interacting with either extracellular or intracellular domains. For instance, heme was found to bind to the intracellular domains of Slo1, K_{ATP}, and Kv1.4 potassium channels [21–23], while a positively charged porphyrin derivative blocked the pores of neuronal Kv1 channels, possibly by binding to the extracellular domains [24]. We herein investigated, through an in silico method, the passive translocation of the three porphyrin derivatives across the cell membrane. All three compounds showed a high potential for permeating the cell membrane, since positive values for their intrinsic permeability coefficients through bilayer membranes (logP_{BLM}) were obtained. The highest logP_{BLM} value was observed for the free base porphyrin derivative (2.02), followed by the copper (1.11) and zinc (0.93) complexes. Moreover, the free energies of binding to the membrane were −8.54 kcal/mol for the free porphyrin, −7.50 kcal/mol for the copper complex, and −7.44 kcal/mol for the zinc complex, showing that the diffusion of the free porphyrin ligand was characterized by a lower transfer energy. The energy profiles along the bilayer normal are shown in Figure 4. Therefore, the assessed free base porphyrin and corresponding metalloporphyrins have the potential to translocate inside the cell and bind to various biological targets, including ion channels and pumps.

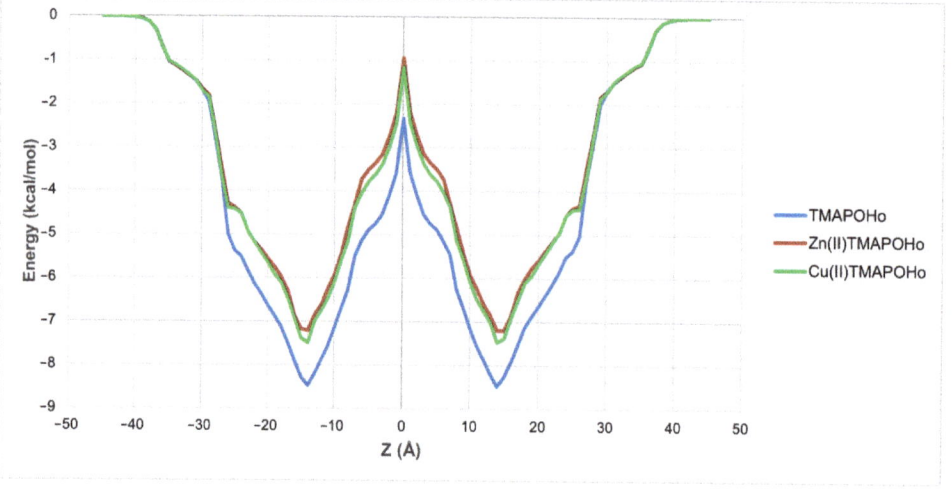

Figure 4. Variation of free energy of binding in relation to the distance from the membrane center (Z).

2.3. Prediction of Full-Length Target Structures of SERCA2b, Slo1 and SUR2

Protein structure modeling was used to generate full sequence structures of the proposed molecular targets. Models were generated with YASARA, SWISS-MODEL, and were also retrieved from the AlphaFold database. The quality parameters determined using MolProbity are shown in Table 1.

For the full-length structure of the calcium pump SERCA2b, the modeling with YASARA yielded the most qualitative structure, characterized by the lowest MolProbity score (0.94), corresponding to a low-resolution crystal structure. Moreover, the model had 97.40 residues in the most favored regions, and only 0.19% residues in disallowed regions. In the case of the Slo1 potassium channel, SWISS-MODEL and YASARA generated structures with similar quality scores. The structure generated with SWISS-MODEL had the lowest MolProbity score (1.40), 94.06% residues in the most favored regions, and 0.73% residues in disallowed regions. On the other hand, the SUR2 model predicted with AlphaFold was significantly more qualitative than the structures modelled with YASARA and SWISS-MODEL, showing a considerably lower MolProbity score (0.71), 97.87% residues in optimal regions, 0.45% residues in disallowed regions, and only 0.88% poor rotamers. Superpositions of the predicted structures on the experimentally determined templates and Ramachandran plots for the selected models are shown in Figure 5.

Table 1. Quality parameters of predicted full-length structures of target proteins.

Target	Method	MolProbity Score	Ramachandran Distribution Z-Score	Residues in Most Favored Regions (%)	Residues in Disallowed Regions (%)	Favored Rotamers (%)	Poor Rotamers (%)
SERCA2b	SWISS-MODEL	1.35	−1.61 ± 0.24	96.70	0.19	91.37	2.95
	YASARA	0.94	−1.06 ± 0.24	97.40	0.19	96.75	1.57
	AlphaFold	1.02	0.36 ± 0.25	97.98	0.48	95.74	1.57
Slo1	SWISS-MODEL	1.40	−1.59 ± 0.23	94.06	0.73	93.05	2.70
	YASARA	1.65	−1.94 ± 0.11	94.48	0.36	97.68	1.15
	AlphaFold	1.77	−1.75 ± 0.22	84.93	8.27	94.17	2.59
SUR2 (K_{ATP})	SWISS-MODEL	1.27	−1.14 ± 0.19	94.42	0.65	93.49	2.07
	YASARA	1.24	−0.70 ± 0.19	96.19	0.32	95.58	2.43
	AlphaFold	0.71	0.15 ± 0.19	97.87	0.45	97.42	0.88

SERCA2b—sarco/endoplasmic reticulum Ca^{2+}-ATPase isoform 2b; Slo1—large conductance calcium-activated potassium channel (KCa1.1); SUR2—sulfonylurea receptor isoform 2; K_{ATP}—ATP-sensitive potassium channel.

2.4. Prediction of Interaction Models between Porphyrin Derivatives and SERCA2b, Slo1 and SUR2 through Molecular Docking

The full-length structure models with the most optimal quality parameters were further used for molecular docking experiments, in order to investigate the potential mechanisms of action of the assessed porphyrin derivatives. Firstly, positive controls were docked into the binding sites of SERCA2b, Slo1, and the SUR2 subunit of K_{ATP}, as a means to validate the docking procedure. BHQ, an inhibitor of the sarco/endoplasmic reticulum Ca^{2+}-ATPase was docked into the binding pocket of the SERCA2b isoform in closed E2 state conformation, and its binding mode was compared with the experimentally determined BHQ-SERCA1 complex (PDB ID: 2AGV [25]). Similar to the experimental structure, BHQ formed a hydrogen bond with Pro308 and hydrophobic interactions with Leu61, Val62, and Pro312 (Figure 6a,b). However, the predicted conformation did not form hydrogen bonds with Asp59, possibly due to the larger distance between the hydroxyl moiety and Asp59 carboxyl group, which can be reduced in reality since the binding pocket has the potential to adapt its conformation to promote stronger interactions with the ligand. The binding energy of BHQ was −7.195 kcal/mol, the predicted Kd value being 5.321 µM. Heme was docked into the corresponding heme-binding motifs of Slo1 and the SUR2 subunit of the K_{ATP} channel. Heme successfully bound to the cytochrome c-like motif (CXXCH) in the region between the RCK1 and RCK2 domains of Slo1, forming a coordinative bond with Cys680 (Figure 6c,d). Moreover, the porphyrinic carboxyl groups formed hydrogen bonds and attractive charge interactions with the positively charged residues Lys688 and Arg689, but also an unfavorable negative-negative contact with Asp683. Interestingly, one vinyl moiety was in close contact with Cys693, showing the potential to react covalently. Previous studies have suggested that heme vinyl groups can covalently bind to cysteine residues within proteins [26]. The predicted binding energy for the heme-Slo1 interaction was −6.295 kcal/mol, with a Kd value of 24.305 µM. Heme also formed a metal bond with His651 within the $CXXHX_{16}H$ motif of the SUR2 loop, that links the first nucleotide binding domain (NBD1) and the first transmembrane domain (Figure 6e,f). The heme carboxyl moieties interacted with Arg745 through hydrogen bonding, salt bridges, and attractive charges. Moreover, one carboxyl moiety was also involved in hydrogen bonding with Thr744 and formed a carbon hydrogen bond with Glu742. Pi-sigma interactions were formed between Glu742 and one pyrrole ring, while Phe741 interacted with heme through pi-alkyl, pi-sigma, and pi-pi stacked interactions. The binding energy was −8.653 kcal/mol, and the predicted Kd was 0.454 µM.

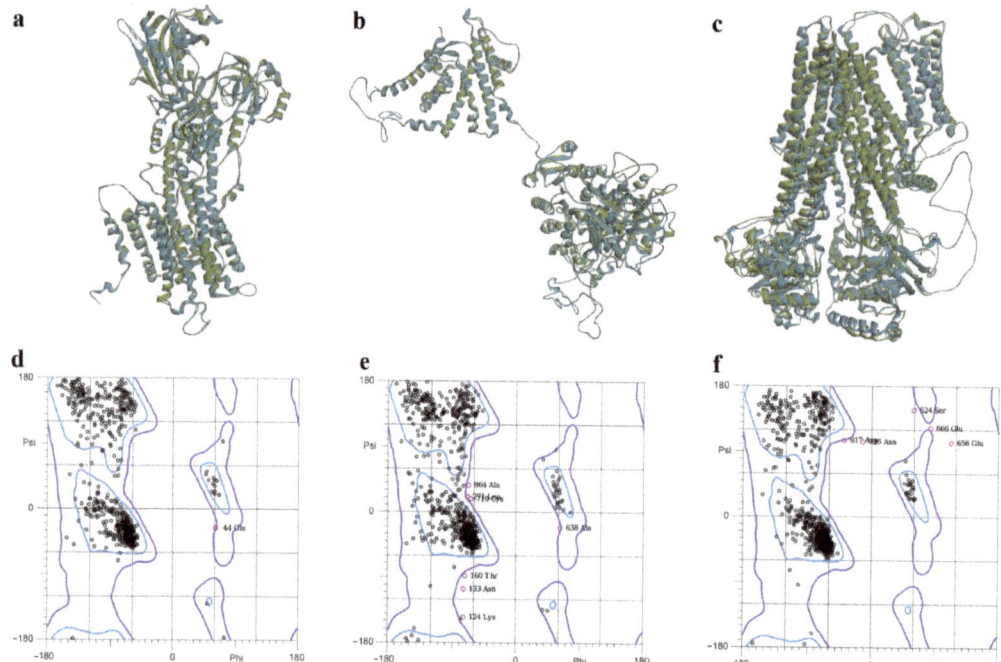

Figure 5. (**a**)—Superposition of SERCA2b model generated with YASARA (blue) on experimental (green) structure (PDB ID: 6LN9); (**b**)—superposition of Slo1 model (monomer) generated with SWISS-MODEL (blue) on experimental (green) structure (PDB ID: 6V35); (**c**)—superposition of SUR2 model generated with AlphaFold (blue) on experimental (green) structure (PDB ID: 7MIT); (**d**)—Ramachandran plot for SERCA2b model; (**e**)—Ramachandran plot for Slo1 model; (**f**)—Ramachandran plot for SUR2 model.

Further, the studied asymmetric porphyrin derivatives were docked on a binding pocket of SERCA2b that overlaps with the BHQ binding site. The lowest binding energy was observed for the free base porphyrin (−12.962 kcal/mol), followed by the copper complex (−12.917 kcal/mol), and zinc complex (−12.679 kcal/mol), the corresponding Kd values being 0.315 nM, 0.340 nM, and 0.508 nM, respectively. The free ligand formed five conventional hydrogen bonds with residues within the binding site, including Asp59, and three carbon-hydrogen bonds (Figure 7a,b). Leu61 and Pro312 were also involved in ligand binding, forming pi-sigma and pi-alkyl interactions. Moreover, pi-cation interactions were established between Arg246 and a phenyl radical, while pi-anion interactions were observed between Asp59, Asp254, and pyrrole rings. The zinc and copper metalloporphyrins formed coordinative bonds with two aspartic acid residues: Asp59 and Asp254. Both metal complexes adopted similar conformations in the binding pocket as the free ligand, forming pi-sigma interactions with Leu61. The zinc complex formed three hydrogen bonds and two carbon-hydrogen bonds (Figure 7c,d), while the copper metalloporphyrin formed three hydrogen bonds and three carbon-hydrogen bonds (Figure 7e,f). The docking results indicate that both the free base porphyrin derivative and its corresponding metalloporphyrins have the potential to inhibit SERCA2b by interacting with the same binding site. The free porphyrin can form hydrogen bonds with two aspartic acid residues through the two protonated pyrrole rings, while the metal complexes can form metal bonds with the same residues.

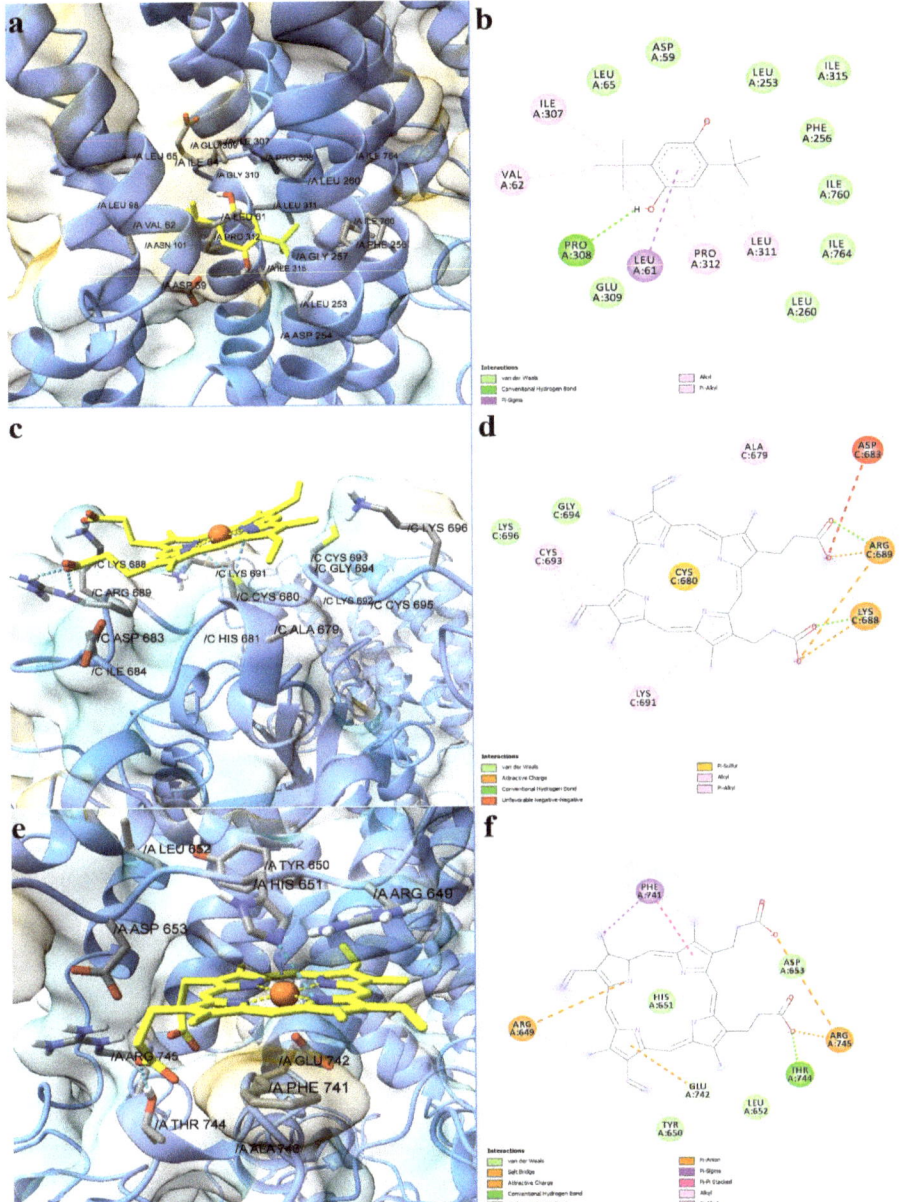

Figure 6. Predicted binding modes between positive controls and full-length structures of SERCA2b, Slo1, and the SUR2 subunit of K_{ATP}. (**a**)—predicted conformation of BHQ-SERCA2b complex; (**b**)—interaction diagram for BHQ-SERCA2b complex; (**c**)—predicted conformation of heme-Slo1 complex; (**d**)—interaction diagram for heme-Slo1 complex; (**e**)—predicted conformation of heme-SUR2 complex; (**f**)—interaction diagram for heme-SUR2 complex. Metal coordination bonds are represented as purple dashes.

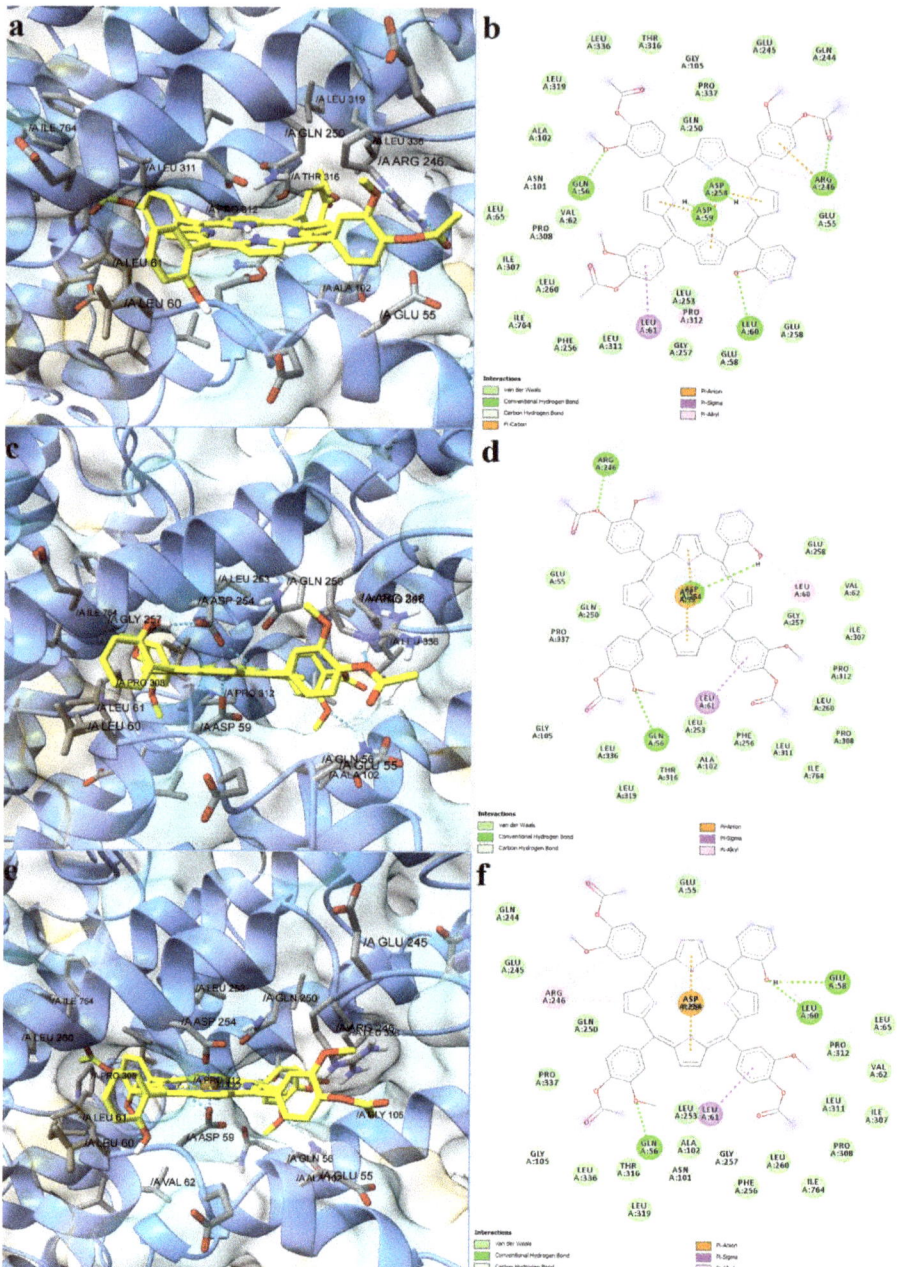

Figure 7. Predicted binding modes between porphyrin derivatives and SERCA2b. (**a**)—predicted conformation of TMAPOHo-SERCA2b complex; (**b**)—interaction diagram for TMAPOHo-SERCA2b complex; (**c**)—predicted conformation of Zn(II)TMAPOHo-SERCA2b complex; (**d**)—interaction diagram for Zn(II)TMAPOHo-SERCA2b complex; (**e**)—predicted conformation of Cu(II)TMAPOHo-SERCA2b complex; (**f**)—interaction diagram for Cu(II)TMAPOHo-SERCA2b complex. Metal coordination bonds are represented as purple dashes.

The three asymmetric porphyrin derivatives were docked into the region corresponding to the CXXCH heme binding motif of Slo1. The free porphyrinic ligand formed three hydrogen bonds with Cys680, Arg689, and Gly694, and a carbon-hydrogen bond with Cys693 (Figure 8a,b). Pi-anion interactions were formed with Asp683, and pi-sulfur interactions with the porphyrin ring. However, it is highly unlikely that the free porphyrin derivative would bind to a heme-binding domain, since its structure lacks a bivalent metal that would engage in stable interactions with Cys680. On the other hand, both zinc and copper metalloporphyrins formed metal bonds with Cys680. The zinc complex formed hydrogen bonds with Asp683, Arg689, and Gly698, and a carbon-hydrogen bond with Cys693 (Figure 8c,d). Moreover, the protein-ligand complex is further stabilized by pi-anion, pi-cation, and pi-alkyl interactions between amino acid residues and pyrrole and phenyl rings. The copper bound porphyrin formed hydrogen bonds with Cys680, Asp682, Asp683, and Arg689, and carbon-hydrogen bonds with Cys693 and Gly694. Pi-anion and pi-alkyl interactions were also present, similar to the zinc metalloporphyrin (Figure 9e,f). The predicted binding energies were -7.060 kcal/mol for the free porphyrin (6.683 µM), -6.630 kcal/mol for the zinc complex (13.808 µM), and -6.4390 kcal/mol for the copper complex (19.061 µM).

Lastly, we simulated the interaction between the porphyrin derivatives and residues within the $CXXHX_{16}H$ heme binding motif of SUR2. For the free porphyrin, the binding energy was -8.621 kcal/mol, the predicted Kd being 0.479 µM. The zinc metalloporphyrin had a binding energy of -7.995 kcal/mol (1.379 µM), while the copper complex had a binding energy of -8.152 kcal/mol (1.058 µM). The free porphyrin ligand formed three hydrogen bonds with Arg659 and His651, pi-anion and pi-cation interactions with Arg745 and Glu743, and pi-pi stacked interactions between Phe741 and three pyrrole rings (Figure 9a,b). In the case of the metalloporphyrins, metal coordination bonds were formed with His651, and pi-pi stacked interactions were formed between Phe741 and all four pyrrole rings. The zinc complex was involved in hydrogen bonding with three residues (Arg649, Leu652, Arg745) and formed a carbon-hydrogen bond with Glu656 (Figure 9c,d). Moreover, pi-cation and pi-anion interactions were formed with Arg745 and Glu742. Four hydrogen bonds and two carbon-hydrogen bonds were formed between the copper complex and SUR2 (Arg649, Arg745, Leu652, Gln657, Glu656). Furthermore, the complex was further stabilized by two pi-cation interactions with Arg649 and Arg745, and one pi-anion interaction with Glu742 (Figure 9e,f). Interestingly, the hydrophobic interactions between Phe741 and the docked ligands were stronger in the case of metalloporphyrins.

Further, we analyzed the electrostatic potential maps of the binding sites for all three target proteins, exemplified for the predicted complexes with the zinc metalloporphyrin derivative. For SERCA2b it can be noted that the binding site is characterized by mostly low electrostatic potential energy values, which is correlated with a large number of negatively charged residues (aspartate and glutamate) with key roles in calcium binding and transport. The liganded zinc is complexed by aspartate residues, the porphyrin ring and the hydroxyl moiety within the porphyrinic structure are involved in electrostatic interactions with the same negatively charged residues. On the other hand, the phenyl rings make contacts with more neutral residues, through non-electrostatic interactions (Figure 10a). The electrostatic potential energy values within the core of the binding site of Slo1 are closer to neutrality, since zinc is complexed by a cysteine. In this case, the porphyrin ring is involved in non-electrostatic interactions with the binding site, such as pi interactions, while the substituted phenyl rings bind to positively and negatively charged residues through both electrostatic and non-electrostatic interactions (Figure 10b). In the case of SUR2, the porphyrin ring and metal atom interact with residues characterized by higher electrostatic potential energy values (histidine), while the substituted phenyls bind with other residues through both electrostatic and non-electrostatic interactions (Figure 10c). Therefore, the three binding pockets have different properties in terms of electrostatic potentials, and the porphyrinic structures can interact with the binding sites via various types of molecular interactions.

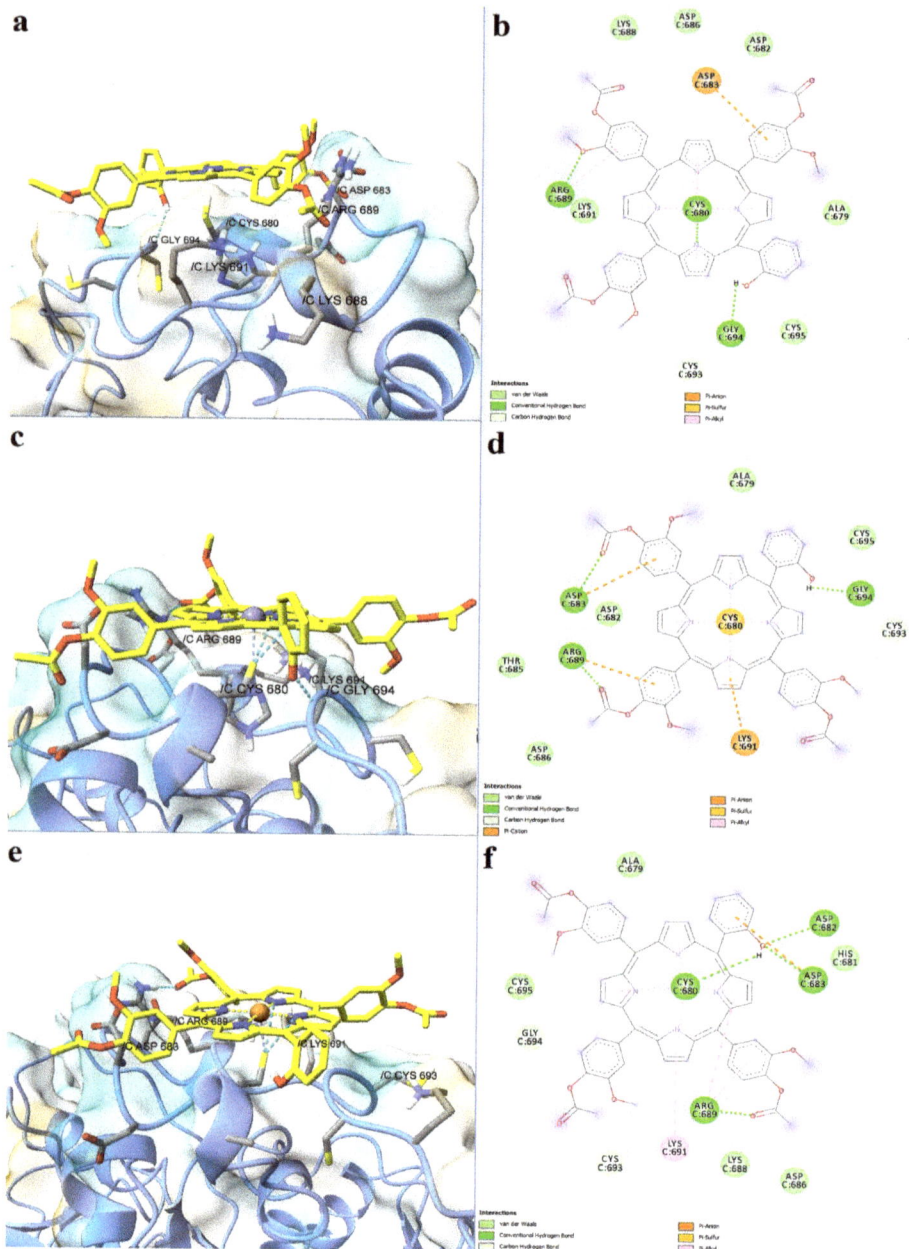

Figure 8. Predicted binding modes between porphyrin derivatives and Slo1. (**a**)—predicted conformation of TMAPOHo-Slo1 complex; (**b**)—interaction diagram for TMAPOHo-Slo1 complex; (**c**)—predicted conformation of Zn(II)TMAPOHo-Slo1 complex; (**d**)—interaction diagram for Zn(II)TMAPOHo-Slo1 complex; (**e**)—predicted conformation of Cu(II)TMAPOHo-Slo1 complex; (**f**)—interaction diagram for Cu(II)TMAPOHo-Slo1 complex. Metal coordination bonds are represented as purple dashes.

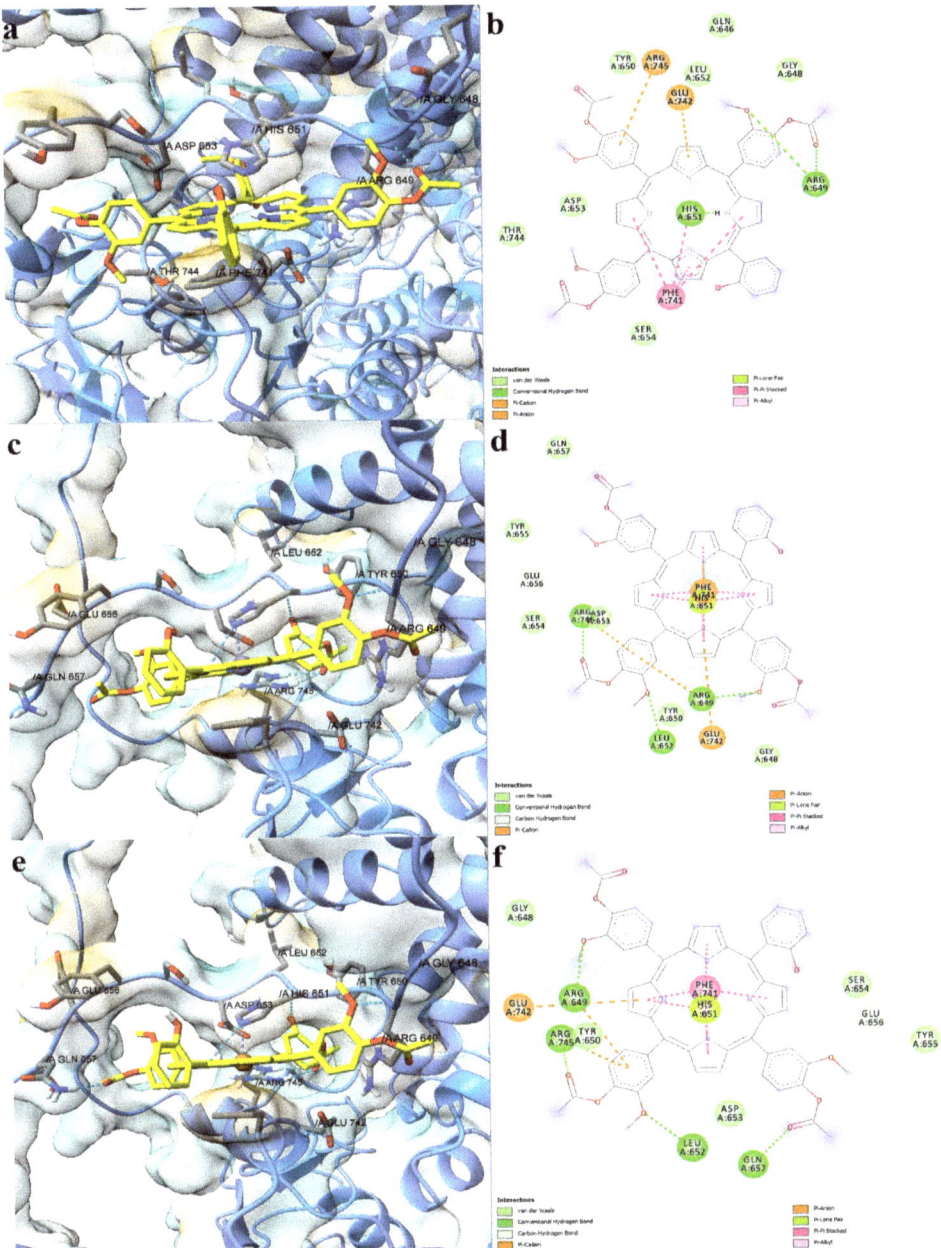

Figure 9. Predicted binding modes between porphyrin derivatives and the SUR2 subunit of K_{ATP}. (**a**)—predicted conformation of TMAPOHo-SUR2 complex; (**b**)—interaction diagram for TMAPOHo-SUR2 complex; (**c**)—predicted conformation of Zn(II)TMAPOHo-SUR2 complex; (**d**)—interaction diagram for Zn(II)TMAPOHo-SUR2 complex; (**e**)—predicted conformation of Cu(II)TMAPOHo-SUR2 complex; (**f**)—interaction diagram for Cu(II)TMAPOHo-SUR2 complex. Metal coordination bonds are represented as purple dashes.

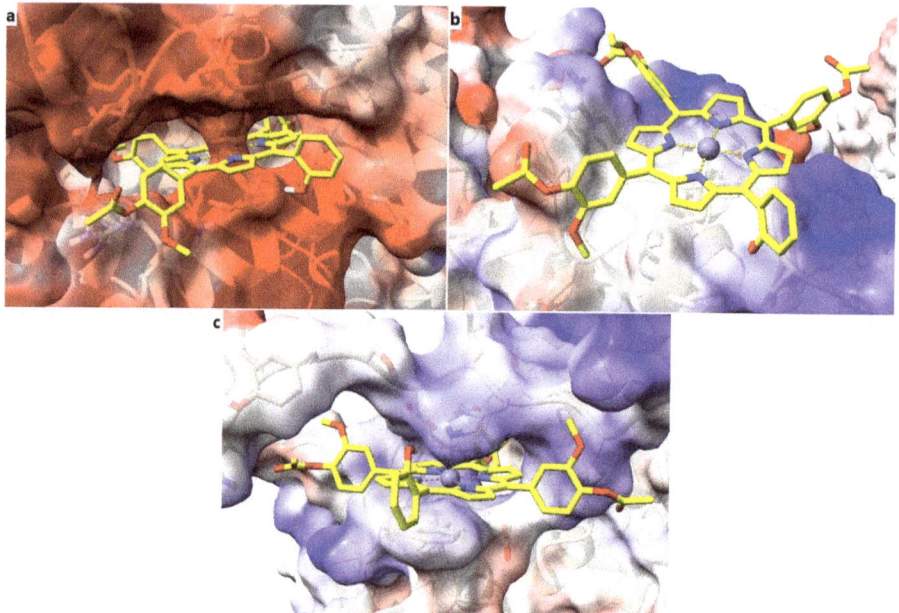

Figure 10. Electrostatic potential maps of the binding sites in complex with Zn(II)TMAPOHo. (**a**) – binding site of SERCA2b; (**b**) – binding site of Slo1; (**c**) – binding site of SUR2.

2.5. MM/PBSA Binding Free Energy

Short MD simulations were carried out to estimate the free energy of binding of docked porphyrins, using the MM/PBSA method. Firstly, phospholipid bilayers representing the cell/endoplasmic membranes were inserted, and a 250 ps simulation was performed to equilibrate the systems. Thereafter, another simulation of 250 ps was carried out to calculate the binding free energies, to assess the stability of the protein-ligand complexes. The docked conformations of the positive controls were also simulated. Both the free porphyrin and metal complexes showed lower energy values than BHQ, highlighting the potential of the assessed porphyrins to form stable complexes with SERCA2b (Table 2). Moreover, both metalloporphyrins had higher affinities for SERCA2b than the free base porphyrin, this observation being in accordance with the experimental results, which indicated higher hyperpolarizing effects for the metalloporphyrins.

Interestingly, the free base porphyrin showed positive values for the binding free energies after simulating its complexes with Slo1 and SUR2, which supports the hypothesis that only metalloporphyrins interact with these targets. For both metalloporphyrins, the highest affinity was observed after the interaction with SERCA2b, followed by Slo1 and SUR2. Heme had an affinity for Slo1 lower than the copper complex, but higher than the zinc metalloporphyrin. On the other hand, the zinc complex showed a higher affinity than heme for SUR2, while the copper complex had higher values for the binding free energy than heme. For the copper metalloporphyrin, the predicted results can be highly correlated with the experimental data, suggesting that at lower concentrations, the metal complex may inhibit SERCA2b, leading to an elevation of cytosolic calcium and activation of calcium-dependent Slo1 potassium channels. At intermediate concentrations, the copper complex could bind to Slo1 and inhibit its activity, reducing the hyperpolarizing effect, while at high concentrations it might bind to the SUR2 subunit of the K_{ATP} channels and further stimulate outward potassium currents. The same hypothesis can be taken into consideration for the zinc metalloporphyrin, although only small differences in binding energies were recorded between the interactions with Slo1 and SUR2. Since the zinc metalloporphyrin diminished the hyperpolarizing effect more efficiently than the copper complex at intermediate concentrations, we can assume that the binding free energy for the interaction between Slo1 and the zinc complex was underestimated. Superpositions of simulated protein-ligand complexes on the original docked conformations are exemplified in Figure 11a–c for the copper metalloporphyrin.

Table 2. Free energy of binding for positive controls and assessed porphyrins after 500 ps MD simulations.

Ligand	Free Energy of Binding (kcal/mol)		
	SERCA2b	Slo1	SUR2
BHQ	−48.892	-	-
Heme	-	−72.996	−42.346
TMAPOHo	−103.677	37.164	31.457
Zn(II)TMAPOHo	−150.943	−68.181	−64.116
Cu(II)TMAPOHo	−157.041	−90.028	−30.083

BHQ (2,5-di-tert-butylbenzene-1,4-diol); TMAPOHo (5-(2-hydroxyphenyl)-10,15,20-tris-(4-acetoxy-3-methoxyphenyl)porphyrin); Zn(II)TMAPOHo (5-(2-hydroxyphenyl)-10,15,20-tris-(4-acetoxy-3-methoxyphenyl)porphyrinatozinc(II)); Cu(II)TMAPOHo (5-(2-hydroxyphenyl)-10,15,20-tris-(4-acetoxy-3-methoxyphenyl)porphyrinatocopper(II)); SERCA2b (sarco/endoplasmic reticulum Ca^{2+}-ATPase isoform 2b); Slo1 (large conductance calcium-activated potassium channel (KCa1.1); SUR2 (sulfonylurea receptor isoform 2).

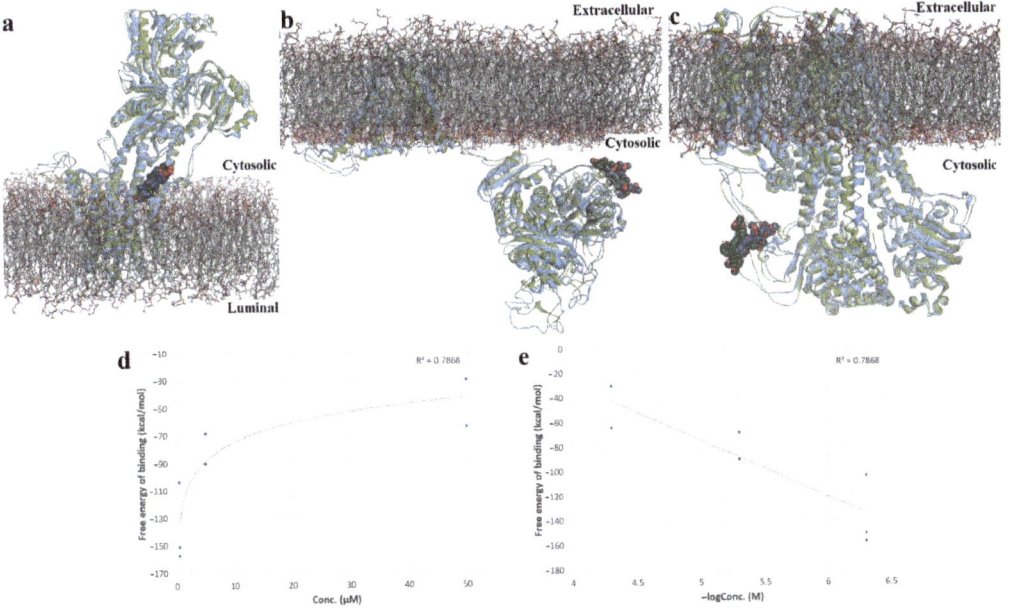

Figure 11. (a)—Superposition of simulated conformation (green) of Cu(II)TMAPOHo-SERCA2b complex on original structure (blue); (b)—superposition of simulated conformation (green) of Cu(II)TMAPOHo-Slo1 complex on original structure (blue); (c)—superposition of simulated conformation (green) of Cu(II)TMAPOHo-SUR2 complex on original structure (blue); (d)—exponential relationship between free binding energy values and tested porphyrin concentrations; (e)—correlation diagram between free binding energies and negative logarithmic values of tested porphyrin concentrations. Solvent molecules are hidden for clarity.

The Pearson statistical test was applied, to assess the correlation between the estimated free binding energies and tested porphyrin concentrations. As observed in Figure 11, a good, statistically significant negative correlation was obtained between the predicted free energy of binding values and the used concentrations (R^2 = 0.7868, p = 0.0078), further supporting our hypothesized mechanism of action (Figure 11d,e).

The root mean square fluctuations (RMSF) of amino acid residues were also analyzed, following the short MD simulations. For SERCA2b, lower fluctuations were observed in the amino acid sequence located within the binding site (residues 55–61, 244–260), when compared to the apo state protein simulation (Figure 12a). However, higher conformational changes could be seen in other regions. Moreover, the structural conformations of both Slo1 and SUR2 were more stable when bound to heme and the other two metalloporphyrins (Figure 12b,c). Lower fluctuations were noticed for residues 679–695, located within the Slo1 binding site. The same observation could be made for the SUR2 binding site, since lower RMSF values were recorded for residues 649–656 and 741–745.

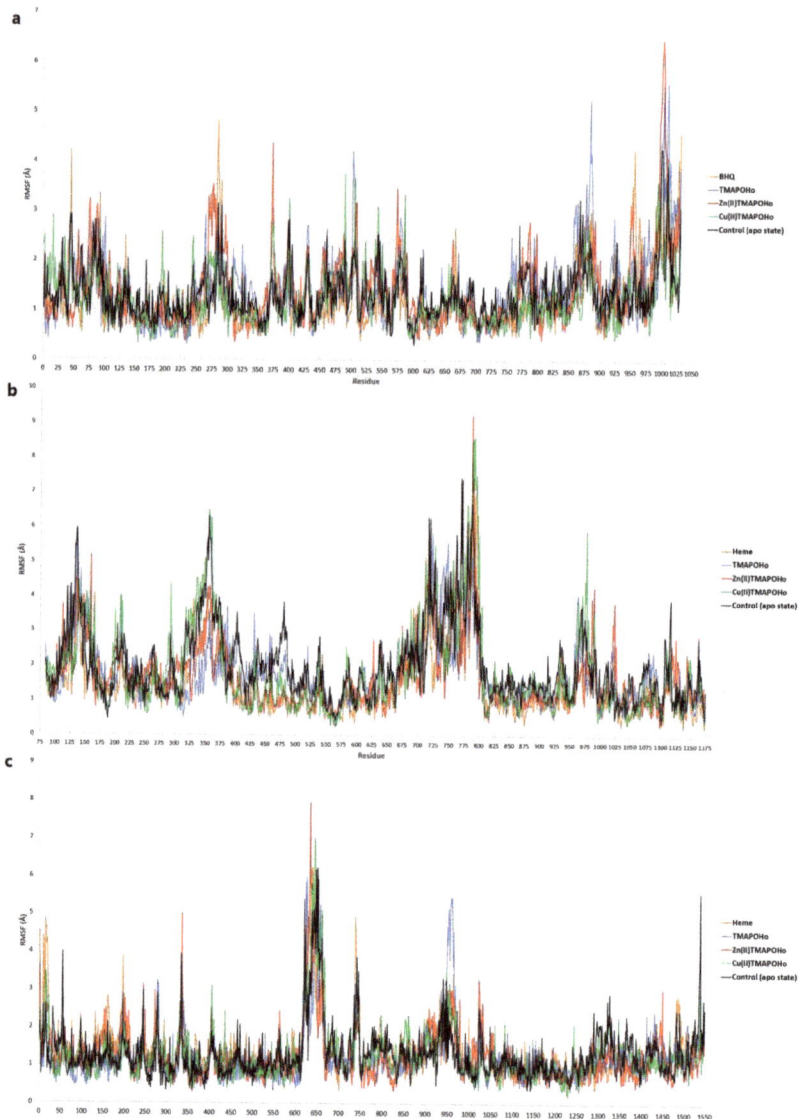

Figure 12. RMSF values per amino acid residue after short MD simulations. (**a**) – SERCA2b; (**b**) – Slo1; (**c**) – SUR2.

3. Materials and Methods

3.1. Evaluation of Transmembrane Potential and Membrane Anisotropy

Porphyrins were obtained according to the previously described procedures [27,28]. Commercially available chemicals and solvents were purchased from Sigma-Aldrich and Merck, both from Germany. Porphyrin stock solutions (50 mM), in dimethyl sulfoxide (DMSO), were kept until use, at room temperature and in the dark, for preventing photodegradation. The final dilution of DMSO in the cell culture was below 1/1000.

For assessing membrane anisotropy, 1-(4-trimethylammoniumphenyl)-6-phenyl-1, 3,5-hexatriene p-toluenesulfonate (TMA-DPH, Molecular Probes, Thermo Fisher Scientific, Waltham, MA, USA) was used as the fluorescence probe. Bis-(1,3-dibutylbarbituric acid)trimethine oxonol (DiBAC4(3), Molecular Probes ThermoFisher Scientific, Waltham, MA, USA) was used as a fluorescence probe for the evaluation of the transmembrane potential changes. The stock solutions (2 mM DiBAC4(3) and 2.5 mM TMA-DPH) in DMSO were kept at $-20\,°C$ until use.

The human pro-monocytic model cell line U937 (CRL-1593.2), deriving from hystiocytic lymphoma, was used for the experiments [29]. The U937 cells (3×10^5 cells/mL) were grown at $37\,°C$, in a 5% CO_2 atmosphere, in RPMI 1640 culture medium (Gibco, Thermo Fisher Scientific, Waltham, MA, USA) supplemented with 2 mM L-glutamine (Sigma-Aldrich, Germany) and 10% fetal bovine serum (Euroclone, Italy). This culture medium will be further designated as complete culture medium. The passage of cells was made at 2–3 days. In experiments, cells at passages 9–11 were used.

The U937 cells in complete culture medium (1×10^6 cells/mL) were incubated for 24 h with the tested compounds (0.5 µM, 5 µM, and 50 µM). After incubation, the cells were washed by centrifugation (1200 rpm, 5 min, $4\,°C$), and were suspended in RMPI 1640 culture medium without phenol red and supplements for fluorescence measurements, at a cellular density of 5×10^5 cells/mL. The cellular viability was assessed by optical microscopy, with the trypan blue exclusion test. Cell cultures with a viability > 95% were used for experiments. Fluorescence measurements were performed in standard 1 cm pathlength cuvettes from polymethyl methacrylate (BrandTech Scientific, Essex, CT, USA) with an LS 50B Perkin Elmer spectrofluorometer (PerkinElmer, Waltham, MA, USA), equipped with a thermostatted cuvette holder, magnetic stirring in cuvette, and analysis system in polarized light.

For assessing the cellular autofluorescence, the autofluorescence spectrum was plotted for each sample in the working range of the DiBAC4(3) probe (493 nm excitation wavelengths in the 500–600 nm spectral range, excitation and emission slits of 5 nm, and a spectrum scanning speed of 500 nm/min).

The measurement of the background signals in fluorescence polarization mode was performed using the Time Drive mode of the spectrofluorometer, using the following parameters: 355 nm excitation, 430 nm emission, 10 nm (excitation) and 10 nm (emission) slits, measurement time 4 s (step 0.02 s). The measurements were carried out for the four positions of the polarizers in the excitation and emission beam: vertical-vertical (I0_vv), vertical-horizontal (I0_vh), horizontal-horizontal (I0_hh), horizontal-vertical (I0_hv).

After measuring the background signals, 2 µL DiBAC4(3) and 2 µL TMA-DPH stock solutions were added to each sample (2 mL volume), for assessing both the stationary fluorescence and polarized fluorescence. Cellular samples were incubated with the probes for 2 min at $37\,°C$ under continuous magnetic stirring, before treating the cells with porphyrinic compounds. At the end of the incubation time of the cells with porphyrins, the emission spectrum was recorded (excitation at 493 nm, between 500 and 600 nm, excitation and emission slits of 5 nm, spectrum scanning speed of 500 nm/min). Measurements in the Time Drive mode were made to obtain the Ivv, Ivh, Ihh, and Ihv values.

For transmembrane potential evaluation, the results were expressed as relative percent change of the intensity signal of the treated cells vs. controls:

$$Ie = 100 \times (1 - Isample/Icontrol) \qquad (1)$$

The calculation of the fluorescence anisotropy (r) was performed using the following equations:

$$r = (I_{vv} - G \cdot I_{vh}) / (I_{vv} + 2 \cdot G \cdot I_{vh}) \quad (2)$$

$$G = I_{hv} / I_{hh} \quad (3)$$

I_{vv}, I_{vh}, I_{hv}, and I_{hh} represent the emitted fluorescence intensity measured for the four positions of the polarizers in the excitation and emission beam (vertical-vertical, vertical-horizontal, horizontal-vertical and horizontal-horizontal) [30].

For evaluation, we used the relative effect, calculated according to the rs (anisotropy value of the sample) and rc (anisotropy value of the control).

$$e_r \% = 100 \times (r_s / r_c - 1) \quad (4)$$

3.2. The Prediction of Permeability across the Cell Membrane

The diffusion through the cell membrane for the porphyrin derivatives was estimated using the Permeability of Molecules across Membranes (PerMM) web server [31]. PerMM is a thermodynamics-based method that uses the anisotropic solvent model of the DOPC (dioleoylphosphatidylcholine) bilayer. Three-dimensional structures of the porphyrin derivatives were generated with OpenBabel v2.4.1 [32], and were energetically minimized using the UFF force field. Simulations were carried out at 298K and physiological pH (7.4). The membrane binding affinities, energy profiles along the bilayer normal, and permeability coefficients were calculated.

3.3. Full-Length Protein Structure Modeling

Three potential protein targets were taken into consideration for the in silico studies. The target proteins were selected based on the in vitro results and on a literature review. We hypothesized that the tested porphyrins might interact with specific ion pumps and ion channels, such as the sarco/endoplasmic reticulum Ca^{2+}-ATPase, isoform 2b (SERCA2b), large conductance calcium-activated potassium channel (KCa1.1/Slo1), and ATP-sensitive potassium channel (K_{ATP}). The Slo1 and K_{ATP} channels were chosen based on a structural analogy with heme (iron protoporphyrin IX), which was previously discovered to negatively modulate Slo1 [10] and activate K_{ATP} [11]. Since heme modulates these potassium channels by binding to disordered regions that could not be solved by cryo-EM, homology modeling was used to predict the full-length structure of the selected proteins. For this purpose, we used the YASARA Structure [33] software and the SWISS-MODEL [34] web-server, which are fully automated resources for protein structure prediction. Moreover, the target human proteins were also retrieved from AlphaFold, since it was shown to have remarkable accuracy in predicting protein structures [35]. The protein sequences were retrieved from the UniProt database [36] in FASTA format (codes P16615-1 for hSERCA2b, Q12791 for hSlo1, and O60706-1 for the SUR2 subunit of hK_{ATP}). Templates were selected manually, based on relevant experimentally determined conformations, such as human SERCA2b in E2 state (PDB ID: 6LN9 [37]), calcium-free human Slo1 in complex with auxiliary protein β4 (PDB ID: 6V35 [38]), and SUR2B subunit of rat K_{ATP} in propeller-like conformation (PDB ID: 7MIT [39]). The qualities of the modeled structures were compared with structures retrieved from the AlphaFold database, and the most optimal models were retained for molecular docking studies [40]. Quality assessment was performed using the MolProbity 4.5.1 web-server [41].

3.4. Molecular Docking

Docking experiments were carried out using the full-length predicted structures of the target proteins, with the most optimal quality parameters. The simulations were performed using the AutoDock Vina v1.1.2 algorithm [42], integrated within the YASARA Structure. In the case of SERCA2b, the docking grid box included a potential binding pocket situated at the interface between the transmembrane and cytosolic domains, overlapping with the binding site of SERCA inhibitor BHQ (2,5-di-tert-butylbenzene-1,4-diol) [25]. The binding pocket was chosen based on previous studies, which suggested that the binding site of BHQ might overlap with the potential binding pocket of hypericin, a photosensitizing agent with a large molecule, similar to porphyrinic derivatives [43]. For potassium channels KCa1.1 and K_{ATP}, the grid box was set around the residues involved in heme binding, which were confirmed through previous mutation studies [21,22,44]. BHQ and heme were used as positive controls.

Both protein and ligand structures were protonated according to the physiological pH. The docking experiments were performed with flexible residues, and 12 docking runs were executed for each ligand. The predicted protein-ligand complexes were refined by minimization with AMBER14 force field. The results were retrieved as the binding energy (ΔG, kcal/mol) and predicted dissociation constant (Kd, μM). The predicted binding poses and molecular interactions were analyzed using ChimeraX v1.4 [45] and BIOVIA Discovery Studio Visualizer (BIOVIA, Discovery Studio Visualizer, Version 17.2.0, Dassault Systèmes, 2016, San Diego, CA, USA).

3.5. MM/PBSA Free Energy of Binding Estimation

Short molecular dynamics (MD) simulations were performed with YASARA, after molecular docking, to evaluate the free binding energies of the investigated porphyrins. Since all three target proteins (SERCA2b, KCa1.1 and K_{ATP}) are membrane proteins, membranes consisting of phosphatidyl-ethanolamine molecules were generated automatically. The simulated protein-ligand complexes were temporarily scaled to 0.9, and clashing membrane lipids were deleted. The scaling was thereafter slowly removed during a short simulation at 298K in vacuo. The proteins were scaled by 1.02 every 200 fs and the membrane atoms were allowed to move. During the simulations, AMBER14 forcefield [46] and Lipid17/GAFF2/AM1BCC [47–49] parameters for non-standard residues were used. Side-chain pKa values were predicted after the protein reached its original size [50]. Protonation states were assigned according to the physiological pH (7.4). The simulation cell was filled with water, 0.9% NaCl and counter ions [51]. Simulations were carried out with the Particle Mesh Ewald algorithm, 0.8 Å cutoff for non-bonded real space forces, 4 fs time-step, constrained hydrogen atoms, and at constant pressure and temperature (isothermal-isobaric ensemble) [52]. An equilibration period of 250 ps was set to prevent water molecules from entering the membrane, and the simulation cell adapted to the pressure exerted by the membrane. A simulation snapshot was obtained after another 250 ps to calculate the free energy of binding (ΔG_{bind}, kcal/mol), using the Poisson–Boltzmann (MM/PBSA), method, excluding the entropic term.

4. Discussion

In this study, three asymmetrical porphyrins were evaluated in vitro, regarding their effects on the transmembrane potential and the membrane anisotropy of human monocytic U937 cells, using DiBAC4(3) and TMA-DPH as fluorescence probes.

The obtained results indicated, for the free base porphyrin, a relatively constant, dose-independent hyperpolarizing effect, but the zinc and copper metalloporphyrins exerted a biphasic dose-dependent hyperpolarizing effect, suggesting a distinctive molecular mechanism of action.

An increased membrane anisotropy was registered in the case of porphyrin-treated cells compared to the untreated control. Since anisotropy is inversely correlated with fluidity, the results suggest that, under the action of the asymmetric porphyrin compounds, the cellular membrane becomes more rigid, hence favoring the interaction of porphyrinic compounds with membrane proteins, through which porphyrinic compounds may alter the cellular functionality.

Another study showed that porphyrinic compounds can induce potassium leakage and cell membrane dysfunction in bovine erythrocytes under photoirradiation, but no changes in potassium homeostasis were observed in the absence of light irradiation [53].

We herein show that a free base porphyrin, and two corresponding metalloporphyrins, can hyperpolarize U937 cell membranes in the absence of photoirradiation. The implications of the hyperpolarizing effects exerted by the investigated porphyrinic compounds are yet to be studied. Previous works indicated that hyperpolarization of the U937 cell membranes promotes cell proliferation and migration [54], these effects being detrimental in oncologic settings. On the other hand, porphyrinic derivatives act as photosensitizers and can photo-inactivate tumor cells.

The potential of the three porphyrins to translocate across the cell membrane was predicted using thermodynamic methods, showing satisfactory permeability coefficients and energy profiles for all compounds These findings support the therapeutic potential of the assessed porphyrins as candidates for further experimental evaluation as photodynamic agents.

Since the U937 cell culture is a pro-monocytic, human histiocytic lymphoma cell line [55], and considering that monocytes and macrophages express BK Slo1 and K_{ATP} potassium channels [56,57], we further investigated, through molecular docking experiments and short MD simulations, the potential of the assessed porphyrins to modulate such channels. Firstly, we predicted the full-length structures of SERCA2b, Slo1, and K_{ATP}. Furthermore, we simulated the interaction of heme (positive control) with the corresponding heme-binding motifs within Slo1 and the SUR2 subunit of K_{ATP}, and generated intermediate interaction models between heme and the regulatory regions. The docked conformations illustrate that heme can indeed form metal coordination bonds with a cysteine residue within Slo1 and a histidine residue within the K_{ATP} channel. Furthermore, the carboxyl moieties of heme formed hydrogen bonds and salt bridges with positively charged residues within the binding sites of both target proteins. Thus, our models further support the previous studies that highlighted the modulatory effect of heme on these potassium channels [10,11].

We hypothesized that the three asymmetrical porphyrins induced a hyperpolarizing effect on the cell membrane by blocking the sarco/endoplasmic reticulum calcium ATPase SERCA2b at 0.5 µM concentrations, thus increasing the levels of cytosolic calcium. SERCA2 is an ion pump involved in the reuptake of Ca^{2+} from the cytosol into the endoplasmic reticulum lumen and has two isoforms, SERCA2a, being expressed in cardiac and skeletal muscles, and SERCA2b, which occurs ubiquitously [58]. Cytosolic calcium would thereafter stimulate the calcium-activated potassium channel BK Slo1 (KCa1.1), inducing potassium leakage to the extracellular space and hyperpolarizing the cell membrane.

Furthermore, we considered that only the metalloporphyrin derivatives would interact with the heme-binding motif of Slo1 and partly counteract membrane hyperpolarization by blocking the calcium-dependent activation of Slo1 at 5 µM concentrations. Moreover, at higher concentrations (50 µM), the metalloporphyrins would potentially bind to the SUR2 subunit of K_{ATP} in the same binding site as heme, and would activate the channel, leading to further hyperpolarizing effects due to increased potassium leakage. This complex, hypothetical mechanism of action was supported by molecular docking simulations and MM/PBSA binding free energy calculations, since the tested concentrations could be correlated with the in silico experiments (Figure 13).

The present study was a preliminary one for investigating some basic properties of some unsymmetrical porphyrinic compounds, such as their incorporation into cell membranes and consequent effects on membrane polarization and anisotropy. Because the transmembrane potential and membrane fluidity are biophysical properties of the cell membrane, with important impacts on cellular homeostasis, further studies are necessary for highlighting the parameters that allow an optimal cellular internalization of the tested porphyrins. Novel porphyrinic derivatives can be designed to increase selectivity on certain ion channels or pumps, in order to further explore the therapeutic potential of such compounds.

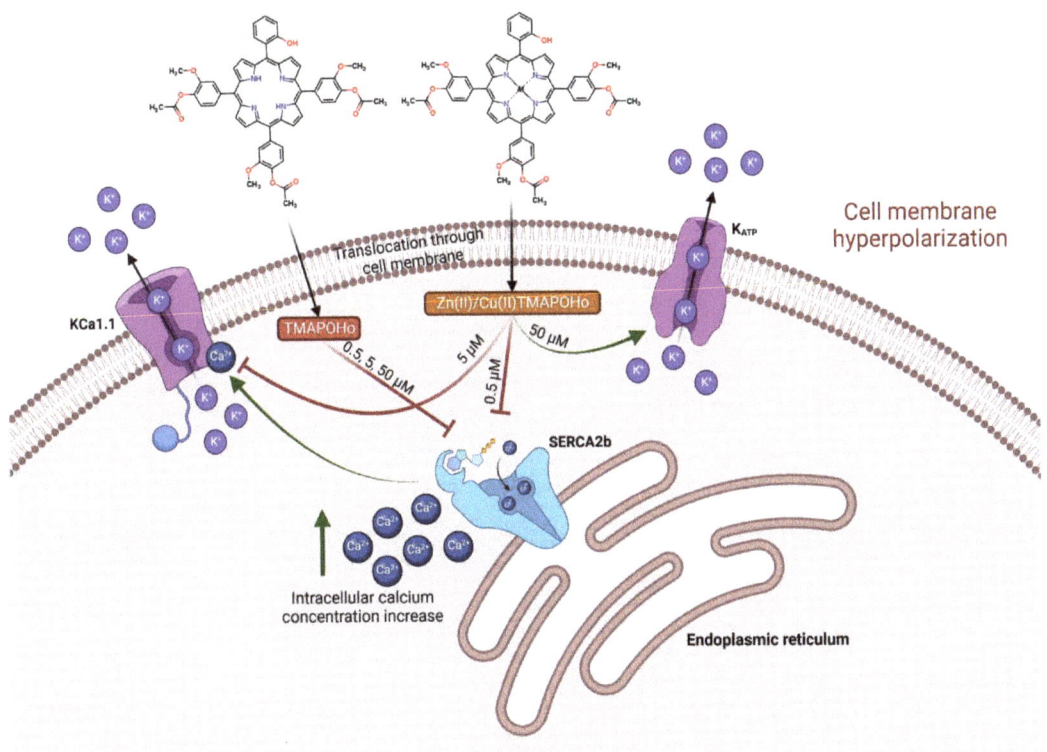

Figure 13. Illustration of hypothesized molecular mechanisms related to the interaction of the investigated porphyrinic compounds with cellular membranes.

5. Conclusions

The effects of three asymmetrical porphyrins, 5-(2-hydroxyphenyl)-10,15,20-tris-(4-acetoxy-3-methoxyphenyl)porphyrin, 5-(2-hydroxyphenyl)-10,15,20-tris-(4-acetoxy-3-methoxyphenyl)porphyrinatozinc(II) and 5-(2-hydroxyphenyl)-10,15,20–tris-(4-acetoxy-3-methoxyphenyl)porphyrinatocopper(II), on transmembrane potential and cell membrane anisotropy was evaluated on human monocytic U937 cells.

All three compounds induced hyperpolarizing effects and increased the anisotropy of the cell membrane. The ability of the three porphyrins to cross the cell membrane was predicted using thermodynamic methods, which evidenced good permeability profiles. In silico studies supported the hypothesis that the tested compounds could potentially interact with the membrane proteins SERCA2b, Slo1, and K_{ATP}. Further studies are needed to experimentally validate the proposed molecular interactions and their consequences on cell homeostasis.

Author Contributions: Conceptualization, G.M., R.B., D.P.M. and A.M.B.; methodology, G.M., R.B., D.P.M., I.V.N. and R.P.S.; software, D.P.M.; writing—original draft preparation, R.B., G.M., D.P.M. and A.M.B.; writing—review and editing, R.B., G.M., G.V., D.P.M., R.P.S.; supervision, G.M. and D.L. All authors have read and agreed to the published version of the manuscript.

Funding: Publication fee of this paper was funded by "Carol Davila" University of Medicine and Pharmacy, Bucharest, Romania, through the institutional program "Publish Not Perish".

Institutional Review Board Statement: Not applicable.

Informed Consent Statement: Not applicable.

Data Availability Statement: Not applicable.

Acknowledgments: Publication of this paper was supported by "Carol Davila" University of Medicine and Pharmacy, Bucharest, Romania, through the institutional program "Publish Not Perish".

Conflicts of Interest: The authors declare no conflict of interest.

References

1. Socoteanu, R.; Boscencu, R.; Hirtopeanu, A.; Manda, G.; Sousa, A.; Ilie, M.; Vieira Ferreir, L.F. Trends in Interdisciplinary Studies Revealing Porphyrinic Compounds Multivalency Towards Biomedical Application. In *Biomedical Engineering—From Theory to Applications*; InTech: London, UK, 2011.
2. Zhang, J.; Jiang, C.; Figueiró Longo, J.P.; Azevedo, R.B.; Zhang, H.; Muehlmann, L.A. An updated overview on the development of new photosensitizers for anticancer photodynamic therapy. *Acta Pharm. Sin. B* **2018**, *8*, 137–146. [CrossRef] [PubMed]
3. Roy Chowdhury, M.; Schumann, C.; Bhakta-Guha, D.; Guha, G. Cancer nanotheranostics: Strategies, promises and impediments. *Biomed. Pharmacother.* **2016**, *84*, 291–304. [CrossRef] [PubMed]
4. Ethirajan, M.; Chen, Y.; Joshi, P.; Pandey, R.K. The role of porphyrin chemistry in tumor imaging and photodynamic therapy. *Chem. Soc. Rev.* **2011**, *40*, 340–362. [CrossRef] [PubMed]
5. Josefsen, L.B.; Boyle, R.W. Unique Diagnostic and Therapeutic Roles of Porphyrins and Phthalocyanines in Photodynamic Therapy, Imaging and Theranostics. *Theranostics* **2012**, *2*, 916–966. [CrossRef] [PubMed]
6. Manda, G.; Hinescu, M.E.; Neagoe, I.V.; Ferreira, L.F.V.; Boscencu, R.; Vasos, P.; Basaga, S.H.; Cuadrado, A. Emerging Therapeutic Targets in Oncologic Photodynamic Therapy. *Curr. Pharm. Des.* **2019**, *24*, 5268–5295. [CrossRef] [PubMed]
7. Razzaq, S.; Minhas, A.M.; Qazi, N.G.; Nadeem, H.; Khan, A.-u.; Ali, F.; Hassan, S.S.U.; Bungau, S. Novel Isoxazole Derivative Attenuates Ethanol-Induced Gastric Mucosal Injury through Inhibition of H$^+$/K$^+$-ATPase Pump. Oxidative Stress and Inflammatory Pathways. *Molecules* **2022**, *27*, 5065. [CrossRef]
8. Das, S.S.; Bharadwaj, P.; Bilal, M.; Barani, M.; Rahdar, A.; Taboada, P.; Bungau, S.; Kyzas, G.Z. Stimuli-Responsive Polymeric Nanocarriers for Drug Delivery, Imaging, and Theragnosis. *Polymers* **2020**, *12*, 1397. [CrossRef]
9. Yutaka, T.; Yoshiaki, O.; Yutaka, S.; Harutoshi, K. Benzodiazepine Receptor Agonists Modulate Thymocyte Apoptosis Through Reduction of the Mitochondrial Transmembrane Potential. *Jpn. J. Pharmacol.* **1999**, *79*, 177–183. [CrossRef]
10. Yanamala, N.; Klein-Seetharaman, J. Allosteric Modulation of G Protein Coupled Receptors by Cytoplasmic, Transmembrane and Extracellular Ligands. *Pharmaceuticals* **2010**, *3*, 3324–3342. [CrossRef]
11. Venugopal, J.; Blanco, G. Ouabain Enhances ADPKD Cell Apoptosis via the Intrinsic Pathway. *Front. Physiol.* **2016**, *7*, 107. [CrossRef]
12. Klapperstück, T.; Glanz, D.; Klapperstück, M.; Wohlrab, J. Methodological aspects of measuring absolute values of membrane potential in human cells by flow cytometry. *Cytom. Part A* **2009**, *75A*, 593–608. [CrossRef]
13. Boscencu, R. Microwave Synthesis Under Solvent-Free Conditions and Spectral Studies of Some Mesoporphyrinic Complexes. *Molecules* **2012**, *17*, 5592–5603. [CrossRef]
14. Boscencu, R.; Oliveira, A.S.; Ferreira, D.P.; Ferreira, L.F.V. Synthesis and Spectral Evaluation of Some Unsymmetrical Mesoporphyrinic Complexes. *Int. J. Mol. Sci.* **2012**, *13*, 8112–8125. [CrossRef]
15. Socoteanu, R.; Manda, G.; Boscencu, R.; Vasiliu, G.; Oliveira, A. Synthesis, Spectral Analysis and Preliminary in Vitro Evaluation of Some Tetrapyrrolic Complexes with 3d Metal Ions. *Molecules* **2015**, *20*, 15488–15499. [CrossRef]
16. Boscencu, R.; Socoteanu, R.P.; Manda, G.; Radulea, N.; Anastasescu, M.; Gama, A.; Machado, I.F.; Ferreira, L.F.V. New A3B porphyrins as potential candidates for theranostic. Synthesis and photochemical behaviour. *Dyes Pigments* **2019**, *160*, 410–417. [CrossRef]
17. Savitskiĭ, V.P.; Zorin, V.P. Selectivity of accumulation of chlorine e6 derivatives in blood leukocytes. *Biofizika* **2003**, *48*, 58–62. [PubMed]
18. Krosl, G.; Korbelik, M.; Dougherty, G.J. Induction of immune cell infiltration into murine SCCVII tumour by photofrin-based photodynamic therapy. *Br. J. Cancer* **1995**, *71*, 549–555. [CrossRef] [PubMed]
19. Cecic, I.; Parkins, C.; Korbelik, M. Induction of systemic neutrophil response in mice by photodynamic therapy of solid tumors. *Photochem. Photobiol.* **2001**, *74*, 712–720. [CrossRef] [PubMed]
20. Soyama, T.; Sakuragi, A.; Oishi, D.; Kimura, Y.; Aoki, H.; Nomoto, A.; Yano, S.; Nishie, H.; Kataoka, H.; Aoyama, M. Photodynamic therapy exploiting the anti-tumor activity of mannose-conjugated chlorin e6 reduced M2-like tumor-associated macrophages. *Transl. Oncol.* **2021**, *14*, 101005. [CrossRef]
21. Tang, X.D.; Xu, R.; Reynolds, M.F.; Garcia, M.L.; Heinemann, S.H.; Hoshi, T. Haem can bind to and inhibit mammalian calcium-dependent Slo1 BK channels. *Nature* **2003**, *425*, 531–535. [CrossRef]
22. Burton, M.J.; Kapetanaki, S.M.; Chernova, T.; Jamieson, A.G.; Dorlet, P.; Santolini, J.; Moody, P.C.E.; Mitcheson, J.S.; Davies, N.W.; Schmid, R.; et al. A heme-binding domain controls regulation of ATP-dependent potassium channels. *Proc. Natl. Acad. Sci. USA* **2016**, *113*, 3785–3790. [CrossRef] [PubMed]
23. Sahoo, N.; Goradia, N.; Ohlenschläger, O.; Schönherr, R.; Friedrich, M.; Plass, W.; Kappl, R.; Hoshi, T.; Heinemann, S.H. Heme impairs the ball-and-chain inactivation of potassium channels. *Proc. Natl. Acad. Sci. USA* **2013**, *110*, E4036–E4044. [CrossRef] [PubMed]

24. Daly, D.; Al-Sabi, A.; Kinsella, G.K.; Nolan, K.; Dolly, J.O. Porphyrin derivatives as potent and selective blockers of neuronal Kv1 channels. *Chem. Commun.* **2015**, *51*, 1066–1069. [CrossRef] [PubMed]
25. Obara, K.; Miyashita, N.; Xu, C.; Toyoshima, I.; Sugita, Y.; Inesi, G.; Toyoshima, C. Structural role of countertransport revealed in Ca 2+ pump crystal structure in the absence of Ca 2+. *Proc. Natl. Acad. Sci. USA* **2005**, *102*, 14489–14496. [CrossRef] [PubMed]
26. Li, T.; Bonkovsky, H.L.; Guo, J. Structural analysis of heme proteins: Implications for design and prediction. *BMC Struct. Biol.* **2011**, *11*, 13. [CrossRef] [PubMed]
27. Vasiliu, G.; Boscencu, R.; Socoteanu, R.; Nacea, V. Complex combinations of some transition metals with new unsymmetrical porphirins. *Rev. Chim.* **2014**, *65*, 998–1001.
28. Boscencu, R.; Socoteanu, R.; Vasiliu, G.; Nacea, V. Synthesis under solvent free conditions of some unsymmetrically substituted porphyrinic compounds. *Rev. Chim.* **2014**, *65*, 888–891.
29. Baudin, B.; Bruneel, A.; Bosselut, N.; Vaubourdolle, M. A protocol for isolation and culture of human umbilical vein endothelial cells. *Nat. Protoc.* **2007**, *2*, 481–485. [CrossRef]
30. Instrumentation for Fluorescence Spectroscopy. *Principles of Fluorescence Spectroscopy*; Springer: Boston, MA, USA, 2006; pp. 27–61.
31. Lomize, A.L.; Hage, J.M.; Schnitzer, K.; Golobokov, K.; LaFaive, M.B.; Forsyth, A.C.; Pogozheva, I.D. PerMM: A Web Tool and Database for Analysis of Passive Membrane Permeability and Translocation Pathways of Bioactive Molecules. *J. Chem. Inf. Model.* **2019**, *59*, 3094–3099. [CrossRef]
32. O'Boyle, N.M.; Banck, M.; James, C.A.; Morley, C.; Vandermeersch, T.; Hutchison, G.R. Open Babel: An Open chemical toolbox. *J. Cheminform.* **2011**, *3*, 33. [CrossRef]
33. Land, H.; Humble, M.S. YASARA: A Tool to Obtain Structural Guidance in Biocatalytic Investigations. *Methods Mol. Biol.* **2018**, *1685*, 43–67.
34. Waterhouse, A.; Bertoni, M.; Bienert, S.; Studer, G.; Tauriello, G.; Gumienny, R.; Heer, F.T.; de Beer, T.A.P.; Rempfer, C.; Bordoli, L.; et al. SWISS-MODEL: Homology modelling of protein structures and complexes. *Nucleic Acids Res.* **2018**, *46*, W296–W303. [CrossRef] [PubMed]
35. Jumper, J.; Evans, R.; Pritzel, A.; Green, T.; Figurnov, M.; Ronneberger, O.; Tunyasuvunakool, K.; Bates, R.; Žídek, A.; Potapenko, A.; et al. Highly accurate protein structure prediction with AlphaFold. *Nature* **2021**, *596*, 583–589. [CrossRef] [PubMed]
36. Bateman, A.; Martin, M.-J.; Orchard, S.; Magrane, M.; Agivetova, R.; Ahmad, S.; Alpi, E.; Bowler-Barnett, E.H.; Britto, R.; Bursteinas, B.; et al. UniProt: The universal protein knowledgebase in 2021. *Nucleic Acids Res.* **2021**, *49*, D480–D489.
37. Zhang, Y.; Inoue, M.; Tsutsumi, A.; Watanabe, S.; Nishizawa, T.; Nagata, K.; Kikkawa, M.; Inaba, K. Cryo-EM structures of SERCA2b reveal the mechanism of regulation by the luminal extension tail. *Sci. Adv.* **2020**, *6*, eabb0147. [CrossRef]
38. Tao, X.; Mackinnon, R. Molecular structures of the human slo1 k+ channel in complex with b4. *Elife* **2019**, *8*, e51409. [CrossRef] [PubMed]
39. Sung, M.W.; Yang, Z.; Driggers, C.M.; Patton, B.L.; Mostofian, B.; Russo, J.D.; Zuckerman, D.M.; Shyng, S.-L. Vascular K ATP channel structural dynamics reveal regulatory mechanism by Mg-nucleotides. *Proc. Natl. Acad. Sci. USA* **2021**, *118*, e2109441118. [CrossRef]
40. Nitulescu, G.; Nitulescu, G.M.; Zanfirescu, A.; Mihai, D.P.; Gradinaru, D. Candidates for Repurposing as Anti-Virulence Agents Based on the Structural Profile Analysis of Microbial Collagenase Inhibitors. *Pharmaceutics* **2021**, *14*, 62. [CrossRef]
41. Williams, C.J.; Headd, J.J.; Moriarty, N.W.; Prisant, M.G.; Videau, L.L.; Deis, L.N.; Verma, V.; Keedy, D.A.; Hintze, B.J.; Chen, V.B.; et al. MolProbity: More and better reference data for improved all-atom structure validation. *Protein Sci.* **2018**, *27*, 293–315. [CrossRef]
42. Trott, O.; Olson, A.J. AutoDock Vina: Improving the speed and accuracy of docking with a new scoring function, efficient optimization, and multithreading. *J. Comput. Chem.* **2010**, *31*, 455–461. [CrossRef]
43. Eriksson, E.S.E.; Eriksson, L.A. Identifying the sarco(endo)plasmic reticulum Ca2+ ATPase (SERCA) as a potential target for hypericin – a theoretical study. *Phys. Chem. Chem. Phys.* **2012**, *14*, 12637. [CrossRef] [PubMed]
44. Hassan, S.S.u.; Abbas, S.Q.; Ali, F.; Ishaq, M.; Bano, I.; Hassan, M.; Jin, H.-Z.; Bungau, S.G. A Comprehensive In Silico Exploration of Pharmacological Properties, Bioactivities, Molecular Docking, and Anticancer Potential of Vieloplain F from *Xylopia vielana* Targeting B-Raf Kinase. *Molecules* **2022**, *27*, 917. [CrossRef] [PubMed]
45. Pettersen, E.F.; Goddard, T.D.; Huang, C.C.; Meng, E.C.; Couch, G.S.; Croll, T.I.; Morris, J.H.; Ferrin, T.E. UCSF ChimeraX: Structure visualization for researchers, educators, and developers. *Protein Sci.* **2021**, *30*, 70–82. [CrossRef] [PubMed]
46. Maier, J.A.; Martinez, C.; Kasavajhala, K.; Wickstrom, L.; Hauser, K.E.; Simmerling, C. ff14SB: Improving the Accuracy of Protein Side Chain and Backbone Parameters from ff99SB. *J. Chem. Theory Comput.* **2015**, *11*, 3696–3713. [CrossRef]
47. Dickson, C.J.; Madej, B.D.; Skjevik, A.A.; Betz, R.M.; Teigen, K.; Gould, I.R.; Walker, R.C. Lipid14: The Amber Lipid Force Field. *J. Chem. Theory Comput.* **2014**, *10*, 865–879. [CrossRef]
48. Wang, J.; Wolf, R.M.; Caldwell, J.W.; Kollman, P.A.; Case, D.A. Development and testing of a general amber force field. *J. Comput. Chem.* **2004**, *25*, 1157–1174. [CrossRef]
49. Jakalian, A.; Jack, D.B.; Bayly, C.I. Fast, efficient generation of high-quality atomic charges. AM1-BCC model: II. Parameterization and validation. *J. Comput. Chem.* **2002**, *23*, 1623–1641. [CrossRef]

50. Krieger, E.; Nielsen, J.E.; Spronk, C.A.E.M.; Vriend, G. Fast empirical pKa prediction by Ewald summation. *J. Mol. Graph. Model.* **2006**, *25*, 481–486. [CrossRef]
51. Krieger, E.; Darden, T.; Nabuurs, S.B.; Finkelstein, A.; Vriend, G. Making optimal use of empirical energy functions: Force-field parameterization in crystal space. *Proteins* **2004**, *57*, 678–683. [CrossRef]
52. Krieger, E.; Vriend, G. New ways to boost molecular dynamics simulations. *J. Comput. Chem.* **2015**, *36*, 996–1007. [CrossRef]
53. Kato, H.; Komagoe, K.; Inoue, T.; Masuda, K.; Katsu, T. Structure—Activity relationship of porphyrin-induced photoinactivation with membrane function in bacteria and erythrocytes. *Photochem. Photobiol. Sci.* **2018**, *17*, 954–963. [CrossRef]
54. Erdogan, A.; Schaefer, C.A.; Most, A.K.; Schaefer, M.B.; Mayer, K.; Tillmanns, H.; Kuhlmann, C.R.W. Lipopolysaccharide-induced proliferation and adhesion of U937 cells to endothelial cells involves barium chloride sensitive hyperpolarization. *J. Endotoxin Res.* **2006**, *12*, 224–230. [CrossRef]
55. Prasad, A.; Sedlářová, M.; Balukova, A.; Ovsii, A.; Rác, M.; Křupka, M.; Kasai, S.; Pospíšil, P. Reactive Oxygen Species Imaging in U937 Cells. *Front. Physiol.* **2020**, *11*, 552569. [CrossRef] [PubMed]
56. Femling, J.K.; Cherny, V.V.; Morgan, D.; Rada, B.; Davis, A.P.; Czirják, G.; Enyedi, P.; England, S.K.; Moreland, J.G.; Ligeti, E.; et al. The Antibacterial Activity of Human Neutrophils and Eosinophils Requires Proton Channels but Not BK Channels. *J. Gen. Physiol.* **2006**, *127*, 659–672. [CrossRef] [PubMed]
57. Ling, M.-Y.; Ma, Z.-Y.; Wang, Y.-Y.; Qi, J.; Liu, L.; Li, L.; Zhang, Y. Up-regulated ATP-sensitive potassium channels play a role in increased inflammation and plaque vulnerability in macrophages. *Atherosclerosis* **2013**, *226*, 348–355. [CrossRef] [PubMed]
58. Park, S.W.; Zhou, Y.; Lee, J.; Lee, J.; Ozcan, U. Sarco(endo)plasmic reticulum Ca^{2+}-ATPase 2b is a major regulator of endoplasmic reticulum stress and glucose homeostasis in obesity. *Proc. Natl. Acad. Sci. USA* **2010**, *107*, 19320–19325. [CrossRef] [PubMed]

Disclaimer/Publisher's Note: The statements, opinions and data contained in all publications are solely those of the individual author(s) and contributor(s) and not of MDPI and/or the editor(s). MDPI and/or the editor(s) disclaim responsibility for any injury to people or property resulting from any ideas, methods, instructions or products referred to in the content.

Article

Biophysical Characterization of the Interaction between a Transport Human Plasma Protein and the 5,10,15,20-Tetra(pyridine-4-yl)porphyrin

Otávio Augusto Chaves [1,*], Bernardo A. Iglesias [2] and Carlos Serpa [1,*]

1. CQC-IMS, Department of Chemistry, University of Coimbra, Rua Larga, 3004-535 Coimbra, Portugal
2. Bioinorganic and Porpyrinoids Materials Lab, Department of Chemistry, Federal University of Santa Maria, Santa Maria 97105-900, Brazil
* Correspondence: otavioaugustochaves@gmail.com (O.A.C.); serpasoa@ci.uc.pt (C.S.)

Abstract: The interaction between human serum albumin (HSA) and the non-charged synthetic photosensitizer 5,10,15,20-tetra(pyridine-4-yl)porphyrin (4-TPyP) was evaluated by in vitro assays under physiological conditions using spectroscopic techniques (UV-vis, circular dichroism, steady-state, time-resolved, synchronous, and 3D-fluorescence) combined with in silico calculations by molecular docking. The UV-vis and steady-state fluorescence parameters indicated a ground-state association between HSA and 4-TPyP and the absence of any dynamic fluorescence quenching was confirmed by the same average fluorescence lifetime for HSA without (4.76 ± 0.11 ns) and with 4-TPyP (4.79 ± 0.14 ns). Therefore, the Stern–Volmer quenching (K_{SV}) constant reflects the binding affinity, indicating a moderate interaction (10^4 M^{-1}) being spontaneous ($\Delta G° = -25.0$ kJ/mol at 296 K), enthalpically ($\Delta H° = -9.31 \pm 1.34$ kJ/mol), and entropically ($\Delta S° = 52.9 \pm 4.4$ J/molK) driven. Binding causes only a very weak perturbation on the secondary structure of albumin. There is just one main binding site in HSA for 4-TPyP ($n \approx 1.0$), probably into the subdomain IIA (site I), where the Trp-214 residue can be found. The microenvironment around this fluorophore seems not to be perturbed even with 4-TPyP interacting via hydrogen bonding and van der Waals forces with the amino acid residues in the subdomain IIA.

Keywords: porphyrin; human serum albumin; spectroscopy; molecular docking; chemical-biological interactions

1. Introduction

Human Serum Albumin (HSA) is the most abundant globular protein in the bloodstream (35–50 g/L). It is synthesized in the liver and is responsible for the transport of both endogenous and exogenous compounds, e.g., fatty acids, hormones, metabolites, and commercial drugs to their target [1,2]. For this reason, from a pharmacological point of view, the interaction between HSA and drugs is crucial for a better understanding of both pharmacokinetics and toxicological profiles [3,4]. The structure of HSA has been elucidated by X-ray analysis and was revealed as an ellipsoid with a heart shape consisting of three domains (I, II, and III), and each domain is divided into two subdomains (A and B) [5,6]. Sudlow and coworkers [7] were one of the first researchers to evaluate the specificity of different drugs in binding with albumin and for this reason, the binding sites I and II, located in subdomains IIA and IIIA, respectively, were known as corresponding Sudlow's sites I and II.

Porphyrin is a class of compounds containing a flat ring of four linked-heterocyclic groups, sometimes with a central metal atom. There are naturally occurring porphyrins, e.g., protoporphyrin IX, chlorophyll, and cobalamin, however, the synthetic ones have attracted attention mainly due to their applicability as photosensitizers in antimicrobial photodynamic therapy (aPDT) or photodynamic therapy (PDT) of malignant tissues [8,9].

In this sense, 5,10,15,20-tetra(pyridine-4-yl)porphyrin (4-TPyP, Figure 1), a simple non-charged synthetic porphyrin, has attracted attention in the design of novel photosensitizers to PDT, mainly due to its high singlet oxygen quantum yield (Φ_Δ = 0.76, in acetonitrile) and phototoxicity in a nanomolar scale under green light irradiation (522 nm) with very low light dose (1.0 J/cm^2) [10]. Recently, binding studies between Bovine Serum Albumin (BSA, a very similar protein compared with HSA, sharing 76% identity and 88% similarity in protein sequence, however, with two tryptophan residues) and 5-phenyl-10,15,20-tri(pyridine-4-yl)porphyrin, a non-charged synthetic porphyrin with structural similarities with 4-TPyP was reported, revealing that the main fluorescence quenching mechanism is static, with binding spontaneous, strong, controlled by electrostatic forces, and the hydrophobicity of the microenvironment around tyrosine (Tyr) and tryptophan (Trp) residues are enhanced in the presence of this porphyrin [11].

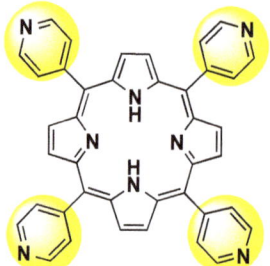

Figure 1. Chemical structure for 5,10,15,20-tetra(pyridine-4-yl)porphyrin.

Since there are not any biophysical reports on the binding capacity of HSA and 4-TPyP (a relevant porphyrin to develop potential leads for PDT), the present study reports this interaction by multiple spectroscopic techniques (UV-vis, circular dichroism, steady-state, time-resolved, synchronous, and 3D-fluorescence) under physiological conditions at pH = 7.4. To offer a molecular-level explanation of the binding HSA: 4-TPyP, molecular docking calculations were also carried out for the three main binding sites of albumin (subdomains IIA, IIIA, and IB).

2. Results

2.1. Experimental Binding Capacity of 4-TPyP to HSA

Absorption spectroscopy is a simple technique to preliminary evaluate qualitatively the binding capacity between HSA and 4-TPyP. From Figure 2A, was noticed that the HSA solution presents two absorption maximums: one at 222 nm due to the $\pi \rightarrow \pi^*$ transition of C=O (from amide group) and the other at 280 nm attributed to $n \rightarrow \pi^*$ transition which is associated with the aromatic amino acid residues tryptophan (Trp), phenylalanine (Phe), and tyrosine (Tyr) [3,12]. On the other hand, the absorption peaks at wavelengths higher than 400 nm is only attributed to the electronic transitions of 4-TPyP (Soret and Q-bands) [10]. Upon the addition of 4-TPyP in the albumin solution, there is a significant hyperchromic effect in the 250–300 nm range (red line in Figure 2A), however, to assess if the observed hyperchromic displacement is in fact a consequence of the HSA:4-TPyP interaction, and not simply a contribution from porphyrin absorption, the HSA:4-TPyP and 4-TPyP absorption spectra were subtracted. The resulting spectrum in the 250–300 nm range (blue line in Figure 2A) continues to present a hyperchromic effect with a small blue-shift, indicating a ground-state association at this concentration [3,12–14]. This phenomenon may be confirmed by the hypochromic effect in the Soret Band ($\approx$400 nm) of 4-TPyP in the presence of HSA. Additionally, there was also evidence of a hyperchromic effect in the 200–250 nm range upon addition of 4-TPyP into the HSA solution, however, after the 4-TPyP absorption spectrum subtraction, this effect was due to the contribution of absorption from 4-TPyP, being the correct effect for this wavelength range as a hypochromic effect that might be considered as preliminary evidence that 4-TPyP can cause some perturbation on the albumin structure [12,15].

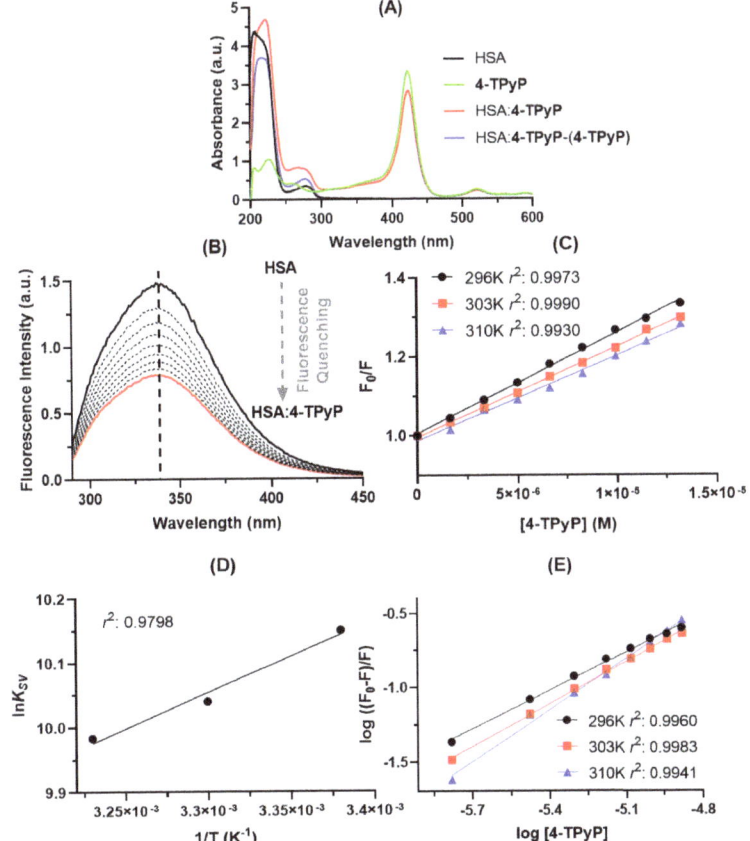

Figure 2. (**A**) UV-vis spectra for HSA (black line, 1.0×10^{-5} M), 4-TPyP (green line, 1.32×10^{-5} M), HSA:4-TPyP (red line), and mathematical subtraction (HSA:4-TPyP) – (4-TPyP) (blue line) in PBS solution (pH = 7.4) at 310 K. (**B**) Steady-state fluorescence emission spectra for HSA without and upon successive additions of 4-TPyP at 310 K (λ_{exc} = 280 nm). (**C**) Stern–Volmer plots for the interaction HSA:4-TPyP corresponding to the steady-state fluorescence data at three different temperatures. (**D**) Van't Hoff plot based on K_{SV} values for HSA:4-TPyP. (**E**) Double logarithmic plots for the interaction HSA:4-TPyP at three different temperatures. The r^2 in each plot is the coefficient of determination. [HSA] = 1.0×10^{-5} M and [4-TPyP] = 0.17, 0.33, 0.50, 0.66, 0.83, 0.99, 1.15, and 1.32×10^{-5} M.

Steady-state fluorescence spectroscopy is a more sensitive technique than UV-vis absorption to study the binding capacity of different small compounds with proteins, including porphyrin binding with HSA [14]. Figure 2B depicts the steady-state fluorescence emission of HSA without and upon successive additions of 4-TPyP, indicating that the porphyrin might interact with albumin without perturbing the microenvironment around the fluorophores due to the lack of blue- or red-shift in the maximum emission wavelength [16]. The steady-state fluorescence emission for 4-TPyP was also determined and as expected, there is not any fluorescence emission in the 290–450 nm range (region corresponding to the albumin fluorescence emission).

Table 1 summarizes the binding parameters obtained by steady-state fluorescence data (Figure 2C–E). The Stern–Volmer quenching (K_{SV}) constant values decrease with increasing temperature and the bimolecular quenching rate (k_q) constant values are 10^{12} M^{-1} s^{-1}, being three orders of magnitude larger than the maximum diffusion rate constant in water ($k_{diff} \approx 7.40 \times 10^9$ M^{-1} s^{-1} at 298 K, according to the Smoluchowski–Stokes–Einstein theory at 298 K) [17], indicating a ground-state association between HSA and 4-TPyP (static quenching mechanism) [15], agreeing with UV-vis results.

Table 1. Binding parameters for the interaction HSA:4-TPyP in PBS (pH = 7.4) at three different temperatures.

T (K)	$K_{SV} \times 10^4$ (M^{-1}) [a]	$k_q \times 10^{12}$ (M^{-1}s^{-1}) [a]	n [b]	$\Delta H°$ (kJ/mol) [c]	$\Delta S°$ (J/molK) [c]	$\Delta G°$ (kJ/mol)
296	2.57 ± 0.05	5.39 ± 0.11	0.857 ± 0.022	−9.31 ± 1.34	52.9 ± 4.4	−25.0
303	2.30 ± 0.03	4.84 ± 0.06	0.948 ± 0.016			−25.3
310	2.17 ± 0.07	4.55 ± 0.14	1.15 ± 0.04			−25.7

[a] Corresponding to Stern–Volmer plots, [b] Corresponding to double-logarithmic plots, [c] Corresponding to van't Hoff plot.

To further confirm the main fluorescence quenching mechanism detected by both UV-vis analysis and Stern–Volmer approximation, time-resolved fluorescence decays were obtained without and with the maximum porphyrin concentration used in the UV-vis absorption and steady-state fluorescence (Figure 3). The HSA decay without porphyrin showed two fluorescence lifetimes: τ_1 = 1.67 ± 0.13 and τ_2 = 5.67 ± 0.11 ns, with relative percentage of 22.7% and 77.3%, respectively, agreeing with the literature [18,19], while the complex HSA:4-TPyP showed τ_1 = 1.80 ± 0.19 and τ_2 = 5.52 ± 0.13, with a relative percentage of 19.7% and 80.3%, respectively. Amiri and coworkers [20] observed that the fluorescence decay for emissive amino acid Trp in HSA yields three lifetimes. The first two lifetimes are already observed for free-Trp (attributed to excited-state Trp substructures) and the third, longer one, results from interactions between the Trp residue and its microenvironment in the protein matrix. The first lifetime is short (sub nanosecond) with a low pre-exponential factor (3%). Thus, due to the instrumental limitations, most reports (as we in the present work) observe solely two lifetimes. Values of the two fluorescence lifetimes and of their pre-exponentials (this is, the relative population of the Trp excited states which are the origin of the fluorescence) may depend on the microenvironment and on the extension and nature of interactions formed upon HAS: ligand complex formation. The measured lifetimes values and the respective percentages do not vary significantly with the formation of the complex, evidence that the microenvironment around the fluorophore is not perturbed very much with the presence of 4-TPyP.

Since the average fluorescence lifetime of HSA without (4.76 ± 0.11 ns) and with 4-TPyP (4.79 ± 0.14 ns) is the same inside the experimental error, time-resolved fluorescence analysis indicates the absence of dynamic quenching mechanisms and confirmed the ground-state association between HSA and 4-TPyP [15,21]. Therefore, the K_{SV} values can also estimate the binding affinity [15,22], being in the order of 10^4 M^{-1}. For this reason, the double-logarithmic approximation was applied only to determine the number of binding sites (n) in the range of 0.857–1.15, which indicates an interaction in the proportion 1:1—one single albumin molecule binds with one molecule of 4-TPyP [23].

The negative $\Delta G°$ values are consistent with the spontaneity of the binding process in all the evaluated temperatures and since $\Delta H°$ and $\Delta S°$ values are negative and positive, respectively, the association HSA:4-TPyP is enthalpically and entropically driven [24].

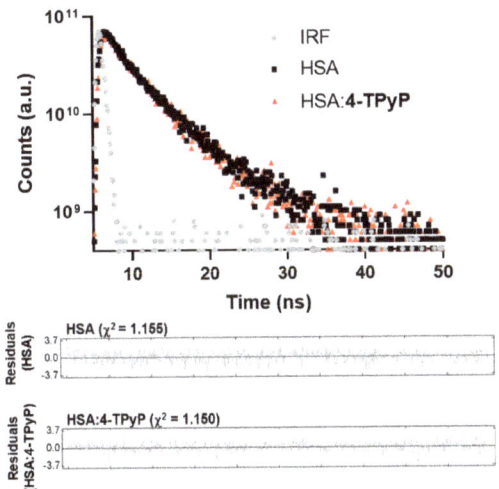

Figure 3. Time-resolved fluorescence decays for HSA without and with 4-TPyP at pH = 7.4 and 296 K using an electrically pumped laser (EPL) with an excitation wavelength of 280 ± 10 nm, pulse width of 850 ps, and a typical average power of 1.8 μW/pulse, monitoring emission at 340 nm. The residuals correspond to the bi-exponential treatment. [HSA] = 1.0×10^{-5} M and [4-TPyP] = 1.32×10^{-5} M.

2.2. Structural and Microenvironment Perturbation of HSA upon 4-TPyP Binding

Comparing the steady-state fluorescence technique with the synchronous fluorescence (SF), the last one has been considered a complementary and more sensitive approach to detect possible perturbations in the microenvironment around the two main fluorophores of albumin (Tyr and Trp residues) after drug binding [25–27]. Figure 4 shows the SF spectra for HSA without and upon successive additions of 4-TPyP at Δλ 15 and 60 nm for Tyr and Trp residues, respectively. For both Δλ, there is a significant decrease in the fluorescence signal upon additions of porphyrin, however, it did not induce any blue- or red-shift, agreeing with steady-state fluorescence data (Figure 2B), which indicated that the binding of 4-TPyP does not induce any significant perturbation on the microenvironment around the fluorophores.

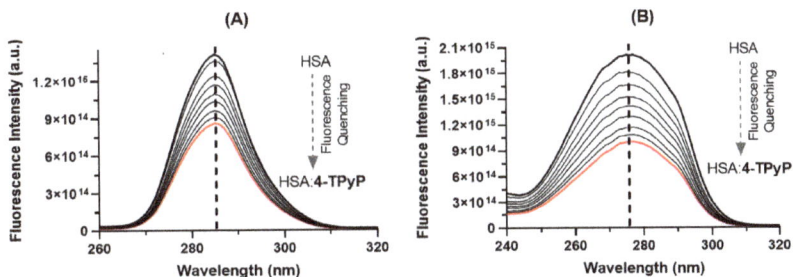

Figure 4. The SF spectra of HSA without and upon successive additions of 4-TPyP at (**A**) Δλ = 15 nm and (**B**) Δλ = 60 nm in pH = 7.4 at room temperature. [HSA] = 1.0×10^{-5} M and [4-TPyP] = 0.17, 0.33, 0.50, 0.66, 0.83, 0.99, 1.15, and 1.32×10^{-5} M.

Circular dichroism (CD) plays an essential role in studying perturbation on the secondary structure of albumin upon drug binding [28,29]. Figure 5A depicts the CD spectra for HSA without and with 4-TPyP, while Figure 5B shows the corresponding secondary structure content. The CD spectra of HSA and HSA:4-TPyP are practically similar in shape and peak position: two minimum peaks, one at 208 nm and the other at 222 nm. Additionally, the secondary structure content did not differ significantly even upon the addition of 4-TPyP in a proportion HSA:4-TPyP of almost 1:13.

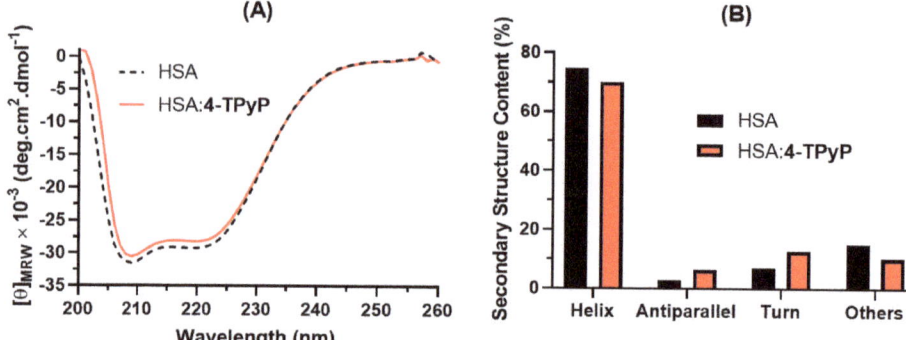

Figure 5. (**A**) Far-UV CD spectra for HSA without and with 4-TPyP in PBS (pH 7.4) at 310K. (**B**) Secondary structure content for HSA without and with 4-TPyP determined by the online server BestSel (Beta Structure Selection http://bestsel.elte.hu/index.php (accessed on 2 June 2022)). [HSA] = 1.0×10^{-6} M and [4-TPyP] = 1.32×10^{-5} M.

Finally, the 3D-fluorescence spectroscopy was applied as additional and conclusive evidence for the experimental data obtained by steady-state fluorescence, SF, and CD spectra [30]. Figure 6 depicts the 3D-fluorescence spectra of HSA and HSA:4-TPyP and their corresponding contour maps, while Table 2 summarizes the fluorescence characteristics of these spectra. Peaks I (λ_{ex} = 280 nm) and II (λ_{ex} = 225 nm) are characteristics of the intrinsic fluorescence spectral behavior of HSA, mainly due to the absorption of the fluorophores Tyr and Trp, however, the peaks "a" ($\lambda_{ex} = \lambda_{em}$) and "b" ($2\lambda_{ex} = \lambda_{em}$) are characteristics of Rayleigh and second-order scattering, respectively [31,32]. The presence of 4-TPyP in the HSA solution did not cause a significant Stokes shift in the position of peaks I and II, as well as the fluorescence intensity was reduced by 10.8% and 16.6% for peaks I and II, respectively, reinforcing the data obtained by the other spectroscopic techniques. It is important to highlight that in this technique, the scattering peaks are so high probably due to the aggregation of 4-TPyP in the PBS medium, indicating that despite the presence of HSA, the non-charged porphyrin 4-TPyP still can form some aggregate.

Table 2. The 3D-fluorescence spectral characteristics for HSA and HSA:4-TPyP in pH = 7.4 at room temperature.

System	Peak	Peak Position ($\lambda_{exc}/\lambda_{em}$ nm/nm)	Intensity $\times 10^3$ (a.u.)
HSA	a	225/225 → 355/355	4.00 → 1.89
	b	230/460	2.01
	I	280/335	2.13
	II	225/330	3.56
HSA:4-TPyP	a	225/225 → 355/355	3.16 → 1.66
	b	230/460	1.66
	I	280/340	1.90
	II	225/335	2.97

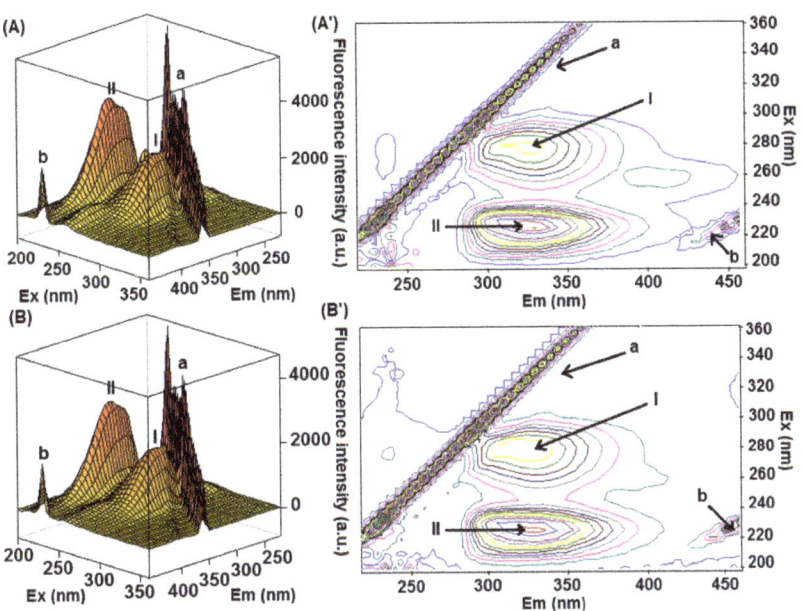

Figure 6. The 3D-fluorescence spectra and the corresponding contour maps for (**A,A'**) HSA and (**B,B'**) HSA:4-TPyP in pH = 7.4 at room temperature. [HSA] = 1.0×10^{-6} M and [4-TPyP] = 1.32×10^{-6} M.

2.3. Molecular-Level Explanation on the Binding Capacity of 4-TPyP to HSA

The 3D structure of HSA has three main binding sites with different specificities: site I, also known as Sudlow's site I, located in the subdomain IIA, site II, also known as Sudlow's site II, located in the subdomain IIIA, and site III located in the subdomain IB [6,33,34]. To suggest the main binding site and to offer a molecular-level explanation of the binding capacity of 4-TPyP to HSA, molecular docking calculations were carried out. The docking score value (dimensionless) for sites I and III was 51.5 and 45.7, respectively, while 4-TPyP did not have any favorable pose for site II. Figure 7A,B depict the superposition of the binding pose of 4-TPyP into subdomains IIA and IB in a cartoon and electrostatic representation, respectively. The 4-TPyP interacts preferentially in a positive electron density pocket being stabilized mainly by hydrogen bonding and van der Waals forces with the amino acid residues of albumin (Figure 7C and Table 3), i.e., not only the main albumin's fluorophore Trp-214 residue interacts with the pyridyl moiety of 4-TPyP via van der Waals forces within a distance of 3.50 Å, but also the amino acid residues His-288, Lys-444, Pro-447, and Val-455 interact by the same intermolecular force (van der Waals) with the pyridyl moieties of 4-TPyP structure within a distance of 3.10, 2.90, 3.70, and 3.20 Å, respectively. On the other hand, the hydrogen atom from the polar group of Lys-195, Arg-218, and Asn-295 is potential donor for hydrogen bonding with 4-TPyP within a distance of 3.40, 3.40, and 3.30 Å, respectively, while the negative charged carboxyl group of Glu-292 is a potential acceptor for hydrogen bonding with the amino group of the tetrapyrrolic core from 4-TPyP structure within distance of 3.40 Å. Finally, molecular docking calculations did not suggest any π-π, π-alkyl, or non-conventional hydrogen bonding interactions between 4-TPyP and the amino acid residues into subdomain IIA.

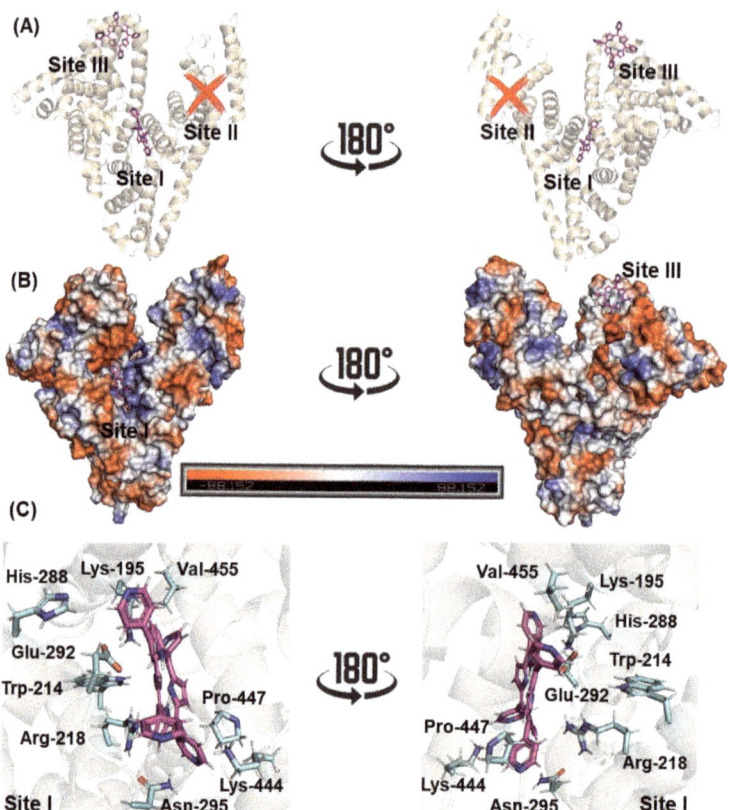

Figure 7. Superposition of molecular docking results for the interaction HSA:4-TPyP into subdomains IIA and IB (sites I and III, respectively) in terms of (**A**) cartoon representation and (**B**) electrostatic potential map for albumin (blue and red for positive and negative electrostatic density, respectively). (**C**) The main amino acid residues that interact with 4-TPyP into site I. Selected amino acid residues and 4-TPyP are represented as sticks in cyan and pink, respectively. Elements' color: hydrogen, nitrogen, and oxygen in white, dark blue, and red, respectively.

Table 3. The main amino acid residues and interactive forces for HSA:4-TPyP in the site I.

Amino Acid Residue	Interaction	Distance (Å)
Lys-195	Hydrogen bonding	3.40
Trp-214	Van der Waals	3.50
Arg-218	Hydrogen bonding	3.40
His-288	Van der Waals	3.10
Glu-292	Hydrogen bonding	3.40
Asn-295	Hydrogen bonding	3.30
Lys-444	Van der Waals	2.90
Pro-447	Van der Waals	3.70
Val-455	Van der Waals	3.20

3. Discussion

The interaction between non-charged or charged porphyrins to albumin has been widely reported as a hot topic by different researchers, however, these studies do not apply the same mathematical approximations and/or methodologies among them [11,14,22,34–40], making a direct comparison with our data difficult. In this sense, for a better interpretation of the binding capacity between HSA and 4-TPyP, reports were selected for porphyrins with both similar structure and experimental/mathematical approaches, more specifically with a non-charged porphyrin (5-phenyl-10,15,20-tri(pyridine-4-yl)-porphyrin, TriPyP) [11] and charged positively porphyrins 5,10,15,20-tetrakis(4-1-benzylpyridinium-4-yl)porphyrin (TBzPyP) [14], 5,10,15,20-tetrakis(1-[Ru(bpy)$_2$Cl]-pyridinium-4-yl)porphyrin (4-RuTPyP) [38], 5,10,15,20-tetrakis(1-methyl-pyridinium-4-porphyrin (4-TMPyP), and 5,10,15,20-tetrakis(1-[Pt(bpy)Cl]-pyridinium-4-yl)porphyrin (4-PtTPyP) [22].

The preliminary binding evaluation was carried out by UV-vis absorption technique, demonstrating hypochromic and hyperchromic effects in the 200–250 and 250–300 nm range, respectively, after the subtraction of the 4-TPyP absorption signal in the UV-vis absorption spectrum of HSA:4-TPyP, indicating that 4-TPyP can cause some perturbation on the albumin structure (analysis in the 200–250 nm range) and there is a ground-state association for HSA:4-TPyP (analysis in the 250–300 nm range) [12,13]. Unfortunately, we did not find any reports for the binding albumin:porphyrin under the same UV-vis approach that we conducted (subtraction of porphyrin absorption signal in the UV-vis absorption profile of albumin:porphyrin) to compare our data with similar porphyrin structures.

The binding of 4-TPyP to HSA does not cause any shift in the maximum fluorescence peak of albumin by both steady-state, synchronous, and 3D-fluorescence techniques, being a clear indication that the evaluated non-charged porphyrin does not perturb the microenvironment around the main fluorophores (Trp, Tyr, and Phe), having the same trend compared to TriPyP [11] and 4-TMPyP [22], however, a different trend considering 4-PtTPyP [22] and 4-RuTPyP [38], indicating that the positive charge in porphyrin structure is not a crucial step in perturbing the microenvironment around the fluorophores, but there is a significant dependence of the steric volume of the groups covalently connected with pyridyl moiety.

The ground-state association for HSA:4-TPyP previously detected by UV-vis analysis was reinforced by both steady-state and time-resolved fluorescence data, showing that the fluorescence quenching mechanism of HSA by 4-TPyP is purely static. Therefore, the double-logarithmic approximation to obtain the K_a values (0.0405, 0.101, and 1.19×10^5 M^{-1} at 296, 303, and 310 K, respectively) is not the best mathematical treatment to estimate the binding affinity of HSA:4-TPyP, reinforcing that in this case, the K_{SV} values besides evaluating the quenching mechanism also determine the binding affinity [15]. As a drug carrier, HSA may aid in the selective delivery of porphyrins to a tumor region and facilitate drug access into the cell via receptor mechanisms (moderate binding affinity for HSA). On the other hand, the same carrier may cause a decrease in the amount of porphyrin available for PDT by its rapid removal from circulation (strong or weak binding affinity for HSA) [37,41]. Since the K_{SV} values are in the order of 10^4 M^{-1}, 4-TPyP binds moderately with albumin, which is favorable to achieving the ideal pharmacokinetic profile for PDT. The same trend was identified in 4-TMPyP and 4-PtTPyP [22], while the porphyrins TBzPyP [14], TriPyP [11], and 4-RuTPyP [38] bind stronger with albumin, indicating that there is not necessarily a charge or steric volume dependence on the binding affinity of pyridyl-porphyrins to albumin but there is a crucial dependence on the nature of the chemical group connected covalently with pyridyl moiety.

In all evaluated temperatures, negative $\Delta G°$ values were obtained, which are consistent with the spontaneity of the binding HSA:4-TPyP and since there are negative and positive values for $\Delta H°$ and $\Delta S°$, respectively, both thermodynamics parameters contribute to the negative $\Delta G°$ value, therefore, the association HSA:4-TPyP is considered enthalpically and entropically driven. According to Ross and Subramanian [42] $\Delta H° < 0$ and $\Delta S° > 0$, indicate that electrostatic forces might contribute significantly to the complex stability, agreeing with in silico data which detected that besides hydrogen bonding, the electrostatic van der Waals interactions play a key intermolecular force for the interaction HSA:4-TPyP into subdomain IIA (site I). Additionally, $\Delta S° > 0$ can also be correlated with the hydrophobic effect governed by desolvation factors upon binding of 4-TPyP into HSA [43]. Interestingly, these results are opposite from those reported for the charged porphyrins 4-RuTPyP [38], 4-

TMPyP, and 4-PtTPyP [22] that identified subdomain IB (site III) as the main binding region, possibly due to the negative electrostatic potential surface of subdomain IB. Unfortunately, there are no in silico or experimental data to identify the main binding site of TBzPyP [14] and TriPyP [11], however, the thermodynamics parameters of TriPyP [11] are like those obtained of 4-TPyP, indicating that the non-charged pyridyl-porphyrins derivatives might interact by the same types of intermolecular forces.

4. Materials and Methods

4.1. Chemicals and Instruments

Commercially available phosphate buffer solution (PBS) and HSA (lyophilized powder, fatty acid-free, globulin free with purity higher than 99%, code A3782-1G) were obtained from Sigma-Aldrich/Merck (St. Louis, MO, USA). One tablet of PBS dissolved in 200 mL of millipore water yields a 0.01 M, 0.0027 M, and 0.137 M of phosphate buffer, potassium chloride, and sodium chloride, respectively, with pH 7.4 at 298 k. Water used in all experiments was Millipore water. Acetonitrile (spectroscopic grade) was obtained from Vetec (Rio de Janeiro, Brazil). The porphyrin 4-TPyP was purchased from PorphyChem (Dijon, France) and a stock solution was prepared in acetonitrile. To increase the solubility of porphyrin in the stock solution the mixture was heated to 323 K and inserted into a home-built ultrasound system. The chemical stability of 4-TPyP under this condition was determined following its corresponding UV-vis profile [10].

The UV-vis, steady-state fluorescence, and circular dichroism (CD) spectra were measured on a Jasco model J-815 optical spectrometer (Easton, MD, USA) and a thermostatic cuvette holder Jasco PFD-425S15F (Easton, MD, USA) was applied to control the temperature in the quartz cell (1.00 cm optical path). All spectra were obtained as the average of three scans with appropriate background corrections. Time-resolved fluorescence measurements were performed on an Edinburgh Instruments fluorimeter model FL920 CD (Edinburgh, UK), equipped with an EPL with excitation wavelength of 280 ± 10 nm, pulse width of 850 ps, and a typical average power of 1.8 µW/pulse, monitoring emission at 340 nm. Synchronous fluorescence (SF) and 3D-fluorescence spectra were performed by the Edinburgh Instruments fluorimeter model Xe900 (Edinburgh, UK).

4.2. Spectroscopic Measurements

The UV-vis spectra were obtained in three different conditions in a quartz cell with 1.00 cm optical path: HSA solution (1.0×10^{-5} M in PBS), 4-TPyP solution (1.32×10^{-5} M in PBS), and HSA:4-TPyP mixture at 310 K in the 200-600 nm range.

The steady-state fluorescence measurements in the 290-450 nm range (λ_{exc} = 280 nm, was used the excitation wavelength of 280 nm instead of 295 nm due to the highest absorption contribution of albumin than porphyrin to minimize the inner filter effect) were carried out for 3.0 mL of HSA solution (1.0×10^{-5} M, in PBS), without and with 4-TPyP (added manually by a micro syringe to achieve final concentrations of 0.17, 0.33, 0.50, 0.66, 0.83, 0.99, 1.15, and 1.32×10^{-5} M) at 296, 303, and 310 K. The maximum concentration of 4-TPyP used in the binding assays corresponds to a stock aliquot of 40 µL of acetonitrile that does not perturb both protein structure and spectroscopic signal [38]. To compensate for the inner filter effect, the maximum steady-state fluorescence intensity values for the system HSA:4-TPyP were corrected by the absorption of porphyrin at excitation (λ = 280 nm) and emission wavelengths (λ = 340 nm), applying Equation (1) [43,44]:

$$F_{cor} = F_{obs} 10^{[\frac{(A_{ex}+A_{em})}{2}]} \tag{1}$$

where F_{cor} and F_{obs} are the corrected and observed steady-state fluorescence intensity values, respectively, while A_{ex} and A_{em} are the absorption value at the excitation (λ = 280 nm, ε = 7537.3 M^{-1}cm^{-1}) and maximum fluorescence emission (λ = 340 nm, ε = 12,659.3 M^{-1}cm^{-1}) wavelengths, respectively.

To obtain quantitative parameters on the binding capacity of HSA:4-TPyP, the maximum fluorescence data after inner filter correction were treated by Stern–Volmer (Equation (2)), double-logarithmic (Equation (3)), van't Hoff (Equation (4)), and Gibbs' free energy (Equation (5)) approximations [15,45–47].

$$\frac{F_0}{F} = 1 + k_q \tau_0 [Q] = 1 + K_{SV}[Q] \quad (2)$$

$$\log\left(\frac{F_0 - F}{F}\right) = n \log [Q] + \log K_a \quad (3)$$

$$\ln K_{SV} = -\frac{\Delta H^\circ}{RT} + \frac{\Delta S^\circ}{R} \quad (4)$$

$$\Delta G^\circ = \Delta H^\circ - T\Delta S^\circ \quad (5)$$

where F_0 and F are the steady-state fluorescence intensities of HSA without and with 4-TPyP, respectively. The $[Q]$, K_{SV}, and k_q are the porphyrin concentration, Stern–Volmer quenching constant, and bimolecular quenching rate constant, respectively. The τ_0 is the obtained experimental average fluorescence lifetime for HSA without 4-TPyP in PBS ($\tau_0 = 4.76 \pm 0.11$ ns), while K_a and n are the binding constant and number of binding sites, respectively. The ΔH°, ΔS°, and ΔG° are the enthalpy, entropy, and Gibbs' free energy change, respectively. Finally, T and R are the temperature (296, 303, and 310 K) and gas constant (8.3145 Jmol^{-1}K^{-1}), respectively.

Time-resolved fluorescence decays were obtained for 3.0 mL of HSA solution (1.0×10^{-5} M, in PBS) without and with 4-TPyP (1.32×10^{-5} M) at room temperature. The instrumental response function (IRF) was collected using a Ludox® dispersion.

The SF spectra were obtained for 3.0 mL of has solution (1.0×10^{-5} M, in PBS) without and upon successive additions of 4-TPyP in the same concentrations of porphyrin used in the steady-state fluorescence studies (0.17, 0.33, 0.50, 0.66, 0.83, 0.99, 1.15, and 1.32×10^{-5} M) at room temperature in the range of 260-320 nm for Tyr ($\Delta\lambda$ = 15 nm) and 240-320 nm for Trp ($\Delta\lambda$ = 60 nm). Finally, the 3D-fluorescence spectra were recorded at the 240–320 nm range for 3.0 mL of HSA solution (1.0×10^{-6} M, in PBS) without and upon addition of 4-TPyP (1.32×10^{-6} M) using λ_{exc} = 200–360 nm and λ_{em} = 200–460 nm, at room temperature.

The CD spectra (200–260 nm) were recorded for HSA without and with 4-TPyP in PBS at 310 K. The HSA concentration was fixed at 1.0×10^{-6} M and porphyrin concentration was those to achieve proportion HSA:4-TPyP of 1:13. The average spectra were obtained from three successive runs and corrected by subtraction of the buffer signal. The CD raw data in ellipticity (Θ_{obs}, in millidegrees) were normalized and expressed as the mean residue weight ($[\Theta]_{MRW}$) in deg. cm^2·dmol^{-1}, defined as $[\Theta]_{MRW} = (\Theta_{obs} \times 10^{-3}) \times 100 \times MW/(l \times c \times NAA)$, where MW is the protein molecular weight, c is the protein concentration in milligrams per milliliter, l is the light path length in centimeters, and NAA is the number of amino acids per protein. The secondary structure content was estimated by analysis of the CD spectra using the online server BestSel (Beta Structure Selection http://bestsel.elte.hu/index.php (accessed on 2 June 2022)).

4.3. Molecular Docking Procedure

The 4-TPyP structure was built and energy-minimized by the Density Functional Theory (DFT) method with the potential B3LYP and basis set 6-31G*, available in the Spartan'14 software (Wavefunction, Inc., Irvine, USA). The crystallographic structure of HSA was obtained in the Protein Data Bank (PDB), with access code 3JRY [48]. Molecular docking calculations were performed with GOLD 2020.2 software (CCDC, Cambridge Crystallographic Data Centre, Cambridge, UK).

Hydrogen atoms were added to HSA according to the data inferred by GOLD 2020.2 software on the ionization and tautomeric states at pH 7.4. Docking interaction cavity in the protein was established with 8 Å radius from the amino acid residues Trp-214, Tyr-411, and Tyr-161 for sites I, II, and III, respectively. The number of genetic operations (crossover, migration, mutation) in each docking run was set to 100,000. The scoring function used was 'ChemPLP', which is the default function of GOLD 2020.2 software. Figures for the docking poses were generated using PyMOL Delano Scientific LLC software (Schrödinger, New York, NY, USA).

Author Contributions: Conceptualization, O.A.C., B.A.I. and C.S.; methodology, O.A.C. and C.S.; software, O.A.C.; validation, O.A.C., B.A.I. and C.S.; formal analysis, O.A.C. and C.S.; investigation, O.A.C.; resources, B.A.I. and C.S.; data curation, O.A.C.; writing—original draft preparation, O.A.C.; writing—review and editing, O.A.C., B.A.I. and C.S.; visualization, O.A.C. and C.S.; supervision, B.A.I. and C.S.; project administration, B.A.I. and C.S.; funding acquisition, B.A.I. and C.S. All authors have read and agreed to the published version of the manuscript.

Funding: This work was financed by national funds through FCT—*Fundação para a Ciência e a Tecnologia, I.P.* (Portugal), under the project UIDB/00313/2020.

Institutional Review Board Statement: Not applicable.

Informed Consent Statement: Not applicable.

Data Availability Statement: All analyzed data are contained in the main text of the article. Raw data are available from the authors upon request.

Acknowledgments: The authors acknowledge Nanci Camara de Lucas Garden from Chemistry Institute at Federal University of Rio de Janeiro (UFRJ, Brazil) for the time-resolved, synchronous, and 3D-fluorescence facilities. O.A. Chaves thanks *Fundação para a Ciência e a Tecnologia* (FCT, Portugal) for his PhD fellowship 2020.07504.BD. B.A. Iglesias also thanks *Conselho Nacional de Desenvolvimento Científico e Tecnológico* (CNPq-Brazil, 409.150/2018-5 and 304.711/2018-7), *Coordenação de Aperfeiçoamento de Pessoal de Nível Superior* (CAPES-Brazil, finance code 001), and *Fundação de Amparo à Pesquisa do Estado do Rio Grande do Sul* (FAPERGS-Brazil, PQ Gaucho-21/2551-0002114-4) for the grants.

Conflicts of Interest: The authors declare no conflict of interest. The funders had no role in the design of the study; in the collection, analysis, or interpretation of data; in the writing of the manuscript; or in the decision to publish the results.

Sample Availability: Not applicable.

References

1. Lee, P.; Wu, X. Review: Modifications of human serum albumin and their binding effect. *Curr. Pharm. Des.* **2015**, *21*, 1862–1865. [CrossRef]
2. Merlot, A.M.; Kalinowski, D.S.; Richardson, D.R. Unraveling the mysteries of serum albumin—More than just a serum protein. *Front. Physiol.* **2014**, *5*, 299. [CrossRef] [PubMed]
3. Chen, B.; Luo, H.; Chen, W.; Huang, Q.; Zheng, K.; Xu, D.; Li, S.; Liu, A.; Huang, L.; Zheng, Y.; et al. Pharmacokinetics, tissue distribution, and human serum albumin binding properties of delicaflavone, a novel anti-tumor candidate. *Front. Pharmacol.* **2021**, *12*, 761884. [CrossRef] [PubMed]
4. Zhivkova, Z.D. Studies on drug-human serum albumin binding: The current state of the matter. *Curr. Pharm. Des.* **2015**, *21*, 1817–1830. [CrossRef] [PubMed]
5. Carter, D.C.; Ho, J.X. Structure of serum albumin. *Adv. Protein Chem.* **1994**, *45*, 153–203.
6. Wardell, M.; Wang, Z.; Ho, J.X.; Robert, J.; Ruker, F.; Ruble, J.; Carter, D.C. The atomic structure of human methemalbumin at 1.9 A. *Biochem. Biophys. Res. Commun.* **2002**, *291*, 913–918. [CrossRef]
7. Sudlow, G.; Birkett, D.J.; Wade, D.N. Further characterization of specific drug binding sites on human serum albumin. *Mol. Pharmacol.* **1976**, *12*, 1052–1061.
8. Kou, J.; Dou, D.; Yang, L. Porphyrin photosensitizers in photodynamic therapy and its applications. *Oncotarget* **2017**, *8*, 81591–81603. [CrossRef]
9. Amos-Tautua, B.M.; Songca, S.P.; Oluwafemi, O.S. Application of porphyrins in antibacterial photodynamic therapy. *Molecules* **2019**, *24*, 2456. [CrossRef]
10. Tasso, T.T.; Tsubone, T.M.; Baptista, M.S.; Mattiazzi, L.M.; Acunha, T.V.; Iglesias, B.A. Isomeric effect on the properties of tetraplatinated porphyrins showing optimized phototoxicity for photodynamic therapy. *Dalton Trans.* **2017**, *46*, 11037–11045. [CrossRef]
11. Chen, X.; Cai, Y.; Zhao, Y.; Ma, H.; Wu, D.; Du, B.; Wei, Q. Quenching and binding mechanism of the intrinsic fluorescence of bovine serum albumin by 5-phenyl-10,15,20-tri-(4-pyridyl)-porphyrin. *J. Porphyr. Phthalocyanines* **2009**, *13*, 933–938. [CrossRef]

12. Jokar, S.; Pourjavadi, A.; Adeli, M. Albumin–graphene oxide conjugates; carriers for anticancer drugs. *RSC Adv.* **2014**, *4*, 33001. [CrossRef]
13. Zhang, W.; Wang, F.; Xiong, X.; Ge, Y.; Liu, Y. Spectroscopic and molecular docking studies on the interaction of dimetridazole with human serum albumin. *J. Chil. Chem. Soc.* **2013**, *58*, 1717–1721. [CrossRef]
14. Bordbar, A.K.; Eslami, A.; Tangestaninejad, S. Spectral investigations of the solution properties of 5,10,15,20-tetrakis(4-N-benzylpyridyl)porphyrin (TBzPyP) and its interaction with human serum albumin (HSA). *J. Porphyr. Phthalocyanines* **2002**, *6*, 225–232. [CrossRef]
15. Lakowicz, J.R. *Principles of Fluorescence Spectroscopy*, 3rd ed.; Springer: New York, NY, USA, 2006.
16. Chaves, O.A.; da Silva, V.A.; Sant'Anna, C.M.R.; Ferreira, A.B.B.; Ribeiro, T.A.N.; de Carvalho, M.G.; Cesarin-Sobrinho, D.; Netto-Ferreira, J.C. Binding studies of lophirone B with bovine serum albumin (BSA): Combination of spectroscopic and molecular docking techniques. *J. Mol. Struct.* **2017**, *1128*, 606–611. [CrossRef]
17. Montalti, M.; Credi, A.; Prodi, L.; Gandolfi, M.T. *Handbook of Photochemistry*, 3rd ed.; CRC Press; Taylor & Francis: Boca Raton, FL, USA, 2006.
18. Sun, H.; Liu, Y.; Li, M.; Han, S.; Yang, X.; Liu, R. Toxic effects of chrysoidine on human serum albumin: Isothermal titration calorimetry and spectroscopic investigations. *Luminescence* **2016**, *31*, 335–340. [CrossRef]
19. Chaves, O.A.; Soares, M.A.G.; de Oliveira, M.C.C. Monosaccharides interact weakly with human serum albumin. Insights for the functional perturbations on the binding capacity of albumin. *Carbohydr. Res.* **2021**, *501*, 108274. [CrossRef]
20. Amiri, M.; Jankeje, K.; Albani, J.R. Origin of fluorescence lifetimes in human serum albumin. Studies on native and denatured protein. *J. Fluoresc.* **2010**, *20*, 651–656. [CrossRef]
21. Chaves, O.A.; Cesarin-Sobrinho, D.; Sant'Anna, C.M.R.; de Carvalho, M.G.; Suzart, L.R.; Catunda-Junior, F.E.A.; Netto-Ferreira, J.C.; Ferreira, A.B.B. Probing the interaction between 7-O-β-D-glucopyranosyl-6-(3-methylbut-2-enyl)-5,4′-dihydroxyflavonol with bovine serum albumin (BSA). *J. Photochem. Photobiol. A* **2017**, *336*, 32–41. [CrossRef]
22. Chaves, O.A.; Acunha, T.V.; Iglesias, B.A.; Jesus, C.S.H.; Serpa, C. Effect of peripheral platinum(II) bipyridyl complexes on the interaction of tetra-cationic porphyrins with human serum albumin. *J. Mol. Liq.* **2020**, *301*, 112466. [CrossRef]
23. Safarnejad, A.; Shaghaghi, M.; Dehghan, G.; Soltani, S. Binding of carvedilol to serum albumins investigated by multi-spectroscopic and molecular modeling methods. *J. Lumin.* **2016**, *176*, 149–158. [CrossRef]
24. Caruso, I.P.; Filho, J.M.B.; de Araújo, A.S.; de Souza, F.P.; Fossey, M.A.; Cornélio, M.L. An integrated approach with experimental and computational tools outlining the cooperative binding between 2-phenylchromone and human serum albumin. *Food Chem.* **2016**, *196*, 935–942. [CrossRef]
25. Karthiga, D.; Chandrasekaran, N.; Mukherjee, N. Spectroscopic studies on the interactions of bovine serum albumin in presence of silver nanorod. *J. Mol. Liq.* **2017**, *232*, 251–257. [CrossRef]
26. Hashempour, S.; Shahabadi, N.; Adewoye, A.; Murphy, B.; Rouse, C.; Salvatore, B.A.; Stratton, C.; Mahdavian, E. Binding studies of AICAR and human serum albumin by spectroscopic, theoretical, and computational methodologies. *Molecules* **2020**, *25*, 5410. [CrossRef] [PubMed]
27. Byadagi, K.; Meti, M.; Nandibewoor, S.; Chimatadar, S. Investigation of binding behaviour of procainamide hydrochloride with human serum albumin using synchronous, 3D fluorescence and circular dichroism. *J. Pharm. Anal.* **2017**, *7*, 103–109. [CrossRef] [PubMed]
28. Chaves, O.A.; de Oliveira, C.H.C.S.; Ferreira, R.C.; Pereira, R.P.; de Melos, J.L.R.; Rodrigues-Santos, C.E.; Echevarria, A.; Cesarin-Sobrinho, D. Investigation of interaction between human plasmatic albumin and potential fluorinated anti-trypanosomal drugs. *J. Fluor. Chem.* **2017**, *199*, 103–112. [CrossRef]
29. Bertucci, C.; de Simone, A.; Pistolozzi, M.; Rosini, M. Reversible human serum albumin binding of lipocrine: A circular dichroism study. *Chirality* **2011**, *23*, 827–832. [CrossRef] [PubMed]
30. Buddanavar, A.T.; Nandibewoor, S.T. Multi-spectroscopic characterization of bovine serum albumin upon interaction with atomoxetine. *J. Pharm. Sci.* **2017**, *7*, 148–155. [CrossRef]
31. Nasruddin, A.N.; Feroz, S.R.; Mukarram, A.K.; Mohamad, S.B.; Tayyab, S. Fluorometric and molecular docking investigation on the binding characteristics of SB202190 to human serum albumin. *J. Lumin.* **2016**, *174*, 77–84. [CrossRef]
32. Sharma, A.S.; Anandakumar, S.; Ilanchelian, M. A combined spectroscopic and molecular docking study on site selective binding interaction of Toluidine blue O with Human and Bovine serum albumins. *J. Lumin.* **2014**, *151*, 206–218. [CrossRef]
33. Johansson, E.; Nielsen, A.D.; Demuth, H.; Wiberg, C.; Schjødt, C.B.; Huang, T.; Chen, J.; Jensen, S.; Petersen, J.; Thygesen, P. Identification of binding sites on human serum albumin for somapacitan, a long-acting growth hormone derivative. *Biochemistry* **2020**, *59*, 1410–1419. [CrossRef] [PubMed]
34. Da Silveira, C.H.; Chaves, O.A.; Marques, A.C.; Rosa, N.M.O.; Costa, L.A.S.; Iglesias, B.A. Synthesis, photophysics, computational approaches, and biomolecule interactive studies of metalloporphyrins containing pyrenyl units: Influence of the metal center. *Europ. J. Inorg. Chem.* **2022**, *12*, e202200075. [CrossRef]
35. Rinco, O.; Brenton, J.; Douglas, A.; Maxwell, A.; Henderson, M.; Indrelie, K.; Wessels, J.; Widin, J. The effect of porphyrin structure on binding to human serum albumin by fluorescence spectroscopy. *J. Photochem. Photobiol. A* **2009**, *208*, 91–96. [CrossRef]
36. Yang, L.; Ji, Z.; Peng, Z.; Cheng, G. A novel low symmetry sulfur containing porphyrazine: Synthesis and its interaction with serum albumin. *J. Porphyr. Phthalocyanines* **2003**, *7*, 420–425. [CrossRef]

37. Cohen, S.; Margalit, R. Binding of porphyrin to human serum albumin. Structure–activity relationships. *Biochem. J.* **1990**, *270*, 325–330. [CrossRef]
38. Chaves, O.A.; Menezes, L.B.; Iglesias, B.A. Multiple spectroscopic and theoretical investigation of meso-tetra-(4-pyridyl)porphyrin ruthenium(II) complexes in HSA binding studies. Effect of Zn(II) in protein binding. *J. Mol. Liq.* **2019**, *294*, 111581. [CrossRef]
39. Ding, Y.; Lin, B.; Huie, C.W. Binding studies of porphyrins to human serum albumin using affinity capillary electrophoresis. *Electrophoresis* **2001**, *22*, 2210–2216. [CrossRef]
40. Kubát, P.; Lang, K.; Anzenbacher, P. Modulation of porphyrin binding to serum albumin by pH. *Biochim. Biophys. Acta (BBA)-Gen. Subj.* **2004**, *1670*, 40–48. [CrossRef]
41. Naveenraj, S.; Anandan, S. Binding of serum albumins with bioactive substances—Nanoparticles to drugs. *J. Photochem. Photobiol. C* **2013**, *14*, 53–71. [CrossRef]
42. Ross, P.D.; Subramanian, S. Thermodynamics of protein association reactions: Forces contributing to stability. *Biochemistry* **1981**, *20*, 3096–3102. [CrossRef]
43. Chaves, O.A.; Jesus, C.S.H.; Cruz, P.F.; Sant'Anna, C.M.R.; Brito, R.M.M.; Serpa, C. Evaluation by fluorescence, STD-NMR, docking and semi-empirical calculations of the o-NBA photo-acid interaction with BSA. *Spectrochim. Acta A Mol. Biomol. Spectrosc.* **2016**, *169*, 175–181. [CrossRef]
44. Matei, I.; Hillebrand, M. Interaction of kaempferol with human serum albumin: A fluorescence and circular dichroism study. *J. Pharm. Biomed. Anal.* **2010**, *51*, 768–773. [CrossRef]
45. Chaves, O.A.; Amorim, A.P.O.; Castro, L.H.E.; de Sant'Anna, C.M.R.; de Oliveira, M.C.C.; Cesarin-Sobrinho, D.; Netto-Ferreira, J.C.; Ferreira, A.B.B. Fluorescence and docking studies of the interaction between human serum albumin and pheophytin. *Molecules* **2015**, *20*, 19526–19539. [CrossRef] [PubMed]
46. Zaidi, N.; Ajmal, M.R.; Rabbani, G.; Ahmad, E.; Khan, R.H. A comprehensive insight into binding of hippuric acid to human serum albumin: A study to uncover its impaired elimination through hemodialysis. *PLoS ONE* **2013**, *8*, e71422. [CrossRef] [PubMed]
47. Grossweiner, L.I. A note on the analysis of ligand binding by the double-logarithmic plot. *J. Photochem. Photobiol. B* **2000**, *58*, 175–177. [CrossRef]
48. Hein, K.L.; Kragh-Hansen, U.; Morth, J.P.; Jeppesen, M.D.; Otzen, D.; Moller, J.V.; Nissen, P. Crystallographic analysis reveals a unique lidocaine binding site on human serum albumin. *J. Struct. Biol.* **2010**, *171*, 353–360. [CrossRef]

Article

8(*meso*)-Pyridyl-BODIPYs: Effects of 2,6-Substitution with Electron-Withdrawing Nitro, Chloro, and Methoxycarbonyl Groups

Caroline Ndung'U [1], Petia Bobadova-Parvanova [2], Daniel J. LaMaster [1], Dylan Goliber [2], Frank R. Fronczek [1] and Maria da Graça H. Vicente [1,*]

[1] Department of Chemistry, Louisiana State University, Baton Rouge, LA 70803, USA; cndung2@lsu.edu (C.N.); daniel.j.lamaster@gmail.com (D.J.L.); ffroncz@lsu.edu (F.R.F.)
[2] Department of Chemistry and Fermentation Sciences, Appalachian State University, Boone, NC 28608, USA; bobadovap@appstate.edu (P.B.-P.); goliberda@appstate.edu (D.G.)
* Correspondence: vicente@lsu.edu

Abstract: The introduction of electron-withdrawing groups on 8(*meso*)-pyridyl-BODIPYs tends to increase the fluorescence quantum yields of this type of compound due to the decrease in electronic charge density on the BODIPY core. A new series of 8(*meso*)-pyridyl-BODIPYs bearing a 2-, 3-, or 4-pyridyl group was synthesized and functionalized with nitro and chlorine groups at the 2,6-positions. The 2,6-methoxycarbonyl-8-pyridyl-BODIPYs analogs were also synthesized by condensation of 2,4-dimethyl-3-methoxycarbonyl-pyrrole with 2-, 3-, or 4-formylpyridine followed by oxidation and boron complexation. The structures and spectroscopic properties of the new series of 8(*meso*)-pyridyl-BODIPYs were investigated both experimentally and computationally. The BODIPYs bearing 2,6-methoxycarbonyl groups showed enhanced relative fluorescence quantum yields in polar organic solvents due to their electron-withdrawing effect. However, the introduction of a single nitro group significantly quenched the fluorescence of the BODIPYs and caused hypsochromic shifts in the absorption and emission bands. The introduction of a chloro substituent partially restored the fluorescence of the mono-nitro-BODIPYs and induced significant bathochromic shifts.

Keywords: BODIPY; pyridyl; nitro; chloro; methyl ester

1. Introduction

Over the last three decades, boron dipyrromethene, abbreviated BODIPY, dyes have been synthesized and investigated [1–3] for a wide range of applications. These include their use as photosensitizers for photodynamic therapy, as biological labels, as imaging agents [4,5], as fluorescence switches, and as laser dyes [6–8]. These diverse BODIPY applications are due to their remarkable properties, which include large molar absorption coefficients [7–10], high photochemical stability, narrow spectral band widths, generally high fluorescence quantum yields, and low cytotoxicity [1–3]. In addition, due to their tunable properties, BODIPY applications have continued to attract researchers' attention.

BODIPYs are easily functionalized at all the carbon atoms and at the boron center, enabling the fine tuning of their chemical and photophysical properties for a particular application [11,12]. In particular, the introduction of water-solubilizing groups, including phosphates [13], sulfonates [14], carboxylates [15], carbohydrates [16], and oligoethylene glycol chains [17], has allowed the development of water-soluble BODIPYs for biological applications. Among the cationic BODIPY derivatives, 8(*meso*)-pyridyl-substituted BODIPYs have received special attention. This is due to their ability to coordinate with metals and the easy protonation or alkylation of the pyridyl groups, which has led to their use as pH sensors [18,19], as mitochondria-specific probes [20,21], as photosensitizers for antimicrobial photodynamic inactivation [22,23], as metal ion sensors [8,24], as G-series nerve agent sensors [25], and as photocatalysts for hydrogen production [26,27]. We have recently reported that the functionalization of 8(*meso*)-pyridyl-BODIPYs with moderate electron-withdrawing chlorine atoms at the 2,6-positions induces bathochromic shifts in the absorption and emission wavelengths of the resulting BODIPYs, increases their reduction potentials, and enhances their relative fluorescence quantum yields [28,29]. We hypothesized that stronger electron-withdrawing groups at the 2,6-positions of 8(*meso*)-pyridyl-BODIPYs would have stronger effects on their photophysical and electrochemical properties. Furthermore, mono-2-functionalized BODIPY derivatives and the introduction of two different electron-withdrawing groups at the 2 and 6 positions induce molecular asymmetry and can, therefore, also affect their properties. Herein, we report the synthesis and investigation of 8(*meso*)-pyridyl-BODIPYs bearing electron-withdrawing 2,6-methylester substituents and the asymmetric 2-nitro and 2-chloro-6-nitro BODIPY derivatives. While the nitro and chlorine groups can be directly introduced by functionalization of the BODIPY via electrophilic substitution, the 2,6-methoxycarbonyl-8-pyridyl-BODIPYs were prepared from 2,4-dimethyl-3-methoxycarbonyl-pyrrole and the corresponding pyridine carboxaldehyde. The presence of the methyl groups at the 1,7-positions of these 8(*meso*)-pyridyl-BODIPYs prevents the rotation of the *meso*-pyridyl groups, enhancing their fluorescence properties and inducing nearly perpendicular dihedral angles between the pyridyl group and the BODIPY core [11,27,28].

2. Results and Discussion

2.1. Synthesis

The syntheses of **2PyCO$_2$Me**, **3PyCO$_2$Me**, and **4PyCO$_2$Me** were accomplished as shown in Scheme 1. The 2,4-dimethyl-3-methoxycarbonyl-pyrrole was prepared from *tert*-butyl acetoacetate and methyl acetoacetate in three steps using the Knorr pyrrole synthesis [30]. The condensation of 2,4-dimethyl-3-methoxycarbonyl-pyrrole with 2-, 3-, or 4-formylpyridine in the presence of TFA followed by oxidation with DDQ (2,3-dichloro-5,6-dicyano-1,4-benzoquinone) and boron complexation using boron trifluoride diethyl etherate produced the corresponding functionalized BODIPYs in 45%, 40%, and 28% yields, respectively. The lower yield obtained for **4PyCO$_2$Me** is due to its lower stability on silica gel, which made its purification with chromatography more challenging. The structures of the three isomers were confirmed with ^{1}H-NMR and X-ray crystallography. Interestingly, these molecules can adopt different conformations depending on the orientation of the carbonyl groups on the methyl esters relative to the BODIPY core (see Section 2.2).

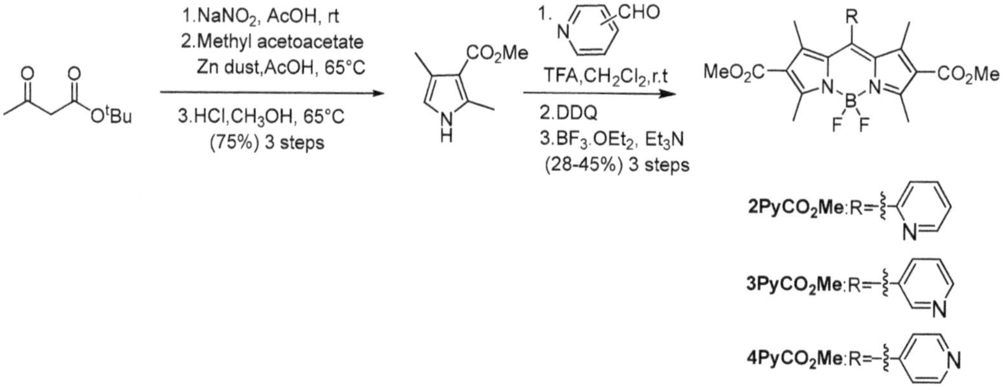

Scheme 1. Synthesis of **2PyCO$_2$Me**, **3PyCO$_2$Me**, and **4PyCO$_2$Me**.

The mono-nitrated derivatives **2PyNO₂**, **3PyNO₂**, and **4PyNO₂** were obtained by nitration of the corresponding 8(*meso*)-pyridyl-BODIPYs using NO_2BF_4 in dichloroethane [31] in 82%, 82%, and 77% yield, respectively, as shown in Scheme 2. The single pyrrolic proton of these BODIPY derivatives was clearly observed in the ^{1}H-NMR spectrum at approximately 6.5 ppm (see Supplemental Materials). Chlorination of these three compounds using trichloroisocyanuric acid (TCCA) in dichloromethane [28,29] gave the corresponding **2PyNO₂Cl**, **3PyNO₂Cl**, and **4PyNO₂Cl** BODIPY derivatives in nearly quantitative yields. The structures of all BODIPYs were confirmed with ^{1}H, ^{13}C NMR, HRMS, and X-ray crystallography (see Supplemental Materials).

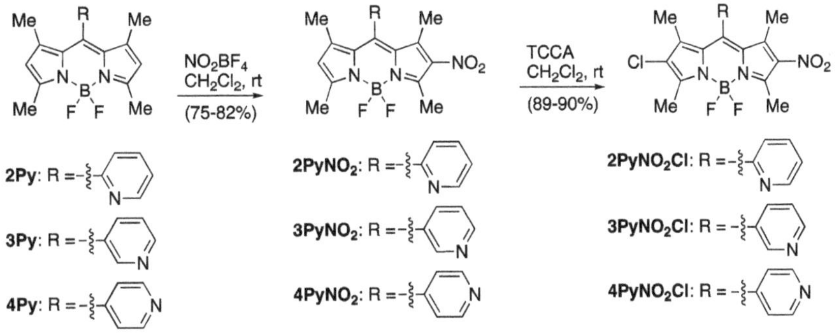

Scheme 2. Direct functionalization of **2Py**, **3Py**, and **4Py** with nitration and chlorination.

2.2. X-ray and Computational Structural Analysis

Crystals suitable for X-ray analysis were obtained for **2PyCO₂Me**, **4PyCO₂Me**, **2PyNO₂**, **3PyNO₂**, **4PyNO₂**, **2PyNO₂Cl**, **3PyNO₂Cl**, and **4PyNO₂Cl** with slow evaporation of dichloromethane and hexanes. The structures are shown in Figure 1. For **2PyCO₂Me**, the 12-atom BODIPY core is fairly planar with a mean deviation of 0.029 Å. The 2-pyridyl substituent is disordered by a 180° rotation, switching the 2 and 6 positions. Only one orientation is shown in Figure 1. The 2-pyridyl and the BODIPY core planes form a dihedral angle of 84.7°, and the two CO₂Me planes form dihedral angles of 16.0° and 22.1° with the BODIPY core. For the **4PyCO₂Me**, the central six-membered C_3N_2B ring has an envelope distortion from planarity with the B atom lying 0.133 Å out of the plane of the other five atoms. The 12-atom BODIPY core thus has a bowed conformation with the C atoms at the 2 and 6 positions carrying the ester substituents lying 0.127 Å and 0.128 Å out of the core plane both on the opposite side from the B atom. The 4-pyridyl ring forms a dihedral angle of 84.0° with the BODIPY core, and the CO₂Me planes form dihedral angles of 8.7° and 11.7° with the BODIPY core. The C=O groups are nearly in the planes of their respective pyrroles with the O atoms lying only 0.023 Å and 0.045 Å out of the planes.

The performed computational studies are in agreement with the above experimental findings. We investigated three different conformations computationally with the two C=O bonds oriented toward the *meso*-pyridyl ring with the two C=O bonds oriented away from the *meso*-pyridyl ring and with one C=O bond toward and one away from the *meso*-pyridyl ring (Figure S1a, Supplemental Materials). The energy differences between the three possible structures are very small (between 0.1 and 0.7 kcal/mol); therefore, all of them likely exist in solution. The pyridyl rings are nearly perpendicular to the BODIPY core with dihedral angles of 86° to 87°. The two CO₂Me groups are not co-planar with the BODIPY core, forming dihedral angles of 18° to 19° for **2PyCO₂Me**, **3PyCO₂Me**, and **4PyCO₂Me**.

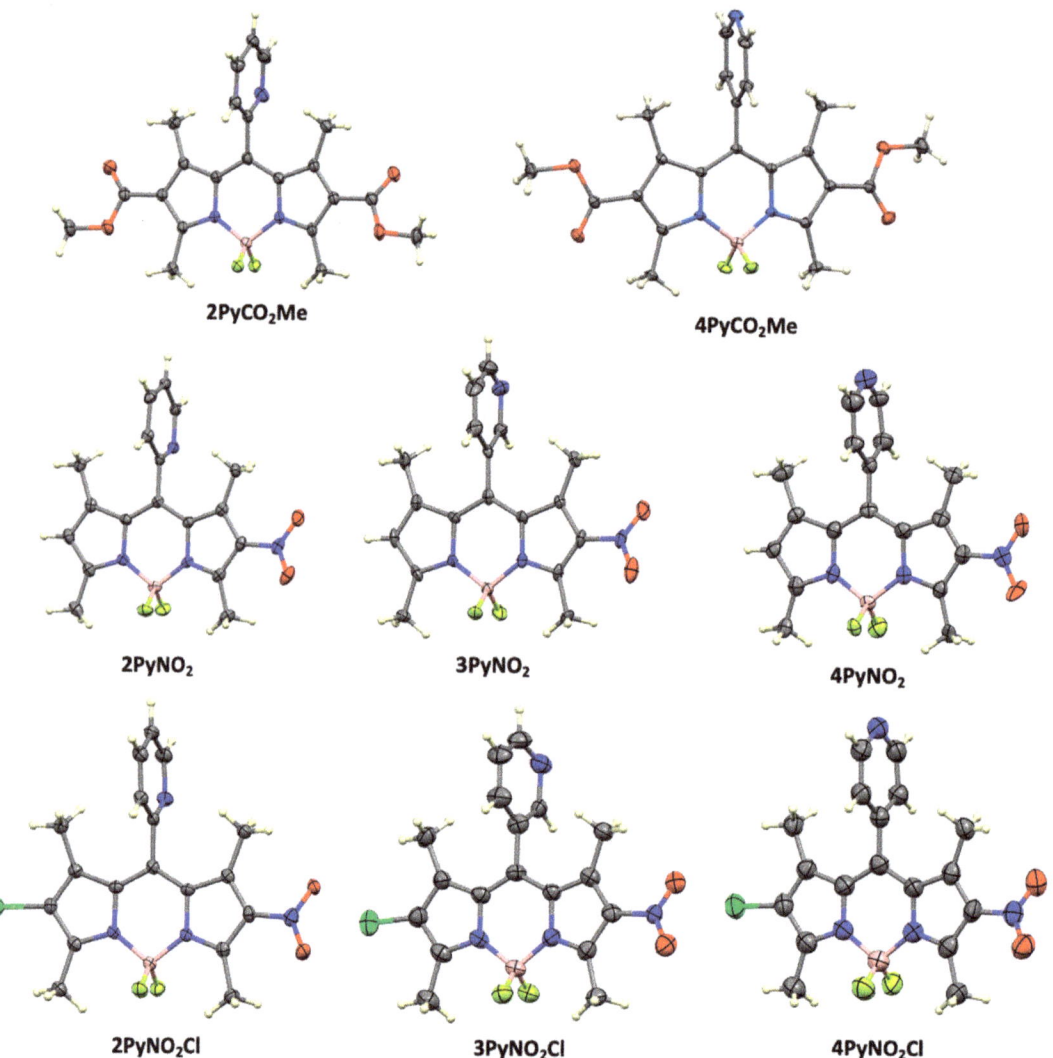

Figure 1. Crystal structures of **2PyCO₂Me**, **4PyCO₂Me**, **2PyNO₂**, **3PyNO₂**, **4PyNO₂**, **2PyNO₂Cl**, **3PyNO₂Cl**, and **4PyNO₂Cl** with 50% ellipsoids.

For the **2PyNO₂**, the 12-atom BODIPY core is fairly planar with a mean deviation of 0.027 Å. The 2-pyridyl ring forms a dihedral angle of 84.9° with the BODIPY core, and the nitro group makes a dihedral angle of 19.6° with the BODIPY core. In the case of **3PyNO₂**, the molecular structure is very similar to that of the 2-pyridyl isomer with a mean deviation of the 12 BODIPY core atoms of 0.015 Å, the 3-pyridyl ring having an 88.8° dihedral angle with it, and the nitro group having a 23.7° dihedral angle with it. Similarly, for the **4PyNO₂**, the molecular structure as the hexane solvate is also similar to those of the 2- and 3-pyridyl isomers. The mean deviation from the BODIPY plane is 0.031 Å. The 4-pyridyl ring makes a dihedral angle of 86.3° with it. The nitro group appears less tilted out of the BODIPY plane, however, having a 4.4° dihedral angle with the BODIPY core. As previously observed [29], the dihedral angle of the *meso*-pyridyl group with the BODIPY core is slightly lower in the case of the 2-pyridyl compared with the 3- and 4-pyridyl analogs, although the difference is small (1–4°).

The computational modeling predicts similar structures and small differences between **2PyNO$_2$**, **3PyNO$_2$**, and **4PyNO$_2$** (Figure S1b, Supplemental Materials). The pyridyl rings form dihedral angles of 81°, 80°, and 82°, respectively. The nitro group is oriented at 15°, 16°, and 17°, respectively. In the case of **3PyNO$_2$** and **4PyNO$_2$**, two possible structures were investigated, i.e., with the pyridyl nitrogen oriented away from the viewer or with the pyridyl nitrogen oriented toward the viewer. The energy differences are very small (0.1–0.2 kcal/mol); therefore, both orientations are possible in solution.

For the **2PyNO$_2$Cl**, the molecule is disordered in the crystal with the nitro and chloro substituents swapped approximately 13% of the time. Only one orientation is shown in Figure 1. The 12-atom BODIPY core is fairly planar with a 0.039 Å mean deviation. The 2-pyridyl group forms a dihedral angle of 84.4° with the BODIPY core. The nitro group in this compound forms a dihedral angle of 21.0° (weighted average of two) with the BODIPY core. The structure of **3PyNO$_2$Cl** has three independent molecules, two of which lie on mirror planes in the crystal. There is no disorder between the nitro and chloro groups as was observed in the 2-pyridyl isomer; however, for the mirror-symmetric molecules, the 3-pyridyl group, which lies across the mirror, is necessarily disordered. In the mirror-symmetric molecules, the BODIPY cores are planar by symmetry, the nitro groups are coplanar with the BODIPY core, and the 3-pyridyl rings are exactly perpendicular by symmetry to the BODIPY cores. For the asymmetric molecule, the BODIPY core is nearly planar with a 0.019 Å mean deviation, the 3-pyridyl ring forms a dihedral angle of 80.0° with the core, and the nitro group has a 16.1° dihedral angle with the core. Since the displacement parameters are large for this structure determination, there may be unresolved disorder that affects the above values. For the **4PyNO$_2$Cl**, there are two independent molecules with nearly identical conformations with the BODIPY core being nearly planar. Averaged over the two molecules, the mean core deviation is 0.019 Å, the 4-pyridyl dihedral angle with the BODIPY core is 84.8°, and the nitro group dihedral angle is 18.6°.

Computational modelling predicts a pyridyl-BODIPY dihedral angle of 81°, 79°, and 81° for **2PyNO$_2$Cl**, **3PyNO$_2$Cl**, and **4PyNO$_2$Cl**, respectively, and a nitro-BODIPY dihedral angle of 16° for all three molecules (Figure S1c, Supplemental Materials). The two possible orientations of the pyridyl ring (pyridyl nitrogen toward or away from the viewer) showed very small energy differences (0.1–0.2 kcal/mol) in the case of **3PyNO$_2$Cl** and **4PyNO$_2$Cl**; therefore, both orientations are possible in solution.

2.3. Spectroscopic Properties

The absorption and emission spectra of the BODIPYs were obtained at room temperature in acetonitrile, methanol, and toluene, and the results are summarized in Tables 1 and S1 (Supplemental Materials). The computationally modeled spectroscopic and electronic properties are given in Table 2. The results obtained for the 2,6-unsubstituted 8(*meso*)-pyridyl-BODIPYs [28] are also given for comparison purposes. All BODIPYs exhibit a characteristic strong absorption peak attributed to the S_0-S_1 (π-π^*) transition that appears at similar maximum absorption and emission wavelength values within each individual series of pyridyl-BODIPYs. The performed calculations show that this is the dominant transition contributing approximately 70% to the total absorption or emission. The next excited singlet state is more than 0.9 eV higher in energy.

Table 1. Spectroscopic properties of *meso*-pyridyl-BODIPYs in CH_3CN and CH_3OH.

Solvent	BODIPY	λ_{abs} (nm)	λ_{em} (nm)	Stokes Shift (nm)	Φ_f [a]	ε (M^{-1} cm^{-1})
CH_3CN	2Py [b]	502	514	12	0.04	87,500
	3Py [b]	502	514	12	0.43	92,900
	4Py [c]	501	515	14	0.31	72,100
	2PyCO$_2$Me	501	515	14	0.09	64,160
	3PyCO$_2$Me	501	512	11	0.61	87,690
	4PyCO$_2$Me	500	510	10	0.43	84,500
	2PyNO$_2$	491	509	18	0.08	55,810
	3PyNO$_2$	491	507	16	0.26	54,740
	4PyNO$_2$	490	508	18	0.25	43,450
	2PyNO$_2$/Cl	512	527	15	0.13	23,790
	3PyNO$_2$/Cl	512	526	14	0.36	23,850
	4PyNO$_2$/Cl	511	525	14	0.28	16,510
CH_3OH	2PyCO$_2$Me	502	516	14	0.21	88,200
	3PyCO$_2$Me	501	514	13	0.61	70,760
	4PyCO$_2$Me	501	516	15	0.39	72,880
	2PyNO$_2$	491	511	20	0.09	64,750
	3PyNO$_2$	491	510	19	0.13	587,240
	4PyNO$_2$	491	510	19	0.12	52,160
	2PyNO$_2$/Cl	511	530	19	0.20	37,730
	3PyNO$_2$/Cl	511	529	18	0.35	51,110
	4PyNO$_2$/Cl	511	529	18	0.32	41,050

[a] Calculated using rhodamine 6G ($\Phi_f = 0.86$) in methanol at $\lambda_{exc} = 473$ nm as the standard. [b] Previous work from this laboratory [29]. [c] Previous work from this laboratory [28].

Table 2. CAM-B3LYP/6-31+G(d,p) calculated spectroscopic and electronic properties of pyridyl BODIPYs (in vacuum). All transitions are S_0-S_1.

BODIPY	λ_{abs} (nm)	Oscillator Strength	HOMO (eV)	LUMO (eV)	HOMO-LUMO Gap (eV)	Dipole Moment (D)	λ_{em} (nm)	Stokes Shift (nm)
2Py [a]	416	0.542	−6.77	−1.62	5.14	6.05	430	14
3Py [a]	416	0.542	−6.92	−1.78	5.14	3.94	429	13
4Py [a]	415	0.543	−6.96	−1.82	5.15	2.13	428	12
2PyCO$_2$Me	413	0.797	−7.21	−2.02	5.19	3.00	436	23
3PyCO$_2$Me	413	0.793	−7.36	−2.18	5.18	1.89	430	17
4PyCO$_2$Me	412	0.794	−7.40	−2.21	5.19	1.05	434	22
2PyNO$_2$	408	0.655	−7.44	−2.22	5.21	9.51	436	28
3PyNO$_2$	407	0.656	−7.58	−2.37	5.21	8.02	429	22
4PyNO$_2$	406	0.660	−7.62	−2.40	5.22	7.54	431	25
2PyNO$_2$Cl	417	0.673	−7.55	−2.40	5.14	8.14	443	26
3PyNO$_2$Cl	416	0.675	−7.69	−2.55	5.14	6.44	438	21
4PyNO$_2$Cl	416	0.676	−7.73	−2.58	5.15	5.66	440	24

[a] Previous work from this laboratory [29].

The similarity in the observed absorption and emission wavelengths correlates with the similarity in the calculated absorption and emission wavelengths and the almost identical HOMO–LUMO gaps observed within a given series (see Table 2). For the entire series of 8(*meso*)-pyridyl-BODIPYs, the HOMO is almost entirely localized on the BODIPY core, while the LUMO partially involves the pyridyl group (Figures 2, A1 and A2). This is consistent with our previous findings for the symmetric 2,6-unsubstituted and 2,6-dichloro-substituted 8(*meso*)-pyridyl-BODIPYs [29]. The shapes of the LUMO orbitals reflect the electron-withdrawing effect of the methyl ester and nitro substituents with the nitro effect being much more significant.

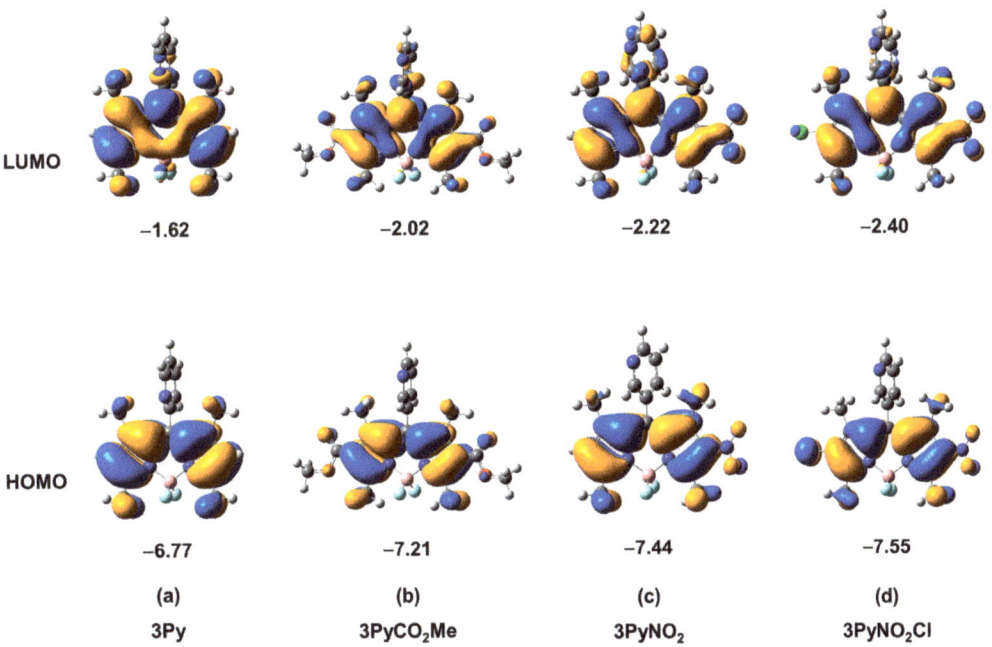

Figure 2. Frontier orbitals of (**a**) **3Py** [29], (**b**) **3PyCO$_2$Me**, (**c**) **3PyNO$_2$**, and (**d**) **3PyNO$_2$Cl** BODIPYs. Orbital energies in eV. The frontier orbitals for the entire series are given in Appendix A.

The absorption and emission bands for **2PyCO$_2$Me**, **3PyCO$_2$Me**, and **4PyCO$_2$Me** appear at approximately 501 and 514 nm, respectively, in polar solvents, similar to those observed for the 2,6-unsubstituted 8(*meso*)-pyridyl-BODIPYs [29]. In the case of the mono-nitro BODIPYs **2PyNO$_2$**, **3PyNO$_2$**, and **4PyNO$_2$**, the absorption and emission bands are blue-shifted, appearing at approximately 491 and 510 nm, respectively, in polar solvents. This result was confirmed with computational modeling (see Table 2) and is due to the slightly larger HOMO–LUMO gap induced by the strong electron-withdrawing nitro substituent. The introduction of a chlorine group on the mono-nitro BODIPYs causes significant red-shifted absorption and emission bands by ca. 20 nm and reduces the molar absorptivity of the absorption bands, as previously observed [29]. The absorption and emission bands of **2PyNO$_2$Cl**, **3PyNO$_2$Cl**, and **4PyNO$_2$Cl** appear at approximately 511 and 527 nm in polar solvents. This is consistent with the smaller HOMO–LUMO gaps for these compounds compared to the mono-nitro BODIPYs.

The Stokes shifts observed in the polar solvents, methanol, and acetonitrile are similar, indicating similar changes in geometry following excitations within a given 2,6-substituted series. The mono-nitro BODIPYs display the largest Stokes shifts of all the studied compounds. These observed shifts agree with the performed computational studies (see Table 2) and are likely due to greater geometry changes in the case of the nitro-BODIPYs excited states. These effects are currently under investigation in our laboratory.

As previously observed in the 2,6-unsubstituted 8(*meso*)-pyridyl-BODIPYs [29], their fluorescence properties largely depend on the relative position of the nitrogen atom on the pyridine ring. The 2-pyridyl BODIPY derivatives consistently show the lowest fluorescence quantum yields due to the lower rotational barrier of the 2-pyridyl group and the closer proximity of the nitrogen atom to the BODIPY core in these compounds. On the other hand, the 3-pyridyl BODIPY derivatives display the largest fluorescence quantum yields closely followed by the 4-pyridyl BODIPY derivatives, as previously observed [29].

The presence of two methyl ester substituents at the 2,6-positions of the 8(*meso*)-pyridyl-BODIPYs increased their relative fluorescence quantum yields in acetonitrile and methanol. This is due to the ester groups decreasing the electron density on the BODIPY core. However, the introduction of a single nitro substituent on the BODIPY core significantly decreases the relative fluorescence quantum yields for all 8(*meso*)-pyridyl-BODIPYs, particularly in polar solvents, probably due to non-radiative deactivation pathways that are more pronounced in polar media. The presence of a single strongly electron-withdrawing nitro group dramatically increases the calculated dipole moment of the molecule, as shown in Table 2. On the other hand, when a chlorine atom is introduced into the mono-nitro BODIPYs, the fluorescence is partially restored due to the reduction in the polarity of the molecule.

3. Materials and Methods

3.1. Synthesis and Characterization

3.1.1. General

Commercially available reagents and solvents were used as received from VWR or Sigma Aldrich unless noted otherwise. All reactions were monitored with thin-layer chromatography (TLC) using 0.2 mm silica gel plates (with UV indicator, polyester backed, 60 Å, pre-coated). Liquid chromatography was performed on preparative TLC plates or via silica gel column chromatography (60 Å, 230–400 mesh). NMR spectra were measured on 400 or 500 MHz for ^{1}H, 400 MHz for ^{11}B NMR, and 500 MHz for ^{13}C spectrometer. Chemical shifts (δ) are given in parts per million (ppm) in CDCl$_3$ (7.27ppm for ^{1}H NMR, 77.0 ppm for ^{13}C NMR) or (CD$_3$)$_2$CO (2.05 ppm for ^{1}H NMR, 206.68 and 29.92 ppm for ^{13}C NMR) or CD$_2$Cl$_2$ (5.32 ppm for ^{1}H NMR, 53.5 ppm for ^{13}C NMR); coupling constants (J) are given in hertz. BF$_3$·OEt$_2$ was used as the reference (0.00 ppm) for ^{11}B NMR spectra. High-resolution mass spectra (HRMS) were obtained using an Agilent 6230-B ESI-TOF mass spectrometer.

2,4-Dimethyl-3-methoxycarbonyl-pyrrole [32] and BODIPYs **2Py** [18], **3Py** [27], and **4Py** [33] were prepared as previously reported, and their spectroscopic data agree with the literature reports.

3.1.2. 1,3,5,7-Tetramethyl-2,6-dimethoxycarbonyl-8-(2-pyridyl)-BODIPY (**2PyCO$_2$Me**)

To a solution of 2,4-dimethyl-3-methyl carbonyl pyrrole in dichloromethane (0.984 g, 6.42 mmol) in a 250 mL round-bottomed flask under nitrogen was added 2-pyridyl carboxaldehyde (0.31 mL, 3.21 mmol) and 2–5 drops of TFA. The reaction mixture was stirred at rt for 40 h. DDQ (0.729 g, 3.21 mmol) was added to the reaction mixture. After 2 h, Et$_3$N (4.48 mL, 32.12 mmol) was added to the reaction mixture followed by BF$_3$·OEt$_2$ (3.96 mL, 32.12 mmol). The reaction was stirred for another 48 h. The reaction mixture was washed with water, and NaHCO$_3$ and the organic layers were extracted with dichloromethane. The combined organic layers were then washed with brine, dried over Na$_2$SO$_4$, and the solvent evaporated under reduced pressure. Purification with column chromatography using 20–40% ethyl acetate/hexanes for elution gave 0.567 g, 45% of the titled BODIPY as a reddish solid. ^{1}H NMR (500 MHz, CDCl$_3$) δ 8.85 (d, *J* = 4.8 Hz, 1H, Ar–H), 7.92 (t, *J* = 7.7 Hz, 1H, Ar–H), 7.57–7.49 (m, 1H, Ar–H), 7.46 (d, *J* = 7.7 Hz, 1H, Ar–H), 3.83 (s, 6H, CO$_2$CH$_3$), 2.85 (s, 6H, CH$_3$), 1.60 (s, 6H, CH$_3$). ^{13}C NMR (126 MHz, CDCl$_3$) δ 164.6, 160.4, 153.1, 150.5, 147.4, 137.6, 131.5, 124.6, 124.4, 51.4, 15.3, 13.2, 13.1. ^{11}B NMR (128 MHz, CD$_2$Cl$_2$) δ 0.72 (t, *J* = 32.0 Hz). HRMS (ESI) *m/z* calcd (%) for C$_{22}$H$_{22}$BF$_2$N$_3$O$_4$: 441.1783 [M + H]$^+$; found 442.1746.

3.1.3. 1,3,5,7-Tetramethyl-2,6-dimethoxycarbonyl-8-(3-pyridyl)-BODIPY (3PyCO₂Me)

This BODIPY was prepared as described above for **2PyCO₂Me** using 2,4-dimethyl-3-methoxycarbonylpyrrole (0.307 g, 2.00 mmol), 3-pyridinecarboxaldehyde (0.1 mL, 1.00 mmol), TFA (5 drops), DDQ (0.454 g, 2.00 mmol), Et$_3$N (2.09 mL, 15.01 mmol), and BF$_3$·OEt$_2$ (1.85 mL, 15.01 mmol). Purification with column chromatography using dichloromethane/ethyl acetate/hexanes 3:1:6 afforded the title BODIPY 0.353 g, 40% yield. ^{1}H NMR (500 MHz, CDCl$_3$) δ 8.89–8.83 (m, 1H, Ar–H), 8.60 (s, 1H, Ar–H), 7.72 (d, J = 7.8 Hz, 1H, Ar–H), 7.63–7.57 (m, 1H, Ar–H), 3.82 (s, 6H, CO$_2$CH$_3$), 2.84 (s, 6H, CH$_3$), 1.65 (s, 6H, CH$_3$). ^{13}C NMR (126 MHz, CDCl$_3$) δ 164.4, 160.6, 147.3, 137.6, 131.5, 124.9, 123.1, 122.9, 51.6, 15.2, 15.1, 14.5, 14.4. ^{11}B NMR (128 MHz, CDCl$_3$) δ 0.66 (t, J = 31.7 Hz). HRMS (ESI) m/z calcd (%) for C$_{22}$H$_{22}$BF$_2$N$_3$O$_4$: 441.1783 [M + H]$^+$; found 442.1771.

3.1.4. 1,3,5,7-Tetramethyl-2,6-dimethoxycarbonyl-8-(4-pyridyl)-BODIPY (4PyCO₂Me)

This BODIPY was prepared as described above for **2PyCO₂Me** using 2,4-dimethyl-3-methoxycarbonylpyrrole (1.630 g, 10.64 mmol), 4-pyridinecarboxaldehyde (0.5 mL, 5.32 mmol), TFA (5 drops), DDQ (1.2078 g, 5.32 mmol), triethylamine (11.12 mL, 79.81 mmol), and BF$_3$·OEt$_2$ (9.85 mL, 79.81 mmol). Purification with column chromatography using 60% ethyl acetate/hexanes 3:1:6 afforded the title BODIPY in 0.232 g, 28% yield. ^{1}H NMR (500 MHz, CDCl$_3$) δ 8.90 (d, 4.8Hz, 2H, Ar–H), 7.48 (d, J = 5.0 Hz, 2H, Ar–H), 3.82 (s, 6H, CO$_2$CH$_3$), 2.84 (s, 6H, CH$_3$), 1.67 (s, 6H, CH$_3$). ^{13}C NMR (126 MHz, CD$_2$Cl$_2$) δ 164.7, 160.5, 151.5, 147.7, 143.1, 142.4, 130.9, 123.4, 123.3, 51.2, 14.8, 13.9, 13.6, 13.5. ^{11}B NMR (128 MHz, CD$_2$Cl$_2$) δ 0.65 (t, J = 32.1 Hz). HRMS (ESI) m/z calcd (%) for C$_{22}$H$_{22}$BF$_2$N$_3$O$_4$: 441.1783 [M + H]$^+$; found 442.1743.

3.1.5. 2-Nitro-1,3,5,7-tetramethyl-8-(2-pyridyl)-BODIPY (2PyNO₂)

1,3,5,7-Tetramethyl-8-(2-pyridyl)-BODIPY (20 mg, 0.0615 mmol) was dissolved in 10 mL of dry dichloromethane in an oven-dried 50 mL round-bottomed flask. Nitronium tetrafluoroborate (0.12 mL, 0.06 mmol) was then added, and the mixture was stirred for 5 h at rt. The solvent was evaporated under reduced pressure, and the crude product was dissolved in methyl tert-butyl methyl ether and then washed with water. The organic layer was dried over Na$_2$SO$_4$, and the solvent was evaporated under reduced pressure. The reaction was purified via preparative TLC using 40% ethyl acetate/hexanes, yielding 18.72 mg, 82% yield of the title BODIPY as a reddish solid. ^{1}H NMR (500 MHz, acetone-d_6) δ 8.85 (dt, J = 4.9, 1.4 Hz, 1H, Ar–H), 8.09 (dt, J = 7.7, 1.7 Hz, 1H, Ar–H), 7.76 (dt, J = 7.8, 1.2 Hz, 1H, Ar–H), 7.66 (ddd, J = 7.7, 4.9, 1.2 Hz, 1H, Ar–H), 6.49 (s, 1H,pyrrolic-H), 2.80 (s, 3H, CH$_3$), 2.65 (s, 3H, CH$_3$), 1.56 (s, 3H, CH$_3$), 1.42 (s, 3H, CH$_3$). ^{13}C NMR (126 MHz, acetone-d_6) δ 153.2, 151.5, 149.9, 142.2, 138.7, 126.4, 126.3, 126.3, 126.3, 125.9, 125.6, 15.4, 14.5, 14.3, 11.9. ^{11}B NMR (128 MHz, acetone-d_6) δ 0.57 (t, J = 31.5 Hz). HRMS (ESI) m/z calcd (%) for C$_{18}$H$_{17}$BF$_2$N$_4$O$_2$: 371.1489 [M + H]$^+$; found 371.1494.

3.1.6. 2-Nitro-1,3,5,7-tetramethyl-8-(3-pyridyl)-BODIPY (3PyNO₂)

This BODIPY was prepared as described above for **2PyNO₂** using 1,3,5,7-tetramethyl-8-(3-pyridyl)-BODIPY [25] (40mg, 0.12 mmol) and NO$_2$BF$_4$ (0.25 mL, 0.12 mmol). The residue was purified via preparative TLC with ethyl acetate/hexanes 1:1, yielding 37.44 mg, 82% yield of the title BODIPY as a reddish solid. ^{1}H NMR (500 MHz, acetone-d_6) δ 8.88 (dd, J = 4.9, 1.7 Hz, 1H, Ar–H), 8.75 (d, J = 2.2 Hz, 1H, Ar–H), 8.02 (dt, J = 7.8, 2.0 Hz, 1H, Ar–H), 7.71 (dd, J = 7.8, 4.9 Hz, 1H, Ar–H), 6.54 (s, 1H, pyrrolic-H), 2.80 (s, 3H, CH$_3$), 2.66 (s, 3H, CH$_3$), 1.68 (s, 3H, CH$_3$), 1.50 (s, 3H, CH$_3$). ^{13}C NMR (126 MHz, acetone-d_6) δ 151.9, 149.1, 137.1, 130.9, 126.7, 126.7, 125.1, 15.8, 15.4, 14.3, 12.9. ^{11}B NMR (128 MHz, acetone-d_6) δ 0.53 (t, J = 31.5 Hz). HRMS (ESI) m/z calcd (%) for C$_{18}$H$_{17}$BF$_2$N$_4$O$_2$: 371.1489 [M + H]$^+$; found 371.1490.

3.1.7. 2-Nitro-1,3,5,7-tetramethyl-8-(4-pyridyl)-BODIPY (**4PyNO$_2$**)

This compound was prepared as described above for **2PyNO$_2$** using 1,3,5,7-tetramethyl-8-(4-pyridyl)-BODIPY (0.030 g, 0.09 mmol) and NO$_2$BF$_4$ (0.18 mL, 0.09 mmol). Purification with preparative TLC using dichloromethane/ethyl acetate/hexanes 3:2:5 solvent afforded 0.023 g, 77% yield of the title BODIPY as a reddish solid. ^{1}H NMR (500 MHz, acetone-d_6) δ 8.90 (d, J = 5.0 Hz, 2H, Ar–H), 7.65 (d, J = 4.9 Hz, 2H, Ar–H), 6.54 (s, 1H, pyrrolic-H), 2.80 (s, 3H, CH$_3$), 2.66 (s, 3H, CH$_3$), 1.71 (s, 3H, CH$_3$), 1.55 (s, 3H, CH$_3$). ^{13}C NMR (126 MHz, CDCl$_3$) δ 166.8, 147.6, 147.3, 137.5, 134.1, 133.9, 126.4, 125.1, 22.8, 15.6, 12.9, 11.2. ^{11}B NMR (128 MHz, acetone-d_6) δ 0.49 (t, J = 31.3 Hz). HRMS (ESI) m/z calcd (%) for C$_{18}$H$_{17}$BF$_2$N$_4$O$_2$: 371.1489 [M + H]$^+$; found 371.1492.

3.1.8. 2-Chloro-6-nitro-1,3,5,7-tetramethyl-8-(2-pyridyl)-BODIPY (**2PyNO$_2$Cl**)

BODIPY **2PyNO$_2$** (11 mg, 0.0297 mmol) was dissolved in dry degassed dichloromethane in a 10 mL oven-dried round-bottomed flask. TCCA (0.003 g, 0.01 mmol) in dry dichloromethane was added dropwise to the solution. The mixture was stirred at room temperature for 2 h and purified with preparative TLC using 40% ethyl acetate/hexanes for elution to afford 0.010 g, 90% yield of the title BODIPY as a pink reddish product. ^{1}H NMR (500 MHz, CDCl$_3$) δ 8.88 (d, J = 4.9 Hz, 1H, Ar–H), 8.05–7.99 (m, 1H, Ar–H), 7.65–7.61 (m, 1H, Ar–H), 7.51 (t, J = 5.8 Hz, 1H, Ar–H), 2.88 (s, 3H, CH$_3$), 2.69 (s, 3H, CH$_3$), 1.60 (s, 3H, CH$_3$), 1.36 (s, 3H, CH$_3$). ^{13}C NMR (126 MHz, CDCl$_3$) δ 160.9, 152.4, 151.4, 150.8, 141.4, 141.2, 137.8, 136.2, 132.9, 131.1, 128.9, 127.4, 125.1, 124.5, 14.7, 14.3, 12.2, 12.1. ^{11}B NMR (128 MHz, acetone-d_6) δ 0.37 (t, J = 30.9 Hz). HRMS (ESI) m/z calcd (%) for C$_{18}$H$_{16}$BClF$_2$N$_4$O$_2$: 403.106 [M + H]$^+$; found 404.1130.

3.1.9. 2-Chloro-6-nitro-1,3,5,7-tetramethyl-8-(3-pyridyl)-BODIPY (**3PyNO$_2$Cl**)

This BODIPY was prepared as described above for **2PyNO$_2$Cl** using **3PyNO$_2$** (12.7 mg, 0.0343 mmol) and TCCA (0.004 g, 0.02 mmol). The reaction was stirred for 30 min and purified with preparative TLC using dichloromethane/ethyl acetate/hexanes 3:2:5 for elution to afford 0.013 g, 89% yield of the title BODIPY as a pink reddish product. ^{1}H NMR (500 MHz, acetone-d_6) δ 8.89 (dd, J = 4.9, 1.7 Hz, 1H, Ar–H), 8.77 (d, J = 2.3 Hz, 1H, Ar–H), 8.04 (dt, J = 7.8, 2.0 Hz, 1H, Ar–H), 7.75–7.70 (m, 1H, Ar–H), 2.80 (s, 3H, CH$_3$), 2.67 (s, 3H, CH$_3$), 1.68 (s, 3H, CH$_3$), 1.47 (s, 3H, CH$_3$). ^{13}C NMR (126 MHz, acetone-d_6) δ 159.9, 151.2, 148.1, 142.1, 141.7, 136.2, 135.9, 132.9, 129.8, 124.3, 13.5, 12.4, 12.3, 12.2. ^{11}B NMR (128 MHz, acetone-d_6) δ 0.34 (t, J = 30.8 Hz). HRMS (ESI) m/z calcd (%) for C$_{18}$H$_{16}$BClF$_2$N$_4$O$_2$: 403.106 [M + H]$^+$; found 404.1134.

3.1.10. 2-Chloro-6-nitro-1,3,5,7-tetramethyl-8-(4-pyridyl)-BODIPY (**4PyNO$_2$Cl**)

This BODIPY was prepared as described above for **2PyNO$_2$Cl** using **4PyNO$_2$** (15 mg, 0.0405 mmol) and TCCA (0.003 g, 0.01 mmol). The reaction was stirred for 2 h. Column chromatography was performed using dichloromethane/ethyl acetate/hexanes 3:2:5 for elution to afford 0.013 g, 89% yield of the title BODIPY as a pink reddish product. ^{1}H NMR (400 MHz, CDCl$_3$) δ 8.90 (d, J = 6.1 Hz, 2H, Ar–H), 7.39 (d, J = 6.0 Hz, 2H, Ar–H), 2.88 (s, 3H, CH$_3$), 2.70 (s, 3H, CH$_3$), 1.72 (s, 3H, CH$_3$), 1.46 (s, 3H, CH$_3$). ^{13}C NMR (126 MHz, CDCl$_3$) δ 161.8, 151.9, 148.7, 140.5, 139.1, 135.9, 131.7, 126.1, 124.3, 14.6, 14.3, 13.5, 13.1. ^{11}B NMR (128 MHz, acetone-d_6) δ 0.37 (t, J = 30.9 Hz). HRMS (ESI) m/z calcd (%) for C$_{18}$H$_{16}$BClF$_2$N$_4$O$_2$: 403.106 [M + H]$^+$; found 404.1134.

3.2. Spectroscopy Methods

UV–vis absorption spectra were collected on a Varian Cary 50 Bio spectrophotometer. Emission spectra were obtained on a PerkinElmer LS55 spectrophotometer at room temperature. Spectrophotometric grade solvents and quartz cuvettes (1 cm path length) were used. Relative fluorescence quantum yields (Φ_f) were calculated using rhodamine 6G (Φ_f = 0.86 in methanol) as the reference using the following equation: $\Phi_x = \Phi_{st} \times \text{Grad}_x/\text{Grad}_{st} \times (\eta_x/\eta_{st})^2$, where Φ_X and Φ_{ST} are the quantum yields of the sample and standard, Grad$_X$ and Grad$_{ST}$ are the gradients from the plot of integrated fluorescence intensity vs. absorbance, and η represents the refractive index of the solvent (x is for the sample and st standard).

3.3. X-ray Crystallography

The structures were determined using data collected at low temperature on a Bruker Kappa ApexII DUO diffractometer with CuKα radiation for **2PyCO₂Me** (90 K), **2PyNO₂** (100 K), **4PyNO₂** (100 K), **2PyNO₂Cl** (100 K), **3PyNO₂Cl** (100 K), and **4PyNO₂Cl** (100 K), or with MoKα radiation for **4PyCO₂Me** (120 K) and **3PyNO₂** (100 K). Disorder was present in several of the structures, and disordered solvent contribution was removed using the SQUEEZE procedure for **3PyNO₂**. Data in CIF format have been deposited with the Cambridge Crystallographic Data Centre as CCDC 2249642-2249649 in the order shown in Figure 1.

3.4. Computational Methods

The geometries of the ground states of all compounds were optimized without symmetry constraints at the -b3lyp/6-31+G(d,p) level in dichloromethane. The solvent effects were considered using the Polarized Continuum Model (PCM). The potential energy minima were confirmed with frequency calculations. The absorption and emission data were calculated using the TD-DFT/6-31+G(d,p) method in vacuum. This method has been shown to correctly reproduce the experimental trends [28]. The first three singlet excitations were considered, and the lowest-energy excited singlet state was optimized to calculate the maximum emission wavelengths. All calculations were performed using the Gaussian 09 program package [34].

4. Conclusions

Three series of 8(*meso*)-pyridyl-BODIPYs bearing a 2-, 3-, or 4-pyridyl group and electron-withdrawing groups at either the 2- or 2,6-positions were synthesized, and their structural and spectroscopic properties were investigated. These BODIPYs were prepared either with direct electrophilic nitration or chlorination of the pyridyl-BODIPY core or by total synthesis from *tert*-butyl acetoacetate. Eight of the new BODIPYs were characterized with X-ray crystallography, and their structures were modeled computationally. All 8(*meso*)-pyridyl rings are nearly perpendicular to the BODIPY core with dihedral angles between 80° and 90°, slightly lower in the case of the 2- versus the 3- and 4-pyridyl derivatives.

The 2,6-methyl ester groups in **2PyCO₂Me**, **3PyCO₂Me**, and **4PyCO₂Me** were observed to increase the relative fluorescence quantum yields of these derivatives compared with the corresponding 2,6-unsubstituted analogs due to the electron-withdrawing effect of the methyl ester groups. On the other hand, the introduction of a 2-nitro substituent on the 8(*meso*)-pyridyl-BODIPYs drastically increases the calculated dipole moment of the molecules, induces significant hypsochromic shifts, and decreases their relative fluorescence quantum yields due to non-radiative deactivation processes. Introduction of a chlorine atom in the series **2PyNO₂Cl**, **3PyNO₂Cl**, and **4PyNO₂Cl** reduces the polarity of the molecules relative to the mono-nitro compounds, induces pronounced bathochromic shifts, and partially restores the fluorescence.

Supplementary Materials: The following supporting information can be downloaded at: https://www.mdpi.com/article/10.3390/molecules28124581/s1, Figure S1, conformations and relative energies of BODIPYs studied; Figures S2 and S3, normalized absorption (a,c,e) and emission (b,d,f) spectra of BODIPYs in acetonitrile (a, b, c, d) and water (e, f); Table S1, spectroscopic properties determined in toluene using rhodamine-6G (Φ = 0.86), λ_{exc} = 473 nm; Figures S4–S30, ^{1}H, ^{13}C, and ^{11}B NMR spectra of BODIPYs.

Author Contributions: Conceptualization, C.N., D.J.L., P.B.-P. and M.d.G.H.V.; methodology, C.N., P.B.-P., D.J.L., F.R.F. and M.d.G.H.V.; software, P.B.-P., D.G. and M.d.G.H.V.; validation, C.N., D.G. and P.B.-P.; formal analysis, C.N., D.G. and F.R.F.; investigation, C.N., D.J.L., P.B.-P., D.G. and M.d.G.H.V.; resources, P.B.-P. and M.d.G.H.V.; data curation, C.N., D.J.L. and F.R.F.; writing—original draft preparation, C.N., P.B.-P. and M.d.G.H.V.; writing—review and editing, C.N., D.J.L., P.B.-P., F.R.F. and M.d.G.H.V.; visualization, C.N., P.B.-P., F.R.F. and M.d.G.H.V.; supervision, P.B.-P. and M.d.G.H.V.; project administration, P.B.-P. and M.d.G.H.V.; funding acquisition, P.B.-P. and M.d.G.H.V. All authors have read and agreed to the published version of the manuscript.

Funding: This research was funded by the National Science Foundation, grant number CHE-2055190.

Institutional Review Board Statement: Not applicable.

Informed Consent Statement: Not applicable.

Data Availability Statement: Not applicable.

Acknowledgments: The authors are thankful to the Louisiana State University High Performance Computing Center (http://www.hpc.lsu.edu) for use of its computational resources in conducting this research.

Conflicts of Interest: The authors declare no conflict of interest. The funders had no role in the design of the study; in the collection, analyses, or interpretation of data; in the writing of the manuscript; or in the decision to publish the results.

Sample Availability: Not applicable.

Appendix A

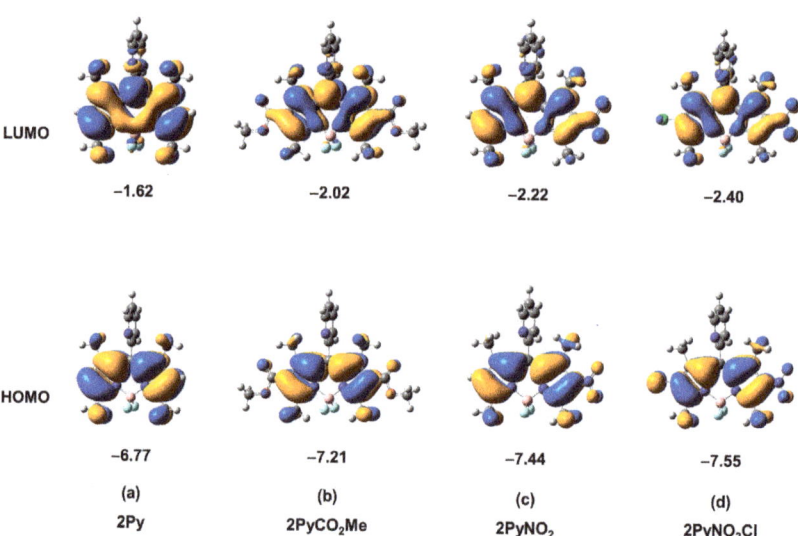

Figure A1. Frontier orbitals of (**a**) **2Py** [29], (**b**) **2PyCO$_2$Me**, (**c**) **2PyNO$_2$**, and (**d**) **2PyNO$_2$Cl** BODIPYs. Orbital energies in eV.

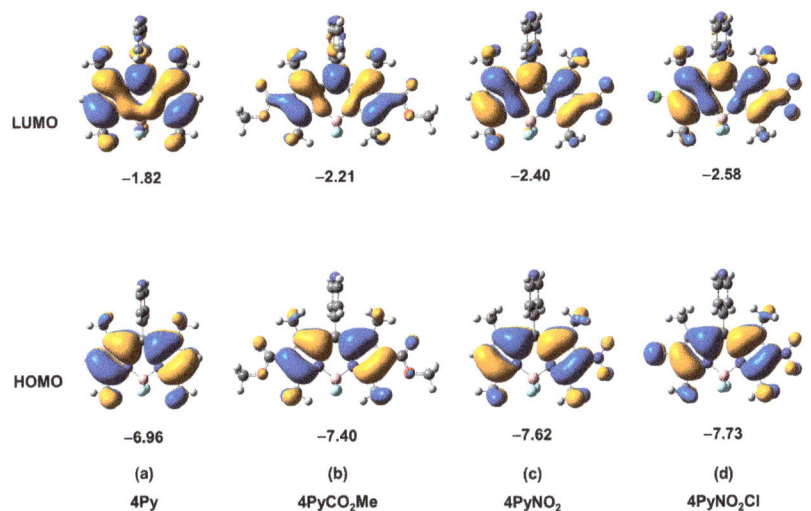

Figure A2. Frontier orbitals of (**a**) 4Py [29], (**b**) 4PyCO$_2$Me, (**c**) 4PyNO$_2$, and (**d**) 4PyNO$_2$Cl BODIPYs. Orbital energies in eV.

References

1. Krumova, K.; Cosa, G. Bodipy Dyes with Tunable Redox Potentials and Functional Groups for Further Tethering: Preparation, Electrochemical, and Spectroscopic Characterization. *J. Am. Chem. Soc.* **2010**, *132*, 17560–17569. [CrossRef]
2. Clarke, R.G.; Hall, M.J. Chapter Three—Recent Developments in the Synthesis of the BODIPY Dyes. *Adv. Heterocycl. Chem.* **2019**, *128*, 181–261. [CrossRef]
3. Ulrich, G.; Ziessel, R.; Harriman, A. The Chemistry of Fluorescent Bodipy Dyes: Versatility Unsurpassed. *Angew. Chem. Int. Ed.* **2008**, *47*, 1184–1201. [CrossRef] [PubMed]
4. Kaur, P.; Singh, K. Recent Advances in the Application of BODIPY in Bioimaging and Chemosensing. *J. Mater. Chem. C* **2019**, *7*, 11361–11405. [CrossRef]
5. Gurubasavaraj, P.M.; Sajjan, V.P.; Muñoz-Flores, B.M.; Jiménez Pérez, V.M.; Hosmane, N.S. Recent Advances in BODIPY Compounds: Synthetic Methods, Optical and Nonlinear Optical Properties, and Their Medical Applications. *Molecules* **2022**, *27*, 1877. [CrossRef]
6. Kaufman, N.E.; Meng, Q.; Griffin, K.E.; Singh, S.S.; Dahal, A.; Zhou, Z.; Fronczek, F.R.; Mathis, J.M.; Jois, S.D.; Vicente, M.G.H. Synthesis, Characterization, and Evaluation of near-IR Boron Dipyrromethene Bioconjugates for Labeling of Adenocarci-Nomas by Selectively Targeting the Epidermal Growth Factor Receptor. *J. Med. Chem.* **2019**, *62*, 3323–3335. [CrossRef]
7. Liu, Z.; Jiang, Z.; Yan, M.; Wang, X. Recent Progress of BODIPY Dyes With Aggregation-Induced Emission. *Front. Chem.* **2019**, *7*, 712. [CrossRef]
8. Poddar, M.; Misra, R. Recent Advances of BODIPY Based Derivatives for Optoelectronic Applications. *Coord. Chem. Rev.* **2020**, *421*, 213462. [CrossRef]
9. Ray, C.; Schad, C.; Moreno, F.; Maroto, B.L.; Bañuelos, J.; Arbeloa, T.; García-Moreno, I.; Villafuerte, C.; Muller, G.; de la Moya, S. BCl3-Activated Synthesis of COO-BODIPY Laser Dyes: General Scope and High Yields under Mild Conditions. *J. Org. Chem.* **2020**, *85*, 4594–4601. [CrossRef]
10. Loudet, A.; Burgess, K. Bodipy Dyes and Their Derivatives: Syntheses and Spectroscopic Properties. *Chem. Rev.* **2007**, *107*, 4891–4932. [CrossRef]
11. Zhang, G.; Wang, M.; Ndung'U, C.; Bobadova-Parvanova, P.; Fronczek, F.R.; Smith, K.M.; Vicente, M.G.H. Synthesis and Investigation of BODIPYs with Restricted Meso-8-Aryl Rotation. *J. Porphyr. Phthalocyanines* **2020**, *24*, 869–877. [CrossRef]
12. Ortiz, M.J.; Garcia-Moreno, I.; Agarrabeitia, A.R.; Duran-Sampedro, G.; Costela, A.; Sastre, R.; Arbeloa, F.L.; Prieto, J.B.; Arbeloa, I.L. Red-Edge-Wavelength Finely-Tunable Laser Action from New BODIPY Dyes. *Phys. Chem. Chem. Phys.* **2010**, *12*, 7804–7811. [CrossRef] [PubMed]
13. Bura, T.; Ziessel, R. Water-Soluble Phosphonate-Substituted BODIPY Derivatives with Tunable Emission Channels. *Org. Lett.* **2011**, *13*, 3072–3075. [CrossRef] [PubMed]
14. Kim, J.; Kim, Y. A Water-Soluble Sulfonate-BODIPY Based Fluorescent Probe for Selective Detection of HOCl/OCl$^-$ in Aqueous Media. *Analyst* **2014**, *139*, 2986–2989. [CrossRef] [PubMed]
15. Komatsu, T.; Urano, Y.; Fujikawa, Y.; Kobayashi, T.; Kojima, H.; Terai, T.; Hanaoka, K.; Nagano, T. Development of 2,6-Carboxy-Substituted Boron Dipyrromethene (BODIPY) as a Novel Scaffold of Ratiometric Fluorescent Probes for Live Cell Imaging. *Chem. Commun.* **2009**, *45*, 7015–7017. [CrossRef]

16. Nguyen, A.L.; Griffin, K.E.; Zhou, Z.; Fronczek, F.R.; Smith, K.M.; Vicente, M.G.H. Syntheses of 1,2,3-Triazole-BODIPYs Bearing up to Three Carbohydrate Units. *New J. Chem.* **2018**, *42*, 8241–8246. [CrossRef]
17. Zhu, S.; Zhang, J.; Vegesna, G.; Luo, F.-T.; Green, S.A.; Liu, H. Highly Water-Soluble Neutral BODIPY Dyes with Controllable Fluorescence Quantum Yields. *Org. Lett.* **2010**, *13*, 438–441. [CrossRef]
18. Wang, Y.-W.; Li, M.; Shen, Z.; You, X.-Z. Meso-Pyridine Substituted Boron-Dipyrromethene (BDP) Dye as a PH Probe: Syn-Thesis, Crystal Structure and Spectroscopic Properties. *Chin. J. Inorg. Chem.* **2008**, *24*, 1247–1252.
19. Zhou, Z.; Maki, T. Ratiometric Fluorescence Acid Probes Based on a Tetrad Structure Including a Single BODIPY Chromo-Phore. *J. Org. Chem.* **2021**, *86*, 17560–17566. [CrossRef]
20. Zhang, S.; Wu, T.; Fan, J.; Li, Z.; Jiang, N.; Wang, J.; Dou, B.; Sun, S.; Song, F.; Peng, X. A BODIPY-Based Fluorescent Dye for Mitochondria in Living Cells, with Low Cytotoxicity and High Photostability. *Org. Biomol. Chem.* **2013**, *11*, 555–558. [CrossRef]
21. Raza, M.K.; Gautam, S.; Howlader, P.; Bhattacharyya, A.; Kondaiah, P.; Chakravarty, A.R. Pyriplatin-Boron-Dipyrrome-Thene Conjugates for Imaging and Mitochondria-Targeted Photodynamic Therapy. *Inorg. Chem.* **2018**, *57*, 14374–14385. [CrossRef] [PubMed]
22. Carpenter, B.L.; Situ, X.; Scholle, F.; Bartelmess, J.; Weare, W.W.; Ghiladi, R.A.A. Antifungal and Antibacterial Activities of a BODIPY-Based Photosensitizer. *Molecules* **2015**, *20*, 10604–10621. [CrossRef] [PubMed]
23. Durantini, A.M.; Heredia, D.A.; Durantini, J.E.; Durantini, E.N. BODIPYs to the Rescue: Potential Applications in Photo-Dynamic Inactivation. *Eur. J. Med. Chem.* **2018**, *144*, 651–661. [CrossRef]
24. Xie, H.-R.; Gu, Y.-Q.; Liu, L.; Dai, J.-C. A H-Aggregating Fluorescent Probe for Recognizing Both Mercury and Copper Ions Based on a Dicarboxyl-Pyridyl Bifunctionalized Difluoroboron Dipyrromethene. *New J. Chem.* **2020**, *44*, 19713–19722. [CrossRef]
25. Kim, Y.; Jang, Y.J.; Lee, D.; Kim, B.-S.; Churchill, D.G. Real Nerve Agent Study Assessing Pyridyl Reactivity: Selective Fluo-Rogenic and Colorimetric Detection of Soman and Simulant. *Sens. Actuators B* **2017**, *238*, 145–149. [CrossRef]
26. Luo, G.-G.; Fang, K.; Wu, J.-H.; Dai, J.-C.; Zhao, Q.-H. Noble-Metal-Free BODIPY-Cobaloxime Photocatalysts for Visible-Light-Driven Hydrogen Production. *Phys. Chem. Chem. Phys.* **2014**, *16*, 23884–23894. [CrossRef]
27. Shen, X.-F.; Watanabe, M.; Takagaki, A.; Song, J.T.; Ishihara, T. Pyridyl-Anchored Type BODIPY Sensitizer-TiO$_2$ Photocata-Lyst for Enhanced Visible Light-Driven Photocatalytic Hydrogen Production. *Catalysts* **2020**, *10*, 535. [CrossRef]
28. LaMaster, D.J.; Kaufman, N.E.M.; Bruner, A.S.; Vicente, M.G.H. Structure based modulation of electron dynamics in meso-(4-pyridyl)-BODIPYs: A computational and synthetic approach. *J. Phys. Chem. A* **2018**, *122*, 6372–6380. [CrossRef]
29. Ndung'U, C.; LaMaster, D.; Dhingra, S.; Michel, N.H.; Bobadova-Parvanova, P.; Fronczek, F.R.; Elgrishi, N.; Vicente, M.G.H. A Comparison of the Photophysical and Electrochemical Properties of Meso-(2-, 3-, and 4-Pyridyl)-BODIPYs and Their Derivatives. *Sensors* **2022**, *22*, 5121. [CrossRef]
30. Mula, S.; Ray, A.K.; Banerjee, M.; Chaudhuri, T.; Dasgupta, K.; Chattopadhyay, S. Design and Development of a New Pyrromethene Dye with Improved Photostability and Lasing Efficiency: Theoretical Rationalization of Photophysical and Photochemical Properties. *J. Org. Chem.* **2008**, *73*, 2146–2154. [CrossRef]
31. Smith, N.W.; Dzyuba, S.V. Efficient Nitration of Meso-Tetraphenylporphyrin with Nitronium Tetrafluoroborate. *Arkivoc* **2010**, *2010*, 10–18. [CrossRef]
32. Sun, L.; Liang, C.; Shirazian, S.; Zhou, Y.; Miller, T.; Cui, J.; Fukuda, J.Y.; Chu, J.-Y.; Nematalla, A.; Wang, X.; et al. Discovery of 5-[5-Fluoro-2-oxo-1,2- dihydroindol-(3Z)-ylidenemethyl]-2,4- dimethyl-1H-pyrrole-3-carboxylic Acid (2-Diethylaminoethyl)amide, a Novel Tyrosine Kinase Inhibitor Targeting Vascular Endothelial and Platelet-Derived Growth Factor Receptor Tyrosine Kinase. *J. Med. Chem.* **2003**, *46*, 1116–1119. [CrossRef] [PubMed]
33. Bartelmess, J.; Weare, W.W.; Latortue, N.; Duong, C.; Jones, D.S. Meso-Pyridyl BODIPYs with Tunable Chemical, Optical and Electrochemical Properties. *New J. Chem.* **2013**, *37*, 2663–2668. [CrossRef]
34. Frisch, M.J.; Trucks, G.W.; Schlegel, H.B.; Robb, M.A.; Cheeseman, J.R.; Scalmani, G.; Barone, V.; Mennucci, B.; Petersson, G.A.; Nakatsuji, H.; et al. *Gaussian 09, Revision D.01, Computational Chemistry Software*; Gaussian, Inc.: Wallingford, CT, USA, 2009.

Disclaimer/Publisher's Note: The statements, opinions and data contained in all publications are solely those of the individual author(s) and contributor(s) and not of MDPI and/or the editor(s). MDPI and/or the editor(s) disclaim responsibility for any injury to people or property resulting from any ideas, methods, instructions or products referred to in the content.

Article

Effects of Substituents on the Photophysical/Photobiological Properties of Mono-Substituted Corroles

Vitória Barbosa de Souza [1], Vinícius N. da Rocha [2], Paulo Cesar Piquini [2], Otávio Augusto Chaves [3] and Bernardo A. Iglesias [1,*]

[1] Bioinorganic and Porphyrinoids Materials Laboratory, Department of Chemistry, Federal University of Santa Maria, Santa Maria 97105-900, RS, Brazil
[2] Department of Physics, Federal University of Santa Maria, Santa Maria 97105-900, RS, Brazil
[3] CQC-IMS, Department of Chemistry, University of Coimbra, Rua Larga, 3004-535 Coimbra, Portugal
* Correspondence: bernardopgq@gmail.com or bernardo.iglesias@ufsm.br

Abstract: The *trans*-A_2B-corrole series was prepared starting with 5-(pentafluorophenyl)dipyrromethene, which was then reacted with respective aryl-substituted aldehyde by Gryko synthesis. It was further characterized by HRMS and electrochemical methods. In addition, we investigated experimental photophysical properties (absorption, emission by steady-state and time-resolved fluorescence) in several solvents and TDDFT calculations, aggregation, photostability and reactive oxygen species generation (ROS), which are relevant when selecting photosensitizers used in photodynamic therapy and many other photo-applications. In addition, we also evaluated the biomolecule-binding properties with CT-DNA and HSA by spectroscopy, viscometry and molecular docking calculations assays.

Keywords: corroles; *trans*-A_2B-corroles; photophysics; photobiology

1. Introduction

Tetrapyrrole macrocycles such as corroles are known as photosensitizers for use in photobiological processes and have recently drawn attention for their great capacity to generate reactive oxygen species (ROS), such as singlet oxygen, low cytotoxicity, desirable photostability and advantageous use of light at long absorption wavelengths up to near-infrared, which facilitates the application of these derivatives in photodynamic processes [1–3].

Corroles have a corrin-like skeleton with a direct bond between two pyrrole rings, as well as an 18-electron porphyrin-like π system, i.e., a hybrid structure between the two tetrapyrrole macrocycles. With recent synthetic improvements, the corrole moiety has attracted significant attention as a new photosensitizer [4,5]. These corroles have been extensively investigated by spectroscopic techniques (absorption/emission), theoretical molecular orbitals calculations, as well as by electrochemical studies. Understanding the general photophysics of corrole derivatives is also necessary for their application in photobiological processes such as photodynamic therapy (PDT) and antimicrobial photodynamic therapy (aPDT), as photophysical parameters such as excited state lifetime, intersystem crossing (ISC) rates and different quantum yields of the sensitizers have specific effects on the functionality of corroles, including their ability to generate ROS [6–8]. The presence of specific substituents in the *meso*-aryl positions of corrole can directly affect these parameters, meaning these compounds have different photophysical properties [9,10].

Interactive studies into Human Serum Albumin (HSA), the main carrier protein in the human bloodstream, are important as it is used in preliminary laboratory evaluations of pharmacokinetic parameters (offering interesting information before clinical evaluations) due to its ability to transport endogenous and exogenous compounds [11–15]. In the case of DNA interaction assays, the possible formation of a corrole-DNA adducts through intercalation or by major/minor groove interactions may induce structural changes in DNA and possible cell destruction by apoptosis or necrosis [16–20].

Citation: de Souza, V.B.; da Rocha, V.N.; Piquini, P.C.; Chaves, O.A.; Iglesias, B.A. Effects of Substituents on the Photophysical/Photobiological Properties of Mono-Substituted Corroles. *Molecules* **2023**, *28*, 1385. https://doi.org/10.3390/molecules28031385

Academic Editor: Artur M. S. Silva

Received: 30 December 2022
Revised: 25 January 2023
Accepted: 26 January 2023
Published: 1 February 2023

Copyright: © 2023 by the authors. Licensee MDPI, Basel, Switzerland. This article is an open access article distributed under the terms and conditions of the Creative Commons Attribution (CC BY) license (https://creativecommons.org/licenses/by/4.0/).

In this manuscript, four corrole derivatives **1–4** (Figure 1), which have different substituents at the *meso*-aryl position, were used, and their photophysical properties were investigated in combination with theoretical approaches (TDDFT). In addition, photobiological parameters and the interaction with biomacromolecules (DNA and HSA) were previously investigated by considering the effect of different types of substituents on the corrole macrocycle. The corrole structural change from phenyl (**1**), naphthyl (**2**), 4-(hydroxy)phenyl (**3**) or 4-(thiomethyl)phenyl (**4**) moiety in the *meso* position of the *trans*-C_6F_5 corrole was also evaluated in terms of its impact on the binding affinity with biomolecules.

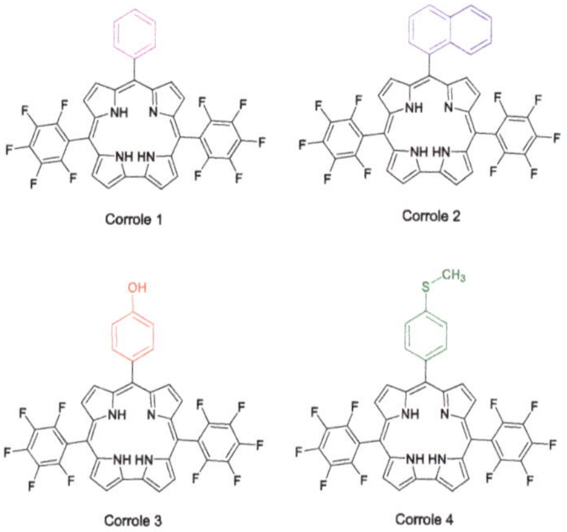

Figure 1. Chemical representative structure of mono-substituted corroles **1–4**.

2. Results

2.1. Corroles

Corroles **1–4** were prepared according to Gryko's synthesis [21–24] and used as molecules to evaluate their photophysical/photobiological processes and interaction with biomolecules (CT-DNA and HSA). The HRMS-ESI(+) mass spectra of derivatives are presented in the Supplementary Materials (Figures S1–S4).

2.2. Photophysical Properties

The absorption spectra in the UV-Vis region of corroles **1–4** in several solvents (DCM, ACN, MeOH and DMSO) are shown in Figure 2 and the data referring to molar absorptivity (ε) and wavelengths of the main transitions (λ_{abs}) are listed in Table 1.

Table 1. Photophysical data of corroles **1–4**.

	DCM					
Corrole	λ_{Abs} (ε; M^{-1} cm^{-1} × 10^5) [a]	λ_{Em} (QY; %) [b]	SS (nm/cm^{-1}) [c]	τ_f (ns) [d]/X^2	k_r (×10^8 s^{-1}) [e]	k_{nr} (×10^8 s^{-1}) [e]
1	411 (1.47), 563 (0.24), 614 (0.14)	659, 717 (15.0)	248/9155	3.66/1.05815	0.41	2.32
2	412 (0.96), 562 (0.14), 614 (0.07)	658, 719 (9.0)	246/9075	4.20/0.97026	0.21	2.16
3	411 (0.86), 563 (0.15), 615 (0.09)	662, 718 (8.0)	251/9225	3.70/1.16247	0.21	2.49
4	411 (1.16), 562 (0.18), 612 (0.11)	657, 719 (12.0)	246/9110	3.87/1.14992	0.31	2.27

Table 1. Cont.

	ACN					
Corrole	λ_{Abs} (ε; M^{-1} cm^{-1} × 10^5) [a]	λ_{Em} (QY; %) [b]	SS (nm/cm^{-1}) [c]	τ_f (ns)[d]/X^2	k_r (×10^8 s^{-1}) [e]	k_{nr} (×10^8 s^{-1}) [e]
1	419 (0.94), 577 (0.11), 625 (0.20)	635, 693 (11.0)	216/8120	4.04/1.05903	0.27	2.20
2	420 (1.01), 580 (0.10), 624 (0.25)	634, 692 (17.0)	214/8035	4.24/1.15462	0.40	1.96
3	419 (0.76), 583 (0.09), 627 (0.20)	637, 694 (8.0)	218/8165	3.68/1.13217	0.22	2.50
4	420 (0.79), 583 (0.07), 623 (0.22)	633, 692 (11.0)	213/8010	4.07/1.17485	0.27	2.18
	MeOH					
Corrole	λ_{Abs} (ε; M^{-1} cm^{-1} × 10^5) [a]	λ_{Em} (QY; %) [b]	SS (nm/cm^{-1}) [c]	τ_f (ns)[d]/X^2	k_r (×10^8 s^{-1}) [e]	k_{nr} (×10^8 s^{-1}) [e]
1	409 (0.82), 571 (0.15), 617 (0.11)	658 (11.0)	249/9250	3.77/1.13135	0.29	2.36
2	410 (1.21), 567 (0.20), 613 (0.18)	661 (6.0)	251/9260	3.78/1.01780	0.16	2.48
3	410 (1.17), 568 (0.21), 613 (0.15)	666 (22.0)	256/9375	3.56/1.07740	0.62	2.19
4	409 (1.39), 567 (0.24), 613 (0.18)	658 (10.0)	249/9250	3.66/1.05354	0.27	2.46
	DMSO					
Corrole	λ_{Abs} (ε; M^{-1} cm^{-1} × 10^5) [a]	λ_{Em} (QY; %) [b]	SS (nm/cm^{-1}) [c]	τ_f (ns) [d]/X^2	k_r (×10^8 s^{-1}) [e]	k_{nr} (×10^8 s^{-1}) [e]
1	425 (1.56), 580 (0.22), 627 (0.40)	637, 692 (38.0)	212/7830	4.26/1.00523	0.89	1.45
2	427 (1.05), 587 (0.11), 628 (0.31)	640, 697 (26.0)	213/7795	4.20/1.10980	0.62	1.76
3	426 (1.00), 579 (0.14), 630 (0.25)	643, 701 (22.0)	217/7920	4.12/1.08084	0.53	1.89
4	427 (1.37), 585 (0.15), 627 (0.38)	639, 696 (27.0)	212/7770	4.35/1.09784	0.62	1.68

[a] [ ] = 10 µM; [b] [ ] = 5.0 µM, using TPP as standard (QY = 11%, DMF solution); [c] SS = Stokes shifts = $\lambda_{em} - \lambda_{Soret}$ (nm) = $1/\lambda_{Soret} - 1/\lambda_{em}$ (cm^{-1}); [d] Excitation by NanoLED source at 441 nm; [e] According ref. [20].

In general, all corroles in both solvents showed electronic transitions already predicted for this type of derivative; in this case, the Soret band and two Q bands (Figure 2). For both corroles in ACN and DMSO solutions, the Soret band splitting and a difference in intensities of Q-bands were noted when compared to the DCM or MeOH solution. This fact has been reported in the literature by several authors and can be attributed to the presence of possible tautomeric species in a solution, occurring by the possibility of pyrrole nitrogen deprotonation or hydrogen bond interactions between pentafluorophenyl-corroles and the solvent [25,26]. On this occasion, different tautomeric types of the studied corroles are predicted in polar solvents such as ACN and DMSO, thus it should be expected that a possible thermodynamic equilibrium in the presence of both tautomeric species would be reached. Polar solvents can interact with the molecules by solvation and stabilize them, mainly through possible intermolecular interactions such as hydrogen bonds or dipole–dipole forces. This case is no longer observed in MeOH solution, as it is protic and acidic enough to keep the H atoms in the N-pyrrole macrocycle ring.

The steady-state emission fluorescence of corroles 1–4 in both solvents (λ_{exc} at Soret band and 600–800 nm range) and the photophysical data are reported in Table 1 and fluorescence emission spectra of corroles are shown in Figure 3. The emission quantum yield values (φ_F) were calculated according to the reference tetra(phenyl)porphyrin standard in DMF solution (TPP; φ_F = 11.0%). In general, the values of the emission quantum yield of the studied corrole derivatives agree with the predicted structure (Table 1). More notable differences are observed in the DMSO solution, where the compounds in general showed the highest φ$_F$ values, probably due to a greater stabilization of the excited states by the more polar solvent. The differences in the fluorescence data come from the presence of the substituents attached at the *meso*-10-aryl position of the corrole ring. In the MeOH solution, the change in the spectrum profile is also noticeable, with only one transition in the excited state, a fact that can be attributed to interactions by hydrogen bond interactions by the protic solvent (Figure 3). Data referring to HOMO-LUMO theoretical calculations by TDDFT analysis of these corroles will help a better interpretation of the presented results to be made.

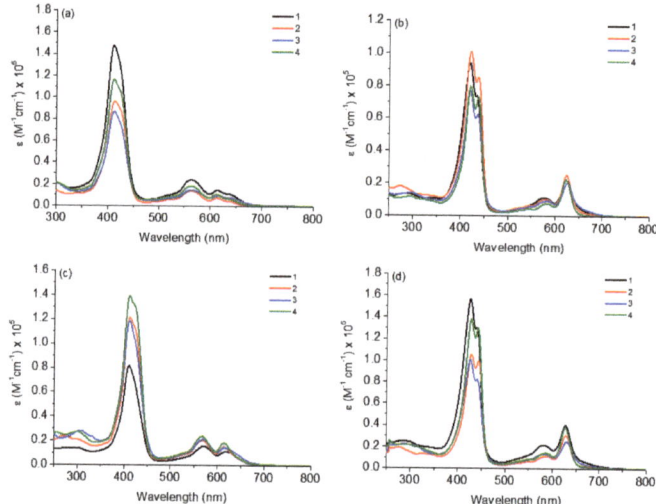

Figure 2. Absorption UV-Vis spectra of corroles **1–4** in (**a**) DCM, (**b**) ACN, (**c**) MeOH and (**d**) DMSO, at concentration of 10 µM.

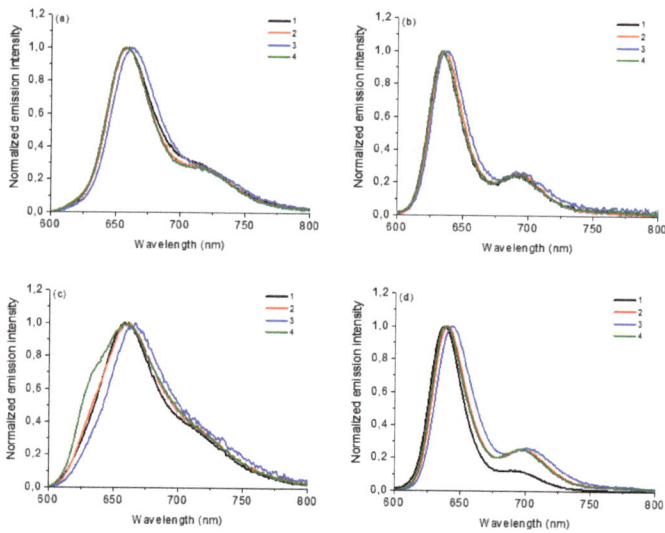

Figure 3. Normalized steady-state fluorescence emission spectra of corroles **1–4** in (**a**) DCM, (**b**) ACN, (**c**) MeOH and (**d**) DMSO, at concentration of 5.0 µM (λ_{exc} at Soret transition band).

With regard to the lifetime (τ_f) values of the corroles **1–4**, no major changes were observed according to the polarity of the solvent, with lifetimes between the 3.0 and 5.0 ns range and non-radiative (k_{nr}) rates higher than the radiative (k_r) ones. The fluorescence decay plots of the compounds are shown in the Supplementary Materials (Figures S5–S8).

2.3. TDDFT Analysis

Table 2 summarizes the theoretical results for the absorption wavelengths, in nm, and oscillator strengths, f, of the main peaks of the Soret and Q-bands in DCM for all the studied corroles. The natural transition orbitals relating to the lowest energy electronic excitation in the DCM solvent are also shown in Table 2. These data were obtained at the ground state equilibrium geometry of each compound. The theoretical optical absorption spectra and TDDFT data of two tautomeric states of corroles in DMSO are given in the Supplementary Materials (Figures S9–S16 and Table S1). Differences in the absorption spectra due to the implicit solvent environment are seen only for corrole **1** (Ph moiety), with the spectrum for DMSO showing longer absorption wavelengths for the Q bands and shorter wavelengths for the Soret band, compared to the spectrum in DCM. For compounds **2** to **4**, the calculated absorption spectra show practically the same profile in both solvents. Previous studies on corrole-like systems show the same behavior for these two solvents [5]. The theoretical absorption wavelengths for the Soret and Q bands are blue-shifted in DMSO as well as in DCM, compared to the experimental results. The reason for the deviations is related to the fact that the energies of the excited states are calculated at the ground state geometry.

Table 2. HOMO-LUMO plots for the lowest energy electronic transitions ($S_0 \rightarrow S_1$), transition energies in nm, and oscillator strengths (in parentheses), f, for the main peak at the Soret band, and the two main peaks of the Q-band in DCM, for corroles **1–4**. The data for the Soret peak was calculated as the average for the two theoretically intense peaks (see Supplementary Materials). This was done to allow a better comparison with the experimental results.

Corrole	HOMO Plot	LUMO Plot	Soret Band (f)	Q Bands (f)
1			390.98 (1.1256)	535.89 (0.1006); 555.81 (0.2818)
2			367.33 (1.3793)	547.84 (0.1095); 564.03 (0.2953)
3			371.05 (1.3838)	554.66 (0.0884); 575.63 (0.3254)
4			371.99 (1.4419)	553.48 (0.0835); 573.47 (0.3171)

To better understand the threshold for the optical transitions predicted by TDDFT calculations, the natural transition orbitals (NTOs) are shown in Table 2. These orbitals are associated with the lowest energy Q-band absorption peak for the DCM solvent. These NTOs for the DMSO solvent are very similar to those for the DCM solvent. For all studied cases, the NTOs associated with the S_0 to S_1 excitation show electronic transitions involving π-like orbitals distributed on the macrocyclic ring. The $\pi \rightarrow \pi^*$ absorption peaks in this specific energy range indicate that these compounds can be good candidates for photodynamic therapy. An analysis of the excited states shows that there are two triplet states, T_1 and T_2, lower in energy than the first excited state, S_1, that can be involved in intersystem crossing processes for all the studied corroles, as show in Figures S17–S18 in the Supplementary information. Furthermore, these two triplet states have energy gaps relative to the ground state that are larger than 0.98 eV, satisfying the energetic requirement to the generation of $^1O_2$ species [27].

2.4. Aggregation, Stability in Solution and Photostability Assays

In the aggregation experiments, we observed a very low tendency of aggregation of the studied corroles **1–4** in both solvents, including in the DMSO (5%)/Tris-HCl pH 7.4 mixture buffered solution (for biological applications). A linear increase was observed in the absorption spectra as a function of the concentration variation from 2.0 to 30 µM. Both observations indicate that aggregation is not present in all assays. The UV-Vis spectra of corroles **1–4** in DCM, ACN, MeOH, DMSO and DMSO (5%)/Tris-HCl pH 7.4 buffered mixture solutions are presented in the Supplementary Materials (Figures S19–S38).

Regarding the stability of derivatives **1–4** in the solution, they were monitored by UV-Vis spectroscopy for a period of seven days (see Supplementary Materials—Figures S39–S46). The experiments were only conducted in DMSO solution and DMSO (5%)/Tris-HCl pH 7.4 buffered mixture due to their use in biological medium. In this way, we can conclude that these compounds are stable and can be used safely in this period of time.

Anticipating photodynamic applications, corroles **1–4** were submitted to photostability tests under white-light LED irradiation (fluence rate of 50 mW cm^{-2} and a total light dosage of 90 J cm^{-2}) for a period of 20 min. Both corroles were tested in DMSO solution and DMSO (5%)/Tris-HCl pH 7.4 mixture. It is possible to note from the graphs in Figure 4 that in DMSO solution, the studied corroles are more susceptible to photodegradation processes than in the DMSO (5%)/Tris-HCl pH 7.4 media (Figure 4). This fact can be explained by the fact that in the absence of an aqueous medium, the derivatives can generate more ROS, thus increasing the photodegradation power of the derivatives. Another fact that may be related to this behavior is the fact that these reactive oxygen species have shorter lifetimes in water [28], thus favoring the integrity of the corroles.

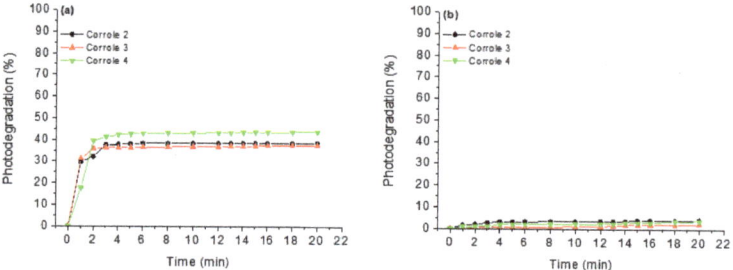

Figure 4. Photostability assays under white-light LED irradiation (fluence rate of 50 mW cm^{-2} and a total light dosage of 90 J cm^{-2}) for a period of 20 min, where (**a**) DMSO and (**b**) DMSO(5%)/Tris-HCl pH 7.4 mixture buffered solution.

2.5. Redox Analysis

The electrochemical profile by cyclic voltammetry of corroles **1–4** shows redox processes between −2.00 V and +2.00 V versus SHE in dry DCM solution at a scan rate of 100 mV s^{-1}, using TBAPF$_6$ as the supporting electrolyte. All CVs of the studied corroles are presented in the Supplementary Materials (Figures S47–S50) and the assignments of the redox peaks are presented in the Table 3. In the positive region, the oxidation processes of studied corroles exhibit three redox peak potentials for the scan between 0.0 V and +2.00 V (Table 3). The oxidation peaks of corroles are assigned to the formation of the mono, second and third mono-electronic oxidation peaks, generating π-cation radical, cation and di-cations, respectively [4,5]. Only in the case of corrole **4**, which contains a SCH$_3$

substituent, can the first oxidation process be attributed to an oxidation of the S atom to a sulfoxide species (S=O). In the negative region and the cathodic region, only one reduction process was observed in both cases between the 0.0 V and −1.50 V potential range. The reduction process can be attributed to the π-anion radical species for corroles (Table 3).

Table 3. Redox potentials of corroles **1–4** in dry DCM solution (*E* versus SHE).

Corrole	E_{ox1}	E_{ox2}	E_{ox3}	E_{red1}	E_{HOMO} [d]	E_{LUMO} [e]	ΔE
1 *	+0.65 V [a]	+0.78 V [c]	+0.95 V [c]	−1.29 V [b]	−5.45	−3.51	1.94
2	+0.80 V [a]	+1.32 V [c]	-	−1.18 V [b]	−5.60	−3.62	1.98
3	+0.60 V [a]	+1.09 V [c]	+1.93 V [c]	−1.21 V [b]	−5.40	−3.59	1.81
4	+0.21 V [a]	+0.95 V [c]	+1.62 V [c]	−1.37 V [b]	−5.01	−3.43	1.58

[a] Anodic peak = E_{pa}; [b] Cathodic peak = E_{pc}; [c] $E_{1/2} = E_{pa} + E_{pc}/2$; [d] $E_{HOMO} = -[4.8 + E_{ox1}$ (versus SHE)]eV; [e] $E_{LUMO} = -[4.8 + E_{red1}$ (versus SHE)]eV; * Ref. [5].

2.6. Photobiological Assays

For singlet oxygen ($^1O_2$) generation, corroles **1–4** are suitable for phototherapeutic application (Table 4). All DPBF photooxidation UV-Vis spectra are compiled in the Supplementary Materials (Figures S51–S53). Compared to the standard *meso*-tri(phenyl)corrole (TPhCor) in DMSO solution, the presence of substituents in the *meso*-aryl position can interfere in $^1O_2$ production, probably by intersystem crossing processes, mainly due to the presence of chemical groups attached in the corrole structure.

Table 4. Photobiological parameters of corroles **1–4**.

Corrole	k_{po} (M^{-1} s^{-1}) [a]	φ_Δ (%) [b]	k_{SO} (min^{-1}) [c]	log P_{OW} [d]
1 *	0.322	33.0	0.641	+2.948
2	0.504	51.0	0.540	+3.542
3	0.444	45.0	0.913	+2.431
4	0.390	40.0	0.513	+2.135
TPhCor **	0.650	67.0	—	—

[a] DPBF photooxidation constant in DMSO solution; [b] Singlet oxygen quantum yields in DMSO solution; [c] Superoxide formation constant in DMSO solution; [d] Water-octanol partition coefficients; * Ref. [5]; ** Ref. [5].

The capacity of corroles **1–4** to generate superoxide species ($O_2^{\bullet-}$) was investigated in DMSO solution. For this application, solutions of corroles containing NBT and the reducing agent NADH were irradiated with a white-light LED source (fluence rate of 50 mW cm^{-2} and a total light dosage of 90 J cm^{-2}) in aerobic conditions, at a period of 20 min. The reaction of NBT with superoxide radical species produced diformazan, which can be monitored following the absorption band of this product (Supplementary Materials, Figure S54). The superoxide generation constant (k_{SO}) by the NBT reduction assays is shown in Table 4. These results indicate that corroles **1–4**, after white-light irradiation conditions, can possibly form $O_2^{\bullet-}$ species. The generation of superoxide is also dependent on the substituent inserted in corrole, since electron donating groups favor the formation of these species in a solution (Table 4).

Finally, the partition coefficients (log P_{OW}) were measured for each corrole derivative and the values found for the neutral corroles containing different substituents are in accordance with the literature [29], with corroles **1** and **2** showing a more hydrophobic character when compared to the hydroxy-phenyl **3** and thio(methyl)phenyl **4** derivatives. This fact is attributed to the presence of more polar groups in the periphery of corroles, leaving compounds **3** and **4** with a more hydrophilic character (Table 4).

3. Biomolecule-Binding Assays

3.1. Binding Properties with DNA by UV-Vis Analysis

In order to evaluate the interaction between DNA and studied corroles **2–4**, UV-Vis absorption analysis was carried out. The UV-Vis spectra for the corroles in the absence and presence of successive additions of CT-DNA concentrations are shown in Figure 5, using compound **4** as an example, and the DNA-binding properties are listed in Table 5. All UV-Vis spectra of corroles **2** and **3** are presented in the Supplementary Materials (Figures S55–S57).

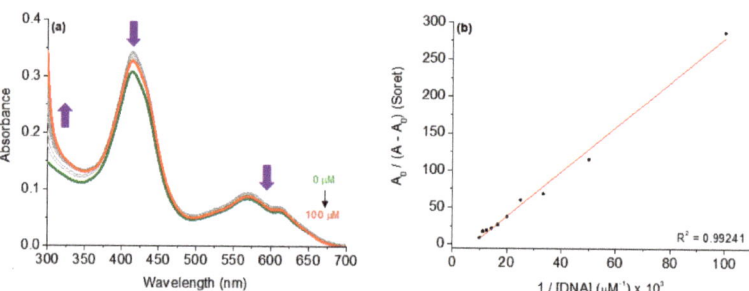

Figure 5. (**a**) UV-Vis spectra of the corrole **4** upon successive additions of CT-DNA concentrations (0 to 100 µM) in DMSO(5%)/Tris-HCl pH 7.4 mixture buffered solution. (**b**) Benesi-Hidelbrandt plots of $A_0/(A - A_0)$ vs. $1/[CT\text{-}DNA]$.

Table 5. DNA-binding properties of corroles **1–4** by UV-Vis and fluorescence emission analysis.

		UV-Vis Analysis			
Corrole	$H(\%)$ [a]	$\Delta\lambda$ (nm) [b]	K_b ($\times 10^5$; M^{-1}) [c]	$\Delta G°$ (kcal mol^{-1}) [d]	
1 *	6.10	0.0	1.49 ± 0.02	−7.05	
2	10.6	0.0	3.21 ± 0.07	−7.50	
3	3.80	0.0	1.75 ± 0.10	−7.15	
4	10.7	0.0	3.32 ± 0.03	−7.55	
		AO:DNA by Emission			
Corrole	$Q(\%)$ [e]	$K_{SV}(\times 10^3;$ $M^{-1})$ [f]	$k_q(\times 10^{12};$ $M^{-1}s^{-1})$ [g]	$K_b(\times 10^3;$ $M^{-1})$ [h]	$\Delta G°$ (kcal mol^{-1}) [d]
1 *	25.0	3.43 ± 0.13	2.02	—	—
2	7.40	0.80 ± 0.01	0.47	1.16 ± 0.18	−4.18
3	33.3	4.27 ± 0.07	2.51	3.41 ± 0.12	−4.82
4	34.2	5.00 ± 0.10	2.94	2.71 ± 0.06	−4.68
		DAPI:DNA by Emission			
Corrole	$Q(\%)$ [e]	$K_{SV}(\times 10^3;$ $M^{-1})$ [f]	$k_q(\times 10^{12};$ $M^{-1}s^{-1})$ [i]	$K_b(\times 10^3;$ $M^{-1})$ [h]	$\Delta G°$ (kcal mol^{-1}) [d]
1 *	37.4	5.81 ± 0.11	2.64	—	—
2	21.2	1.96 ± 0.02	0.89	3.90 ± 0.09	−4.90
3	33.8	3.43 ± 0.03	1.56	3.42 ± 0.13	−4.82
4	41.5	6.47 ± 0.02	2.94	3.29 ± 0.16	−4.80

[a] $H(\%) = (A_0 - A)/A \times 100\%$; [b] Red-shift; [c] Binding constant by Benesi-Hidelbrandt equation; [d] Determined by Gibbs free-energy equation; [e] $Q(\%) = (F_0 - F)/F \times 100\%$; [f] Determined by Stern–Volmer quenching constant; [g] Determined by K_{SV}/τ_0 ratio, where τ_0 = 1.70 ns (AO:DNA); [h] Determined by double-logarithm equation; [i] Determined by K_{SV}/τ_0 ratio, where τ_0 = 2.20 ns (DAPI:DNA).

In this analysis, the successive additions of CT-DNA to corrole compound solutions caused hypochromic effects at the Soret and Q bands without a red or blue shift, indicating that the studied corroles can be interacted with CT-DNA, probably via secondary interactions by major or minor grooves around the biomacromolecule. Although neutral derivatives lack the potential for cationic–anionic electrostatic binding with DNA phosphate units, the corroles under study demonstrated an ability to interact with nucleic acids, corroborating previous studies that reported the activity of non-charged tetrapyrrolic macrocycles [13].

In this way, the binding constant (K_b) values are in the 10^5 M^{-1} range (Table 5), indicating that both corroles interact strongly with CT-DNA and the presence of substituents can interfere in the binding affinity of CT-DNA. Thermodynamic analysis via Gibbs free energy $\Delta G°$ values (Table 5) indicated that all corroles interacted spontaneously with CT-DNA, thus reinforcing the results observed by K_b values. In the next section, the possibilities of interaction between corroles in terms of intercalation or via grooves are investigated.

3.2. Competitive Binding Assays with DNA by Steady-State Fluorescence Emission

The steady-state fluorescence emission spectra involving the competition assays for the binding between corroles and DNA:dyes adducts are presented, using corrole 4 as an example (Figure 6). The fluorescence Stern–Volmer (K_{SV}), bimolecular rate quenching (k_q), binding (K_b) and $\Delta G°$ parameters for DNA:dye:corroles are listed in Table 5.

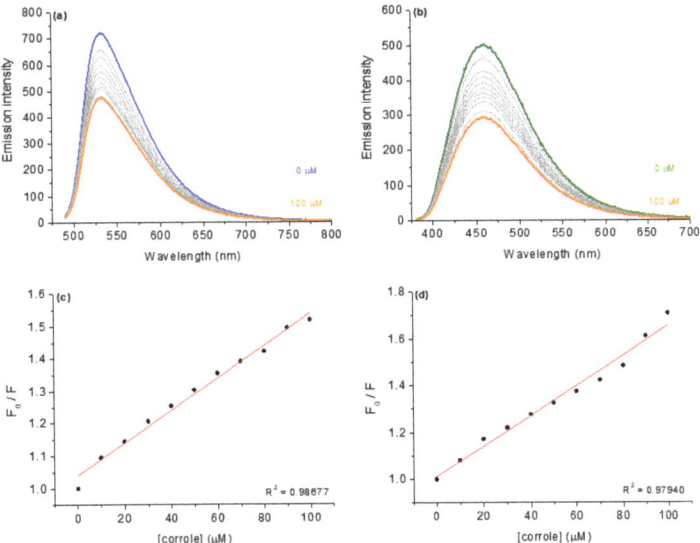

Figure 6. Steady-state fluorescence emission spectra for (**a**) AO:DNA and (**b**) DAPI:DNA without and in the presence of corrole **4**, in in DMSO(5%)/Tris-HCl pH 7.4 mixture buffered solution. Graphs (**c**,**d**) shows the plot F_0/F versus [corrole]. [corrole] = 0–100 µM.

As an example, Figure 6a depicts the fluorescence emission spectra for AO bound to CT-DNA (fluorescence emission at 534 nm when excited at 480 nm) in the absence and presence of corrole **4**. When corroles were added to CT-DNA pre-treated with acridine orange dye, corrole **4** induced a decrease in the fluorescence intensity of CT-AO:DNA adduct, indicating a displacement of AO from DNA, which can be assigned as a viable competition between AO and corroles for CT-DNA strands (specifically to the region rich in adenine and thymine). The Stern–Volmer (K_{SV}), bimolecular quenching rate (k_q) and binding (K_b) constant values in the presence of each derivative are summarized in Table 5.

To evaluate whether the possible interaction of substituted corroles occurs via the groove site, the minor groove binder DAPI dye was used for the steady-state fluorescence emission quenching assays (Figure 6b). In independent experiments, it was possible to observe significant fluorescence quenching of DAPI:DNA upon successive additions of corrole **4**. Comparing both K_{SV} and K_b values for competitive binding assays into AO:DNA and DAPI:DNA adducts, it can be inferred that there is a significant variation in the fluorescence quenching constants, mainly in the presence of the studied corroles (Table 5). Overall, the K_{SV} and K_b data variation can be attributed to a preference for corrole interaction by the minor groove and not only by an intercalation phenomenon, agreeing with the molecular docking calculations (described in the next section) and the literature [13,20].

Moreover, the bimolecular quenching rate constant ($k_q \sim 10^{12}$ M^{-1} s^{-1}) for the corroles, being higher than the diffusion rate constant according to Smoluchowski–Stokes–Einstein theory [30], indicates a ground state association between the corroles and nucleobases in the DNA strands (probably static fluorescence quenching mechanism). Additionally, the fluorescence emission spectra for the other corroles into AO:DNA and DAPI:DNA solutions are presented in the Supplementary Materials (Figures S58–S59).

3.3. Viscosity Measurements with DNA

It is known that viscosity assays are sensitive to the change in the length of the DNA double helix. In the absence of other structural techniques, viscosity measurements are considered an important method to determine intercalation or non-intercalation binding of compounds to DNA nucleobases. The results of the viscosity measurements are shown in Figure S60 in the Supplementary Materials. The DNA viscosity remains almost unchanged upon the addition of corroles **1–4**, with an increase in the ratio [corrole]/[CT-DNA]. These results indicate that the studied corroles do not intercalate between the DNA bases, corroborating the fluorescence emission DAPI experiments (probably binds to minor grooves). This is an expected result, since these corrole derivatives have a steric volume that is not conducive to promoting some intercalation phenomenon.

3.4. Binding Properties with HSA by UV-Vis and Emission Analysis

In addition to the UV-Vis analysis for DNA, the interaction of the studied corroles with HSA was also tested and the absorption spectra are listed in the Supplementary Materials (Figures S61–S63). In general, the values found for interactions such as K_b and $\Delta G°$ are lower than for the interaction with CT-DNA, as can be seen in Table 6.

Table 6. HSA-binding properties of corroles **1–4** by UV-Vis and fluorescence emission analysis.

		UV-Vis Analysis			
Corrole	$H(\%)$ [a]	$\Delta\lambda$ (nm) [b]	K_b ($\times 10^3$; M^{-1}) [c]	$\Delta G°$ (kcal mol^{-1}) [d]	
1	17.4	0.0	4.83 ± 0.04	−5.00	
2	7.25	0.0	4.76 ± 0.04	−5.00	
3	16.0	0.0	3.30 ± 0.07	−4.80	
4	29.0	0.0	16.4 ± 0.07	−5.75	

			Emission Analysis				
Corrole	$Q(\%)$ [e]	$K_{SV}(\times 10^4$; M^{-1}) [f]	$k_q(\times 10^{12}$; M^{-1}s^{-1}) [g]	$K_a(\times 10^4$; M^{-1}) [h]	$K_b(\times 10^4$; M^{-1}) [i]	n [j]	$\Delta G°$ (kcal mol^{-1}) [d]
1 *	—	1.41 ± 0.01	2.49	2.29 ± 0.13	-	-	−5.85
2	65.7	1.91 ± 0.05	3.37	1.00 ± 0.11	5.22 ± 0.19	1.23	−6.45
3	88.6	7.60 ± 0.18	13.4	8.75 ± 0.23	8.63 ± 0.42	1.95	−6.75
4	77.2	2.10 ± 0.06	3.70	9.13 ± 0.25	6.29 ± 0.39	1.45	−6.75

[a] $H(\%) = (A_0 - A)/A \times 100\%$; [b] Red-shift; [c] Binding constant by Benesi-Hidelbrandt equation; [d] Determined by Gibbs free-energy equation; [e] $Q(\%) = (F_0 - F)/F \times 100\%$; [f] Determined by Stern–Volmer quenching constant; [g] Determined by K_{SV}/τ_0 ratio, where τ_0 = 5.67 ns (HSA); [h] Determined by modified Stern–Volmer equation; [i] Determined by double-logarithm equation, where n is the number of binding sites.

For HSA albumin interactions, the fluorescence emission was monitored by gradually increasing the corrole derivative concentration in DMSO (5%)/Tris-HCl pH 7.4 mixture buffered solution; for example, compound **3** is presented in Figure 7. The HSA fluorescence emission quenching spectra of all corroles are shown in the Supplementary Materials (Figures S64–S65).

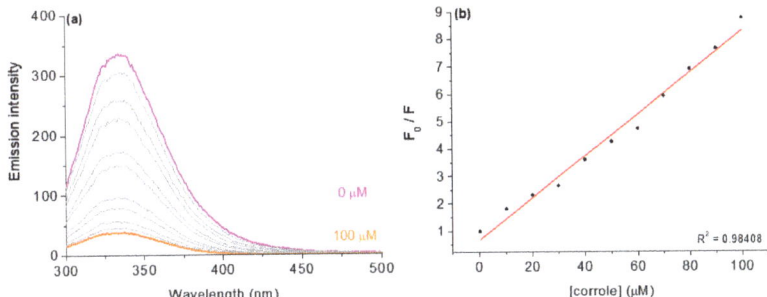

Figure 7. Steady-state fluorescence emission spectra for (**a**) HSA without and in the presence of corrole **3**, in DMSO(5%)/Tris-HCl pH 7.4 mixture buffered solution. Graph (**b**) shows the plot F_0/F versus [corrole]. [corrole] = 0–100 µM.

To examine the nature of the fluorescence quenching mechanism induced by these corrole derivatives, the Stern–Volmer equation was applied. The results showed a good linear relationship and the Stern–Volmer quenching constants (K_{SV} and k_q) (Table 6). The values observed for the bimolecular quenching rate constant ($k_q \sim 10^{12}$ M^{-1} s^{-1}) are three orders of magnitude larger than the diffusion rate constants according to Smoluchowski–Stokes–Einstein [30], corroborating that the nature of quenching is a possible static mechanism. The association constant (K_a), binding constant (K_b), and the number of binding sites (n) were obtained using the modified Stern–Volmer and double-logarithm equations [13,20]. The K_a and K_b constants correlated well with the values of the Stern–Volmer quenching constants (K_{SV}) and were obtained from the emission assays for the interactions of derivatives with HSA. In addition, since the n values are variable (1.23 to 1.95), the derivatives probably bind to HSA in different sites; hence, a SF and time-resolved analysis was made to check this possibility (see next section).

3.5. Synchronous Fluorescence (SF) Analysis

Comparing the steady-state fluorescence technique with the synchronous fluorescence (SF), the latter has been considered a complementary and more sensitive approach to detect possible perturbations in the microenvironment around the two main fluorophores of albumin (tyrosine and tryptophan residues) after drug binding [31]. Figures S66–S68 in the Supplementary Materials show the SF spectra for HSA without and upon successive additions of corroles **1–4** at $\Delta\lambda$ 15 nm and 60 nm for Tyr and Trp residues, respectively. In general, for both $\Delta\lambda$ there is a significant decrease in the fluorescence signal upon additions of corrole compounds; however, it did not induce any blue- or red-shift, agreeing with the steady-state fluorescence data (Section 3.4), which indicated that the binding of corroles does not induce any significant perturbation on the microenvironment around the fluorophores.

3.6. Time-Resolved Fluorescence with HSA and Corroles **1–4**

In order to identify the main fluorescence quenching mechanism (static or dynamic), time-resolved fluorescence decays were obtained for HSA without and in the presence of corroles **1–4** in DMSO(5%)/Tris-HCl pH 7.4 buffered mixture solution and lifetime plots; values are presented in the Supplementary Materials (Figure S69 and Table S3). There was a decrease in the τ_f value of HSA upon corrole addition, indicating that at high derivative concentration a combined static and dynamic fluorescence quenching mechanism is feasible for the interaction HSA:corroles.

3.7. Molecular Docking Analysis with DNA and HSA

Molecular docking calculations are interesting tools to provide the molecular aspects of the interaction between small compounds and biomacromolecules [32,33]. Thus, to determine the binding modes and binding sites of the studied corroles **2–4** with both DNA and HSA, a blind docking was performed in the major and minor grooves of DNA and in the three main binding sites of albumin. Molecular docking analysis between DNA and HSA with corrole **1** has been previously described by Acunha and co-workers [13].

Firstly, Table 7 and Figure 8 depict the docking score value (dimensionless) and docking pose for DNA:corroles. The GOLD 2020.2 software ranked the ten best docking poses for each system (biomacromolecule-ligand) and the best result was treated as shown in Figure 8. Since the docking score value of each pose is considered as the negative value of the sum of energy terms from the mechanical-molecular type component, which includes intramolecular tensions in the ligand and intermolecular interactions in the biomacromolecule-ligand association, the more positive the score value (dimensionless), the better the interaction. According to the computational results, corroles **2–4** can be accommodated mainly in the minor groove of DNA (highest docking score value, e.g., 61.19 and 40.37 for DNA:**2** in the minor and major grooves, respectively—Table 7), which is in agreement with experimental competitive binding assays.

Table 7. Molecular docking score values (dimensionless) for the interaction between DNA/HSA and each corrole derivative under study into the corresponding main binding site.

Corrole	DNA		HSA		
	Minor Groove	Major Groove	Site I	Site II	Site III
2	61.19	40.37	83.03	30.30	74.09
3	63.75	39.87	53.84	30.90	72.20
4	65.60	43.86	69.31	30.88	76.72

In the case of HSA there are three main binding pockets for endogenous and exogenous compounds: subdomain IIA (site I), located in a hydrophobic binding pocket where Trp-214 residues can be found, subdomain IIIA (site II), also located in a hydrophobic binding pocket, and subdomain IB (site III), located on the surface of the albumin (Figure 9). Table 7 shows the docking score value (dimensionless) for the three main binding sites. Since the highest docking score value for the corrole derivatives was obtained for site III, molecular docking results suggested that the external pocket (subdomain IB) is the main binding region for compounds **3** and **4**, while site I was the main binding site for derivative **2**. Although all corroles under study possess a high steric volume that might impact the preference for binding in the external region of albumin, it is probable that corrole **2** interacts mainly in the internal pocket of albumin due to the high lipophilicity of the compound. It has been previously described that fluorinated-phenyl- and pyrenyl-corroles bind mainly into the site I of albumin [13], indicating that the presence of hetero-atoms in the phenyl moieties of the corroles under study changes the polarity of the corrole and the capacity of interaction with an external pocket.

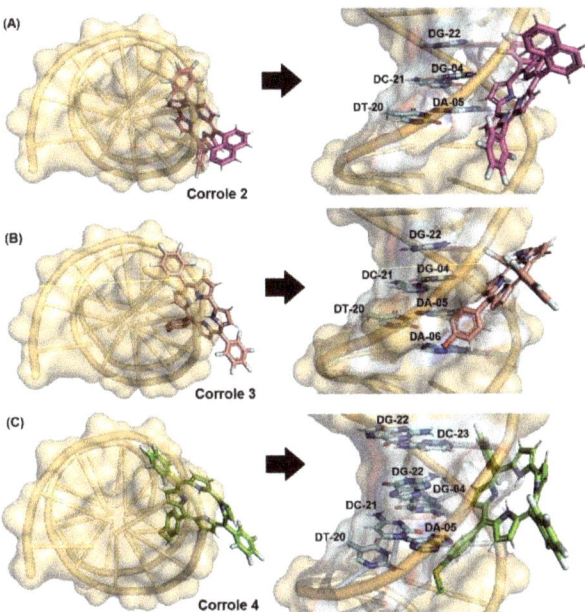

Figure 8. Best docking pose for (**A**) DNA:**2**, (**B**) DNA:**3**, and (**C**) DNA:**4** in the minor groove. Selected nitrogenated bases and corroles are in stick representation in cyan, pink, beige, green, and brown, respectively. Elements' color: hydrogen: white; oxygen: red; nitrogen: dark blue, fluorine: light blue, and sulfur: yellow.

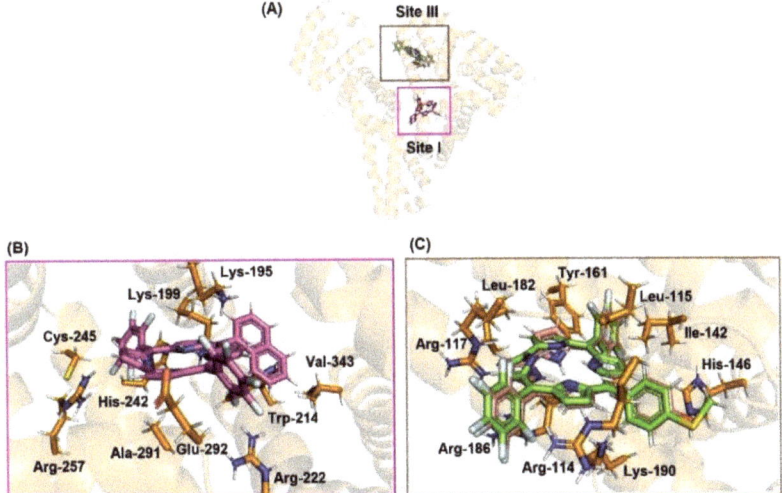

Figure 9. (**A**) Superposition of the best docking pose for HSA:corroles in the Sites I and III. Best docking pose for (**B**) HSA:**2** into site I and (**C**) HSA:**3/4** into site III. Selected amino acid residue and corroles are in stick representation in orange, pink, beige, green, and brown, respectively. Elements' color: hydrogen: white; oxygen: red; nitrogen: dark blue, fluorine: light blue, and sulfur: yellow.

Molecular docking results suggested van der Waals and hydrogen bonding as the main binding forces responsible for both interactions DNA:corroles and HSA:corroles (Tables S3–S4 in the Supplementary Materials, and Figures 8 and 9). In silico calculations suggested that the replacement of hydroxy-phenyl (corrole 3) to thio(methyl)phenyl (corrole 4) moieties did not change the binding conformation to albumim.

4. Materials and Methods

4.1. General

All chemical reagents were of analytical grade and purchased from Sigma-Aldrich® and Oakwood Chemical® (Estill, SC, USA) without any further purification. The calf-thymus acid desoxyribonucleic (CT-DNA) and human serum albumin (HSA) was lyophilized powder and fatty acid-free (Sigma-Aldrich®, St. Louis, MO, USA; purity ≥99%). The concentration of the stock solutions of biomolecules was confirmed by UV-Vis analysis through the Beer–Lambert equation with a molar absorptivity (ε) value of 6600 M^{-1} cm^{-1} for CT-DNA at 260 nm (per nucleic acid) and 35,700 M^{-1} cm^{-1} for HSA at 280 nm in Tris-HCl buffer (pH 7.4) solution; the water used in all experiments was milliQ grade. Compounds were analyzed using a high-resolution mass spectrometer with electrospray ionization (HRMS-ESI) in the positive mode using a micrOTOF-QII mass spectrometer (Bruker Daltonics, Billerica, MA, USA). Mass spectra were recorded for each sample in the methanolic solution (concentration of 500 ppb) with a flow of 5.0 µL/min and capillarity of 6000 V.

4.2. Photophysical Measurements

Absorption UV-Vis spectra were obtained using the Shimadzu UV-2600 spectrophotometer (1.0 cm optical path length) and measured in the 250–800 nm region for studied corroles in several solvents such as acetonitrile (ACN), dichloromethane (DCM), methanol (MeOH), dimethyl sulfoxide (DMSO) and DMSO(5%)/Tris-HCl pH 7.4 mixture buffered solution, with fixed concentrations of 10 µM. For the determination of the steady-state fluorescence emission spectra, we employed a fluorimeter Horiba Fluoromax Plus, where the corroles were dissolved at a fixed concentration of 5.0 µM, in a 1.0 cm optical path length cuvette, excited at Soret transition band (slit 5.0; Em/Exc).

The fluorescence quantum yields (ϕ_f) of the corrole derivatives were measured according to the literature, using tetra(phenyl)porphyrin (TPP) in DMF solution as the standard molecule [34]. Fluorescence lifetimes (τ_f) were recorded using the Time Correlated Single Photon Counting (TCSPC) method with DeltaHub controller in conjunction with Horiba fluorimeter. Data were processed with DAS6 and Origin® 8.5 software (Northampton, MA, USA) using mono-exponential fitting of raw data. NanoLED (Horiba) source (1.0 MHz, 441 nm excitation wavelength) was used as an excitation source. The instrumental response function (IRF) was collected using a Ludox® dispersion (Sigma-Aldrich®, St. Louis, MO, USA). Radiative (k_r) and non-radiative (k_{nr}) values were calculated by equations according to the literature [20].

4.3. TDDFT Calculations

The electronic, structural, and optical properties of the studied compounds were determined using the Density Functional Theory (DFT) and its time-dependent extension (TDDFT). The ground state geometrical structures were optimized through conjugated gradient techniques. The wB97XD functional was used to represent the exchange and correlation potential [35], while the molecular orbitals were described by linear combinations of 6-31G(d,p) quality basis sets. We employed the polarizable continuum model (PCM) [36] to calculate the molecular properties either in dimethyl sulfoxide (DMSO) or dichloromethane (DCM) (ε = 46.826 and ε = 8.930, respectively). Natural Transitions Orbitals (NTOs) were calculated for the molecular orbitals involved in the lowest energy electronic transitions [37]. All calculations have been made using the Gaussian 09 quantum chemistry package [38].

4.4. Electrochemical Analysis

Cyclic voltammograms were recorded with a potenciostat/galvanostat AutoLab Eco Chemie PGSTAT 128N system at room temperature and under argon atmosphere in dry dichloromethane (DCM) solution. Electrochemical grade tetrabutylammonium hexafluorophosphate (TBAPF$_6$, 0.1 M) was used as a supporting electrolyte. Employing a standard three-component system these CV experiments were carried out with: a glassy carbon working electrode; a platinum wire auxiliary electrode; and a platinum wire *pseudo*-reference electrode. To monitor the reference electrode, the Fc/Fc$^+$ redox couple was used as an internal reference [39].

4.5. Photobiological Parameters

Aggregation by UV-Vis absorption analysis was conducted as a function of successive increase in corrole concentration (2.0 to 30 μM) in all solvents and changes in the λ_{Soret} in the 250–700 nm range were monitored according to the related literature [5]. The stability experiments in pure dimethyl sulfoxide (DMSO) and in DMSO(5%)/Tris-HCl pH 7.4 mixture buffered solution of studied corrole was also monitored by absorption UV–Vis measurements at several days (one to seven days). All experiments were performed in duplicate and independently.

The photostability assays in dimethyl sulfoxide (DMSO) and in DMSO(5%)/Tris-HCl pH 7.4 mixture buffered solution of related corrole derivatives was also monitored by absorption UV-Vis measurements at different exposure times (0 to 30 min) under the white-light LED array system (400 to 800 nm) at a fluence rate of 50 mW cm^{-2} and a total light dosage of 90 J cm^{-2}. All experiments were performed in duplicate and independently, according to equations described in the literature [5].

Singlet oxygen ($^1O_2$) production was recorded according to typical 1,3-diphenylisobenzofuran (DPBF) photooxidation assays; the maximum volume of 1.0 mL which contained 100 μM DPBF in DMSO was mixed with 0.5 mL (50 μM) of corroles 1–4. The flask was then filled with 2.0 mL of DMSO to a final volume of 3.5 mL. In order to measure singlet oxygen generation (φ_Δ), absorption UV-Vis spectra of each solution were recorded at different exposure times (0 to 600 s, using red-light diode laser; Thera Laser DMC—São Paulo; potency of 100 mW) and φ_Δ were calculated according to the literature [4].

The superoxide radical ($O_2^{\bullet-}$) species by NBT reduction assays was used and this approach was carried out at the same conditions stated in the literature, using Nitro Bule Tetrazolium (NBT) and NADH in DMSO solution [40]. Control experiments were performed in the absence of corroles and all derivatives were irradiated under aerobic conditions with a white-light LED source (fluence rate of 50 mW cm^{-2} and a total light dosage of 90 J cm^{-2}) at a period of 20 min. The progress of the reaction was monitored by following the increase in the absorbance close to 560 nm. The superoxide generation constant (k_{SO}) values can be obtained according to the literature cited above.

The partition coefficient (log P_{OW}) of corroles 1–4 was determined using *n*-octanol (3.0 mL) and water (3.0 mL), according to the literature [20]. The corrole concentrations as well as their respective absorbances were determined by the UV-Vis absorption at 250 to 800 nm range, and the Soret band was chosen for monitoring and calculating the log P_{OW} values.

4.6. Biomacromolecule Interactive Studies

UV-Vis absorption analysis for each corrole without and in the presence of successive additions of CT-DNA or HSA solution were obtained at 298.15 K in DMSO(5%)/Tris-HCl pH 7.4 mixture buffered solution in the 250 to 700 nm range. The corrole concentration was fixed in 5.0 μM and CT-DNA or HSA was in the 0 to 100 μM range. The hyperchromicity (H%), red-shift ($\Delta\lambda$), binding constant (K_b), and Gibb's free-energy ($\Delta G°$) values of the corroles 1–4 were calculated according to the literature, through Benesi-Hildebrand and free-energy equations [13,20]. An interactive DNA study using absorption analysis with corrole 1 has been previously described by Acunha and co-workers [13].

Competitive binding assays between CT-DNA:dyes and corroles by steady-state fluorescence emission analysis are recorded and corroles **2–4** in DMSO(5%)/Tris-HCl pH 7.4 mixture buffered solution (0 to 100 μM) were gradually added in a fixed concentration of acridine orange (AO; A-T rich intercalator; 10 μM ; λ_{exc}= 490 nm, λ_{em}= 500–800 nm) and 4′,6-diamidino-2-phenylindole (DAPI; minor groove binder; 10 μM; λ_{exc}= 359 nm, λ_{em}= 380–700 nm) and CT-DNA (10 μM) in DMSO(5%)/Tris-HCl pH 7.4 mixture buffered solution. The DNA:dye adducts were incubated for 3 min after corrole addition for each measurement. The Stern–Volmer quenching (K_{SV}) and bimolecular quenching rate (k_q) constants of corroles were calculated according to the DNA:dye fluorescence quenching using a plot of F_0/F versus [corrole] and a ratio of K_{SV}/τ_0, where the τ_0 denote the fluorescence lifetime of DNA:dye (AO = 2.20 ns and DAPI = 1.70 ns), respectively [13,20]. Binding (K_b) constant and free-energy interaction (ΔG°) values are obtained by double-logarithm and Gibb's equation according to the literature [20]. Competitive DNA:dye study by emission analysis with corrole **1** has been previously described by Acunha and co-workers [13].

Additionally, viscosity analysis was carried out using an Ostwald viscometer immersed in a water bath maintained at 298.15 K, according to the literature, with some modifications [41]. The CT-DNA concentration was kept constant in all compounds, while the corrole concentration was increased in DMSO (5%)/Tris-HCl pH 7.4 mixture buffered solution. The flow time was measured at least three times with a digital stop-watch (Casio®, Shibuya City, Tokyo, Japan) and the mean value was calculated. Data are presented as $(\eta/\eta^0)^{1/3}$ versus the ratio [corrole]/[CT-DNA], where η and η^0 are the specific viscosity of CT-DNA in the presence and absence of the corroles **1–4**, respectively. The values of η and η^0 were calculated by use of the expression $(t - t_b)/t_b$, where t is the observed flow time and t_b is the flow time of solution alone. The relative viscosity of the CT-DNA was calculated from η/η^0.

For HSA-binding assays by steady-state fluorescence emission analysis to obtain quantitative parameters on the binding capacity of HSA:corroles **1–4**, the maximum fluorescence data after inner filter correction [31] were obtained via the Stern–Volmer (K_{SV}) quenching constant, binding (K_b) constant, and Gibbs' free energy (ΔG°) values by Stern–Volmer and double-logarithmic equations, according to the literature [20]. All emission spectra were obtained by measurements at 298.15 K in DMSO(5%)/Tris-HCl pH 7.4 mixture buffered solution in the 300 to 500 nm range ([HSA] = 10 μM; [corroles] = 0 to 100 μM).

The synchronized fluorescence (SF) spectra were recorded for HSA (10 μM) without and in the presence of corroles **1–4** (concentration ranging from 0 to 100 μM) at room temperature (298.15 K). Spectra were recorded in the 240–320 nm range by setting Δλ = 15 nm (for tyrosine residue) and Δλ = 60 nm (for tryptophan residue).

Fluorescence lifetime decays (τ_f) were recorded using Time Correlated Single Photon Counting (TCSPC) method with DeltaHub controller in conjunction with Horiba Jobin-Yvon Fluoromax Plus spectrofluorometer. Data were processed with DAS6 and Origin® 8.5 software using mono-exponential fitting of raw data. NanoLED (Horiba) source (1.0 MHz, Pulse width < 1.2 ns at 284 nm excitation wavelength) was used as an excitation source. The HSA and corroles concentration were fixed in 10 μM each in DMSO (5%)/Tris-HCl pH 7.4 mixture buffered solution. Additionally, interactive binding properties with HSA with corrole **1** by spectroscopic methods has been previously described by Acunha and co-workers [13].

4.7. Molecular Docking Procedure with DNA and HSA

The crystallographic structure for DNA and HSA was obtained from the Protein Data Bank with access code 1BNA and 1N5U, respectively [39,40]. The chemical structure for corroles **2**, **3** and **4** was built and minimized in terms of energy by Density Functional Theory (DFT) calculations under B3LYP potential with basis set 6–31G*, available in the Spartan'18 software (Wavefunction, Inc., Irvine, CA, USA) [42–44]. The molecular docking calculations were performed with GOLD 2020.2 software (Cambridge Crystallographic Data Centre, Cambridge, CB2 1EZ, UK) [45].

Hydrogen atoms were added to the biomacromolecules following tautomeric states and ionization data inferred by GOLD 2020.2 software at pH 7.4. For DNA structure, the 10 Å radius around the center of mass of DT-20 was analyzed to explore the two main possible binding sites (major and minor grooves) [46]. On the other hand, for HSA structure 10 Å radius around the selected center of mass from the amino acid residue present in one of the three main binding pockets, more specifically Trp-214, Tyr-411, and Tyr-161 residues, for sites I, II, and III, respectively, was delimitated for molecular docking calculations. These amino acid residues were chosen according to the crystallographic structure of each site probe into HSA (warfarin, ibuprofen, and camptothecin) [47,48]. The number of genetic operations (crossing, migration, mutation) during the search procedure was set as 100,000. The software optimizes the geometry for hydrogen bonding by allowing the rotation of hydroxyl and amino groups contained in the biomacromolecules. The side chain rotamers have been defined according to the availability of the library. For all biomacromolecules, ChemPLP was used as the scoring function due to the lowest Root Mean Square Deviation (RMSD) obtained in previous work for small organic compounds, including porphyrins [26,49–51]. The figures for the best docking pose were generated with the PyMOL Delano Scientific LLC software (Schrödinger, New York, NY, USA) [52].

5. Conclusions

In this study, we investigated and studied the photophysical/photobiological properties of substituted corroles containing different groups at *meso*-10-position (phenyl, naphthyl, 4-hydroxy-phenyl and 4-thio(methyl)phenyl) in some solvents. Photobiological parameters such as ROS generation and photostability were evaluated and it was found that these compounds are promising for use in photoinduced processes. Furthermore, the interactive properties of corroles **1–4** against biomacromolecules such as DNA and HSA were evaluated, and the corrole derivatives had a preference for interacting in the minor grooves of the DNA due to secondary forces, which are more evident in site III of the albumin.

Supplementary Materials: The following supporting information can be downloaded at: https://www.mdpi.com/article/10.3390/molecules28031385/s1, Supplementary Materials: Figures S1–S4: HRMS mass spectra of corroles; Figures S5–S8: Normalized fluorescence decays of corroles; Figures S9–S16: Theoretical optical absorption spectra of corroles; Figures S17 and S18: Total energies of the S1, T1 and T2 excited states; Figures S19–S38: Aggregation behavior by UV-Vis analysis; Figures S39–S46: stability in solution assays; Figures S47–S50: Cyclic voltammetry analysis of corroles; Figures S51–S53: DPBF photo-oxidation assays with corroles; Figure S54: NBT reduction assay of corroles; Figures S55–S57: UV-Vis spectra of corroles in the presence of DNA; Figures S58 and S59: Steady-state fluorescence emission analysis of AO-DNA and DAPI-DNA adducts in the presence of corroles; Figure S60: Viscosimetry analysis of DNA in the presence of corroles; Figures S61–S63: UV-Vis spectra of corroles in the presence of HSA; Figures S64 and S65: Steady-state fluorescence emission analysis of HSA in the presence of corroles; Figures S66–S68: SF fluorescence emission spectra between HSA and corroles; Figure S69: Normalized fluorescence decay plots of corroles, in the presence of HSA; Tables S1–S2: TD-DFT theoretical parameters; Table S3: Fluorescence decay data of corroles in the presence of HSA; Tables S4–S5: Molecular docking results of corroles with DNA and HSA biomolecules.

Author Contributions: Conceptualization, V.B.d.S. and B.A.I.; methodology, V.B.d.S. and B.A.I.; software, B.A.I.; validation, V.B.d.S. and B.A.I.; formal analysis, V.B.d.S., V.N.d.R., P.C.P., O.A.C. and B.A.I.; investigation, O.A.C.; resources, B.A.I.; data curation, O.A.C.; writing—original draft preparation, V.B.d.S.; writing—review and editing, O.A.C., B.A.I. and P.C.P.; visualization, V.B.d.S. and B.A.I.; supervision, B.A.I. and O.A.C.; project administration, B.A.I.; funding acquisition, B.A.I. All authors have read and agreed to the published version of the manuscript.

Funding: This work was financed by national funds through FCT—*Fundação para a Ciência e a Tecnologia*, I.P. (Portugal), under the project UIDB/00313/2020. B.A. Iglesias also to thanks the CNPq (Universal—403210/2021-6 and PQ—305458/2021-3), CAPES (Finance code 001) and FAPERGS grants (PQ Gaucho—21/2551-0002114-4). P.C. Piquini also to thanks CNPq (312388/2018-7), CAPES and FAPERGS.

Institutional Review Board Statement: Not applicable.

Informed Consent Statement: Not applicable.

Data Availability Statement: All analyzed data are contained in the main text of the article. Raw data are available from the authors upon request.

Acknowledgments: The authors acknowledge Thiago Barcellos at University of Caxias do Sul (UCS, Brazil) for the HRMS mass spectra facilities. O.A.C. thanks Fundação para a Ciência e a Tecnologia (FCT—Portuguese Agency for Scientific Research) for his PhD fellowship 2020.07504.BD. B.A.I. and P.C.P. thanks to CNPq, CAPES and FAPERGS.

Conflicts of Interest: The authors declare no conflict of interest. The funders had no role in the design of the study; in the collection, analysis, or interpretation of data; in the writing of the manuscript; or in the decision to publish the results.

References

1. Teo, R.D.; Hwang, J.Y.; Termini, J.; Gross, Z.; Gray, H.B. Fighting Cancer with Corroles. *Chem. Rev.* **2017**, *117*, 2711–2729. [CrossRef] [PubMed]
2. Jiang, X.; Liu, R.-X.; Liu, H.-Y.; Chang, C.K. Corrole-based photodynamic antitumor therapy. *J. Chin. Chem. Soc.* **2019**, *66*, 1090–1099. [CrossRef]
3. Lopes, S.M.M.; Pineiro, M.; Pinho e Melo, T.M.V.D. Corroles and Hexaphyrins: Synthesis and Application in Cancer Photodynamic Therapy. *Molecules* **2020**, *25*, 3450. [CrossRef]
4. Pivetta, R.C.; Auras, B.L.; de Souza, B.; Neves, A.; Nunes, F.S.; Cocca, L.H.Z.; De Boni, L.; Iglesias, B.A. Synthesis, photophysical properties and spectroelectrochemical characterization of 10-(4-methyl-bipyridyl)-5,15-(pentafluorophenyl)corrole. *J. Photochem. Photobiol. A Chem.* **2017**, *332*, 306–315. [CrossRef]
5. Acunha, T.V.; Victória, H.F.V.; Krambrock, K.; Marques, A.C.; Costa, L.A.S.; Iglesias, B.A. Photophysical and electrochemical properties of two trans-A_2B-corroles: Differences between phenyl or pyrenyl groups at the meso-10 position. *Phys. Chem. Chem. Phys.* **2020**, *22*, 16965–16977. [CrossRef]
6. Anusha, P.T.; Swain, D.; Hamad, S.; Giribabu, S.; Prashant, T.S.; Tewari, S.P.; Rao, S.V. Ultrafast Excited-State Dynamics and Dispersion Studies of Third-Order Optical Nonlinearities in Novel Corroles. *J. Phys. Chem. C* **2012**, *116*, 17828–17837. [CrossRef]
7. Shao, W.; Wang, H.; He, S.; Shi, L.; Peng, K.; Lin, Y.; Zhang, L.; Ji, L.; Liu, H. Photophysical Properties and Singlet Oxygen Generation of Three Sets of Halogenated Corroles. *J. Phys. Chem. B* **2012**, *116*, 14228–14234. [CrossRef] [PubMed]
8. Mahammed, A.; Gross, Z. Corroles as triplet photosensitizers. *Coord. Chem. Rev.* **2019**, *379*, 121–132. [CrossRef]
9. Shi, L.; Liu, H.-Y.; Shen, H.; Hu, J.; Zhang, G.-L.; Wang, H.; Ji, L.-N.; Chang, C.-K.; Jiang, H.-F. Fluorescence properties of halogenated mono-hydroxyl corroles: The heavy-atom effects. *J. Porphyr. Phthalocyanines* **2009**, *13*, 1221–1226. [CrossRef]
10. Zhao, F.; Zhan, X.; Lai, S.-H.; Zhang, L.; Liu, H.-Y. Photophysical properties and singlet oxygen generation of meso-iodinated free-base corroles. *RSC Adv.* **2019**, *9*, 12626–12634. [CrossRef]
11. Lee, P.; Wu, X. Modifications of human serum albumin and their binding effect. *Curr. Pharm. Des.* **2015**, *21*, 1862–1865. [CrossRef]
12. Mahammed, A.; Gray, H.B.; Weaver, J.J.; Sorasaenee, K.; Gross, Z. Amphiphilic Corroles Bind Tightly to Human Serum Albumin. *Bioconjugate Chem.* **2004**, *15*, 738–746. [CrossRef]
13. Acunha, T.V.; Chaves, O.A.; Iglesias, B.A. Fluorescent pyrene moiety in fluorinated C_6F_5-corroles increases the interaction with HSA and CT-DNA. *J. Porphyr. Phthalocyanines* **2021**, *25*, 75–94. [CrossRef]
14. Zhang, Y.; Wen, J.-Y.; Mahmood, M.H.R.; Wang, X.-L.; Lv, B.-B.; Ying, X.; Wang, H.; Ji, L.-N.; Liu, H.-L. DNA/HSA interaction and nuclease activity of an iron(III) amphiphilic sulfonated corrole. *Luminescence* **2015**, *30*, 1045–1054. [CrossRef]
15. Iglesias, B.A.; Barata, J.F.B.; Pereira, P.M.R.; Girão, H.; Fernandes, R.; Tomé, J.P.C.; Neves, M.G.P.M.S.; Cavaleiro, J.A.S. New platinum(II)–bipyridyl corrole complexes: Synthesis, characterization and binding studies with DNA and HAS. *J. Inorg. Biochem.* **2015**, *153*, 32–41. [CrossRef]
16. Gershman, Z.; Goldberg, I.; Gross, Z. DNA Binding and Catalytic Properties of Positively Charged Corroles. *Angew. Chem. Int. Ed.* **2007**, *119*, 4398–4402. [CrossRef]
17. Na, N.; Zhao, D.-Q.; Li, H.; Jiang, N.; Wen, J.-Y.; Liu, H.-Y. DNA Binding, Photonuclease Activity and Human Serum Albumin Interaction of a Water-Soluble Freebase Carboxyl Corrole. *Molecules* **2016**, *21*, 54. [CrossRef] [PubMed]
18. Wang, Y.-G.; Zhang, Z.; Wang, H.; Liu, H.-Y. Phosphorus(V) corrole: DNA binding, photonuclease activity and cytotoxicity toward tumor cells. *Bioorg. Chem.* **2016**, *67*, 57–63. [CrossRef]
19. Liu, L.-G.; Sun, Y.-M.; Liu, Z.-Y.; Liao, Y.-H.; Zeng, L.; Ye, Y.; Liu, H.-Y. Halogenated Gallium Corroles:DNA Interaction and Photodynamic Antitumor Activity. *Inorg. Chem.* **2021**, *60*, 2234–2245. [CrossRef]
20. Acunha, T.V.; Rodrigues, B.M.; da Silva, J.A.; Galindo, D.D.M.; Chaves, O.A.; da Rocha, V.N.; Piquini, P.C.; Köhler, M.H.; De Boni, L.; Iglesias, B.A. Unveiling the photophysical, biomolecule binding and photo-oxidative capacity of novel Ru(II)-polypyridyl corroles: A multipronged approach. *J. Mol. Liq.* **2021**, *340*, 117223. [CrossRef]
21. Gryko, D.T.; Jadach, K. A Simple and Versatile One-Pot Synthesis of meso-Substituted trans-A_2B-Corroles. *J. Org. Chem.* **2001**, *66*, 4267–4275. [CrossRef]

22. Koszarna, B.; Gryko, D.T. Efficient Synthesis of meso-Substituted Corroles in a H$_2$O−MeOH Mixture. *J. Org. Chem.* **2006**, *71*, 3707–3717. [CrossRef]
23. Gryko, D.T. Adventures in the synthesis of meso-substituted corroles. *J. Porphyr. Phthalocyanines* **2008**, *12*, 906–917. [CrossRef]
24. Orłowski, R.; Gryko, D.; Gryko, D.T. Synthesis of Corroles and Their Heteroanalogs. *Chem. Rev.* **2017**, *117*, 3102–3137. [CrossRef] [PubMed]
25. Kruk, M.M.; Ngo, T.H.; Verstappen, P.; Starukhin, A.S.; Hofkens, J.; Dehaen, W.; Maes, W. Unraveling the Fluorescence Features of Individual Corrole NH Tautomers. *J. Phys. Chem. A* **2012**, *116*, 10695. [CrossRef]
26. Ivanova, Y.B.; Savva, V.A.; Mamardashvili, N.Z.; Starukhin, A.S.; Ngo, T.H.; Dehaen, W.; Maes, W.; Kruk, M.M.; Corrole, N.H. Tautomers: Spectral Features and Individual Protonation. *J. Phys. Chem. A* **2012**, *116*, 10683. [CrossRef]
27. Quina, F.H.; Silva, G.T.M. The photophysics of photosensitization: A brief overview. *J. Photochem. Photobiol.* **2021**, *7*, 100042. [CrossRef]
28. Egorov, S.Y.; Kamalov, V.F.; Koroteev, N.I.; Krasnovsky, A.A., Jr.; Toleutaev, B.N.; Zinukov, S.V. Rise and decay kinetics of photosensitized singlet oxygen luminescence in water. Measurements with nanosecond time-correlated single photon counting technique. *Chem. Phys. Lett.* **1989**, *163*, 421–424. [CrossRef]
29. Barata, J.F.B.; Zamarrón, A.; Neves, M.G.P.M.S.; Faustino, M.A.F.; Tomé, A.C.; Cavaleiro, J.A.S.; Röder, B.; Juarranz, A.; Sanz-Rodríguez, F. Photodynamic effects induced by meso-tris(pentafluorophenyl)corrole and its cyclodextrin conjugates on cytoskeletal components of HeLa cells. *Eur. J. Med. Chem.* **2015**, *92*, 135–144. [CrossRef]
30. Lakowicz, J.R. *Principles of Fluorescence Spectroscopy*, 3rd ed.; Springer: New York, NY, USA, 2006.
31. Chaves, O.A.; Menezes, L.B.; Iglesias, B.A. Multiple spectroscopic and theoretical investigation of meso-tetra-(4-pyridyl)porphyrin ruthenium(II) complexes in HSA binding studies. Effect of Zn(II) in protein binding. *J. Mol. Liq.* **2019**, *294*, 111581. [CrossRef]
32. Da Silveira, C.H.; Chaves, O.A.; Marques, A.C.; Rosa, N.M.O.; Costa, L.A.S.; Iglesias, B.A. Synthesis, photophysics, computational approaches, and biomolecule interactive studies of metalloporphyrins containing pyrenyl units: Influence of the metal center. *Eur. J. Inorg. Chem.* **2022**, *12*, e2022000075. [CrossRef]
33. Chaves, O.A.; Amorim, A.P.O.; Castro, L.H.E.; de Sant'Anna, C.M.R.; de Oliveira, M.C.C.; Cesarin-Sobrinho, D.; Netto-Ferreira, J.C.; Ferreira, A.B.B. Fluorescence and docking studies of the interaction between human serum albumin and pheophytin. *Molecules* **2015**, *20*, 19526–19539. [CrossRef] [PubMed]
34. Engelmann, F.M.; Losco, P.; Winnischofer, H.; Araki, K.; Toma, H.E. Synthesis, electrochemistry, spectroscopy and photophysical properties of a series of meso-phenylpyridylporphyrins with one to four pyridyl rings coordinated to [Ru (bipy) 2 Cl]+ groups. *J. Porphyr. Phthalocyanines* **2002**, *6*, 33. [CrossRef]
35. Chai, J.-D.; Head-Gordon, M. Long-range corrected hybrid density functionals with damped atom-atom dispersion corrections. *Phys. Chem. Chem. Phys.* **2008**, *10*, 6615–6620. [CrossRef] [PubMed]
36. Tomasi, J.; Mennucci, B.; Cammi, R. Quantum mechanical continuum solvation models. *Chem. Rev.* **2005**, *105*, 2999–3094. [CrossRef]
37. Martin, R.L. Natural transition orbitals. *J. Chem. Phys.* **2003**, *118*, 4775–4777. [CrossRef]
38. Revision, C.M.J.; Frisch, G.W.; Trucks, H.B.; Schlegel, G.E.; Scuseria, M.A.; Robb, J.R.; Cheeseman, G.; Scalmani, V.; Barone, G.A.; Petersson, H.; et al. *Gaussian 09*; Gaussian, Inc.: Wallingford, CT, USA, 2016.
39. Santos, F.S.; da Silveira, C.H.; Nunes, F.S.; Ferreira, D.C.; Victória, H.F.V.; Krambrock, K.; Chaves, O.A.; Rodembusch, F.S.; Iglesias, B.A. Photophysical, photodynamical, redox properties and BSA interactions of novel isomeric tetracationic peripheral palladium(ii)-bipyridyl porphyrins. *Dalton Trans.* **2020**, *49*, 16278–16295. [CrossRef]
40. Santamarina, S.C.; Heredia, D.A.; Durantini, A.M.; Durantini, E.N. Antimicrobial Photosensitizing Material Based on Conjugated Zn(II) Porphyrins. *Antibiotics* **2022**, *11*, 91. [CrossRef]
41. Navarro, M.; Cisneros-Fajardo, E.J.; Sierralta, A.; Fernández-Mestre, M.; Silva, P.; Arrieche, D.; Marchán, E. Design of copper DNA intercalators with leishmanicidal activity. *J. Biol. Inorg. Chem.* **2003**, *8*, 401–408. [CrossRef]
42. Drew, H.R.; Wing, R.M.; Takano, T.; Broka, C.; Tanaka, S.; Itakura, K.; Dickerson, R.E. Structure of a B-DNA dodecamer: Conformation and dynamics. *Proc. Natl. Acad. Sci. USA* **1981**, *78*, 2179. [CrossRef]
43. Wardell, M.; Wang, Z.; Ho, J.X.; Robert, J.; Rucker, J.; Ruble, J.; Carter, D.C. The Atomic Structure of Human Methemalbumin at 1.9 Å. *Biochem. Biophys. Res. Commun.* **2002**, *291*, 813–819. [CrossRef] [PubMed]
44. Available online: https://www.wavefun.com/ (accessed on 30 November 2022).
45. Available online: http://www.ccdc.cam.ac.uk/solutions/csd-discovery/components/gold/ (accessed on 30 November 2022).
46. Bessega, T.; Chaves, O.; Martins, F.; Acunha, T.; Back, D.; Iglesias, B.; de Oliveira, G. Coordination of Zn(II), Pd(II) and Pt(II) with ligands derived from diformylpyridine and thiosemicarbazide: Synthesis, structural characterization, DNA/BSA binding properties and molecular docking analysis. *Inorg. Chim. Acta* **2019**, *496*, 119049. [CrossRef]
47. Ghuman, J.; Zunszain, P.; Petitpas, I.; Bhattacharya, A.; Otagiri, M.; Curry, S. Structural basis of the drug-binding specificity of human serum albumin. *J. Mol. Biol.* **2005**, *353*, 38–52. [CrossRef]
48. Wang, Z.; Ho, J.; Ruble, J.; Rose, J.; Ruker, F.; Ellenburg, M.; Murphy, R.; Click, J.; Soistman, E.; Wilkerson, L.; et al. Structural studies of several clinically important oncology drugs in complex with human serum albumin. *Biochim. Biophys. Acta* **2013**, *1830*, 5356–5374. [CrossRef] [PubMed]

49. Chaves, O.; Santos, M.; de Oliveira, M.; Sant'Anna, C.; Ferreira, R.; Echevarria, A.; Netto-Ferreira, J. Synthesis, tyrosinase inhibition and transportation behavior of novel β-enamino thiosemicarbazide derivatives by human serum albumin. *J. Mol. Liq.* **2018**, *254*, 280–290. [CrossRef]
50. Câmara, V.; Chaves, O.; de Araújo, B.; Gonçalves, P.; Iglesias, B.; Ceschi, M.; Rodembusch, F. Photoactive homomolecular bis(n)-Lophine dyads: Multicomponent synthesis, photophysical properties, theoretical investigation, docking and interaction studies with biomacromolecules. *J. Mol. Liq.* **2022**, *349*, 118084. [CrossRef]
51. Chaves, O.; Acunha, T.; Iglesias, B.; Jesus, C.; Serpa, C. Effect of peripheral platinum(II) bipyridyl complexes on the interaction of tetra-cationic porphyrins with human serum albumin. *J. Mol. Liq.* **2020**, *301*, 112466. [CrossRef]
52. Available online: https://pymol.org/2/ (accessed on 30 November 2022).

Disclaimer/Publisher's Note: The statements, opinions and data contained in all publications are solely those of the individual author(s) and contributor(s) and not of MDPI and/or the editor(s). MDPI and/or the editor(s) disclaim responsibility for any injury to people or property resulting from any ideas, methods, instructions or products referred to in the content.

Article

Syntheses and Electrochemical and EPR Studies of Porphyrins Functionalized with Bulky Aromatic Amine Donors

Mary-Ambre Carvalho [1], Khalissa Merahi [1], Julien Haumesser [1], Ana Mafalda Vaz Martins Pereira [1], Nathalie Parizel [1], Jean Weiss [1], Maylis Orio [2], Vincent Maurel [3], Laurent Ruhlmann [1,*], Sylvie Choua [1,*] and Romain Ruppert [1,*]

1. Institut de Chimie, UMR CNRS 7177, Université de Strasbourg, Institut Le Bel, 4 rue Blaise Pascal, 67000 Strasbourg, France
2. Campus of St Jérôme, Aix-Marseille University, CNRS, Centrale Marseille, iSm2, CEDEX 20, 13397 Marseille, France
3. SyMMES, UMR 5819 CEA Grenoble/CNRS/Université Grenoble-Alpes, CEA Grenoble, 17 rue des Martyrs, CEDEX 9, 38054 Grenoble, France
* Correspondence: lruhlmann@unistra.fr (L.R.); sylvie.choua@unistra.fr (S.C.); rruppert@unistra.fr (R.R.)

Abstract: A series of nickel(II) porphyrins bearing one or two bulky nitrogen donors at the *meso* positions were prepared by using Ullmann methodology or more classical Buchwald–Hartwig amination reactions to create the new C-N bonds. For several new compounds, single crystals were obtained, and the X-ray structures were solved. The electrochemical data of these compounds are reported. For a few representative examples, spectroelectrochemical measurements were used to clarify the electron exchange process. In addition, a detailed electron paramagnetic resonance (EPR) study was performed to estimate the extent of delocalization of the generated radical cations. In particular, electron nuclear double resonance spectroscopy (ENDOR) was used to determine the coupling constants. DFT calculations were conducted to corroborate the EPR spectroscopic data.

Keywords: porphyrinoids; phenothiazine; radical cation; EPR spectroscopy; ENDOR; HYSCORE; spectroelectrochemistry

Citation: Carvalho, M.-A.; Merahi, K.; Haumesser, J.; Pereira, A.M.V.M.; Parizel, N.; Weiss, J.; Orio, M.; Maurel, V.; Ruhlmann, L.; Choua, S.; et al. Syntheses and Electrochemical and EPR Studies of Porphyrins Functionalized with Bulky Aromatic Amine Donors. *Molecules* 2023, 28, 4405. https://doi.org/10.3390/molecules28114405

Academic Editors: M. Amparo F. Faustino, Carlos J. P. Monteiro and Carlos Serpa

Received: 10 May 2023
Revised: 24 May 2023
Accepted: 26 May 2023
Published: 29 May 2023

Copyright: © 2023 by the authors. Licensee MDPI, Basel, Switzerland. This article is an open access article distributed under the terms and conditions of the Creative Commons Attribution (CC BY) license (https://creativecommons.org/licenses/by/4.0/).

1. Introduction

The selective peripheral functionalization of the porphyrin macrocycle has a long history [1,2]. The introduction of some substituents remained a synthetic challenge until recently. More specifically, introducing nitrogen nucleophiles at the periphery of porphyrin was mainly developed in the last twenty years. Several approaches were proposed, with or without metal-catalyzed reactions [3–7], and, for example, the classical palladium-catalyzed Buchwald–Hartwig amination reaction worked well for nucleophilic amines [8–11]. Although the introduction of many functional groups was considered a solved problem, adding bulky and/or less-nucleophilic aromatic amines with good yields at the *meso* positions of porphyrins remained a synthetic challenge, and this problem was really tackled over the last decade. The so-called Pd-PEPPSI complexes were chosen by the group of Osuka to prepare porphyrins bearing aromatic amines at their *meso* positions, which could later be fused to the aromatic core of the porphyrin [12–15]. Additionally, inexpensive copper-catalyzed Ullmann couplings were used to make these C-N connections [16]. In this particular case, porphyrins could be substituted once or twice with phenoxazine or carbazole units in good yields [17]. Such reactions are particularly useful to introduce electron donors and/or acceptors to build new molecular dyads and triads for dye-sensitized solar cells (DSSCs) or to prepare model compounds mimicking elementary steps in natural photosynthetic systems [18–22]. Mixed-valence compounds, with porphyrins as large π-bridges and aromatic amines as redox centers, were also studied [23,24]. The compounds described in this manuscript are shown in Figure 1 and were studied by electrochemistry,

EPR spectroscopy, and DFT calculations to obtain insight into the electronic delocalization of the generated radical cation of the amines on the π-systems of the nickel(II) porphyrins.

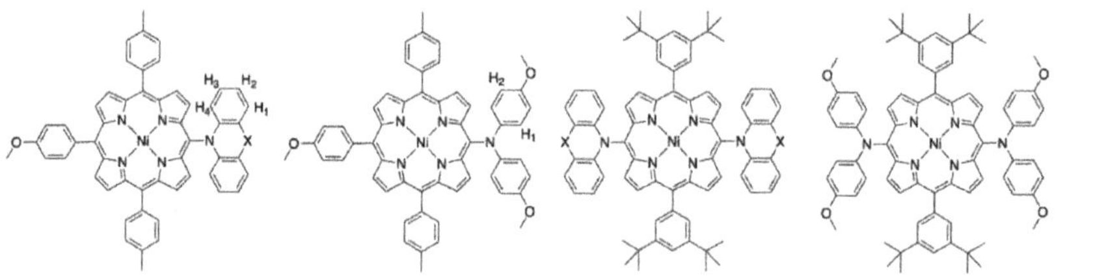

Figure 1. Chemical structures of the nickel(II) porphyrins studied.

2. Results and Discussion

2.1. Syntheses and Characterization

Some compounds described in this study were already described earlier, but their preparation is shortly included in the synthetic procedures presented. The phenoxazine and phenothiazine-substituted porphyrins were obtained by the inexpensive Ullmann coupling between brominated porphyrin precursors and these amines (see Scheme 1) [17]. The dianisylamines were introduced by using a classical palladium-catalyzed Buchwald–Hartwig amination reaction [5].

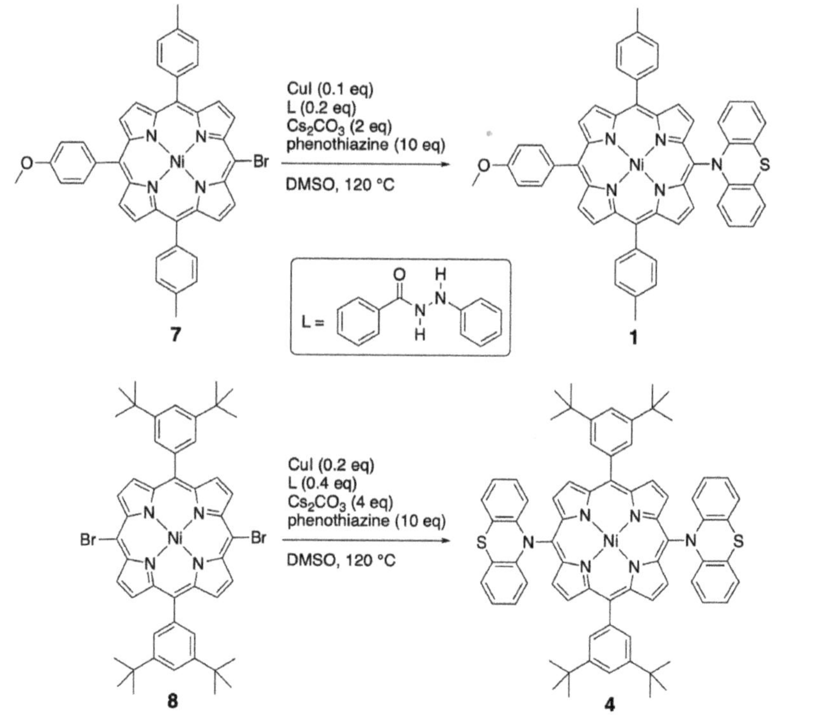

Scheme 1. Preparation of the nickel(II) porphyrins bearing one or two *meso*-phenothiazine substituents.

All compounds were characterized by standard spectroscopic techniques. Single crystals were obtained for some of them, and the structures were solved. The X-ray structure of the bis-phenothiazine-substituted nickel(II) porphyrin is presented in Figure 2. As noticed before in the solid-state structures of the 5,15-bis-carbazole- or 5,15-bis-phenoxazine-substituted nickel(II) porphyrins, the porphyrin plane is almost perfectly planar, despite the fact that the metal ion inside the aromatic cavity is nickel(II). Indeed, the vast majority of the nickel(II) porphyrin structures reported in the literature are strongly ruffled because the nickel(II) ion is too small for the inner porphyrin cavity [25].

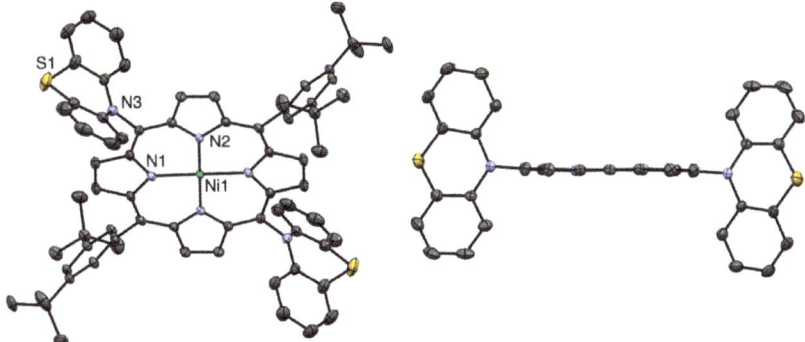

Figure 2. Two views of the nickel(II) porphyrin **4** X-ray structure (all hydrogen atoms are omitted for clarity). **Right**: View showing the planarity of the porphyrin plane (two *meso*-aryl groups omitted for clarity).

The same planarity of the porphyrin macrocycle was observed for the 5,15-bis(dianisylamine)nickel(II) porphyrin (see Figure 3).

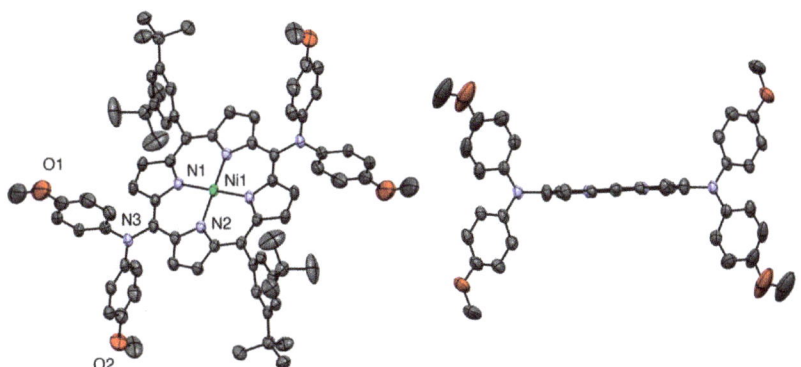

Figure 3. Two views of the nickel(II) porphyrin **6** X-ray structure (all hydrogen atoms are omitted for clarity). **Right**: View showing the planarity of the porphyrin plane (two *meso*-aryl groups omitted for clarity).

The related nickel(II) porphyrin bearing only one phenoxazine donor was described earlier, but the X-ray structure was missing. In this case (see Figure 4), the macrocycle was slightly ruffled, and the Ni-N distances were shorter than in the two other examples (1.94–1.95 Å instead of 1.95–1.97 Å).

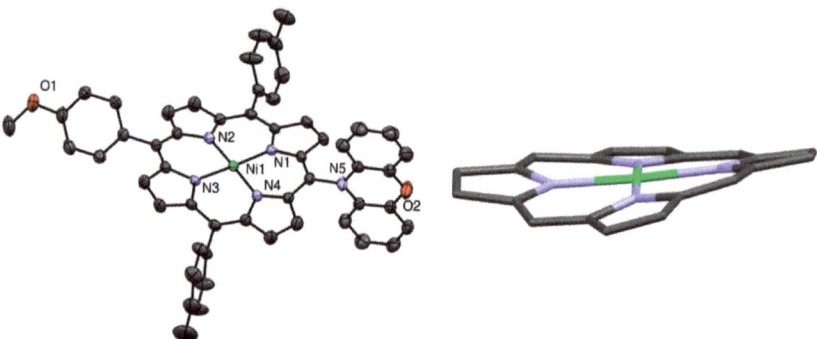

Figure 4. X-ray structure of nickel(II) porphyrin **2** (all hydrogen atoms are omitted for clarity). **Right**: view showing the ruffled nickel(II) porphyrin core.

As expected for very bulky donors (phenoxazine or phenothiazine), the aromatic amines are not in the plane of the porphyrin and are almost orthogonal to this plane. The *meso*-carbon to nitrogen distances were all in the same range (1.43 to 1.44 Å). The main difference between dianisylamine and the two other donors resides in the flexibility of the two aromatic groups attached to the nitrogen, whereas in the case of phenoxazine and phenothiazine, the linking oxygen or sulfur atoms prevent rotation of the two phenyl groups.

2.2. Electrochemical Studies

Compounds **1**, **2**, **3**, **4**, **5**, and **6** were studied by cyclic voltammetry, and the results are summarized in Table 1.

Table 1. Electrochemical data for the six compounds [a].

Compound	E_{red2} (Volts)	E_{red1} (Volts)	E_{ox1} (Volts)	ΔE_p (mV)	E_{ox2} (Volts)	ΔE_p (mV)	E_{ox3} (Volts)	E_{ox4} (Volts)
1	−2.18 [b]	−1.64 (1)	+0.34 (1)	60	+0.85 (2)	100	+1.24 [b]	-
4	−2.20 [b]	−1.62 (1)	+0.34 (1)	60	+0.40 (1)	60	+0.98 (2)	+1.25 [b]
2	-	−1.70 (1)	+0.36 (1)	80	+0.84 (2)	100	-	-
5	-	−1.64 (1)	+0.37 (1)	60	+0.43 (1)	75	+0.99 (2)	-
3	-	−1.09 [b]	+0.58 (1)	133	+0.95 (2)	122	-	-
6	-	−1.09 [b]	+0.57 (2)	100	+1.06 [b]	-	-	-

[a]. Cyclic voltammetry measurements: 1 mM dichloromethane solutions, with NBu_4PF_6 (0.1 M) as the supporting electrolyte, scan rate 0.1 V/s, 298 K. The number of exchanged electrons (indicated in parentheses) was determined by rotating disk voltammetry (RDV). Potentials are referenced vs. the Fc+/Fc redox couple, except for compounds **3** and **6** measured with the AgCl/Ag reference electrode. [b]. Irreversible reduction or oxidation steps.

The reduction and oxidation steps of these molecules can be localized on the porphyrinic core and/or on the aromatic amine *meso*-substituents. In the case of nickel(II) porphyrins, the redox processes are localized on the aromatic moiety, and the nickel(II) ion remains formally in oxidation state 2. Generally, for nickel(II) porphyrins, two reduction and two oxidation states are present, and sometimes a third oxidation step can be observed [26]. For compounds **1**–**6**, one or two additional oxidation states were expected due to the presence of the aromatic amines. The electrochemical data for compounds **2** and **5** were described earlier [17]. The cyclic voltammetry of compounds **1** and **4** is shown in Figure 5. For both compounds, two reduction waves corresponding to the formation of the porphyrinic radical anion and dianion were observed. In oxidation, four and five oxidation steps were, respectively, observed for compounds **1** and **4**.

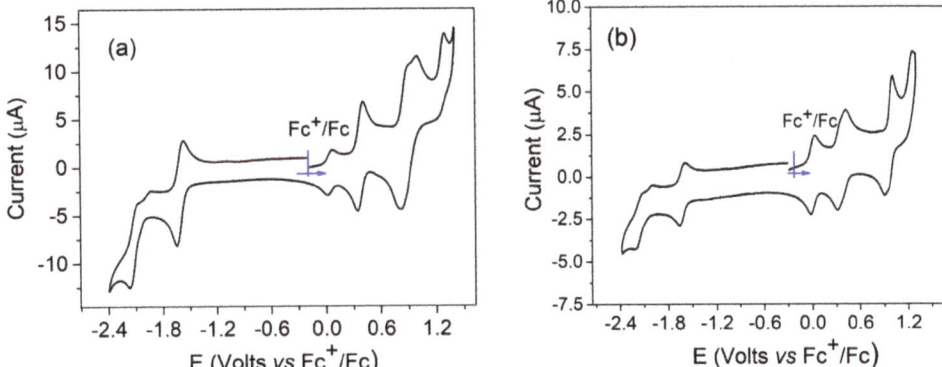

Figure 5. Cyclic voltammetry of compounds **1** (**a**) and **4** (**b**), recorded in dichloromethane, NBu$_4$PF$_6$ (0.1 M), scan rate = 0.1 V/s.

To ascertain the electronic states of the oxidized species of the compounds, spectroelectrochemical studies were carried out. For the monosubstituted porphyrin **1**, the first oxidation led to a new compound with almost no changes in the Soret band (from 415 to 417 nm) or the Q bands (see Figure 6). Three new bands appeared at 716, 789, and 897 nm. The very small bathochromic shift of the Soret band clearly indicated that the porphyrin ring was not involved in this redox process, and the new bands were attributed to a radical cation located on the substituent. Under the same conditions, increasing the oxidation potential to a value higher than the second oxidation modified the spectrum drastically. The intensity of the Soret band decreased drastically, showing that now the formation of the porphyrinic π-cation radical was occurring. The same behavior was observed for compound **2** (see Figure S10).

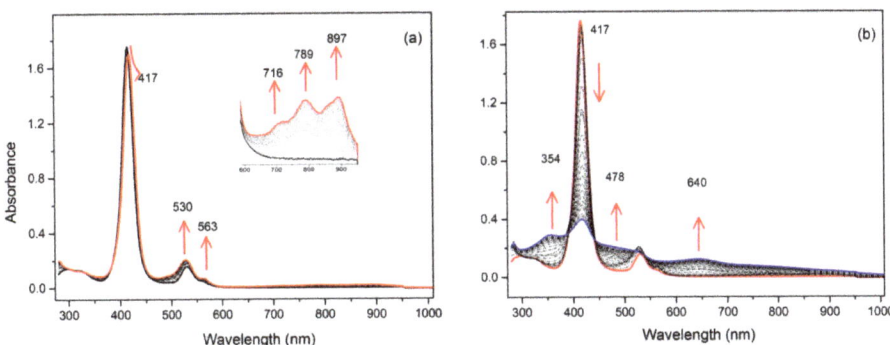

Figure 6. UV/Vis spectra during the stepwise electrochemical oxidation of **1**: (**a**) at the first oxidation potential; (**b**) at the second oxidation potential (recorded in dichloromethane, 0.1 M NBu$_4$PF$_6$).

Spectroelectrochemical studies of compound **3** under similar conditions led to a different result. While for compounds **1** and **2** the intensities of the Soret bands remained quasi-identical after the first oxidation step, the intensity of the Soret band of compound **3** dropped markedly after the first oxidation step (see Figure 7). In this case, the generation of the radical cation of the *meso*-donor nitrogen atom was partly delocalized over the aromatic porphyrinic moiety. For the three compounds bearing only one *meso*-donor group, the second oxidation step led to a large decrease in the Soret intensity, showing that the second oxidation steps were localized on the porphyrin aromatic ring.

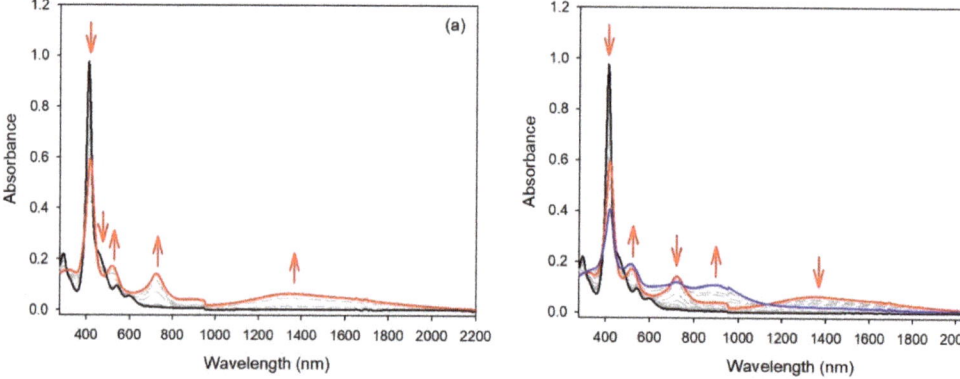

Figure 7. UV/Vis spectra during the stepwise electrochemical oxidation of **3**: (**a**) at the first oxidation potential; (**b**) at the second oxidation potential (recorded in dichloromethane, 0.1 M NBu$_4$PF$_6$).

The cyclic voltammetry of compound **4** (see Figure 5 right) was similar to the results described earlier for compound **5** and a bis-carbazole-substituted nickel(II) porphyrin, with small differences in potential values [17]. Again, compound **6**, bearing two *meso*-bis-anisylamine donor groups, showed a different electrochemical behavior. The first oxidation step was reversible, but the following oxidation steps proved to be irreversible. The spectroelectrochemical study showed again that the aromatic porphyrinic core was involved in the oxidation processes because the Soret band intensity dropped markedly (see Figure 8). To confirm these observations, an EPR study and DFT calculations were carried out (vide infra).

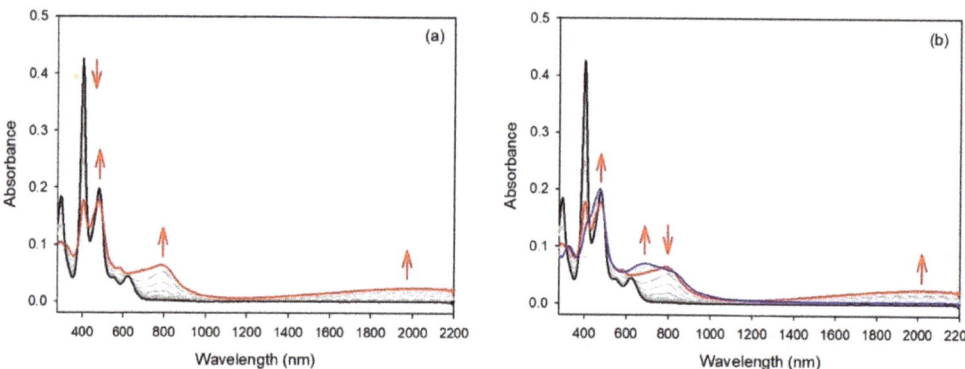

Figure 8. UV/Vis spectra during the stepwise electrochemical oxidation of **6**: (**a**) at the first oxidation potential; (**b**) at the second oxidation potential (recorded in dichloromethane, 0.1 M NBu$_4$PF$_6$).

2.3. EPR Studies

2.3.1. CW-EPR/ENDOR Measurements

The X-band EPR spectra of the fluid solutions of the radicals from mono-substituted nickel(II) porphyrins generated by chemical oxidation with one equivalent of AgSbF$_6$ in CH$_2$Cl$_2$ at room temperature are shown in Figure 1 and in Figure S1 in SI. They exhibit a well-resolved three-line pattern centered at g = 2.003 with nitrogen hyperfine coupling constants A$_{iso}$(^{14}N) of 19, 23, and 19 MHz for **1**, **2**, and **3**, respectively. Additionally, each of the three lines is resolved for **1** and **2** due to further hyperfine coupling interactions with magnetically active nuclei. To gain better information about the small spin density distributions within all compounds, we used CW electron nuclear double resonance spectroscopy (ENDOR) as an alternative tool for studying radicals with many overlapping and/or incompletely resolved EPR lines. Three, four, and two additional hyperfine coupling interactions centered around the nuclear frequency of the proton are observed for **1**, **2**, and **3**, respectively (see Figure 9, Figures S11 and S12, and Table 2). These patterns arise from the contributions of the known additional couplings described for the corresponding radical cations of phenothiazine and phenoxazine [27,28]. From these data, the EPR spectra were best simulated with the hyperfine coupling constants reported in Table 2, which were in close agreement with the DFT-computed values. Oxidations were also performed in situ electrochemically in the EPR cavity. The EPR spectra observed after a few minutes of electrolysis were identical to the spectra obtained after chemical oxidations.

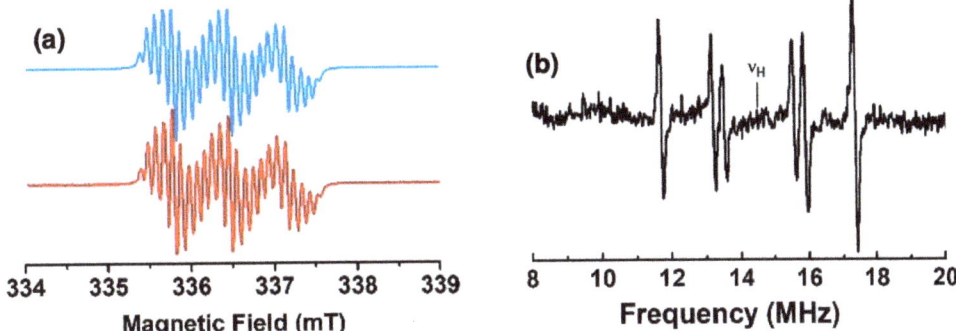

Figure 9. (a) X-band EPR spectrum of the radical cation of **1** in CH_2Cl_2 fluid solution at room temperature (blue) and its simulation (red); (b) ^{1}H ENDOR spectrum in CH_2Cl_2 fluid solution at 200 K.

Table 2. Isotropic hyperfine coupling constants obtained from DfT computations compared to experimental values from CW-EPR and ENDOR spectra (values in MHz).

Compound		N	H_1	H_2	H_3	H_4	g_{iso}
1	Exp	19	-	5.6	2.0	2.2	2.0034
	DFT	15	1.2	6.2	2.2	2.3	2.0060
2	Exp	23	1.3	8.3	2.1	4.3	2.0035
	DFT	18	1.9	8.2	2.0	3.3	2.0037
3	Exp	19	4.3	1.8	-	-	2.0035
	DFT	15	4.9	1.5	-	-	2.0033

2.3.2. HYSCORE Results

Pulse EPR experiments were conducted using the 2D-Hyperfine sublevel correlation experiment (HYSCORE) to obtain information about the electronic structure and the delocalization of the unpaired electron all over the compounds. The use of HYSCORE spectroscopy affords the accurate assignment of various couplings with a large number of nuclei. For **1** and **2**, the HYSCORE spectra revealed in the (+, +) quadrant two distinct ridges centered at 14.5 MHz (the Larmor nuclear frequency of the ^{1}H) with no symmetric cross-peak in the (−, +) quadrant (Figures 10 and S13). These results demonstrated the presence of weak hyperfine constants ($|A_{iso}| < 2|\nu_I|$) in agreement with the CW ^{1}H-ENDOR experiments. At low frequency, the (+,+) quadrants showed nitrogen features (ν_I = 1.06 MHz), which correspond also to a weak coupling situation ($|A_{iso}| < 2|\nu_I|$) with two pairs of cross peaks at 2.1 and 3.0 MHz for **1**, and are attributed to the porphyrin nitrogen atoms. In the (−, +) quadrant, a more complex pattern (i.e., the strong coupling situation where ($|A_{iso}| > 2|\nu_I|$) is present and indicative of a hyperfine coupling with a more important anisotropy probably from a nitrogen nucleus. To interpret the low frequency part of the

HYSCORE spectra, the ^{14}N hyperfine and quadrupole data were computed, and these values were used as a starting point for the simulation. The corresponding ^{14}N hyperfine and nuclear quadrupole couplings are given in Table 3 with the simulations shown in Figure S14 and support the presence of a second nitrogen atom with a hyperfine tensor and a quadrupole contribution very close to nitrogen from porphyrin with the phenothiazine and phenoxazine substituents. In contrast, the HYSCORE spectrum of **3** recorded at 80 K is rather different (Figure S13). A similar shape is observed at low frequency in the (+, +) quadrant, but the pattern visible in the (−, +) quadrant is different. This signal could originate from nitrogen atoms of the porphyrin with close hyperfine coupling values, as suggested by the DFT calculation. From this information, the following trends in the spin density distributions have emerged. The spin distribution over the Ni porphyrin is slightly different when species **1** and **2** are compared to **3**. For the latter, the computed values (see DFT calculations in SI) predict a non-zero spin density on the four ^{14}N porphyrin, whereas only the two nitrogen atoms close to the *meso* group bear spin density for **1** and **2**. These results reflect the enhanced electron donating ability of the bis(*p*-anisyl)-amine substituent and the flexibility of this amine compared to the rigid phenoxazine and phenothiazine. Consequently, enhanced radical delocalization was observed in compound **3** when compared to compounds **1** and **2**.

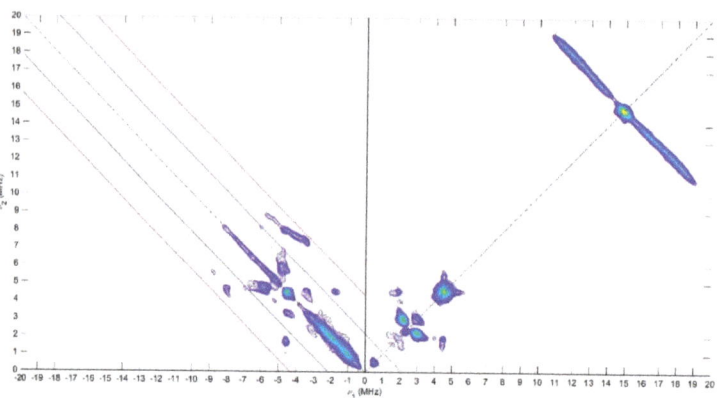

Figure 10. X-Band HYSCORE spectrum of **1** recorded at 70 K and collected at 349.4 mT, microwave frequency 9.70 GHz, τ = 136 ns.

2.3.3. Magnetic Properties

The reaction of **4** and **5** with two equivalents of AgSbF$_6$ in CH$_2$Cl$_2$ fluid solution yielded EPR spectra with five transitions consistent with two spin species and a ^{14}N hyperfine coupling constant approximatively equal to the half of the hyperfine coupling constant of the monoradical species (see Figures 11 and S16). This situation corresponds to the strong exchange limit and means that the exchange coupling is much higher than the hyperfine coupling constant. In contrast, one line was observed for **6** with a linewidth of 10 G and thus also suggested the formation of the biradical. The EPR spectra of the biradicals could also be obtained upon electrochemical generation if the electrolyses were carried out around the second oxidation potential. For compound **4**, a spectrum corresponding to a mixture of the mono- and biradical species (see Figure S17) was obtained. The very small differences between the first and second oxidation potentials for compounds **5** and **6** led directly to the doubly oxidized species, with EPR spectra identical to those obtained by chemical oxidation.

Table 3. ^{14}N hyperfine and quadrupolar coupling parameters obtained from DFT computations and experimental HYSCORE spectra (values in MHz) [a].

	Compound 1				Compound 2				Compound 3			
	Exp	DFT	Exp	DFT	Exp	DFT	Exp	DFT	Exp	DFT	Exp	DFT
	^{14}N$_{porphyrin}$		^{14}N$_{phenothiazine}$		^{14}N$_{porphyrin}$		^{14}N$_{phenoxazine}$		^{14}N$_{porphyrin}$		^{14}N$_{anisyl}$	
T (MHz) ± 0.1	0.2	0.1 [b]	14 [a]	14	0.1	0.1 [b]	13 [a]	17 [a]	0.5	1.1 [c]	-	12
|A$_{iso}$| (MHz) ± 0.1	0.6	0.35 [b]	19	15	0.8	0.5 [b]	19 [c]	18	1.2	1.2 [c]	-	19
|e^{2}qQ/h| (MHz) ± 0.1	1.8	2.1 [b]	3.5	3.2	1.8	2.2 [b]	2.8	2.8	2.0	2.0 [c]	-	3.6
η ± 0.1	0.2	0.1 [b]	0.1	0.1	0.2	0.13 [b]	0.1	0.0	0.1	0.1 [c]	-	0

[a] Values are in agreement with hyperfine coupling originating from field-swept EPR spectra (see Figure S5 and Table S1 in the Supporting Information). [b] Average value considering the two porphyrin nitrogen atoms featuring some spin density. [c] Average value for the four porphyrin nitrogen atoms. Simulations are shown in Figures S14 and S15 in SI. T is the dipolar contribution, e^{2}qQ/h is the quadrupolar coupling constant, and η is the electric field gradient asymmetry parameter.

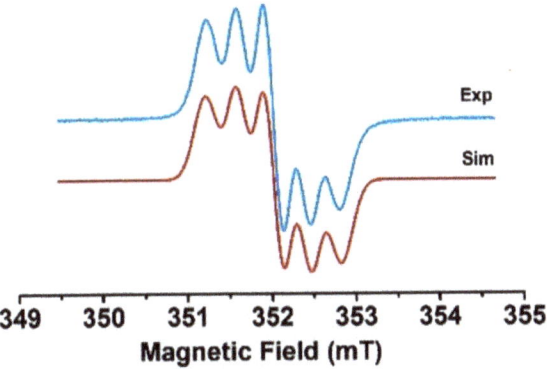

Figure 11. Experimental EPR spectrum of the dication of compound **4** in CH$_2$Cl$_2$ fluid solution at room temperature and its simulation.

The frozen solution EPR spectra of three bis compounds recorded at 4 K exhibit only one intense symmetrical line (ΔB$_{PP}$ = 12 G) centered on g = 2.00 without any fine structure characteristic of the triplet state, but a forbidden ΔMs = ±2 transition was detected at half field. This definitely indicates a biradical nature. To compare the exchange interaction, the temperature dependence of the EPR signal intensity (I) was measured. The intensity is known to be proportional to the magnetic susceptibility. Note that the EPR susceptibility is assessed as usual with the integrated intensity for the ΔMs = ±1 line with a high S/N ratio. This is correct as long as the line shape does not vary. In all three cases, the product IT decreases when the temperature is below 20 K, indicating the presence of an antiferromagnetic coupling (see Figure 12 for **6** and Figure S18). To obtain the exchange coupling constant (J/k$_B$ where k$_B$ is the Boltzmann constant), the observed curves were adjusted with a Bleaney–Bowers law with the additional contribution of a Curie law to account for the presence of the single radical. The estimated J/k$_B$ values of −8 K for **4** and −8 K for **5** were very close, while the value of **6** was in the order of −30 K and showed much better magnetic communication between the two bis(*p*-anisyl)amine *meso*-substituents.

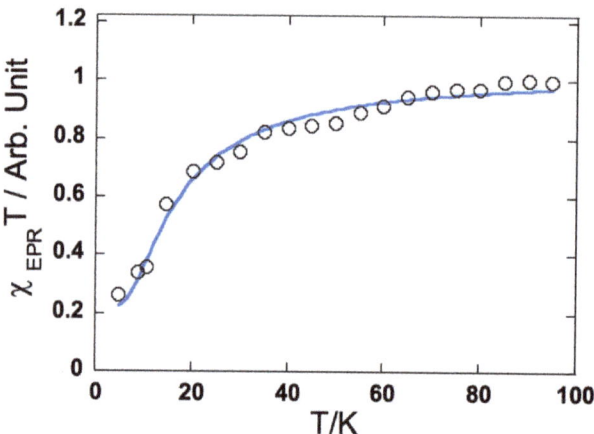

Figure 12. Temperature dependence of the EPR susceptibility (χT product) in CH_2Cl_2 frozen solution for the dication of **6**.

3. Materials and Methods

3.1. Materials

All commercial reagents (Aldrich, St. Louis, MI, USA; Fisher Scientific, Hampton, NH, USA; Merck, Rahway, NJ, USA) were used as received. All solvents were dried and freshly distilled before use. Column chromatography was performed using silica gel (Merck 0.40–0.63 nm).

3.2. Preparation of the Compounds

5,15-Diarylporphyrin and 5,10,15-triarylporphyrin were prepared by well-established porphyrin synthetic procedures, followed by bromination of the remaining free *meso* positions to obtain **7** and **8**. The syntheses and characterizations of these starting materials were described previously. Similarly, the syntheses of the porphyrins **2** and **5** bearing one or two phenoxazine donors at the *meso* positions were described previously [17].

MonoPhenothiazine 1. A degassed solution of brominated nickel(II) porphyrin **7** (53 mg, 72 µmol), phenothiazine (136 mg, 10 eq.), CuI (2 mg, 0.1 eq.), N'-phenylbenzoylhydrazine (3 mg, 0.2 eq.), Cs_2CO_3 (45 mg, 2 eq.) in DMSO (1 mL) was heated under argon at 120 °C for 16 h. After cooling, the reaction mixture was diluted in dichloromethane (100 mL), washed twice with a saturated $NaHCO_3$ aqueous solution (50 mL), and finally twice with water. The organic phase was dried over Na_2SO_4, and the solvents evaporated. After chromatographic purification (silica gel, CH_2Cl_2 then $CHCl_3$) and crystallization (CH_2Cl_2, n-hexane, MeOH), the desired compound **1** was isolated as a purple solid (18 mg, 21 µmol, 29%). ^{1}H NMR (300 MHz, $CDCl_3$, 25 °C), δ (ppm) = 9.29 (d, *J* = 5.0 Hz, 2H, pyrr.), 8.77 (d, *J* = 5.0 Hz, 2H, pyrr.), 8.74 (m, 4H, pyrr.), 7.91 (d, *J* = 8.6 Hz, 2H, $H_{o\text{-Anis}}$), 7.87 (d, *J* = 7.9 Hz, 4H, $H_{o\text{-tolyl}}$),7.45 (d, *J* = 7.9 Hz, 4H, $H_{m\text{-tolyl}}$), 7.20 (d, *J* = 8.6 Hz, 2H, $H_{m\text{-Anis}}$), 7.12 (dd, *J* = 7.6 and 1.5 Hz, 2H, H_a), 6.73 (ddd, *J* = 7.6, 7.6, and 1.0 Hz, 2H, H_b), 6.46 (ddd, *J* = 8.2, 7.6, and 1.5 Hz, 2H, H_c), 5.79 (dd, *J* = 8.2 and 1.0 Hz, 2H, H_d), 4.03 (s, 3H, OCH_3), 2.62 (s, 6H, CH_3). MS (Maldi-TOF, dithranol): Calcd for $[C_{53}H_{37}N_5NiOS]^{+\bullet}$: 849.21; found 849.19. UV-Vis. (CH_2Cl_2): λ_{max} = 416 nm (ε = 24,5000 $M^{-1}.cm^{-1}$), 530 (19,000). Anal. Calcd for $C_{53}H_{37}N_5NiOS\bullet 1/2\ CH_3OH$: C 74.14, H 4.54, N 8.08; found C 74.16, H 4.51, N 8.00.

BisPhenothiazine 4. A degassed solution of dibrominated nickel(II)porphyrin **8** (101 mg, 112 µmol), phenothiazine (220 mg, 10 eq.), CuI (4 mg, 0.2 eq.), N'-phenylbenzoylhydrazine (9 mg, 0.4 eq.), and Cs_2CO_3 (144 mg, 4 eq.) in DMSO (4 mL) was heated under argon at 120 °C for 16 h. After cooling, the reaction mixture was diluted in dichloromethane (100 mL), washed twice with a saturated $NaHCO_3$ aqueous solution (50 mL), and finally twice with water. The organic phase was dried over Na_2SO_4, and the solvents evaporated.

After chromatographic purification (silica gel, CH_2Cl_2 then $CHCl_3$) and crystallization (CH_2Cl_2, n-hexane, MeOH), the desired compound **4** was isolated as a purple solid (32 mg, 28 μmol, 25%). ^{1}H NMR (400 MHz, $CDCl_3$, 25 °C), δ (ppm) = 9.38 (d, J = 5.0 Hz, 4H, pyrr.), 8.81 (d, J = 5.0 Hz, 4H, pyrr.), 7.87 (d, J = 1.9 Hz, 4H, H_{o-Ar}), 7.74 (t, J = 1.9 Hz, 2H, H_{p-Ar}), 7.15 (m, 4H, H_a), 6.75 (br, 4H, H_b), 6.50 (br, 4H, H_c), 5.80 (br, 4H, H_d), 1.47 (s, 36 H, CH_3). MS (Maldi-TOF, dithranol): Calcd for $[C_{72}H_{66}N_6NiS_2]^{+\bullet}$: 1136.41; found 1136.40. UV-Vis. (CH_2Cl_2): λ_{max} = 414 nm (ε = 248,000 $M^{-1}.cm^{-1}$), 533 (19,000), 566 (11,000). Anal. Calcd for $C_{72}H_{66}N_6NiS_2$: C 75.98, H 5.84, N 7.38; found C 75.70, H 5.90, N 7.32. Crystal data. From CH_2Cl_2-CH_3OH, $C_{74}H_{70}Cl_4N_6NiS_2$, M = 1307.99 g.mol^{-1}, 0.25 × 0.20 × 0.10 red prisms, triclinic, space group P-1, a = 9.581(5) Å, b = 12.043(5) Å, c = 15.226(5) Å, α = 69.453° (5), β = 84.502° (5), γ = 83.241° (5), V = 1630.8 (12) Å^3, Z = 1, T = 173 K, MoKα = 0.71073, 1.81 < θ < 30.05, 35,851 reflections measured, 9530 unique reflections, R_1 = 0.0579, wR_2 = 0.1505, GoF = 1.033. CCDC Nr 2261554.

Monophenoxazine 2. The preparation and characterization of this compound were described previously [17]. Crystal data. $C_{53}H_{37}N_5NiO_2$, M = 834.58 g.mol^{-1}, 0.500 × 0.380 × 0.160 mm red, triclinic, space group P-1, a = 9.3928 (6) Å, b = 16.1279 (11) Å, c = 16.7580 (11) Å, α = 74.871° (2), β = 83.3960° (10), γ = 87.470° (2), V = 2434.1 (3) Å^3, Z = 2, T = 173 K, MoKα = 0.71073, 1.308 < θ < 28.125, 56,323 reflections measured, 11,860 unique reflections, R_1 = 0.0515, wR_2 = 0.1166, GoF = 1.048. CCDC Nr 2261556.

Compound 3. A Schlenk tube filled with bis(4-methoxyphenyl)-amine (298 mg, 1.3 mmol), tBuOK (1.5 g, 13.5 mmol), Pd(OAc)$_2$ (14 mg, 0.06 mmol), rac-BINAP (58 mg, 0.09 mmol), and 18-crown-6 (15 mg, 0.06 mmol) was dried under vacuum for two hours and backfilled with argon. A degassed solution of 5-iodo-10,20-ditolyl-15-(p-anisyl)-nickel(II) porphyrin (100 mg, 0.13 mmol) in freshly distilled THF (60 mL) was added to the previously dried solids, and the reaction mixture was stirred at 70 °C under argon. After 16 h, the mixture was cooled to room temperature and the solvent evaporated. Chromatographic purification (silica gel, CH_2Cl_2/C_6H_{12} 3/7) followed by crystallization (CH_2Cl_2, MeOH) afforded the dehalogenated porphyrin (30 mg, 35%) and the desired product **3** (35 mg, 0.04 mmol, 31%). ^{1}H NMR (500 MHz, $CDCl_3$, 25 °C), δ (ppm) = 9.04 (d, J = 5.0 Hz, 2H, pyrr.), 8.67 (d, J = 5.0 Hz, 2H, pyrr.), 8.66 (d, J = 5.0 Hz, 2H, pyrr.), 8.64 (d, J = 5.0 Hz, 2H, pyrr.), 7.87 (d, J = 6.7 Hz, 2H, H_{o-Anis}), 7.82 (d, J = 8.0 Hz, 4H, $H_{o-Tolyl}$), 7.42 (d, J = 8.0 Hz, 4H, $H_{m-Tolyl}$), 7.18 (d, J = 6.7 Hz, 2H, H_{m-Anis}), 7.12 (d, J = 6.7 Hz, 8H, H_{N-Anis}), 6.70 (d, J = 6.7 Hz, 8H, H_{N-Anis}), 4.02 (s, 4H, OCH_3), 3.68 (s, 6H, OCH_3), 2.60 (s, 6H, CH_3). ^{13}C (125 MHz, $CDCl_3$): δ (ppm) = 159.3, 153.7, 164.1, 145.2, 143.2, 143.1, 142.0, 137.7, 137.4, 134.7 (CH), 133.6 (CH), 133.2, 133.1 (CH), 132.1 (CH), 132.0 (CH), 130.8 (CH), 127.6 (CH), 123.0 (CH), 122.0, 119.1, 119.0, 114.5 (CH), 112.4 (CH), 55.6 (OCH_3), 55.5 (OCH_3), 21.5 (CH_3). UV-Vis. (CH_2Cl_2): λ_{max} = 410 nm (ε = 152,000 $M^{-1}.cm^{-1}$), 470 (28,000sh), 541 (12,800), 595 (7000). Anal. Calcd for $C_{55}H_{43}N_5NiO_3$: C 75.01, H 4.92, N 7.95; found C 74.62, H 4.93, N 7.76.

Compound 6. A Schlenk tube filled with bis(4-methoxyphenyl)-amine (415 mg, 1.80 mmol), tBuOK (2.0 g, 18 mmol), Pd(OAc)$_2$ (21 mg, 0.09 mmol), rac-BINAP (78.5 mg, 0.13 mmol), and 18-crown-6 (20 mg, 0.08 mmol) was dried under vacuum for one hour and backfilled with argon. A degassed solution of 5,15-diiodo-10,20-di-(3,5-di-tBu-phenyl)-nickel(II) porphyrin (180 mg, 0.18 mmol) in freshly distilled THF (60 mL) was added to the previously dried solids, and the reaction mixture was stirred at 70 °C under argon. After 15 h, the mixture was cooled to room temperature and the solvent evaporated. Chromatographic purification (silica gel, CH_2Cl_2/C_6H_{12} 3/7) followed by crystallization (CH_2Cl_2, n-hexane, MeOH) afforded green crystals (49 mg, 0.041 mmol, 23%). ^{1}H NMR (500 MHz, $CDCl_3$, 25 °C), δ (ppm) = 9.00 (d, J = 5.0 Hz, 4H, pyrr.), 8.59 (d, J = 5.0 Hz, 4H, pyrr.), 7.75 (d, J = 1.7 Hz, 4H, H_{o-Ar}), 7.65 (t, J = 1.7 Hz, 2H, H_{p-Ar}), 7.12 (d, J = 9.1 Hz, 8H, H_{Anis}), 6.71 (d, J = 9.1 Hz, 8H, H_{Anis}), 3.68 (s, 12H, OCH_3), 1.42 (s, 36H, CH_3). ^{13}C (125 MHz, $CDCl_3$): δ (ppm) = 153.6, 148.9, 146.0, 145.3, 142.4, 139.5, 133.4 (CH), 130.6 (CH), 128.4 (CH), 122.9 (CH), 122.2, 121.3 (CH), 120.3, 114.5 (CH), 55.5 (OCH_3), 35.0, 31.7 (CH_3). MS (ESI-TOF-HRMS): Calcd for $[C_{76}H_{78}N_6NiO_4]^{+\bullet}$: 1196.5433; found 1196.5439. UV-Vis. (CH_2Cl_2):

λ_{max} = 401 nm (ε = 115,000 M^{-1}·cm^{-1}), 477 (53,000), 545 (10,500), 622 (11,300). Anal. Calcd for C$_{76}$H$_{78}$N$_6$NiO$_4$·H$_2$O: C 75.06, H 6.63, N 6.91; found C 75.11, H 6.48, N 6.93. Crystal data. From chlorobenzene and isopropanol, C$_{76}$H$_{78}$Cl$_4$N$_6$NiO$_4$, M = 1198.15 g.mol^{-1}, 0.500 × 0.500 × 0.500 red, triclinic, space group P-1, a = 10.7094 (10) Å, b = 14.9585 (16) Å, c = 20.634 (2) Å, α = 97.122° (5), β = 90.811° (5), γ = 99.452° (6), V = 3233.4 (6) Å^3, Z = 2, T = 173 K, CuKα = 1.54178, 3.932 < θ < 66.975, 44,205 reflections measured, 11,200 unique reflections, R$_1$ = 0.088, wR$_2$ = 0.2262, GoF = 1.030. CCDC Nr 2261547.

3.3. Electrochemistry

Electrochemical measurements were carried out in CH$_2$Cl$_2$ containing NBu$_4$PF$_6$ (0.1 M) in a classical three-electrode cell by cyclic voltammetry (CV) and rotating-disk voltammetry (RDV). The working electrode was a glassy carbon disk (3 mm in diameter), the auxiliary electrode was a Pt wire, and the pseudo-reference electrode was a Pt wire. The cell was connected to an Autolab PGSTAT30 potentiostat (Eco Chemie, Utrecht, Holland) driven by GPSE software running on a personal computer. All potentials are given vs. Fc$^+$/Fc used as internal standards and are uncorrected from ohmic drop. Spectroelectrochemical investigations were carried out with a MCS600 Carl Zeiss photodiodes array spectrometer (MCS 601 UV-vis and MCS 611 NIR2.2 modules) controlled by the Aspect Plus software. A home-made Optical Transparent Thin Layer Electrode (OTTLE) spectroelectrochemical cell was placed in the light beam. The three electrodes of the cell (a platinum grid as the working electrode, platinum wire as the counter electrode, and silver/silver chloride as the reference electrode) were connected to an Autolab 302 potentiostat/galvanostat controlled by the GPES software running on a PC computer.

3.4. EPR Measurements

CW EPR experiments were performed using an X-band (9–10 GHz) Bruker ESP300 spectrometer equipped with a standard rectangular cavity (TE102). A Bruker Elexis E580 spectrometer operating at the X-band (9–10 GHz) equipped with an Oxford instrument CF935 cryostat, an ITC4 temperature controller, and a 1 kW TWT amplifier was used for pulsed EPR measurements. 2D HYSCORE (hyperfine sublevel correlation spectroscopy) spectra were measured by using a pulse sequence (π/2–τ–π/2–t1–π–t2–π/2–τ echo), with pulse lengths of 16 and 32 ns for the π/2 and π pulses, respectively. The experiments were performed with two different delay values τ (136 and 200 ns). Numerous unwanted pulse echoes were removed by four-step phase cycling. The intensity of the echo was measured after the fourth pulse with variable t1 and t2 values, keeping τ constant. The collected ESEEM and HYSCORE data were Fourier transformed by using the spectrum manipulation routines available within the Bruker program: subtraction of the relaxation decay, apodization (Hamming window), and zero filling. The HYSCORE spectra were recorded with a magnetic field corresponding to the maximum intensity of the radical signal in the two-pulse field-swept EPR spectra. The EPR and HYSCORE spectra were simulated using Easyspin V5.2.28 using the MATLAB R2020a version [29].

3.5. Computational Methods

The ORCA program package was used to perform all theoretical calculations based on density functional theory (DFT) [30]. Starting from the experimentally determined solid-state structures, all complexes were subjected to full geometry optimization using the B3LYP functional [31,32] in combination with the 6–31 g* basis sets [33–35]. Increased integration grids and tight SCF convergence criteria were used in the calculations. Cartesian coordinates of the DFT-optimized structures are provided in the SI. Solvent effects were accounted for according to the experimental conditions, and dichloromethane was used as a solvent (ε = 80) within the framework of the conductor-like polarizable continuum model, COSMO [36]. Single point calculations were conducted to compute electronic structures and predict EPR parameters using the B3LYP functional together with the CP(PPP) basis set for the metal center [37] and the EPR-II basis set for other atoms [38]. Optimized geometries as well as electronic structures were visualized using the program Chemdraft.

4. Conclusions

This work demonstrates that the introduction of aromatic amines on the free *meso* positions of diarylporphyrins via either Buchwald or Ullmann methods is very efficient. Radicals can be generated on the nitrogen atoms by electrochemistry and chemical oxidation. The degree of communication between the radicals through the π-electron density of the porphyrin ring strongly depends on the aromatic amines' ability to contribute to their delocalization. The higher global flexibility of amine substituents able to freely rotate along both the N-C$_{meso}$ and the N-Aryl bonds allows a more efficient delocalization of the radicals.

Supplementary Materials: The following supporting information can be downloaded at: https://www.mdpi.com/article/10.3390/molecules28114405/s1, Figure S1: ^{1}H NMR spectrum of compound **1**; Figure S2: MALDI-TOF MS of **1** (top) simulation (bottom); Figure S3: ^{1}H NMR spectrum of compound **4**; Figure S4: MALDI-TOF MS of **4** (top) simulation (bottom); Figure S5: ^{1}H NMR of compound **3** (top) and aromatic area (bottom); Figure S6: ^{13}C NMR of compound **3** (top) and DEPT (bottom); Figure S7: ^{1}H NMR of compound **6**; Figure S8: HRMS (ESI-TOF) of compound **6** (top) and simulation (bottom); Figure S9: ^{13}C NMR of compound **6** (top) and DEPT (bottom); Figure S10: UV/Vis spectra during the stepwise electrochemical oxidation of **2**: (a) at the first oxidation potential; (b) at the second oxidation potential (recorded in dichloromethane, 0.1 M NBu$_4$PF$_6$); Figure S11: (a) X-band EPR spectrum of **2** in CH$_2$Cl$_2$ fluid solution at room temperature (blue) and its simulation (red) (b) ^{1}H-ENDOR spectrum of **2** in CH$_2$Cl$_2$ fluid solution at 200K; Figure S12: (a) X-band EPR spectrum of **3** in CH$_2$Cl$_2$ fluid solution at room temperature (b) ^{1}H-ENDOR spectrum of **3** in CH$_2$Cl$_2$ fluid solution at 200 K; Figure S13: (++) and (+ −) quadrants of the ^{1}H and ^{14}N of X-band HYSCORE spectrum at 80 K of (a) **2** (b) **3** showing the location of ^{14}N cross-peaks weakly coupled nitrogen nuclei, respectively, and ^{1}H cross-ridges. Microwave frequency of 9.71 and 9.74 GHz respectively, magnetic field 350.0 and 352.5 mT respectively and time τ of 136 ns; Figure S14: Experimental (blue) and simulated (red) X-band ^{14}N–HYSCORE spectra of (a) **1**, (b) **2** and (c) **3**. The simulations are carried out using parameters given in Table 2 in the main text; Figure S15: Field-swept EPR spectra at 80 K of (a) **1** (b) **2** and (c) **3**. Experimental (black) and simulated (red); Figure S16: Experimental EPR spectrum of (a) **5** and (b) **6** in CH$_2$Cl$_2$ fluid solution at room temperature; Figure S17: Experimental EPR spectrum of mixture **2** and **4** in CH$_2$Cl$_2$ fluid solution generated by electrolysis at room temperature. Simulated spectrum was obtained by an admixture of 23% of monoradical and 77% of biradical; Figure S18: Temperature dependence of the EPR susceptibility (χT product) in CH$_2$Cl$_2$ frozen solution for (a) **5** and (b) **4**; Table S1: ^{14}N hyperfine coupling parameters in MHz obtained from simulations and experimental Field-swept EPR spectra; and DFT calculations for compounds **1**, **2**, and **3**.

Author Contributions: Conceptualization, J.W. and R.R.; syntheses, M.-A.C., J.H. and A.M.V.M.P.; electrochemistry, L.R.; EPR studies, K.M., N.P., V.M. and S.C.; DFT, M.O.; writing—original draft preparation, S.C. and R.R.; writing—review and editing, J.W., S.C., L.R. and R.R.; funding acquisition, S.C. and R.R. All authors have read and agreed to the published version of the manuscript.

Funding: This research was funded by IR Infrananalytics FR205.

Acknowledgments: We thank L. Karmazin and C. Bailly from the X-ray facilities of the University of Strasbourg for solving the structures. We thank the CNRS and the University of Strasbourg for continuous financial support. We thank also the FRC-Labex CSC (PhD fellowship to MAC) and the Fondation Recherche Chimie (PhD fellowship to KM). Financial support from the IR Infrananalytics FR2054 for conducting the research is gratefully acknowledged.

Conflicts of Interest: The authors declare no conflict of interest.

Sample Availability: Samples of all compounds are available from the authors.

References

1. Kadish, K.M.; Smith, K.M.; Guilard, R. (Eds.) *The Porphyrin Handbook*; Academic Press: San Diego, CA, USA, 2000; Volume 1, pp. 1–400.
2. Hiroto, S.; Miyake, Y.; Shinokubo, H. Synthesis and functionalization of porphyrins through organometallic methodologies. *Chem. Rev.* **2017**, *117*, 2910–3043. [CrossRef]
3. Balaban, M.C.; Chappaz-Gillot, C.; Canard, G.; Fuhr, O.; Roussel, C.; Balaban, T.S. Metal catalyst-free amination of *meso*-bromoporphrins: An entry to supramolecular porphyrinoid frameworks. *Tetrahedron* **2009**, *65*, 3733–3739. [CrossRef]
4. Devillers, C.H.; Hebié, S.; Lucas, D.; Cattey, H.; Clément, S.; Richeter, S. Aromatic nucleophilic substitution (S_NAr) of *meso*-nitroporphyrin with azide and amines as an alternative metal catalyst free synthetic approach to obtain *meso*-N-substituted porphyrins. *J. Org. Chem.* **2014**, *79*, 6424–6434. [CrossRef] [PubMed]
5. Ruiz-Castillo, P.; Buchwald, S.L. Applications of palladium-catalyzed C-N cross-coupling reactions. *Chem. Rev.* **2016**, *116*, 12564–12649. [CrossRef] [PubMed]
6. Imahori, H.; Matsubara, Y.; Iijima, H.; Umeyama, T.; Matano, Y.; Ito, S.; Niemi, M.; Tkachenko, N.V.; Lemmetyinen, H. Effects of *meso*-diarylamino group of porphyrins as sensitizers in dye-sensitized solar cells on optical, electrochemical, and photovoltaic properties. *J. Phys. Chem. C* **2010**, *114*, 10656–10665. [CrossRef]
7. Nowak-Krol, A.; Gryko, D.T. Oxidative aromatic coupling of *meso*-arylamino-porphyrins. *Org. Lett.* **2013**, *15*, 5618–5621. [CrossRef] [PubMed]
8. Fields, K.B.; Ruppel, J.V.; Snyder, N.L.; Zhang, X.P. *Handbook of Porphyrin Science*; Kadish, K.M., Smith, K.M., Guilard, R., Eds.; World Scientific: Singapore, 2010; Volume 3, pp. 367–427.
9. Kawano, S.; Kawada, S.; Kitagawa, Y.; Teramoto, R.; Nakano, M.; Tanaka, K. Near-infrared absorption by intramolecular charge-transfer transition in 5,10,15,20-tetra-(N-carbazolyl)porphyrin through protonation. *Chem. Commun.* **2019**, *55*, 2992–2995. [CrossRef] [PubMed]
10. Kawano, S.; Kawada, S.; Matsubuchi, A.; Tanaka, K. Metalloporphyrins substituted with N-carbazolyl groups quadruply at meso positions. *J. Porphyr. Phthalocyanines* **2022**, *26*, 140–146. [CrossRef]
11. Pawlicki, M.; Hurej, K.; Kwiecinska, K.; Szterenberg, L.; Latos-Grazynski, L. A fused *meso*-aminoporphyrin: A switchable near-IR chromophore. *Chem. Commun.* **2015**, *51*, 11362–11365. [CrossRef]
12. Fukui, N.; Cho, W.Y.; Lee, S.; Tokuji, S.; Kim, D.; Yorimitsu, H.; Osuka, A. Oxidative fusion reactions of *meso*-(diarylamino) porphyrins. *Angew. Chem. Int. Ed.* **2013**, *52*, 9728–9732. [CrossRef]
13. Susuki, Y.; Fukui, N.; Murakami, K.; Yorimitsu, H.; Osuka, A. Amination of *meso*-bromoporphyrins and haloanthracenes with diarylamines catalyzed by a palladium-PEPPSI complex. *Asian J. Org. Chem.* **2013**, *2*, 1066–1071. [CrossRef]
14. Fukui, N.; Lee, S.K.; Kato, K.; Shimizu, D.; Tanaka, T.; Lee, S.; Yorimitsu, H.; Kim, D.; Osuka, A. Regioselective phenylene-fusion reactions of Ni(II)-porphyrins controlled by an electron-withdrawing *meso*-substituent. *Chem. Sci.* **2016**, *7*, 4059–4066. [CrossRef] [PubMed]
15. Wang, K.; Osuka, A.; Song, J. Pd-catalyzed cross coupling strategy for functional porphyrin arrays. *ACS Cent. Sci.* **2020**, *6*, 2159–2178. [CrossRef] [PubMed]
16. Haumesser, J.; Gisselbrecht, J.-P.; Weiss, J.; Ruppert, R. Carbene Spacers in bis-porphyrinic scaffolds. *Chem. Commun.* **2012**, *48*, 11653–11655. [CrossRef] [PubMed]
17. Haumesser, J.; Pereira, A.M.V.M.; Gisselbrecht, J.-P.; Merahi, K.; Choua, S.; Weiss, J.; Cavaleiro, J.A.S.; Ruppert, R. Inexpensive and efficient Ullmann methodology to prepare donor-substituted porphyrins. *Org. Lett.* **2013**, *15*, 6282–6285. [CrossRef]
18. Esdaile, L.J.; Senge, M.O.; Arnold, D.P. New palladium catalysed reactions of bromoporphyrins: Synthesis and crystal structures of nickel(II)complexes of primary 5-aminoporphyrin, 5,5′-bis(porphyrinyl) secondary amine, and 5-hydroxyporphyrin. *Chem. Commun.* **2006**, 4192–4194. [CrossRef]
19. Pereira, A.M.V.M.; Neves, M.G.P.M.S.; Cavaleiro, J.A.S.; Jeandon, C.; Gisselbrecht, J.-P.; Choua, S.; Ruppert, R. Diporphyrinylamines: Synthesis and electrochemistry. *Org. Lett.* **2011**, *13*, 4742–4745. [CrossRef] [PubMed]
20. Kawamata, M.; Sugai, T.; Minoura, M.; Maruyama, K.; Furukawa, K.; Holstrom, C.; Nemykin, V.N.; Nakano, H.; Matano, Y. Nitrogen-bridged metallodiazaporphyrin dimers: Synergistic effects of nitrogen bridges and *meso*-nitrogen atoms on structure and properties. *Chem. Asian J.* **2017**, *12*, 816–821. [CrossRef]
21. Shimizu, D.; Fujimoto, K.; Osuka, A. Stable diporphyrinylaminyl radical and nitrenium ion. *Angew. Chem. Int. Ed.* **2018**, *57*, 9434–9438. [CrossRef]
22. Yella, A.; Lee, H.W.; Tsao, H.N.; Yi, C.; Chandiran, A.K.; Nazeeruddin, M.K.; Diau, E.W.G.; Yeh, C.Y.; Zakeeruddin, S.M.; Grätzel, M. Porphyrin-sensitized solar cells with cobalt(II/III)-based redox electrolyte exceed 12 percent efficiency. *Science* **2011**, *334*, 629–634. [CrossRef]
23. Sakamoto, R.; Sasaki, T.; Honda, N.; Yamamura, T. 5,15-Bis(di-p-anisylamino)-10,20-diphenylporphyrin: Distant and intense electronic communication between two amine sites. *Chem. Commun.* **2009**, *45*, 5156–5158. [CrossRef]
24. Sakamoto, R.; Nishikawa, M.; Yamamura, T.; Kume, S.; Nishihira, H. A new special pair model comprising *meso*-di-p-anisylaminoporphyrin: Enhancement of visible-light absorptivities and quantification of electronic communication in mixed valent cation radical. *Chem. Commun.* **2010**, *46*, 2028–2030. [CrossRef] [PubMed]
25. Senge, M.O.; Davis, M. 5,15-dianthracen-9-yl-10,20-dihexylporphynato)nickel(II): A planar nickel(II) porphyrin. *Acta Crystallogr. Sect. E Struct. Rep. Online* **2010**, *E66*, m790. [CrossRef]

26. Kadish, K.M.; Royal, G.; Van Caemelbecke, E.; Gueletti, L. *The Porphyrin Handbook*; Kadish, K.M., Smith, K.M., Guilard, R., Eds.; Academic Press: San Diego, CA, USA, 2000; Volume 9, pp. 1–219.
27. Neugebauer, F.A.; Bamberger, S. Diarylamine radical cations. *Angew. Chem. Int. Ed. Engl.* **1971**, *10*, 71–72. [CrossRef]
28. Kennedy, D.E.; Dalal, N.S.; McDowell, C.A. Endor of protons and determination of chlorine hyperfine couplings in neutral free radicals of biologically interesting compounds. *Chem. Phys. Lett.* **1974**, *29*, 521–525. [CrossRef]
29. Stoll, S.; Schweiger, A. EasySpin, a comprehensive software package for spectral simulation and analysis in EPR. *J. Magn. Reson.* **2006**, *178*, 42–55. [CrossRef] [PubMed]
30. Neese, F. The ORCA program system. *WIREs Comput. Mol. Sci.* **2012**, *2*, 73–78. [CrossRef]
31. Becke, A.D. Density functional thermochemistry III. The role of exact exchange. *J. Chem. Phys.* **1993**, *98*, 5648–5652. [CrossRef]
32. Lee, C.T.; Yang, W.T.; Parr, R.G. Development of the Colle-Salvetti Correlation-Energy Formula into a Functional of the Electron Density. *Phys. Rev. B* **1988**, *37*, 785–789. [CrossRef]
33. Csonka, G.I. Proper basis set for quantum mechanical studies of potential energy surfaces of carbohydrates. *J. Mol. Struct. THEOCHEM.* **2002**, *584*, 1–4. [CrossRef]
34. Boese, A.D.; Martin, J.M.L.; Handy, N.C. The role of the basis set: Assessing density functional theory. *J. Chem. Phys.* **2003**, *119*, 3005–3014. [CrossRef]
35. Mackie, I.D.; DiLabio, G.A. Accurate dispersion interactions from standard density-functional theory methods with small basis sets. *Phys. Chem. Chem. Phys.* **2010**, *12*, 6092–6098. [CrossRef] [PubMed]
36. Klamt, A.; Schüürmann, G. COSMO: A new approach to dielectric screening in solvents with explicit expressions for the screening energy and its gradient. *J. Chem. Soc. Perkin Trans. 2* **1993**, 799–805. [CrossRef]
37. Neese, F. Prediction and interpretation of the 57Fe isomer shift in Mössbauer spectra by density functional theory. *Inorg. Chim. Acta* **2002**, *337*, 181–192. [CrossRef]
38. Barone, V. *Recent Advances in Density Functional Methods, Part I*; Chong, D.P., Ed.; World Scientific Publ Co.: Singapore, 1995; p. 287.

Disclaimer/Publisher's Note: The statements, opinions and data contained in all publications are solely those of the individual author(s) and contributor(s) and not of MDPI and/or the editor(s). MDPI and/or the editor(s) disclaim responsibility for any injury to people or property resulting from any ideas, methods, instructions or products referred to in the content.

Article

Comparison of the Performance of Density Functional Methods for the Description of Spin States and Binding Energies of Porphyrins

Pierpaolo Morgante [1,2] and Roberto Peverati [1,*]

[1] Department of Chemistry and Chemical Engineering, Florida Institute of Technology, 150 W. University Blvd., Melbourne, FL 32901, USA
[2] Department of Chemistry, University at Buffalo, State University of New York, Buffalo, NY 14260, USA
* Correspondence: rpeverati@fit.edu

Abstract: This work analyzes the performance of 250 electronic structure theory methods (including 240 density functional approximations) for the description of spin states and the binding properties of iron, manganese, and cobalt porphyrins. The assessment employs the Por21 database of high-level computational data (CASPT2 reference energies taken from the literature). Results show that current approximations fail to achieve the "chemical accuracy" target of 1.0 kcal/mol by a long margin. The best-performing methods achieve a mean unsigned error (MUE) <15.0 kcal/mol, but the errors are at least twice as large for most methods. Semilocal functionals and global hybrid functionals with a low percentage of exact exchange are found to be the least problematic for spin states and binding energies, in agreement with the general knowledge in transition metal computational chemistry. Approximations with high percentages of exact exchange (including range-separated and double-hybrid functionals) can lead to catastrophic failures. More modern approximations usually perform better than older functionals. An accurate statistical analysis of the results also casts doubts on some of the reference energies calculated using multireference methods. Suggestions and general guidelines for users are provided in the conclusions. These results hopefully stimulate advances for both the wave function and the density functional side of electronic structure calculations.

Keywords: DFT; porphyrin; organometallic; density functionals; transition metals

1. Introduction

Porphyrins are a class of heterocyclic aromatic compounds that bind transition metals to form a broad family of coordination complexes, mostly known as metalloporphyrins. Due to their ubiquitous presence in biology and biochemistry—for example, in the active site of hemoglobin, myoglobin, and the cytochrome P450 family of enzymes—and their broad applicability as biomimetic catalysts, porphyrins have been extensively studied both experimentally and computationally [1–11]. The presence of the metal makes porphyrins challenging for electronic structure calculations due to several low-lying, nearly degenerate spin states [12–18]. Multireference treatments, such as those based on active-space methods such as CAS-SCF/CASPT2 (CAS = complete active space, SCF = self-consistent field, PT2 = Møller-Plesset perturbation theory truncated at second order), are usually necessary to correctly describe porphyrins and related compounds, as shown, for example, by Pierloot et al. [4,5]. Such calculations are not easily affordable due to their high computational cost and are usually limited to small systems. Variants such as the multiconfiguration pair-density functional theory (MC-PDFT) of Truhlar and Gagliardi et al. [19] appear promising for production calculations [9]. Single-reference methods based on Kohn–Sham (KS) density functional theory (DFT) account for static and dynamic correlation—at least in an approximate manner [20–22]—and thus represent a competitive alternative to multiconfigurational methods for large systems. In KS DFT, however, the accuracy of the approximation severely impacts the reliability of the calculated results [23–27], and choosing an appropriate functional might be a daunting task [28–31]. Benchmarking functional approximations is an effective way to understand the reliability of a functional against a wide range of chemical properties [23–27,32,33]. At the time of writing this article, no benchmark study aimed at assessing KS DFT calculations for metalloporphyrins is available in the literature. The main goal of this paper is to provide such a benchmark. This study reports calculations on spin state energy differences and binding properties of different metal porphyrins with 250 electronic structure methods, including 240 exchange-correlation functional approximations (a large majority of the functionals available in most electronic structure software). Suitable recommendations for choosing an appropriate electronic structure method for the computational study of porphyrins are provided at the end of Section 2.

2. Results and Discussion

2.1. Best Performers and General Trends

To concisely present the results for all 250 methods, we assigned grades to each functional based on their percentile ranking, as reported in Table 1. We note in passing that most of the grades obtained for the entire Por21 database are transferable to the PorSS11 and PorBE10 datasets (the individual rankings for the subsets are reported in the Supplementary Materials). We set the threshold for a passing grade of D or better at the 60th percentile, corresponding to an MUE for Por21 of 23.0 kcal/mol. A total of 106 functionals achieved a passing grade, almost equally distributed among the grades A–D. Most of the grade-A functionals are local, either GGAs or meta-GGAs, with the addition of five global hybrids with a low percentage of exact exchange (r^2SCANh, r^2SCANh-D4, B98, APF(D), O3LYP). The GAM functional is the overall best performer for the Por21 database, ranking first for PorSS1 and second for PorBE10. Other notable grade-A functionals are all four parameterizations of HCTH and several revisions of SCAN, namely, rSCAN, r^2SCAN, and r^2SCANh. The three revisions perform better than the original functional, which has a grade of D. In particular, the r^2SCANh and its -D4 variant stood out, with improvements larger than 50% over the errors obtained with SCAN. These results are consistent with other findings in the literature [34]. Three local Minnesota functionals are also in class A: revM06-L, M06-L, and MN15-L. These three functionals—together with r^2SCANh and r^2SCAN mentioned above—currently represent the best compromise between accuracy for general properties and accuracy for porphyrins chemistry and are at the top of our suggestion list (see below). For spin state energy differences, the results obtained in this work are consistent with the accepted knowledge that local functionals tend to stabilize low or intermediate spin states. In contrast, hybrid functionals stabilize higher spin states by including exact exchange [12,35].

Looking at the results for individual systems, we noticed that most functionals (233 out of 250) predict a triplet ground state for the iron porphyrin (FeP) system, while the CASPT2 reference predicts a quintet ground state (*vide infra*). Three top performers (GAM, HISS, and MN15-L) agree with the reference. The other results involving Fe(III) spin states also appear erratic. For the pentacoordinate FePOH system, 60% of the functionals agree with the CASPT2 references, predicting the high-spin state as the ground state. For the hexacoordinate FePNH$_3$OH system, 72% of the functionals predict the ground state to be either the low- or intermediate-spin state, in contrast to the CASPT2 reference data. Finally, 90% of the methods predict the ground state of the d^7 cobalt porphyrin to be a quartet, with spin state energy differences ranging from 1 to 90 kcal/mol (the CASPT2 reference is 0.90 kcal/mol in favor of the high-spin configuration). These results cast some doubts on the reference spin state energies for porphyrin containing Fe(III) and Co(II).

For most functionals, the binding energies of the iron porphyrin with CO and NO appear unproblematic, regardless of the presence or absence of the imidazole ring. 90% of the functionals correctly predict the bound complexes to be more stable than the separate molecular fragments. The description of the binding with molecular oxygen appears to be more challenging: 60% of the methods predict the unbound FeP to be more stable than the FePO$_2$ adduct, while only 25% of the methods predict the porphyrin-imidazole system to be more stable when unbound. This trend is observed for global hybrid functionals and range-separated functionals in both systems, and there is no apparent correlation between the stability of the adducts and the percentage of exact exchange employed in the functional form.

Table 1. List of methods examined in this study and their overall grade based on the MUE for Por21. Functionals are listed in alphabetical order, and grades are assigned based on percentile ranking, corresponding to the following thresholds: A: MUE < 14.3 kcal/mol; B: MUE < 17.1 kcal/mol; C: MUE < 20.0 kcal/mol; D: MUE < 23.0 kcal/mol; F: MUE > 23.0 kcal/mol. The dispersion corrections are covered in [24,36–46], unless noted otherwise. The MUEs were calculated from the CASPT2 reference energies; see the 'Material and Methods' section for further details. See also the Supporting Information for results on the PorSS11 and PorBE10 subsets.

Functional	Grade	Functional	Grade	Functional	Grade
APF [47]	A	HFLYP [48–51]	F	PKZB [52]	D
APFD [47]	A	HFPW92 [48–50,53]	F	PM6 [54]	F
B2PLYP	F	HISS [55]	A	PM7 [56]	F
B2PLYP-D3(0)	F	HSE-HJS [57–59]	B	PW6B95 [60]	D
B2PLYP-D3(BJ)	F	HSE-HJS-D3(0)	B	PW6B95-D2 [61]	C
B2PLYP-D4	F	HSE-HJS-D3(BJ)	B	PW6B95-D3(0)	D
B3LYP [51,62,63]	C	LC-ωPBE08 [64]	F	PW6B95-D3(BJ)	D
B3LYP-D2	C	LC-ωPBE08-D3(0)	F	PW6B95-D3(CSO)	F
B3LYP-D3(0)	C	LC-ωPBE08-D3(BJ)	F	PW91 [65]	F
B3LYP-D3(BJ)	C	LC-ωPBE08-D3M(BJ)	F	PWB6K [60]	F
B3LYP-D3(CSO)	C	LRC-ωPBE [66]	F	PWB6K-D3(0)	F
B3LYP-D3M(BJ)	C	LRC-ωPBEh [66]	F	PWB6K-D3(BJ)	F
B3LYP-D4	C	M05 [67]	D	PWPB95-D3(BJ)	F
B3LYP-NL [68]	C	M05-2X [69]	F	PWPB95-D4	F
B3LYP* [70]	F	M05-2X-D3(0)	F	r++SCAN [71]	B
B3LYP*-D3(0)	F	M05-D3(0)	D	r^{2}SCAN [72]	A
B3LYP*-D3(BJ)	F	M06 [73]	B	r^{2}SCAN-D4 [74]	A
B3P86 [62,63,75]	C	M06-2X [73]	F	r^{2}SCAN0 [76]	C
B3PW91 [62,63,65]	C	M06-2X-D2	F	r^{2}SCAN0-D4 [76]	B
B3PW91-D2	C	M06-2X-D3(0)	F	r^{2}SCANh [76]	A
B3PW91-D3(0)	C	M06-D2	B	r^{2}SCANh-D4 [76]	A
B3PW91-D3(BJ)	C	M06-D3(0)	B	r^{4}SCAN [71]	C
B97 [77]	C	M06-HF [78]	F	regTM [79]	F
B97-1 [80]	C	M06-HF-D3(0)	F	revM06 [81]	F
B97-1-D2 [61]	C	M06-L [82]	A	revM06-L [83]	A
B97-2 [84]	B	M06-L-D2 [61]	A	revM11 [85]	F
B97-2-D2 [61]	B	M06-L-D3(0)	A	revPBE [86]	D
B97-3 [87]	D	M08-HX [88]	F	revPBE-D2	D
B97-3-D2 [61]	D	M08-SO [88]	F	revPBE-D3(0)	D
B97-3c [89]	B	M11 [90]	F	revPBE-D3(BJ)	F
B97-D [91]	A	M11-D3(BJ)	F	revPBE-NL [68]	F
B97-D2 [36]	C	M11-L [92]	F	revPBE0 [57,86,93]	B
B97-D3(0)	C	M11-L-D3(0)	F	revPBE0-D3(0)	B
B97-D3(BJ)	C	mBEEF [94,95]	D	revPBE0-D3(BJ)	B
B97-K [96]	F	MN12-L [97]	F	revPBE0-NL [68]	C
B97M-rV [98]	C	MN12-L-D3(BJ)	F	revTPSS [99]	F
B97M-V [100]	C	MN12-SX [101]	F	revTPSSh [102]	C
B98 [103]	A	MN12-SX-D3(BJ)	F	RPBE [104]	D
BLOC [105]	F	MN15 [25]	F	RPBE-D3(0)	D
BLOC-D3(0) [104]	F	MN15-L [106]	A	RPBE-D3(BJ)	F
BLYP [51,62]	F	mPW91 [65,107]	F	rPW86PBE [57,108]	F
BLYP-D2	F	MS0 [109]	F	rPW86PBE-D3(0)	F

Table 1. Cont.

Functional	Grade	Functional	Grade	Functional	Grade
BLYP-D3(0)	F	MS0-D3(0) [110]	F	rPW86PBE-D3(BJ)	F
BLYP-D3(BJ)	F	MS1 [110]	F	rregTM [111]	F
BLYP-D3(CSO)	F	MS1-D3(0) [110]	F	rSCAN [71,112]	A
BLYP-D3M(BJ)	F	MS2 [110]	F	rSCAN-D4 [74]	B
BLYP-D4	F	MS2-D3(0) [110]	F	rVV10 [113]	F
BLYP-NL [68]	F	MS2h [110]	F	SCAN [114]	D
BMK [96]	F	MS2h-D3(0) [110]	F	SCAN-D3(0) [115]	F
BMK-D2 [61]	F	mTASK [116]	F	SCAN-D3(BJ) [115]	D
BMK-D3(0)	F	MVS [117]	A	SCAN-rVV10 [118]	F
BMK-D3(BJ)	F	MVSh [117]	D	SCAN0 [119]	D
BOP [62,120]	F	N12 [121]	F	SOGGA [57,122]	F
BOP-D3(0)	F	N12-D3(0)	F	SOGGA11-X [123]	D
BOP-D3(BJ)	F	N12-SX [101]	F	SOGGA11-X-D3(BJ)	D
BP86 [62,75]	F	N12-SX-D3(BJ)	F	SPW92 [53,124]	F
BP86-D2	F	O3LYP [125]	A	SVWN5 [124,126]	F
BP86-D3(0)	F	OLYP [51,125]	A	τ-HCTH [127]	A
BP86-D3(BJ)	F	OLYP-D3(0)	B	τ-HCTHh [127]	B
BP86-D3(CSO)	F	OLYP-D3(BJ)	B	TASK [128]	B
BP86-D3M(BJ)	F	OPBE [57,125]	B	TM [129]	F
BPBE [57,62]	D	oTPSS-D3(0) [37]	F	TPSS [130]	F
BPBE-D3(0)	F	oTPSS-D3(BJ) [37]	F	TPSS-D2	F
BPBE-D3(BJ)	F	PBE [57]	F	TPSS-D3(0)	F
CAM-B3LYP [131]	F	PBE-D2	F	TPSS-D3(BJ)	F
CAM-B3LYP-D3(0)	F	PBE-D3(0)	F	TPSS-D3(CSO)	F
CAM-B3LYP-D3(BJ)	F	PBE-D3(BJ)	F	TPSSh [132]	D
DSD-PBEP86-D3(BJ) [133]	F	PBE-D3(CSO)	F	TPSSh-D2 [61]	D
DSD-PBEPBE-D3(BJ) [133]	F	PBE-D3M(BJ)	F	TPSSh-D3(0)	D
GAM [134]	A	PBE-D4	F	TPSSh-D3(BJ)	D
GFN1-xTB [135]	F	PBE0 [93]	B	TPSSh-D4	F
GFN2-xTB [136]	F	PBE0-2 [137]	F	VV10 [138]	F
HCTH/120 [139]	A	PBE0-D2 [61]	B	ωB97 [140]	F
HCTH/120-D3(0)	A	PBE0-D3(0)	B	ωB97M-V [141]	F
HCTH/120-D3(BJ)	A	PBE0-D3(BJ)	B	ωB97M(2) [142]	F
HCTH/147 [139]	A	PBE0-D3(CSO)	B	ωB97X [140]	F
HCTH/407 [143]	A	PBE0-D3M(BJ)	C	ωB97X-D [144]	F
HCTH/93 [80]	A	PBE0-D4	B	ωB97X-D3 [145]	F
HF [48–50]	F	PBEh-3c [146]	F	ωB97X-V [147]	F
HF-3c [148]	F	PBEOP [57,120]	F	ωM05-D [149]	F
HF-D3(0)	F	PBEsol [94]	F	ωM06-D3 [145]	F
HF-D3(BJ)	F	PBEsol-D3(0)	F	X3LYP [150]	B
HF-NL [68]	F	PBEsol-D3(BJ)	F	XYG3 [151]/XYGJ-OS [152]	F

2.2. Results for Most Used and Most Suggested (MUMS) Functionals

To discuss the results in more detail, we selected a set of 25 functionals among the most used or the most suggested (MUMS) approximations for general and transition metal chemistry applications. The results are reported in Table 2. We note that selecting functionals is always subjective, but we tried to pick MUMS for their (unbiased) historical usage in the transition metal chemistry field [28,153,154] or following recent literature suggestions on methods that perform well for a broad range of properties [23–27,32]. When reporting standard functionals in the MUMS results, such as B3LYP or PBE, we included the parameterization without the dispersion corrections because we noticed that they generally worsen the MUEs for Por21. However, the magnitude of the worsening has no significant effect (usually within 5% or less) on the overall error of the method for all cases. As such, the results of standard functionals such as B3LYP and PBE can be transferred to the different dispersion correction 'flavors' (e.g., B3LYP-D3(BJ) or PBE-D2) without loss of generality. This consideration aligns with the fact that most interactions considered in Por21 are purely electronic. For example, dispersion corrections are almost entirely canceled out for the spin states when calculating the energy differences. For the bond energies, only a little overbinding residual remains, resulting in a modest worsening of the overall MUEs for methods that intrinsically overbind (which, in this case, is the vast majority). For functionals defined with dispersion corrections, such as PWPB95-D4 or ωB97M-V, we report their MUEs including such corrections, since this is how the functional was originally developed. Detailed results on each method can also be found in the Supplementary Materials for more granular analysis.

Table 2. General performance of selected functional approximations for the Por21 database for 25 of the "most used, most suggested" (MUMS) functionals in the literature. All values are mean unsigned errors (MUEs) in kcal/mol calculated from the CASPT2 reference energies.

MUMS Functional:	Type [a]	Por21	PorSS11	PorBE10
r^{2}SCANh	GH-mGGA	10.8	7.49	14.4
M06-L	mGGA	11.8	11.9	11.6
MN15-L	mGGA	11.9	17.9	5.26
r^{2}SCAN	mGGA	13.4	13.1	13.6
M06	GH-mGGA	15.1	17.9	12.0
PBE0	GH-GGA	16.1	17.1	15.0
r^{2}SCAN0	GH-mGGA	17.3	17.4	17.1
B3LYP	GH-GGA	19.1	21.1	16.8
B97M-V	mGGA	19.8	20.6	18.9
PW6B95	GH-mGGA	22.2	20.4	24.2
SCAN	mGGA	22.6	21.3	24.1
TPSSh	GH-mGGA	22.9	26.7	18.8
BLYP	GGA	25.6	27.6	23.4
PBE	GGA	26.3	27.8	24.6
TPSS	mGGA	26.7	30.5	22.5
MN15	GH-mGGA	26.8	30.5	22.7
B2PLYP	DH	27.2	21.3	33.6
BP86	GGA	27.5	29.3	25.5
CAM-B3LYP	RSH-GGA	28.5	32.5	24.2
ωB97M-V	RSH-mGGA	31.5	36.2	26.4
M06-2X	GH-mGGA	31.8	36.1	27.2
ωB97X-V	RSH-GGA	36.0	35.4	36.7
PWPB95-D4	DH	37.1	31.5	43.3
ωB97M(2)	DH	42.0	46.7	36.7
B3LYP*	GH-GGA	43.9	20.8	69.4

[a] GGA: generalized gradient approximation; mGGA: meta-GGA; GH: global hybrid; RSH: range-separated hybrid; DH: double-hybrid.

The overall best performer among the MUMS approximations is the r^{2}SCANh functional, with an MUE of 10.8 kcal/mol for the Por21 database, and 7.49 kcal/mol and 14.4 kcal/mol for the PorSS11 and PorBE10 subsets, respectively. The local Minnesota functionals M06-L and MN15-L come next, followed by r^{2}SCAN, for a total of four functionals with MUEs < 15.0 kcal/mol (M06 is only 0.1 kcal/mol above this threshold). MN15-L is by far the best performer for the PorBE10 subset, with an MUE of 5.26 kcal/mol, but, unfortunately, it is only seventh for PorSS11, with an MUE of 17.9 kcal/mol. Looking at historically significant functionals, the PBE0 approximation performs well, followed by B3LYP. Modern transferable functionals such as B97M-V, PW6B95, SCAN, and MN15 position themselves in the middle of the pack, perhaps disappointingly, given their usually accurate performance for main-group elements [23–25]. TPSSh, traditionally regarded as one of the 'gold standards' of transition metal chemistry, ranks similarly with an MUE of 22.9 kcal/mol. On the disappointing side of the results, with MUEs higher than 25.0 kcal/mol, are some of the most popular semilocal functionals for transition metal applications, such as PBE,

BP86, and TPSS. Range-separated hybrid functionals and double-hybrid functionals are even more disappointing, with MUEs higher than 30.0 kcal/mol. The latter class also presents challenges compared to most other functionals due to their computational cost and difficulty converging to the lowest energy solution. The notorious problems of the PT2-like correlation term for systems with multireference character easily explain these difficulties. For this reason, double-hybrid functionals are the only category where we did not explore every possible functional we had access to, but we limited ourselves to the most accurate approximations [142,155]. The B3LYP* is the worst performer among MUMS functionals for Por21, despite being specifically created to target spin state energy differences [70]. The improvements from B3LYP for PorSS11 are evident, but so are the deteriorating performances for PorBE10, where B3LYP is far superior. As already pointed out above, including dispersion corrections does not impact the results much, consistent with previous findings in the literature [35].

2.3. Results for Functionals Divided by "Ingredients"

An additional way to analyze the results is by classifying them according to how the approximation is constructed. For this purpose, we divided functionals into the following seven groups:

- **Group 0:** LDA, HF, LC, SE—local spin density approximations, Hartree-Fock, low-cost methods, and semiempirical methods (15 methods).
- **Group 1:** GGA—generalized gradient approximations and nonseparable gradient approximations (63 methods).
- **Group 2:** mGGA—meta-GGA and meta-NGA functionals (46 methods).
- **Group 3:** GH-GGA—global hybrid GGA and NGA functionals (43 methods).
- **Group 4:** GH-mGGA—global hybrid meta-GGA and meta-NGA functionals (43 methods).
- **Group 5:** RSH—range-separated hybrid functionals (28 methods).
- **Group 6:** DH—double-hybrid functionals (12 methods).

This classification follows well-established paths in the literature [23,28,156], and it provides valuable information to understand which ingredient is necessary for good performance. The distribution of the results for the methods in these groups is collected in the violin plots in Figure 1.

The plot analysis shows that the group of global hybrid GGA functionals (Group 3) is the best, with an average MUE of 19.8 kcal/mol. Groups 1, 2, and 4 perform similarly, with average MUEs of ~24 kcal/mol. The difference between Group 3 and Group 4 is particularly fascinating since global hybrid meta-GGA functionals, in principle, should be superior to global hybrid GGA functionals. To understand this (perhaps surprising) behavior, we report the MUEs of all functionals in both these groups against the percentage of exact exchange in each functional, as shown in Figure 2. From this plot, we notice a strong dependency of the MUE on the percentage of exact exchange, with the best results obtained by functionals with less than 30%, as expected [34,157,158]. As noted previously, B3LYP* is a particularly surprising outlier for the reasons we discussed in the previous section. The average percentage of exact exchange among the functionals in Group 3 is 23.0%, while the average percentage in Group 3 is 34.6%. This difference explains the superior performance of Group 3 functionals for the entire Por21 database compared to Group 4 functionals. If functionals with a percentage of exact exchange higher than 30% are removed from Group 4, then the average MUE for the remaining global hybrid meta-GGA functionals becomes identical to that of Group 3 (~20 kcal/mol). Additionally, we notice from the results for the subsets that the deterioration of the global hybrids meta-GGA results is mainly due to the PorSS11 database. The PorBE10 subset shows instead that Groups 2, 3, 4, and 5 have very similar average MUEs of ~22.0 kcal/mol, confirming that bond energies are far less sensitive to the choice of the method than spin state energy differences.

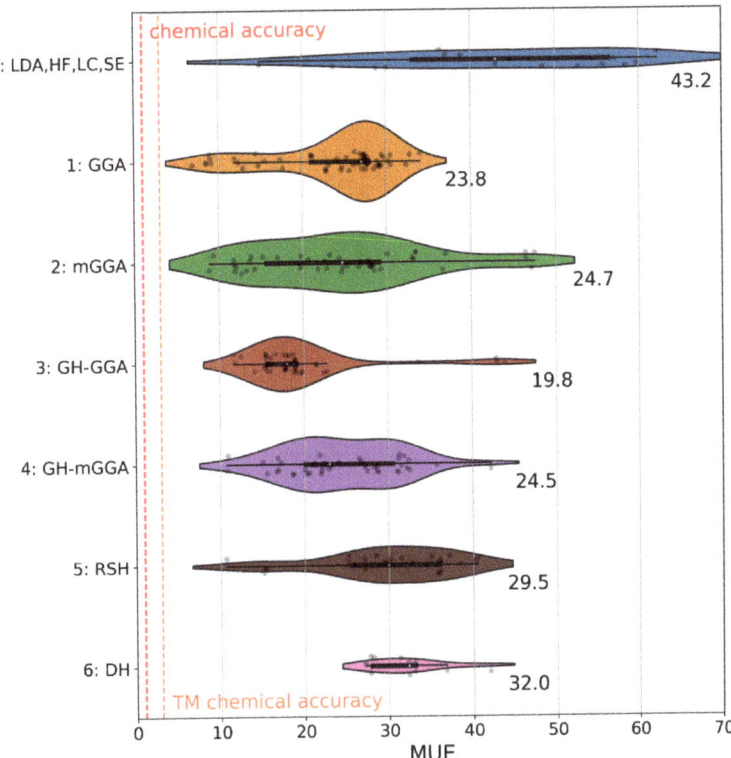

Figure 1. Distribution of the mean unsigned error (MUE) for the Por21 database for electronic structure methods divided into seven groups based on their ingredients (see text). Each group's average MUEs is reported at the bottom-right of each violin. The area of each violin is proportional to the number of functionals in each group. The white dot in the center of the plot shows the median of the distribution. The thicker horizontal bar inside a violin shows the interquartile range of the data. The thinner black bar inside a violin shows 1.5 times the interquartile range of the data. Individual MUEs are also reported as black points within each violin (smoothness was applied; hence some outliers exist). Chemical accuracy and transition metal (TM) chemical accuracy (*vide infra*) are also reported (in red/orange, respectively, vertical dashed lines). All values in this plot are in kcal/mol.

The disappointing performance of range-separated hybrid functionals in Group 5 is also easily rationalized based on a too-high percentage of exact exchange. This rationalization also applies to the double-hybrid functionals in Group 6. DH functionals usually include >50% exact exchange, which, together with the limitations already discussed in the previous section, are highly detrimental to their performance for Por21. Finally, Group 0 is perhaps an oddity since it includes methods with vastly different origins and ingredients. We grouped these methods because they do not belong to any of the other classes. Results reflect this non-uniformity, and given their poor performance, they will not be analyzed further.

When interpreting the results presented here, it is essential to remember that the MUE is a statistical parameter showing *average* performance. In other words, the MUE results of this study can be loosely interpreted as standard deviations (σ) and can be employed to give approximate error bars on metalloporphyrin calculations. In order to do so, however, we need to keep in mind that error bars of 1σ are not sufficiently stringent since individual errors might be (sometimes catastrophically) larger than the average one. This fact is confirmed by a granular analysis of our results, showing that the biggest individual error for one system can be as large as three times the average error. For example, our best performer, GAM, has an MUE of 6.93 kcal/mol, but its largest error is 17.7 kcal/mol. When using the stringent statistical threshold of 3σ (99.7 percentile of the error distribution), even the best performer gives error bars larger than 20 kcal/mol. If the less stringent 2σ threshold (95.5 percentile) is adopted instead, the error bars are smaller but still as significant as ~15 kcal/mol. For example, with a widely popular functional such as B3LYP, the 3σ and 2σ thresholds are ~60 and ~40 kcal/mol, respectively. Even for highly reliable functionals such as ωB97M-V, the 3σ error bars can be as large as ~100 kcal/mol, especially if the system is poorly described by a single determinant. This consideration is not intended as an endorsement in favor of B3LYP and against ωB97M-V but rather as a warning that the choice of the computational method should never be taken lightly [28–31], even if metals such as the closed-shell (d^{10}) Zn(II) might seem less problematic at first glance. It is impossible to know a priori if the studied system requires a multireference treatment.

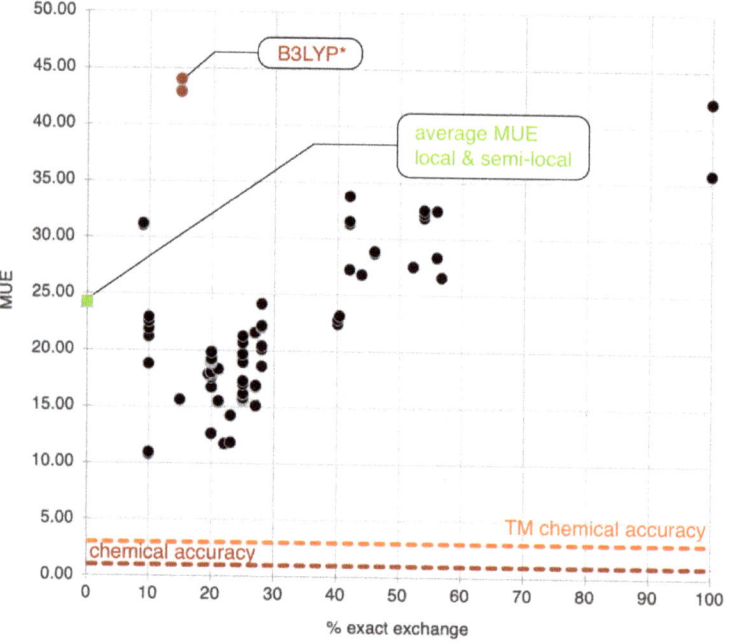

Figure 2. Distribution of the mean unsigned error (MUE) for the Por21 database for global hybrid functionals as a function of the percentage of exact exchange in each functional. The average MUE for the 107 local and semilocal functionals is also reported as a green square at 0%. The B3LYP* family is singled out as unusual outliers in red. Chemical accuracy and transition metal (TM) chemical accuracy are also reported. All values in this plot are in kcal/mol.

2.4. Discussion on Reference Energies and Chemical Accuracy for Transition Metals

The case of iron(II) porphyrin is of particular interest because—despite the ubiquity of this moiety in chemistry and biology—the energetic ordering of its spin states is still highly debated, even in recent literature [9,158,159]. The difficulties in describing FeP arise because the spin states are very close to each other, and even a small change in the geometry of the complex results in a reordering of the energy levels. This situation results in an intricate connection between the Fe–N distance and the predicted ground state, with the quintet state more stable at long distances and the triplet state more stable at short ones [159]. This complex dependence on geometry also complicates the interpretation of the experimental results since it is difficult to exclude the effect of the doming motion of the metal from the porphyrin plane in most experimental conditions. For completely co-planar iron and porphyrin—as in the Por21 geometry—CASPT2 predicts a quintet ground state, which is a reasonable prediction. However, given the complex nature of the energy states and considering the results of other reference calculations, the CASPT2 predicted energy difference of 7.00 kcal/mol between the quintet and the triplet seems suspiciously large.

As emerges from the DFT results already discussed, single-reference methods should stabilize the low-spin states compared to high-spin ones. Pierloot and coworkers reported results from CCSD(T) calculations for FeP alongside the CASPT2 reference. These CCSD(T) results also present stabilization of the triplet, predicting a quintet ground state with a split of only 2.30 kcal/mol from the triplet, compared to the 7.00 kcal/mol predicted by CASPT2 (see Table 3). Most MUMS functionals predict a triplet ground state, except for MN15-L. On the one hand, we notice that the best functionals—r^2SCANh and M06-L—are within 4 kcal/mol from the CCSD(T) results, although more than 8 kcal/mol from the reference CASPT2 results. On the other hand, MN15-L has an error of 5.6 kcal/mol compared to the CASPT2 reference but a difference of more than 10 kcal/mol from the CCSD(T) results. Similar trends, with significant differences between CCSD(T) and CASPT2 reference energies, appear also for the $^5A_{1g} \rightarrow {}^3E_g$ gap of FeP (see last column in Table 3) and the $^6A_{1g} \rightarrow {}^4A_{2g}$ and $^6A_{1g} \rightarrow {}^2E_g$ states of MnP. The $^5A_{1g} \rightarrow {}^3A_{2g}$ of FeP (not considered in Por21 for precisely this reason) and the $^4B_{1g} \rightarrow {}^2A_{1g}$ of CoP (see results in the Supporting Information) appear less problematic.

Table 3. Spin state energy differences for the FeP system with 25 of the "most used, most suggested" (MUMS) functionals in the literature. All values are in kcal/mol. WFT = Wavefunction theory.

MUMS Functional:	Type [a]	FeP^5 → FeP^3	FeP^5 → FeP^1
r^{2}SCANh	GH-mGGA	−1.76	34.9
M06-L	mGGA	−1.35	25.6
MN15-L	mGGA	12.6	48.7
r^{2}SCAN	mGGA	−9.58	26.3
M06	GH-mGGA	−4.49	27.2
PBE0	GH-GGA	−8.69	27.2
r^{2}SCAN0	GH-mGGA	−3.83	33.2
B3LYP	GH-GGA	−14.6	18.6
B97M-V	mGGA	−21.5	10.8
PW6B95	GH-mGGA	−12.4	19.6
SCAN	mGGA	−24.0	5.44
TPSSh	GH-mGGA	−19.7	13.8
BLYP	GGA	−16.4	17.4
PBE	GGA	−15.9	19.1
TPSS	mGGA	−26.2	6.68
MN15	GH-mGGA	−45.2	−9.77
B2PLYP	DH	−19.7	14.3
BP86	GGA	−13.7	19.6
CAM-B3LYP	RSH-GGA	−11.4	24.2
ωB97M-V	RSH-mGGA	−57.5	106.
M06-2X	GH-mGGA	−49.4	−11.8
ωB97X-V	RSH-GGA	−53.9	108.
PWPB95-D3(BJ)	DH	−62.1	93.0
ωB97M(2)	DH	−80.9	−54.0
B3LYP*	GH-GGA	−14.6	18.6
CCSD(T)	WFT	2.30	33.0
CASPT2	WFT, Reference	7.00	39.9

[a] GGA: generalized gradient approximation; mGGA: meta-GGA; GH: global hybrid; RSH: range-separated hybrid; DH: double-hybrid.

Because of the complications highlighted above, these results are hard to interpret. It seems evident that the development of reliable, highly accurate reference electronic structure theory methods is required to address several of these issues. Lowering the cost of methods such as multireference coupled cluster [160], density matrix renormalization group (DMRG) [161], and quantum Monte Carlo (QMC) [162,163] might provide an answer in the future, even if those methods are not immune from interpretational difficulties (see Reference [164] for a recent unsolved controversy). Additional variants of the CAS methodology, such as the Stochastic Generalized Active Space (Stochastic-GAS) by Li Manni et al., also appear promising [158]. Given this situation, we still support the concept of "transition metal chemical accuracy" introduced by Wilson and coworkers [165]. Despite being proposed more than a decade ago, we reiterate the necessity to bring the chemical accuracy threshold to at least 3 kcal/mol for transition-metal-containing systems. The analysis of the results presented here and the available reference energies [9,11,158,159] suggest that a more conservative threshold could be considered at 5 kcal/mol for calculations on some of the most problematic metalloporphyrins. In the words of Wilson et al.: "This targeted accuracy is larger than for energetics of main group species because greater uncertainties are common in the experimental data for transition metal compounds and greater errors are expected with theory due to a number of factors, including increased valence electron space, stronger relativistic effects, and increased complexity of metal–ligand bonding." [166].

2.5. Recommendations for Users and Final Remarks

In this work, we assessed the performance of 250 approximations for calculations of spin state energy differences and binding energies of Mn(II), Co(II), Fe(II), and Fe(III) porphyrins. Unfortunately, the average errors observed for these systems are very far from the chemical accuracy threshold of 1.0 kcal/mol that several functionals achieve for reaction energies of main-group elements [23,24]. These results reflect both an intrinsic difficulty of density functional calculations on metalloporphyrin and the difficulties in obtaining reliable reference energies and experimental results. We support raising such a chemical accuracy threshold to at least 3 kcal/mol for transition metals, as previously suggested by Wilson et al. [165,166].

In light of these underwhelming results, we believe that the purpose of this extensive benchmark should not be the identification of a single best performer but rather to suggest trends and guidelines. In this spirit, we provide the following general recommendations for calculations on porphyrin-containing systems:

1. Avoid LDA, Hartree-Fock, and semiempirical methods (Group 0).
2. Avoid functionals containing PT2-like correlation, such as double-hybrids (Group 6).
3. Avoid range-separated hybrids (Group 5).
4. Prefer semilocal functionals (Group 1 and Group 2) for spin states and other purely electronic properties.
5. Among hybrid functionals in Group 3 and Group 4, prefer those with a percentage of exact exchange below 30%.
6. For all other properties—including thermochemistry—prefer functionals that scored a grade of A or B in this study (Group 3 and Group 4) and are highly transferable to other systems across chemistry, as established in other benchmark studies. Some suitable suggestions are (our grades are in parenthesis, and high transferability is indicated in bold font):

    (1) Semilocal functionals: **MN15-L (A)**, GAM (A), revM06-L (A), M06-L (A), r^2SCAN-D4 (A).

    (2) Global hybrids: **r^2SCANh-D4 (A)**, M06 (B), PBE0-D3(BJ) (B).

The suggestions above are our (current) top recommendations for methods that retain a certain degree of transferability across a broad range of chemical properties.

3. Materials and Methods

3.1. Software and Settings

Most of the calculations in this study are single-point KS DFT calculations performed with the def2-TZVP Gaussian-type basis set [167] and the functionals listed in Table 1. Functionals belonging to all rungs of Perdew's *Jacob's ladder* of density functional approximations [168] were employed. The list of functionals includes almost all the approximations developed to date, spanning more than forty years of functional development. The effect of dispersion corrections was also studied by adding -D2 [36], -D3 (in different "flavors") [24,37–43], or -D4- [44–46] corrections to some common functionals. Selected semiempirical and other low-cost methods were also included for comparison. The calculations with the low-cost methods employed the basis sets accompanying each method (MINIX [148] for HF-3c, def2-mSVP [146] for PBEh-3c, and mTZVP [89] for B97-3c). The majority of the calculations were performed using a development version of the Q-Chem quantum chemistry package [169]. The calculations with the APF(D) [47], B98 [103], HISS [55], and X3LYP [150] functionals and the semiempirical PM6 [54] and PM7 [56] methods were performed using the Gaussian16 program [170]. The xtb program [171] was used for calculations with the GFN1-xTB [135] and GFN2-xTB [136] semiempirical tight binding methods, while the standalone dftd4 program [172] was used to obtain the -D4 dispersion corrections for selected approximations. Among the available semiempirical and low-cost methods, we chose the most accurate according to recent benchmark studies [89,136,173]. A Lebedev grid with 99 radial and 590 angular points was employed to integrate all functionals, and the stabilities of the final solutions were checked to ensure the proper convergence of the electronic energies, allowing for symmetry-breaking when necessary. For double-hybrid functionals, the default frozen-core option for the PT2-like correlation was enabled for computational efficiency.

3.2. The Por21 Database

The database used in this study is called Por21, as defined in the ACCDB collection [174]. The reference energies were obtained with the CASPT2 method by Pierloot, Radón, and coworkers [4,11]. The database includes 11 spin states [11] of Mn(II), Co(II), and differently substituted Fe(II) and Fe(III) porphyrins, and 10 binding energies [4] of the complexes between a model system of the heme group and three different diatomic molecules: NO, CO, and O_2. To analyze the results in more detail, the spin states data have been grouped in a subset called PorSS11, while the bond energies data were grouped in the PorBE10 subset. All the molecular geometries are taken from the original publications [4,11], and were not re-optimized in this work. The geometry for iron porphyrin is optimized at the PBE0/def2-TZVP level [93,167], while all the remaining PorSS11 ones are optimized at the BP86/def2-TZVP level [62,75,167]. The spin multiplicities considered are sextuplet, quadruplet, and doublet for Mn(II) and Fe(III), quartet and doublet for Co(II), and quintet, triplet, and singlet for Fe(II). The binding energies are calculated using the following reaction:

$$FeP(Im) + X \rightarrow FeP(Im)X \qquad (1)$$

where FeP corresponds to the iron porphyrin, FePIm is the iron porphyrin bound to the imidazole ring, and X is either CO, NO, or O_2. The geometries of the reactants in Equation (1) were optimized at the PBE0/def2-TZVP level, while the geometries of the products were optimized at the BP86/def2-TZVP level. The reader is referred to the original publications [4,11] for additional details. A representative example of the systems included in each subset is reported in Figure 3. The geometries used in the calculations are available on our group's GitHub page [175].

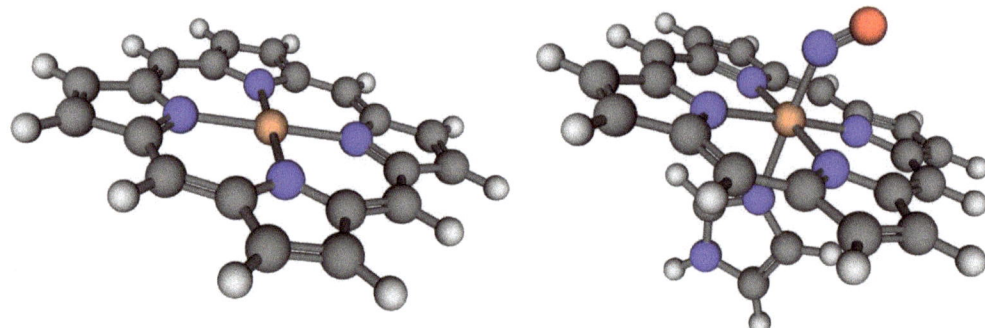

Figure 3. (**Left**) One of the iron-porphyrin rings (FeP) in the PorSS11 dataset. (**Right**) The porphyrin-imidazole-NO complex (FePIm-NO) from the PorBE10 dataset. The imidazole ring is used in place of the full heme active site. Hydrogen atoms are white, carbon atoms are black, nitrogen atoms are blue, the oxygen atom is red, and the iron atom is orange.

Supplementary Materials: The following supporting information can be downloaded at: https://www.mdpi.com/article/10.3390/molecules28083487/s1, The complete results for each data point, including calculated energies, signed and unsigned errors, and grades for each functional against the Por21, PorSS11, and PorBE10 datasets are provided as Excel spreadsheets.

Author Contributions: P.M.: Conceptualization, Methodology, Writing—original draft, Formal analysis, Investigation. R.P.: Conceptualization, Methodology, Writing—review and editing, Conceptualization, Formal analysis, Visualization, Investigation, Supervision, Project administration. All authors have read and agreed to the published version of the manuscript.

Funding: This research received no external funding.

Institutional Review Board Statement: Not applicable.

Informed Consent Statement: Not applicable.

Data Availability Statement: All the data are available within the manuscript and the Supplementary Materials.

Acknowledgments: Some of the calculations were performed on the Florida Tech Blueshark cluster, which is supported by the National Science Foundation under Grant No. CNS 09-23050.

Conflicts of Interest: The authors declare no conflict of interest.

References

1. Meunier, B.; de Visser, S.P.; Shaik, S. Mechanism of Oxidation Reactions Catalyzed by Cytochrome P450 Enzymes. *Chem. Rev.* **2004**, *104*, 3947–3980. [CrossRef]
2. Shaik, S.; Kumar, D.; de Visser, S.P.; Altun, A.; Thiel, W. Theoretical Perspective on the Structure and Mechanism of Cytochrome P450 Enzymes. *Chem. Rev.* **2005**, *105*, 2279–2328. [CrossRef] [PubMed]
3. Shaik, S.; Cohen, S.; Wang, Y.; Chen, H.; Kumar, D.; Thiel, W. P450 Enzymes: Their Structure, Reactivity, and Selectivity—Modeled by QM/MM Calculations. *Chem. Rev.* **2010**, *110*, 949–1017. [CrossRef]
4. Radoń, M.; Pierloot, K. Binding of CO, NO, and O_2 to Heme by Density Functional and Multireference Ab Initio Calculations. *J. Phys. Chem. A* **2008**, *112*, 11824–11832. [CrossRef] [PubMed]
5. Feldt, M.; Phung, Q.M.; Pierloot, K.; Mata, R.A.; Harvey, J.N. Limits of Coupled-Cluster Calculations for Non-Heme Iron Complexes. *J. Chem. Theory Comput.* **2019**, *15*, 922–937. [CrossRef]
6. Sauri, V.; Serrano-Andrés, L.; Shahi, A.R.M.; Gagliardi, L.; Vancoillie, S.; Pierloot, K. Multiconfigurational Second-Order Perturbation Theory Restricted Active Space (RASPT2) Method for Electronic Excited States: A Benchmark Study. *J. Chem. Theory Comput.* **2011**, *7*, 153–168. [CrossRef]
7. Berryman, V.E.J.; Boyd, R.J.; Johnson, E.R. Balancing Exchange Mixing in Density-Functional Approximations for Iron Porphyrin. *J. Chem. Theory Comput.* **2015**, *11*, 3022–3028. [CrossRef] [PubMed]

8. Nguyen, K.A.; Pachter, R. Ground State Electronic Structures and Spectra of Zinc Complexes of Porphyrin, Tetraazaporphyrin, Tetrabenzoporphyrin, and Phthalocyanine: A Density Functional Theory Study. *J. Chem. Phys.* **2001**, *114*, 10757–10767. [CrossRef]
9. Zhou, C.; Gagliardi, L.; Truhlar, D.G. Multiconfiguration Pair-Density Functional Theory for Iron Porphyrin with CAS, RAS, and DMRG Active Spaces. *J. Phys. Chem. A* **2019**, *123*, 3389–3394. [CrossRef]
10. Chen, H.; Lai, W.; Shaik, S. Multireference and Multiconfiguration Ab Initio Methods in Heme-Related Systems: What Have We Learned So Far? *J. Phys. Chem. B* **2011**, *115*, 1727–1742. [CrossRef]
11. Pierloot, K.; Phung, Q.M.; Domingo, A. Spin State Energetics in First-Row Transition Metal Complexes: Contribution of (3s3p) Correlation and Its Description by Second-Order Perturbation Theory. *J. Chem. Theory Comput.* **2017**, *13*, 537–553. [CrossRef]
12. Cramer, C.J.; Truhlar, D.G. Density Functional Theory for Transition Metals and Transition Metal Chemistry. *Phys. Chem. Chem. Phys.* **2009**, *11*, 10757–10816. [CrossRef]
13. Yang, K.; Peverati, R.; Truhlar, D.G.; Valero, R. Density Functional Study of Multiplicity-Changing Valence and Rydberg Excitations of p-Block Elements: Delta Self-Consistent Field, Collinear Spin-Flip Time-Dependent Density Functional Theory (DFT), and Conventional Time-Dependent DFT. *J. Chem. Phys.* **2011**, *135*, 044118. [CrossRef] [PubMed]
14. Janet, J.P.; Kulik, H.J. Predicting Electronic Structure Properties of Transition Metal Complexes with Neural Networks. *Chem. Sci.* **2017**, *8*, 5137–5152. [CrossRef]
15. Wilbraham, L.; Verma, P.; Truhlar, D.G.; Gagliardi, L.; Ciofini, I. Multiconfiguration Pair-Density Functional Theory Predicts Spin-State Ordering in Iron Complexes with the Same Accuracy as Complete Active Space Second-Order Perturbation Theory at a Significantly Reduced Computational Cost. *J. Phys. Chem. Lett.* **2017**, *8*, 2026–2030. [CrossRef] [PubMed]
16. Verma, P.; Varga, Z.; Klein, J.E.M.N.; Cramer, C.J.; Que, L.; Truhlar, D.G. Assessment of Electronic Structure Methods for the Determination of the Ground Spin States of Fe(II), Fe(III) and Fe(IV) Complexes. *Phys. Chem. Chem. Phys.* **2017**, *19*, 13049–13069. [CrossRef]
17. Taylor, M.G.; Yang, T.; Lin, S.; Nandy, A.; Janet, J.P.; Duan, C.; Kulik, H.J. Seeing Is Believing: Experimental Spin States from Machine Learning Model Structure Predictions. *J. Phys. Chem. A* **2020**, *124*, 3286–3299. [CrossRef] [PubMed]
18. Verma, P.; Truhlar, D.G. Status and Challenges of Density Functional Theory. *Trends Chem.* **2020**, *2*, 302–318. [CrossRef]
19. Gagliardi, L.; Truhlar, D.G.; Li Manni, G.; Carlson, R.K.; Hoyer, C.E.; Bao, J.L. Multiconfiguration Pair-Density Functional Theory: A New Way To Treat Strongly Correlated Systems. *Acc. Chem. Res.* **2017**, *50*, 66–73. [CrossRef] [PubMed]
20. Mok, D.K.W.; Neumann, R.; Handy, N.C. Dynamical and Nondynamical Correlation. *J. Phys. Chem.* **1996**, *100*, 6225–6230. [CrossRef]
21. Cohen, A.J.; Handy, N.C. Dynamic Correlation. *Mol. Phys.* **2001**, *99*, 607–615. [CrossRef]
22. Polo, V.; Kraka, E.; Cremer, D. Some Thoughts about the Stability and Reliability of Commonly Used Exchange Correlation Functionals: Coverage of Dynamic and Nondynamic Correlation Effects. *Theor. Chem. Acc.* **2002**, *107*, 291–303. [CrossRef]
23. Mardirossian, N.; Head-Gordon, M. Thirty Years of Density Functional Theory in Computational Chemistry: An Overview and Extensive Assessment of 200 Density Functionals. *Mol. Phys.* **2017**, *115*, 2315–2372. [CrossRef]
24. Goerigk, L.; Hansen, A.; Bauer, C.; Ehrlich, S.; Najibi, A.; Grimme, S. A Look at the Density Functional Theory Zoo with the Advanced GMTKN55 Database for General Main Group Thermochemistry, Kinetics and Noncovalent Interactions. *Phys. Chem. Chem. Phys.* **2017**, *19*, 32184–32215. [CrossRef] [PubMed]
25. Yu, H.S.; He, X.; Li, S.L.; Truhlar, D.G. MN15: A Kohn–Sham Global-Hybrid Exchange–Correlation Density Functional with Broad Accuracy for Multi-Reference and Single-Reference Systems and Noncovalent Interactions. *Chem. Sci.* **2016**, *7*, 5032–5051. [CrossRef]
26. Dohm, S.; Hansen, A.; Steinmetz, M.; Grimme, S.; Checinski, M.P. Comprehensive Thermochemical Benchmark Set of Realistic Closed-Shell Metal Organic Reactions. *J. Chem. Theory Comput.* **2018**, *14*, 2596–2608. [CrossRef]
27. Maurer, L.R.; Bursch, M.; Grimme, S.; Hansen, A. Assessing Density Functional Theory for Chemically Relevant Open-Shell Transition Metal Reactions. *J. Chem. Theory Comput.* **2021**, *17*, 6134–6151. [CrossRef]
28. Rappoport, D.; Crawford, N.R.M.; Furche, F.; Burke, K. Approximate Density Functionals: Which Should I Choose? In *Encyclopedia of Inorganic Chemistry*; King, R.B., Crabtree, R.H., Lukehart, C.M., Atwood, D.A., Scott, R.A., Eds.; John Wiley & Sons, Ltd.: Chichester, UK, 2009; p. ia615. [CrossRef]
29. Morgante, P.; Peverati, R. The Devil in the Details: A Tutorial Review on Some Undervalued Aspects of Density Functional Theory Calculations. *Int. J. Quantum Chem.* **2020**, *120*, e26332. [CrossRef]
30. Bursch, M.; Mewes, J.; Hansen, A.; Grimme, S. Best-Practice DFT Protocols for Basic Molecular Computational Chemistry**. *Angew. Chem. Int. Ed.* **2022**, *61*, e202205735. [CrossRef]
31. Goerigk, L.; Mehta, N. A Trip to the Density Functional Theory Zoo: Warnings and Recommendations for the User. *Aust. J. Chem.* **2019**, *72*, 563. [CrossRef]
32. Najibi, A.; Goerigk, L. The Nonlocal Kernel in van Der Waals Density Functionals as an Additive Correction: An Extensive Analysis with Special Emphasis on the B97M-V and ωB97M-V Approaches. *J. Chem. Theory Comput.* **2018**, *14*, 5725–5738. [CrossRef] [PubMed]
33. Peverati, R.; Truhlar, D.G. Quest for a Universal Density Functional: The Accuracy of Density Functionals across a Broad Spectrum of Databases in Chemistry and Physics. *Phil. Trans. R. Soc. A* **2014**, *372*, 20120476. [CrossRef] [PubMed]
34. Cirera, J.; Ruiz, E. Assessment of the SCAN Functional for Spin-State Energies in Spin-Crossover Systems. *J. Phys. Chem. A* **2020**, *124*, 5053–5058. [CrossRef] [PubMed]

35. Mortensen, S.R.; Kepp, K.P. Spin Propensities of Octahedral Complexes From Density Functional Theory. *J. Phys. Chem. A* **2015**, *119*, 4041–4050. [CrossRef]
36. Grimme, S. Semiempirical GGA-Type Density Functional Constructed with a Long-Range Dispersion Correction. *J. Comput. Chem.* **2006**, *27*, 1787–1799. [CrossRef]
37. Goerigk, L.; Grimme, S. A General Database for Main Group Thermochemistry, Kinetics, and Noncovalent Interactions—Assessment of Common and Reparameterized (Meta-)GGA Density Functionals. *J. Chem. Theory Comput.* **2010**, *6*, 107–126. [CrossRef]
38. Grimme, S.; Antony, J.; Ehrlich, S.; Krieg, H. A Consistent and Accurate Ab Initio Parametrization of Density Functional Dispersion Correction (DFT-D) for the 94 Elements H-Pu. *J. Chem. Phys.* **2010**, *132*, 154104. [CrossRef]
39. Grimme, S.; Ehrlich, S.; Goerigk, L. Effect of the Damping Function in Dispersion Corrected Density Functional Theory. *J. Comput. Chem.* **2011**, *32*, 1456–1465. [CrossRef]
40. Schröder, H.; Creon, A.; Schwabe, T. Reformulation of the D3(Becke–Johnson) Dispersion Correction without Resorting to Higher than C_6 Dispersion Coefficients. *J. Chem. Theory Comput.* **2015**, *11*, 3163–3170. [CrossRef]
41. Smith, D.G.A.; Burns, L.A.; Patkowski, K.; Sherrill, C.D. Revised Damping Parameters for the D3 Dispersion Correction to Density Functional Theory. *J. Phys. Chem. Lett.* **2016**, *7*, 2197–2203. [CrossRef]
42. Goerigk, L.; Grimme, S. Efficient and Accurate Double-Hybrid-Meta-GGA Density Functionals—Evaluation with the Extended GMTKN30 Database for General Main Group Thermochemistry, Kinetics, and Noncovalent Interactions. *J. Chem. Theory Comput.* **2011**, *7*, 291–309. [CrossRef] [PubMed]
43. Goerigk, L. Treating London-Dispersion Effects with the Latest Minnesota Density Functionals: Problems and Possible Solutions. *J. Phys. Chem. Lett.* **2015**, *6*, 3891–3896. [CrossRef] [PubMed]
44. Caldeweyher, E.; Bannwarth, C.; Grimme, S. Extension of the D3 Dispersion Coefficient Model. *J. Chem. Phys.* **2017**, *147*, 034112. [CrossRef] [PubMed]
45. Caldeweyher, E.; Ehlert, S.; Hansen, A.; Neugebauer, H.; Spicher, S.; Bannwarth, C.; Grimme, S. A Generally Applicable Atomic-Charge Dependent London Dispersion Correction. *J. Chem. Phys.* **2019**, *150*, 154122. [CrossRef]
46. Najibi, A.; Goerigk, L. DFT -D4 Counterparts of Leading Meta-Generalized-gradient Approximation and Hybrid Density Functionals for Energetics and Geometries. *J. Comput. Chem.* **2020**, *41*, 2562–2572. [CrossRef]
47. Austin, A.; Petersson, G.A.; Frisch, M.J.; Dobek, F.J.; Scalmani, G.; Throssell, K. A Density Functional with Spherical Atom Dispersion Terms. *J. Chem. Theory Comput.* **2012**, *8*, 4989–5007. [CrossRef]
48. Hartree, D.R. The Wave Mechanics of an Atom with a Non-Coulomb Central Field. Part I. Theory and Methods. *Math. Proc. Camb. Phil. Soc.* **1928**, *24*, 89–110. [CrossRef]
49. Hartree, D.R. The Wave Mechanics of an Atom with a Non-Coulomb Central Field. Part II. Some Results and Discussion. *Math. Proc. Camb. Phil. Soc.* **1928**, *24*, 111–132. [CrossRef]
50. Fock, V. Näherungsmethode zur Lösung des quantenmechanischen Mehrkörperproblems. *Z. Phys.* **1930**, *61*, 126–148. [CrossRef]
51. Lee, C.; Yang, W.; Parr, R.G. Development of the Colle-Salvetti Correlation-Energy Formula Into a Functional of the Electron-Density. *Phys. Rev. B* **1988**, *37*, 785–789. [CrossRef]
52. Perdew, J.P.; Kurth, S.; Zupan, A.; Blaha, P. Accurate Density Functional with Correct Formal Properties: A Step Beyond the Generalized Gradient Approximation. *Phys. Rev. Lett.* **1999**, *82*, 2544–2547. [CrossRef]
53. Perdew, J.P.; Wang, Y. Accurate and Simple Analytic Representation of the Electron-Gas Correlation-Energy. *Phys. Rev. B* **1992**, *45*, 13244–13249. [CrossRef] [PubMed]
54. Stewart, J.J.P. Optimization of Parameters for Semiempirical Methods V: Modification of NDDO Approximations and Application to 70 Elements. *J. Mol. Model.* **2007**, *13*, 1173–1213. [CrossRef]
55. Henderson, T.M.; Izmaylov, A.F.; Scuseria, G.E.; Savin, A. Assessment of a Middle-Range Hybrid Functional. *J. Chem. Theory Comput.* **2008**, *4*, 1254–1262. [CrossRef]
56. Stewart, J.J.P. Optimization of Parameters for Semiempirical Methods VI: More Modifications to the NDDO Approximations and Re-Optimization of Parameters. *J. Mol. Model.* **2013**, *19*, 1–32. [CrossRef] [PubMed]
57. Perdew, J.P.; Burke, K.; Ernzerhof, M. Generalized Gradient Approximation Made Simple. *Phys. Rev. Lett.* **1996**, *77*, 3865–3868. [CrossRef]
58. Krukau, A.V.; Vydrov, O.A.; Izmaylov, A.F.; Scuseria, G.E. Influence of the Exchange Screening Parameter on the Performance of Screened Hybrid Functionals. *J. Chem. Phys.* **2006**, *125*, 224106. [CrossRef]
59. Henderson, T.M.; Janesko, B.G.; Scuseria, G.E. Generalized Gradient Approximation Model Exchange Holes for Range-Separated Hybrids. *J. Chem. Phys.* **2008**, *128*, 194105. [CrossRef]
60. Zhao, Y.; Truhlar, D.G. Design of Density Functionals That Are Broadly Accurate for Thermochemistry, Thermochemical Kinetics, and Nonbonded Interactions. *J. Phys. Chem. A* **2005**, *109*, 5656–5667. [CrossRef]
61. Karton, A.; Gruzman, D.; Martin, J.M.L. Benchmark Thermochemistry of the C_nH_{2n+2} Alkane Isomers (n = 2−8) and Performance of DFT and Composite Ab Initio Methods for Dispersion-Driven Isomeric Equilibria. *J. Phys. Chem. A* **2009**, *113*, 8434–8447. [CrossRef]
62. Becke, A.D. Density-Functional Exchange-Energy Approximation with Correct Asymptotic-Behavior. *Phys. Rev. A* **1988**, *38*, 3098–3100. [CrossRef] [PubMed]

63. Becke, A.D. Density-functional Thermochemistry. III. The Role of Exact Exchange. *J. Chem. Phys.* **1993**, *98*, 5648–5652. [CrossRef]
64. Weintraub, E.; Henderson, T.M.; Scuseria, G.E. Long-Range-Corrected Hybrids Based on a New Model Exchange Hole. *J. Chem. Theory Comput.* **2009**, *5*, 754–762. [CrossRef]
65. Perdew, J.P.; Chevary, J.A.; Vosko, S.H.; Jackson, K.A.; Pederson, M.R.; Singh, D.J.; Fiolhais, C. Atoms, Molecules, Solids, and Surfaces: Applications of the Generalized Gradient Approximation for Exchange and Correlation. *Phys. Rev. B* **1992**, *46*, 6671–6687. [CrossRef]
66. Rohrdanz, M.A.; Martins, K.M.; Herbert, J.M. A Long-Range-Corrected Density Functional That Performs Well for Both Ground-State Properties and Time-Dependent Density Functional Theory Excitation Energies, Including Charge-Transfer Excited States. *J. Chem. Phys.* **2009**, *130*, 054112. [CrossRef] [PubMed]
67. Zhao, Y.; Schultz, N.E.; Truhlar, D.G. Exchange-Correlation Functional with Broad Accuracy for Metallic and Nonmetallic Compounds, Kinetics, and Noncovalent Interactions. *J. Chem. Phys.* **2005**, *123*, 161103. [CrossRef]
68. Hujo, W.; Grimme, S. Performance of the van Der Waals Density Functional VV10 and (Hybrid)GGA Variants for Thermochemistry and Noncovalent Interactions. *J. Chem. Theory Comput.* **2011**, *7*, 3866–3871. [CrossRef]
69. Zhao, Y.; Schultz, N.E.; Truhlar, D.G. Design of Density Functionals by Combining the Method of Constraint Satisfaction with Parametrization for Thermochemistry, Thermochemical Kinetics, and Noncovalent Interactions. *J. Chem. Theory Comput.* **2006**, *2*, 364–382. [CrossRef]
70. Reiher, M.; Salomon, O.; Artur Hess, B. Reparameterization of Hybrid Functionals Based on Energy Differences of States of Different Multiplicity. *Theor. Chem. Acc.* **2001**, *107*, 48–55. [CrossRef]
71. Furness, J.W.; Kaplan, A.D.; Ning, J.; Perdew, J.P.; Sun, J. Construction of Meta-GGA Functionals through Restoration of Exact Constraint Adherence to Regularized SCAN Functionals. *J. Chem. Phys.* **2022**, *156*, 034109. [CrossRef]
72. Furness, J.W.; Kaplan, A.D.; Ning, J.; Perdew, J.P.; Sun, J. Accurate and Numerically Efficient r2SCAN Meta-Generalized Gradient Approximation. *J. Phys. Chem. Lett.* **2020**, *11*, 8208–8215. [CrossRef] [PubMed]
73. Zhao, Y.; Truhlar, D.G. The M06 Suite of Density Functionals for Main Group Thermochemistry, Thermochemical Kinetics, Noncovalent Interactions, Excited States, and Transition Elements: Two New Functionals and Systematic Testing of Four M06-Class Functionals and 12 Other Functionals. *Theor. Chem. Acc.* **2008**, *120*, 215–241.
74. Ehlert, S.; Huniar, U.; Ning, J.; Furness, J.W.; Sun, J.; Kaplan, A.D.; Perdew, J.P.; Brandenburg, J.G. R2SCAN-D4: Dispersion Corrected Meta-Generalized Gradient Approximation for General Chemical Applications. *J. Chem. Phys.* **2021**, *154*, 061101. [CrossRef]
75. Perdew, J.P. Density-Functional Approximation for the Correlation-Energy of the Inhomogeneous Electron-Gas. *Phys. Rev. B* **1986**, *33*, 8822–8824. [CrossRef]
76. Bursch, M.; Neugebauer, H.; Ehlert, S.; Grimme, S. Dispersion Corrected r2SCAN Based Global Hybrid Functionals: R2SCANh, r2SCAN0, and r2SCAN50. *J. Chem. Phys.* **2022**, *156*, 134105. [CrossRef] [PubMed]
77. Becke, A.D. Density-Functional Thermochemistry. V. Systematic Optimization of Exchange-Correlation Functionals. *J. Chem. Phys.* **1997**, *107*, 8554–8560. [CrossRef]
78. Zhao, Y.; Truhlar, D.G. Density Functional for Spectroscopy: No Long-Range Self-Interaction Error, Good Performance for Rydberg and Charge-Transfer States, and Better Performance on Average than B3LYP for Ground States. *J. Phys. Chem. A* **2006**, *110*, 13126–13130. [CrossRef]
79. Patra, A.; Jana, S.; Samal, P. A Way of Resolving the Order-of-Limit Problem of Tao–Mo Semilocal Functional. *J. Chem. Phys.* **2020**, *153*, 184112. [CrossRef]
80. Hamprecht, F.A.; Cohen, A.J.; Tozer, D.J.; Handy, N.C. Development and Assessment of New Exchange-Correlation Functionals. *J. Chem. Phys.* **1998**, *109*, 6264–6271. [CrossRef]
81. Wang, Y.; Verma, P.; Jin, X.; Truhlar, D.G.; He, X. Revised M06 Density Functional for Main-Group and Transition-Metal Chemistry. *Proc. Natl. Acad. Sci. USA* **2018**, *115*, 10257–10262. [CrossRef]
82. Zhao, Y.; Truhlar, D.G. A New Local Density Functional for Main-Group Thermochemistry, Transition Metal Bonding, Thermochemical Kinetics, and Noncovalent Interactions. *J. Chem. Phys.* **2006**, *125*, 194101. [CrossRef] [PubMed]
83. Wang, Y.; Jin, X.; Yu, H.S.; Truhlar, D.G.; He, X. Revised M06-L Functional for Improved Accuracy on Chemical Reaction Barrier Heights, Noncovalent Interactions, and Solid-State Physics. *Proc. Natl. Acad. Sci. USA* **2017**, *114*, 8487–8492. [CrossRef] [PubMed]
84. Wilson, P.J.; Bradley, T.J.; Tozer, D.J. Hybrid Exchange-Correlation Functional Determined from Thermochemical Data and Ab Initio Potentials. *J. Chem. Phys.* **2001**, *115*, 9233–9242. [CrossRef]
85. Verma, P.; Wang, Y.; Ghosh, S.; He, X.; Truhlar, D.G. Revised M11 Exchange-Correlation Functional for Electronic Excitation Energies and Ground-State Properties. *J. Phys. Chem. A* **2019**, *123*, 2966–2990. [CrossRef] [PubMed]
86. Zhang, Y.; Yang, W. Comment on "Generalized Gradient Approximation Made Simple. *Phys. Rev. Lett.* **1998**, *80*, 890. [CrossRef]
87. Keal, T.W.; Tozer, D.J. Semiempirical Hybrid Functional with Improved Performance in an Extensive Chemical Assessment. *J. Chem. Phys.* **2005**, *123*, 121103. [CrossRef]
88. Zhao, Y.; Truhlar, D.G. Exploring the Limit of Accuracy of the Global Hybrid Meta Density Functional for Main-Group Thermochemistry, Kinetics, and Noncovalent Interactions. *J. Chem. Theory Comput.* **2008**, *4*, 1849–1868. [CrossRef]
89. Brandenburg, J.G.; Bannwarth, C.; Hansen, A.; Grimme, S. B97-3c: A Revised Low-Cost Variant of the B97-D Density Functional Method. *J. Chem. Phys.* **2018**, *148*, 064104. [CrossRef]

90. Peverati, R.; Truhlar, D.G. Improving the Accuracy of Hybrid Meta-GGA Density Functionals by Range Separation. *J. Phys. Chem. Lett.* **2011**, *2*, 2810–2817. [CrossRef]
91. Burns, L.A.; Mayagoitia, Á.V.; Sumpter, B.G.; Sherrill, C.D. Density-Functional Approaches to Noncovalent Interactions: A Comparison of Dispersion Corrections (DFT-D), Exchange-Hole Dipole Moment (XDM) Theory, and Specialized Functionals. *J. Chem. Phys.* **2011**, *134*, 084107. [CrossRef]
92. Peverati, R.; Truhlar, D.G. M11-L: A Local Density Functional That Provides Improved Accuracy for Electronic Structure Calculations in Chemistry and Physics. *J. Phys. Chem. Lett.* **2012**, *3*, 117–124. [CrossRef]
93. Adamo, C.; Barone, V. Toward Reliable Density Functional Methods without Adjustable Parameters: The PBE0 Model. *J. Chem. Phys.* **1999**, *110*, 6158–6170. [CrossRef]
94. Perdew, J.P.; Ruzsinszky, A.; Csonka, G.I.; Vydrov, O.A.; Scuseria, G.E.; Constantin, L.A.; Zhou, X.; Burke, K. Restoring the Density-Gradient Expansion for Exchange in Solids and Surfaces. *Phys. Rev. Lett.* **2008**, *100*, 136406. [CrossRef] [PubMed]
95. Wellendorff, J.; Lundgaard, K.T.; Jacobsen, K.W.; Bligaard, T. mBEEF: An Accurate Semilocal Bayesian Error Estimation Density Functional. *J. Chem. Phys.* **2014**, *140*, 144107. [CrossRef] [PubMed]
96. Boese, A.D.; Martin, J.M.L. Development of Density Functionals for Thermochemical Kinetics. *J. Chem. Phys.* **2004**, *121*, 3405–3416. [CrossRef]
97. Peverati, R.; Truhlar, D.G. An Improved and Broadly Accurate Local Approximation to the Exchange-Correlation Density Functional: The MN12-L Functional for Electronic Structure Calculations in Chemistry and Physics. *Phys. Chem. Chem. Phys.* **2012**, *14*, 13171–13174. [CrossRef]
98. Mardirossian, N.; Pestana, L.R.; Womack, J.C.; Skylaris, C.-K.; Head-Gordon, T.; Head-Gordon, M. Use of the rVV10 Nonlocal Correlation Functional in the B97M-V Density Functional: Defining B97M-rV and Related Functionals. *J. Phys. Chem. Lett.* **2017**, *8*, 35–40. [CrossRef]
99. Perdew, J.P.; Ruzsinszky, A.; Csonka, G.I.; Constantin, L.A.; Sun, J. Workhorse Semilocal Density Functional for Condensed Matter Physics and Quantum Chemistry. *Phys. Rev. Lett.* **2009**, *103*, 026403. [CrossRef]
100. Mardirossian, N.; Head-Gordon, M. Mapping the Genome of Meta-Generalized Gradient Approximation Density Functionals: The Search for B97M-V. *J. Chem. Phys.* **2015**, *142*, 074111–074132. [CrossRef]
101. Peverati, R.; Truhlar, D.G. Screened-Exchange Density Functionals with Broad Accuracy for Chemistry and Solid-State Physics. *Phys. Chem. Chem. Phys.* **2012**, *14*, 16187–16191. [CrossRef]
102. Csonka, G.I.; Perdew, J.P.; Ruzsinszky, A. Global Hybrid Functionals: A Look at the Engine under the Hood. *J. Chem. Theory Comput.* **2010**, *6*, 3688–3703. [CrossRef]
103. Schmider, H.L.; Becke, A.D. Optimized Density Functionals from the Extended G2 Test Set. *J. Chem. Phys.* **1998**, *108*, 9624–9631. [CrossRef]
104. Hammer, B.; Hansen, L.B.; Nørskov, J.K. Improved Adsorption Energetics within Density-Functional Theory Using Revised Perdew-Burke-Ernzerhof Functionals. *Phys. Rev. B* **1999**, *59*, 7413–7421. [CrossRef]
105. Constantin, L.A.; Fabiano, E.; Della Sala, F. Meta-GGA Exchange-Correlation Functional with a Balanced Treatment of Nonlocality. *J. Chem. Theory Comput.* **2013**, *9*, 2256–2263. [CrossRef] [PubMed]
106. Yu, H.S.; He, X.; Truhlar, D.G. MN15-L: A New Local Exchange-Correlation Functional for Kohn–Sham Density Functional Theory with Broad Accuracy for Atoms, Molecules, and Solids. *J. Chem. Theory Comput.* **2016**, *12*, 1280–1293. [CrossRef]
107. Adamo, C.; Barone, V. Exchange Functionals with Improved Long-Range Behavior and Adiabatic Connection Methods without Adjustable Parameters: The mPW and mPW1PW Models. *J. Chem. Phys.* **1998**, *108*, 664–675. [CrossRef]
108. Murray, É.D.; Lee, K.; Langreth, D.C. Investigation of Exchange Energy Density Functional Accuracy for Interacting Molecules. *J. Chem. Theory Comput.* **2009**, *5*, 2754–2762. [CrossRef]
109. Sun, J.; Xiao, B.; Ruzsinszky, A. Communication: Effect of the Orbital-Overlap Dependence in the Meta Generalized Gradient Approximation. *J. Chem. Phys.* **2012**, *137*, 051101. [CrossRef]
110. Sun, J.; Haunschild, R.; Xiao, B.; Bulik, I.W.; Scuseria, G.E.; Perdew, J.P. Semilocal and Hybrid Meta-Generalized Gradient Approximations Based on the Understanding of the Kinetic-Energy-Density Dependence. *J. Chem. Phys.* **2013**, *138*, 044113. [CrossRef]
111. Jana, S.; Sharma, K.; Samal, P. Improving the Performance of Tao–Mo Non-Empirical Density Functional with Broader Applicability in Quantum Chemistry and Materials Science. *J. Phys. Chem. A* **2019**, *123*, 6356–6369. [CrossRef]
112. Bartók, A.P.; Yates, J.R. Regularized SCAN Functional. *J. Chem. Phys.* **2019**, *150*, 161101. [CrossRef] [PubMed]
113. Sabatini, R.; Gorni, T.; de Gironcoli, S. Nonlocal van Der Waals Density Functional Made Simple and Efficient. *Phys. Rev. A* **2013**, *87*, 041108. [CrossRef]
114. Sun, J.; Ruzsinszky, A.; Perdew, J.P. Strongly Constrained and Appropriately Normed Semilocal Density Functional. *Phys. Rev. Lett.* **2015**, *115*, 036402. [CrossRef] [PubMed]
115. Brandenburg, J.G.; Bates, J.E.; Sun, J.; Perdew, J.P. Benchmark Tests of a Strongly Constrained Semilocal Functional with a Long-Range Dispersion Correction. *Phys. Rev. B* **2016**, *94*, 115144. [CrossRef]
116. Neupane, B.; Tang, H.; Nepal, N.K.; Adhikari, S.; Ruzsinszky, A. Opening Band Gaps of Low-Dimensional Materials at the Meta-GGA Level of Density Functional Approximations. *Phys. Rev. Mater.* **2021**, *5*, 063803. [CrossRef]

117. Sun, J.; Perdew, J.P.; Ruzsinszky, A. Semilocal Density Functional Obeying a Strongly Tightened Bound for Exchange. *Proc. Natl. Acad. Sci. USA* **2015**, *112*, 685–689. [CrossRef]
118. Peng, H.; Yang, Z.-H.; Perdew, J.P.; Sun, J. Versatile van Der Waals Density Functional Based on a Meta-Generalized Gradient Approximation. *Phys. Rev. X* **2016**, *6*, 041005. [CrossRef]
119. Hui, K.; Chai, J.-D. SCAN-Based Hybrid and Double-Hybrid Density Functionals from Models without Fitted Parameters. *J. Chem. Phys.* **2016**, *144*, 044114. [CrossRef]
120. Tsuneda, T.; Suzumura, T.; Hirao, K. A New One-Parameter Progressive Colle–Salvetti-Type Correlation Functional. *J. Chem. Phys.* **1999**, *110*, 10664–10678. [CrossRef]
121. Peverati, R.; Truhlar, D.G. Exchange–Correlation Functional with Good Accuracy for Both Structural and Energetic Properties While Depending Only on the Density and Its Gradient. *J. Chem. Theory Comput.* **2012**, *8*, 2310–2319. [CrossRef]
122. Zhao, Y.; Truhlar, D.G. Construction of a Generalized Gradient Approximation by Restoring the Density-Gradient Expansion and Enforcing a Tight Lieb–Oxford Bound. *J. Chem. Phys.* **2008**, *128*, 184109. [CrossRef] [PubMed]
123. Peverati, R.; Truhlar, D.G. Communication: A Global Hybrid Generalized Gradient Approximation to the Exchange-Correlation Functional That Satisfies the Second-Order Density-Gradient Constraint and Has Broad Applicability in Chemistry. *J. Chem. Phys.* **2011**, *135*, 191102. [CrossRef] [PubMed]
124. Slater, J.C. A Simplification of the Hartree-Fock Method. *Phys. Rev.* **1951**, *81*, 385–390. [CrossRef]
125. Handy, N.C.; Cohen, A.J. Left-Right Correlation Energy. *Mol. Phys.* **2001**, *99*, 403–412. [CrossRef]
126. Vosko, S.H.; Wilk, L.; Nusair, M. Accurate Spin-Dependent Electron Liquid Correlation Energies for Local Spin-Density Calculations: A Critical Analysis. *Can. J. Phys.* **1980**, *58*, 1200–1211. [CrossRef]
127. Boese, A.D.; Handy, N.C. New Exchange-Correlation Density Functionals: The Role of the Kinetic-Energy Density. *J. Chem. Phys.* **2002**, *116*, 9559–9569. [CrossRef]
128. Aschebrock, T.; Kümmel, S. Ultranonlocality and Accurate Band Gaps from a Meta-Generalized Gradient Approximation. *Phys. Rev. Res.* **2019**, *1*, 033082. [CrossRef]
129. Tao, J.; Mo, Y. Accurate Semilocal Density Functional for Condensed-Matter Physics and Quantum Chemistry. *Phys. Rev. Lett.* **2016**, *117*, 073001. [CrossRef]
130. Tao, J.; Perdew, J.P.; Staroverov, V.N.; Scuseria, G.E. Climbing the Density Functional Ladder: Nonempirical Meta–Generalized Gradient Approximation Designed for Molecules and Solids. *Phys. Rev. Lett.* **2003**, *91*, 146401. [CrossRef]
131. Yanai, T.; Tew, D.P.; Handy, N.C. A New Hybrid Exchange–Correlation Functional Using the Coulomb-Attenuating Method (CAM-B3LYP). *Chem. Phys. Lett.* **2004**, *393*, 51–57. [CrossRef]
132. Staroverov, V.N.; Scuseria, G.E.; Tao, J.; Perdew, J.P. Comparative Assessment of a New Nonempirical Density Functional: Molecules and Hydrogen-Bonded Complexes. *J. Chem. Phys.* **2003**, *119*, 12129–12137. [CrossRef]
133. Kozuch, S.; Martin, J.M.L. Spin-Component-Scaled Double Hybrids: An Extensive Search for the Best Fifth-Rung Functionals Blending DFT and Perturbation Theory. *J. Comput. Chem.* **2013**, *34*, 2327–2344. [CrossRef] [PubMed]
134. Yu, H.S.; Zhang, W.; Verma, P.; He, X.; Truhlar, D.G. Nonseparable Exchange–Correlation Functional for Molecules, Including Homogeneous Catalysis Involving Transition Metals. *Phys. Chem. Chem. Phys.* **2015**, *17*, 12146–12160. [CrossRef] [PubMed]
135. Grimme, S.; Bannwarth, C.; Shushkov, P. A Robust and Accurate Tight-Binding Quantum Chemical Method for Structures, Vibrational Frequencies, and Noncovalent Interactions of Large Molecular Systems Parametrized for All Spd-Block Elements (Z = 1–86). *J. Chem. Theory Comput.* **2017**, *13*, 1989–2009. [CrossRef] [PubMed]
136. Bannwarth, C.; Ehlert, S.; Grimme, S. GFN2-XTB—An Accurate and Broadly Parametrized Self-Consistent Tight-Binding Quantum Chemical Method with Multipole Electrostatics and Density-Dependent Dispersion Contributions. *J. Chem. Theory Comput.* **2019**, *15*, 1652–1671. [CrossRef]
137. Chai, J.-D.; Mao, S.-P. Seeking for Reliable Double-Hybrid Density Functionals without Fitting Parameters: The PBE0-2 Functional. *Chem. Phys. Lett.* **2012**, *538*, 121–125. [CrossRef]
138. Vydrov, O.A.; van Voorhis, T. Nonlocal van Der Waals Density Functional: The Simpler the Better. *J. Chem. Phys.* **2010**, *133*, 244103. [CrossRef]
139. Boese, A.D.; Doltsinis, N.L.; Handy, N.C.; Sprik, M. New Generalized Gradient Approximation Functionals. *J. Chem. Phys.* **2000**, *112*, 1670–1678. [CrossRef]
140. Chai, J.-D.; Head-Gordon, M. Systematic Optimization of Long-Range Corrected Hybrid Density Functionals. *J. Chem. Phys.* **2008**, *138*, 084106. [CrossRef]
141. Mardirossian, N.; Head-Gordon, M. ωB97M-V: A Combinatorially Optimized, Range-Separated Hybrid, Meta-GGA Density Functional with VV10 Nonlocal Correlation. *J. Chem. Phys.* **2016**, *144*, 214110. [CrossRef]
142. Mardirossian, N.; Head-Gordon, M. Survival of the Most Transferable at the Top of Jacob's Ladder: Defining and Testing the ωB97M(2) Double Hybrid Density Functional. *J. Chem. Phys.* **2018**, *148*, 241736. [CrossRef] [PubMed]
143. Boese, A.D.; Handy, N.C. A New Parametrization of Exchange–Correlation Generalized Gradient Approximation Functionals. *J. Chem. Phys.* **2001**, *114*, 5497–5503. [CrossRef]
144. Chai, J.-D.; Head-Gordon, M. Long-Range Corrected Hybrid Density Functionals with Damped Atom–Atom Dispersion Corrections. *Phys. Chem. Chem. Phys.* **2008**, *10*, 6615. [CrossRef] [PubMed]
145. Lin, Y.-S.; Li, G.-D.; Mao, S.-P.; Chai, J.-D. Long-Range Corrected Hybrid Density Functionals with Improved Dispersion Corrections. *J. Chem. Theory Comput.* **2013**, *9*, 263–272. [CrossRef]

146. Grimme, S.; Brandenburg, J.G.; Bannwarth, C.; Hansen, A. Consistent Structures and Interactions by Density Functional Theory with Small Atomic Orbital Basis Sets. *J. Chem. Phys.* **2015**, *143*, 054107. [CrossRef] [PubMed]
147. Mardirossian, N.; Head-Gordon, M. ωB97X-V: A 10-Parameter, Range-Separated Hybrid, Generalized Gradient Approximation Density Functional with Nonlocal Correlation, Designed by a Survival-of-the-Fittest Strategy. *Phys. Chem. Chem. Phys.* **2014**, *16*, 9904–9924. [CrossRef]
148. Sure, R.; Grimme, S. Corrected Small Basis Set Hartree-Fock Method for Large Systems. *J. Comput. Chem.* **2013**, *34*, 1672–1685. [CrossRef]
149. Lin, Y.-S.; Tsai, C.-W.; Li, G.-D.; Chai, J.-D. Long-Range Corrected Hybrid Meta-Generalized-Gradient Approximations with Dispersion Corrections. *J. Chem. Phys.* **2012**, *136*, 154109. [CrossRef]
150. Xu, X.; Goddard, W.A. The X3LYP Extended Density Functional for Accurate Descriptions of Nonbond Interactions, Spin States, and Thermochemical Properties. *Proc. Natl. Acad. Sci. USA* **2004**, *101*, 2673–2677. [CrossRef]
151. Zhang, Y.; Xu, X.; Goddard, W.A. Doubly Hybrid Density Functional for Accurate Descriptions of Nonbond Interactions, Thermochemistry, and Thermochemical Kinetics. *Proc. Natl. Acad. Sci. USA* **2009**, *106*, 4963–4968. [CrossRef]
152. Zhang, I.Y.; Xu, X.; Jung, Y.; Goddard, W.A. A Fast Doubly Hybrid Density Functional Method Close to Chemical Accuracy Using a Local Opposite Spin Ansatz. *Proc. Natl. Acad. Sci. USA* **2011**, *108*, 19896–19900. [CrossRef] [PubMed]
153. Sousa, S.F.; Fernandes, P.A.; Ramos, M.J. General Performance of Density Functionals. *J. Phys. Chem. A* **2007**, *111*, 10439–10452. [CrossRef] [PubMed]
154. Van Noorden, R.; Maher, B.; Nuzzo, R. The Top 100 Papers. *Nature* **2014**, *514*, 550–553. [CrossRef]
155. Mehta, N.; Casanova-Páez, M.; Goerigk, L. Semi-Empirical or Non-Empirical Double-Hybrid Density Functionals: Which Are More Robust? *Phys. Chem. Chem. Phys.* **2018**, *20*, 23175–23194. [CrossRef] [PubMed]
156. Casida, M.E. Jacob's Ladder for Time-Dependent Density-Functional Theory: Some Rungs on the Way to Photochemical Heaven. In *Low-Lying Potential Energy Surfaces*; Hoffmann, M.R., Dyall, K.G., Eds.; ACS Symposium Series; American Chemical Society: Washington, DC, USA, 2002; Volume 828, pp. 199–220. [CrossRef]
157. Ioannidis, E.I.; Kulik, H.J. Towards Quantifying the Role of Exact Exchange in Predictions of Transition Metal Complex Properties. *J. Chem. Phys.* **2015**, *143*, 034104. [CrossRef]
158. Weser, O.; Guther, K.; Ghanem, K.; Li Manni, G. Stochastic Generalized Active Space Self-Consistent Field: Theory and Application. *J. Chem. Theory Comput.* **2022**, *18*, 251–272. [CrossRef] [PubMed]
159. Antalík, A.; Nachtigallová, D.; Lo, R.; Matoušek, M.; Lang, J.; Legeza, Ö.; Pittner, J.; Hobza, P.; Veis, L. Ground State of the Fe(II)-Porphyrin Model System Corresponds to Quintet: A DFT and DMRG-Based Tailored CC Study. *Phys. Chem. Chem. Phys.* **2020**, *22*, 17033–17037. [CrossRef]
160. Drosou, M.; Mitsopoulou, C.A.; Pantazis, D.A. Reconciling Local Coupled Cluster with Multireference Approaches for Transition Metal Spin-State Energetics. *J. Chem. Theory Comput.* **2022**, *18*, 3538–3548. [CrossRef]
161. Baiardi, A.; Reiher, M. The Density Matrix Renormalization Group in Chemistry and Molecular Physics: Recent Developments and New Challenges. *J. Chem. Phys.* **2020**, *152*, 040903. [CrossRef]
162. Konkov, V.; Peverati, R. QMC-SW: A Simple Workflow for Quantum Monte Carlo Calculations in Chemistry. *SoftwareX* **2019**, *9*, 7–14. [CrossRef]
163. Rudshteyn, B.; Coskun, D.; Weber, J.L.; Arthur, E.J.; Zhang, S.; Reichman, D.R.; Friesner, R.A.; Shee, J. Predicting Ligand-Dissociation Energies of 3d Coordination Complexes with Auxiliary-Field Quantum Monte Carlo. *J. Chem. Theory Comput.* **2020**, *16*, 3041–3054. [CrossRef] [PubMed]
164. Al-Hamdani, Y.S.; Nagy, P.R.; Zen, A.; Barton, D.; Kállay, M.; Brandenburg, J.G.; Tkatchenko, A. Interactions between Large Molecules Pose a Puzzle for Reference Quantum Mechanical Methods. *Nat. Commun.* **2021**, *12*, 3927. [CrossRef] [PubMed]
165. DeYonker, N.J.; Peterson, K.A.; Steyl, G.; Wilson, A.K.; Cundari, T.R. Quantitative Computational Thermochemistry of Transition Metal Species. *J. Phys. Chem. A* **2007**, *111*, 11269–11277. [CrossRef] [PubMed]
166. Jiang, W.; DeYonker, N.J.; Determan, J.J.; Wilson, A.K. Toward Accurate Theoretical Thermochemistry of First Row Transition Metal Complexes. *J. Phys. Chem. A* **2012**, *116*, 870–885. [CrossRef] [PubMed]
167. Weigend, F.; Ahlrichs, R. Balanced Basis Sets of Split Valence, Triple Zeta Valence and Quadruple Zeta Valence Quality for H to Rn: Design and Assessment of Accuracy. *Phys. Chem. Chem. Phys.* **2005**, *7*, 3297–3305. [CrossRef] [PubMed]
168. Perdew, J.P.; Schmidt, K. Jacob's Ladder of Density Functional Approximations for the Exchange-Correlation Energy. In *AIP Conference Proceedings*; American Institute of Physics: College Park, MD, USA, 2001; Volume 577, pp. 1–20. [CrossRef]
169. Epifanovsky, E.; Gilbert, A.T.B.; Feng, X.; Lee, J.; Mao, Y.; Mardirossian, N.; Pokhilko, P.; White, A.F.; Coons, M.P.; Dempwolff, A.L.; et al. Software for the Frontiers of Quantum Chemistry: An Overview of Developments in the Q-Chem 5 Package. *J. Chem. Phys.* **2021**, *155*, 084801. [CrossRef]
170. Frisch, M.J.; Trucks, G.W.; Schlegel, H.B.; Scuseria, G.E.; Robb, M.A.; Cheeseman, J.R.; Scalmani, G.; Barone, V.; Petersson, G.A.; Nakatsuji, H.; et al. *Gaussian 16 Revision A.03*; Gaussian, Inc.: Wallingford, CT, USA, 2016.
171. XTB; Grimme Group. Available online: https://github.com/grimme-lab/xtb (accessed on 27 March 2023).
172. DFT-D4; Grimme Group. Available online: https://github.com/grimme-lab/dftd4 (accessed on 27 March 2023).
173. Dral, P.O.; Wu, X.; Spörkel, L.; Koslowski, A.; Thiel, W. Semiempirical Quantum-Chemical Orthogonalization-Corrected Methods: Benchmarks for Ground-State Properties. *J. Chem. Theory Comput.* **2016**, *12*, 1097–1120. [CrossRef]

174. Morgante, P.; Peverati, R. ACCDB: A Collection of Chemistry Databases for Broad Computational Purposes. *J. Comput. Chem.* **2019**, *40*, 839–848. [CrossRef]
175. ACCDB; Peverati Group. Available online: https://github.com/peverati/ACCDB (accessed on 27 March 2023).

Disclaimer/Publisher's Note: The statements, opinions and data contained in all publications are solely those of the individual author(s) and contributor(s) and not of MDPI and/or the editor(s). MDPI and/or the editor(s) disclaim responsibility for any injury to people or property resulting from any ideas, methods, instructions or products referred to in the content.

Article

Chiral Porphyrin Assemblies Investigated by a Modified Reflectance Anisotropy Spectroscopy Spectrometer

Ilaria Tomei [1], Beatrice Bonanni [1], Anna Sgarlata [1], Massimo Fanfoni [1], Roberto Martini [2], Ilaria Di Filippo [2], Gabriele Magna [2], Manuela Stefanelli [2], Donato Monti [3], Roberto Paolesse [2] and Claudio Goletti [1,*]

1 Department of Physics, Università di Roma Tor Vergata, Via della Ricerca Scientifica 1, 00133 Roma, Italy
2 Department of Chemical Science and Technologies, Università di Roma Tor Vergata, Via della Ricerca Scientifica 1, 00133 Rome, Italy
3 Department of Chemistry, Sapienza Università di Roma, Piazzale Aldo Moro 5, 00185 Rome, Italy
* Correspondence: goletti@roma2.infn.it

Abstract: Reflectance anisotropy spectroscopy (RAS) has been largely used to investigate organic compounds: Langmuir–Blodgett and Langmuir–Schaeffer layers, the organic molecular beam epitaxy growth in situ and in real time, thin and ultrathin organic films exposed to volatiles, in ultra-high vacuum (UHV), in controlled atmosphere and even in liquid. In all these cases, porphyrins and porphyrin-related compounds have often been used, taking advantage of the peculiar characteristics of RAS with respect to other techniques. The technical modification of a RAS spectrometer (CD-RAS: circular dichroism RAS) allows us to investigate the circular dichroism of samples instead of the normally studied linear dichroism: CD-RAS measures (in transmission mode) the anisotropy of the optical properties of a sample under right and left circularly polarized light. Although commercial spectrometers exist to measure the circular dichroism of substances, the "open structure" of this new spectrometer and its higher flexibility in design makes it possible to couple it with UHV systems or other experimental configurations. The importance of chirality in the development of organic materials (from solutions to the solid state, as thin layers deposited—in liquid or in vacuum—on transparent substrates) could open interesting possibilities to a development in the investigation of the chirality of organic and biological layers. In this manuscript, after the detailed explanation of the CD-RAS technique, some calibration tests with chiral porphyrin assemblies in solution or deposited in solid film are reported to demonstrate the quality of the results, comparing curves obtained with CD-RAS and a commercial spectrometer.

Keywords: circular dichroism; porphyrins; chirality; chiral layers; supramolecular chirality; reflectance anisotropy spectroscopy

1. Introduction

Reflectance anisotropy spectroscopy (RAS) is a surface-sensitive optical technique with significant peculiarities: it strongly reduces or avoids the contamination or even damage of the sample (when electrons or charged particles are used as probes in other experimental techniques), is utilizable in vacuum, in atmosphere or in transparent media (also in liquids) without limitations due to pressure, allows for the investigation of insulating samples without problems of charging, and gives the possibility to study the structure of surfaces/layers and even buried or immersed interfaces [1–3].

RAS has been applied to clean surfaces of metals and semiconductors in ultra-high vacuum (UHV), boosting a significant development of the knowledge in surface science [4–7], then to low-dimensionality solid state systems (films, wires, dots) [8,9], and finally to organic samples (for example, samples grown by organic molecular beam epitaxy (OMBE) [10,11] or porphyrin and sapphyrin layers deposited by Langmuir–Blodgett and Langmuir–Schaeffer technique) [12,13]. Clean surfaces of metals and semiconductors immersed in solutions and organic and biological layers deposited in liquid have been recently studied, down to the characterization of samples in electrolytes, monitoring, for example, in real time and in situ by RAS and scanning tunneling microscope (STM), the electrochemical reactions at the Cu(110)–liquid interface triggered by voltage cycles applied to the sample [14–16].

In all these experiments, RAS has been limited to the anisotropy of the linear dichroism of matter. As we will show hereinafter, the same technique can be used to investigate the circular dichroism in transmittance by a proper modification in the experimental apparatus, opening intriguing perspectives in the experimental study of chirality. This topic constitutes one of the most active and dynamic research areas related to varied disciplines such as chemistry, physics, biology, and material science [17].

Chirality is a characteristic that pervades the universe, from the small range of the subatomic particles to the immensity of the spiral galaxies: it is defined as the geometric property of a rigid object (or spatial arrangement of points or atoms) of being non-superposable on its mirror image [18]. This term derives from the Ancient Greek word "cheir" ($\chi\varepsilon\ell\rho$) for hand, to give a pictorial sketch of the chirality meaning.

In chemistry, molecules that feature chirality can be spatially arranged into two specular, nonsuperimposable structures called enantiomers [19]. Since enantiomeric pairs are the same chemical species, the different spatial arrangement does not induce physical or chemical changes in the properties of these two isomers unless they are placed or interact with an asymmetrical environment. To have an idea of the importance of chirality, it is sufficient to note that all living systems are inherently chiral and represent an asymmetrical environment since, during evolution, only a single enantiomer has been selected for the synthons of essential biological macromolecules, such as L-aminoacids for proteins and D-sugars for nucleic acids. As a consequence, chirality plays a fundamental role in biological processes, driving the selectivity of most interactions essential for life.

Although conventionally considered important at the molecular level, especially in the pharmaceutical field, chirality is even more pursued at the supramolecular level to produce novel artificial chiral materials suitable for emerging applications such as circular polarizers, chiral chromatography, and (bio)sensor devices. Among the countless combinations of assembling methods and building blocks chosen, porphyrins have been intensively investigated since they offer the possibility to easily modulate molecular interactions, leading to stereospecific chiral systems by means of different approaches that use both chiral and achiral porphyrin units [20]. Additionally, the high absorptions in the ultraviolet (UV) and visible range possessed by these chromophores guarantee strong exciton coupling over large distances that result in diagnostic chiroptical signals unveiling the specific molecular orientation of the macrocycles within the aggregated species [21,22].

We are traditionally interested in characterizing the aggregation mechanisms of chiral metalloporphyrins with an appended proline group in hydroalcoholic solutions [23]. More recently, we have broadened this interest in the preparation of chemical sensors for chiral discrimination by using materials based on the same chiral porphyrins [24]. The enantiorecognition with chemical sensors is particularly challenging since these devices can rely only on a single binding event for chiral recognition. Additionally, controlling the deposition of chiral layers onto the surface of transducers constitutes another critical step since additional factors that may affect the film morphology/chiroptical properties, such as solvent, concentration, and evaporation time should be considered during the formation of sensing films [25].

Here, we investigated the possibility of utilizing a RAS spectrometer in transmission to measure features of porphyrins in solution and in solid films related to circular dichroism (CD), comparing the spectra obtained with the ones produced by commercial extended wavelength CD (ECD) apparatuses [26]. The possibility of using a RAS spectrometer with this new configuration for investigating CD (hereafter indicated as circular dichroism RAS, CD-RAS) could in principle represent just another way of performing experiments that are possible with the existing commercial instruments and then of questionable utility. However, unlike commercial CD spectrometers, the "open structure" of our CD-RAS spectrometer allows for higher flexibility in designing the optical path, making it possible to couple optics with UHV systems, with electrochemical cells, or more generally with other experimental configurations that otherwise should be impossible to configure. The increasing importance of experiments investigating the time domain of the behavior of chiral films (from evolution of the chiral signal after deposition to the investigation of kinetics when exposed to chiral molecules) suggests to go beyond the quite rigid methodology of application of the optical methods normally available in laboratory.

2. Experimental Methods and Techniques

The RAS signal is defined as the ratio $\Delta R/R$ between the difference ΔR of the light intensity R_α and R_β reflected by the sample for a beam polarized into two different and independent polarization states α and β, and their average $R = (R_\alpha + R_\beta)/2$, as a function of photon energy [2,3]:

$$\frac{\Delta R}{R} = 2\frac{R_\alpha - R_\beta}{R_\alpha + R_\beta} \quad (1)$$

In the case of linearly polarized light, the electric field is directed along two orthogonal directions α and β, usually aligned with the main anisotropy axes of the sample surface plane. In this version, a RAS spectrometer is essentially an ellipsometer at near normal incidence, whose experimental set-up strongly simplifies the interpretation of data with respect to traditional ellipsometry. A (sometimes complex) deconvolution from rough data is still necessary to correctly understand and explain the obtained results by modeling [2,3]: but the existence of a non-vanishing anisotropy often represents a meaningful result per se, suitable to characterize a certain physical system, or to follow its dependence upon the variation of definite experimental conditions as temperature, contamination, strain, coverage, pressure, electric field, etc., finally defined as the signature of the existence of a certain, well-defined phase at the surface. Linearly polarized RAS has been widely applied to investigate surfaces and interfaces, more generally 2D systems, exploiting the different symmetries of the bulk/substrate with respect to the surface/top layer [4–7,15]: the signal anisotropy coming from centrosymmetric crystals or amorphous substrates is ideally null, and then, any anisotropy signal measured by RAS is recognized and isolated as due to the surface/top layer [2,3]. In the experiments reported in Figure 1, α and β were parallel to the $[\bar{2}11]$ and $[1\bar{1}0]$ directions on the (111) cleavage plane of a diamond sample 2×1 reconstructed. The evident peak dominating the spectrum is due to optical transitions between electronic surface states of the π-bonded reconstructed surface [27].

In a typical RAS system, light from a source (usually Xe or tungsten lamp, emitting photons in the near UV–visible–near infrared (IR) range, that is 280–1000 nm) is shined and focused into a polarizer, then onto a photoelastic modulator (PEM) and finally onto the sample (properly oriented), eventually passing through a special low-birefringence window if the sample is in ultra-high vacuum (UHV) or in liquid. The PEM (properly driven by an oscillating circuit at the resonance frequency ν_0 of the piezoelectric crystal) introduces a phase shift equal to $\pm\pi$ between light beams propagating along ordinary and extraordinary axes, thus modulating the linear polarization of light between two orthogonal, independent states x and y. The light intensity reflected by the sample for two polarizations (R_x and R_y) is then collected and focused onto a second polarizer (analyzer) and finally into a monochromator. At the exit slit, there is a detector (photomultiplier, photodiode, etc.) chained to a preamplifier and then to a lock-in amplifier to filter the signal, tuned at the correct modulation frequency (exactly $2\nu_0$, that is the second harmonic of the modulated signal). Another version of a RAS spectrometer exists, where the analyzer is not present.

If the PEM introduces a phase shift equal to $\pm\pi/2$, the outgoing light is alternatively polarized between right and left circularly polarized light, with a resulting signal carrying information about the circular dichroism of the shined sample. The high modulation frequency (about 50 kHz) favors the signal stability, eliminating the mechanical noise and the low frequency modes. In Figure 2, we draw a time-sketch of the successive polarization states of circularly polarized light after the passage through the PEM: the oscillation frequency between circularly left and circularly right polarization states is exactly coincident with the oscillation frequency of the PEM. In the signal, a linear polarization contribution also appears (just for one of the two orthogonal independent states, we will call it "α"), at a frequency that is twice the PEM frequency. Just tuning the lock-in at a frequency equal to ν_0, the only contribution due to the circularly polarized light is selected.

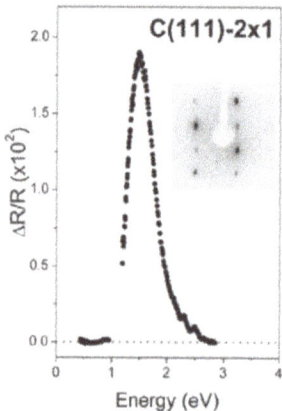

Figure 1. ΔR/R as a function of photon energy for a single domain C(111)-2 × 1 surface, in the energy range from 0.4 to 2.8 eV. The sharp peak (whose sign means light electric field along the 1D chains of the reconstructed surface) is related to optical transitions between the surface electronic bands. The inset shows a low-energy electron diffraction (LEED) picture taken at 70 eV, demonstrating the existence on the reconstructed surface of a largely dominating single domain. Doubling of the spots is due to the regular array of surface steps. More extended experimental details are reported in Ref. [27]. Reproduced from G. Bussetti et al., Europhys. Lett. 2007, 79, 57002, with permission.

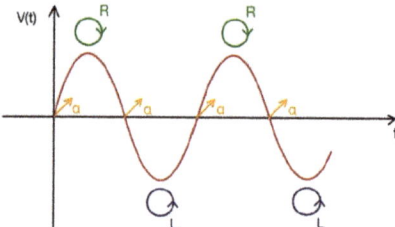

Figure 2. Sketch of the polarization of the light beam after the PEM, for an applied voltage value such that the phase shift is $\Delta\varphi = \pi/2$. The red curve represents (vs. time) the voltage driving the PEM at the frequency ν_0. At V = 0 volt, the polarization is linear, here directed as α (yellow arrows). When the voltage reaches its maximum, the resulting polarization is circular and right handed (green circles). When the voltage reaches its minimum, the resulting polarization is circular and left handed (blue circles). The oscillation frequency of the signal between the two states of circular polarization is evidently ν_0. The frequency of the linearly polarized signal is $2\nu_0$.

The technical equipment for the CD-RAS experiments is sketched in Figure 3. After being transmitted through the sample, the beam is reflected on an oxidized Si(100), perfectly isotropic for linearly polarized light, not introducing any artifacts in the signal. Strictly speaking, the term "reflectance" in the acronym CD-RAS is inappropriate here, as the experiments are conducted in transmission: reflectance anisotropy for circularly polarized light is rigorously null. This can be demonstrated using the formalism of the Jones matrices [28]; however, an immediate and identical conclusion is reached by evaluating the effect on the beam of the surface/layer (that is crossed twice). We have experimentally verified this null result for chiral films whose thicknesses were down to 10 nm.

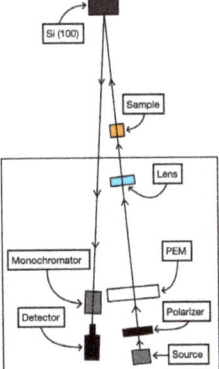

Figure 3. Sketch of the CD-RAS experimental spectrometer. The version here reported is the Safarov/Berkovits-like (see Refs. [1–3,29]), with one polarizer. After passing the sample, the light beam is reflected back by an isotropic Si(100) sample. Although the system was mounted horizontally in this case, it can also work vertically.

The apparatus is highly efficient (as we will show in the next section) in extracting the CD-related features in transmission from solutions and thin solid-state layers deposited onto transparent substrates. For CD-RAS, more properly and more generally, we will then define the signal as:

$$\frac{\Delta I}{I} = 2\frac{I_\alpha - I_\beta}{I_\alpha + I_\beta} \qquad (2)$$

where I_α and I_β are the light intensities (in this case, transmitted) measured after passing through the sample for a beam alternatively polarized into two different and independent circular polarization states α and β. The "atout" of our CD-RAS (with respect to more traditional apparatuses commonly available, such as the JASCO CD spectrometer [26]) is a more flexible design of the optical path that can be tuned (in length), with a careful setting of the whole experimental setup, according to the peculiar necessity of the experiments, to be coupled with UHV systems, electrochemical cells, or transparent recipients larger that the cuvette used in commercial CD spectrometers.

The data obtained by CD-RAS were compared with the ones obtained by a commercial spectrometer [26]. While CD-RAS data represent the intensity modulation $\Delta I/I$ of the signal when passing from the right circularly polarized state to the left circularly polarized state (see Formula (2)), a commercial spectrometer measures the ellipticity due to the sample, which is a composition (generally associated with the elliptic polarization of light) of the two circularly polarized states, with the same intensity when radiation impinges on the sample, but is then differently absorbed. CD spectra are reported as ellipticity, θ, and are measured in units of mdeg. The ellipticity $[\theta]$ is defined as:

$$[\theta] = A\,(\varepsilon_L - \varepsilon_R)\,c\,l \qquad (3)$$

where ε_L and ε_R are the molar extinction coefficients for light left- and right-handed polarized ones, respectively; l is the optical pathway of the sample, and c is its concentration. A is a constant with appropriate dimensions.

It is possible to demonstrate that the ellipticity $[\theta]$ and the $\Delta I/I$ result of a CD-RAS experiment are proportional [26]:

$$[\theta] \sim \Delta I/I \qquad (4)$$

3. Results and Discussion

In this section, the CD-RAS spectra measured in transmission mode for different samples are presented and compared with the spectra recorded by a commercial spectrometer for circular dichroism:

i. A solution of chiral porphyrins with a chiral signature (in terms ellipticity) in the order of some thousands of millideg (samples A1 and A2);
ii. A solution of chiral porphyrins with a chiral signal in the order of some millideg (sample B);
iii. A thin chiral film of porphyrins deposited onto glass substrates, in two enantiomeric configurations (samples C1 and C2).

In cases (i) and (ii), the spectrometer used in CD-RAS experiments was in the Aspnes version, with two polarizers [30]. In case (iii), we used the Safarov–Berkovits version of the CD-RAS spectrometer (without the analyzer), sketched in Figure 3 [29]. In all the experiments, the beam after transmission through the sample was reflected on an oxidized Si(100), perfectly isotropic for linearly polarized light, and then in principle not introducing artifacts in the signal. The spectra were measured in the range 380 nm < λ < 500 nm (where the main molecular contribution was expected in all cases), although the covered possible range is more extended (300 nm < λ < 700 nm).

The CD-RAS spectrometer was first tested on chiral porphyrin aggregates characterized by intense CD-related signals in solution (samples A1 and A2). In Figure 4, the CD-RAS spectrum (red curve) and the JASCO spectrum (black curve) of Sample A1 are reported and compared. The different axes (with different units) are also drawn. The two curves are in excellent agreement. A similar result was obtained with sample A2 (Figure 5). In this case, with a different molecule and different CD activity, the curves exhibit an excellent overall similitude, with the same lineshape at a very good accuracy. The CD-RAS spectrum was here corrected by subtracting the background measured (in the same experimental conditions) with an achiral solution.

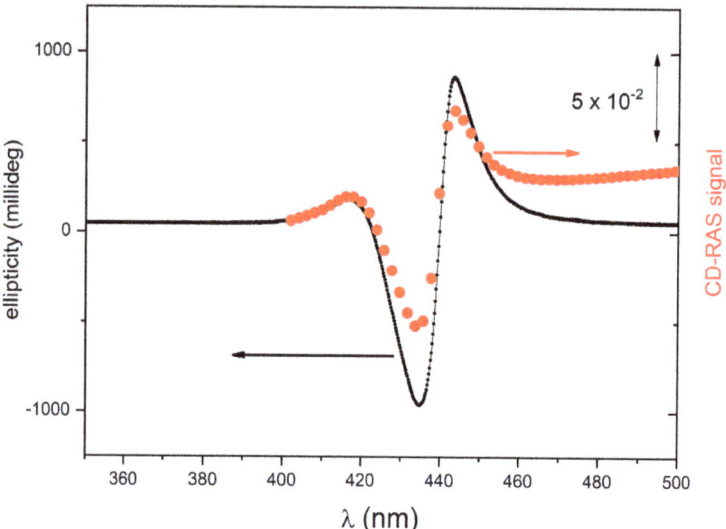

Figure 4. CD-RAS experimental spectrum for sample A1 (red curve, right vertical axis), compared with the ellipticity spectrum measured for the same sample by a commercial spectrometer for circular dichroism (black curve, left vertical axis). The CD-RAS spectrum is reported as measured, without correction. The double-arrow reported in the figure is the unit for the CD-RAS data. Here, sample A1 is a solution containing II-type aggregates of ZnP(L)Pro(−) in hydroalcoholic solution (EtOH:water 25:75, 5 µM).

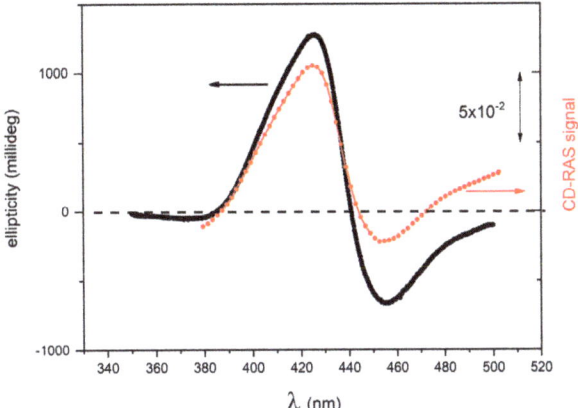

Figure 5. CD-RAS experimental spectrum for sample A2 (red curve, right vertical axis), compared with the ellipticity spectrum measured for the same sample by a commercial spectrometer for circular dichroism (black curve, left vertical axis). The double-arrow reported in the figure is the unit for the CD-RAS data. The CD-RAS spectrum was corrected by subtracting the background measured (in the same experimental conditions) with an achiral solution. Here, sample A2 is a solution containing aggregates of $H_2P(L)Pro(-)$ in hydroalcoholic solution (EtOH:water 25:75, 10 µM).

The system was then tested on sample B to check the performance with a sample exhibiting three orders of magnitude lower CD-related signals in the millideg range. Given this low value of the CD, the CD-RAS spectrum reported in Figure 6 was obtained by a longer integration time (2 s/spectral point) to gain a higher signal-to-noise ratio, and from the resulting curve, the optical background of the glass cuvette containing the solution (being a quite larger signal due to the birefringence of the vessel) was subtracted. The good agreement of the two curves in this case allows us to demonstrate the sensitivity of CD-RAS at a higher sensitivity scale.

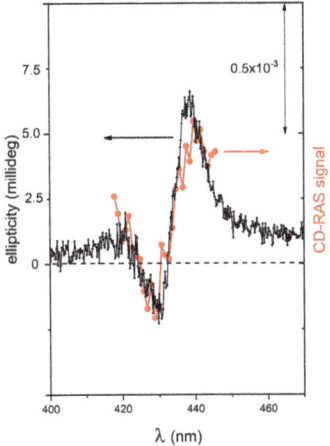

Figure 6. CD-RAS experimental spectrum for sample B (red curve), compared with the ellipticity spectrum measured for the same sample by a commercial spectrometer for circular dichroism (black curve, left vertical axis). The double-arrow reported in the figure is the unit for the CD-RAS data. The CD-RAS spectrum was corrected by subtracting the optical background of the glass cuvette containing an achiral solution (measured in the same experimental conditions). Here, sample B is a solution containing I-type aggregates of $ZnP(D)Pro(-)$ in hydroalcoholic solution (EtOH:water 25:75, 5 µM).

Finally, chiral porphyrin films (samples C1 and C2) deposited onto a glass substrate were measured in transmission mode. The films resulting from letting one 50 µL droplet dry out were approximately 10 nm thick (estimated from atomic force microscope images). In Figure 7, the CD-RAS spectra (red and black curves in panel A) and the JASCO spectra (red and black curves in panel B) are shown for both enantiomers, deposited separately following the same protocol. The background spectrum measured in the same configuration for a clean glass substrate identical to the one used for deposition was always subtracted from the CD-RAS data, to eliminate the spurious anisotropy signal due to the experimental configuration. The final agreement is fully satisfactory, with an excellent signal-to-noise ratio in the CD-RAS curve (integration time: 4 s/spectral point). The broader structures in panel A are due to the larger bandpass of the monochromator used in the CD-RAS spectrometer.

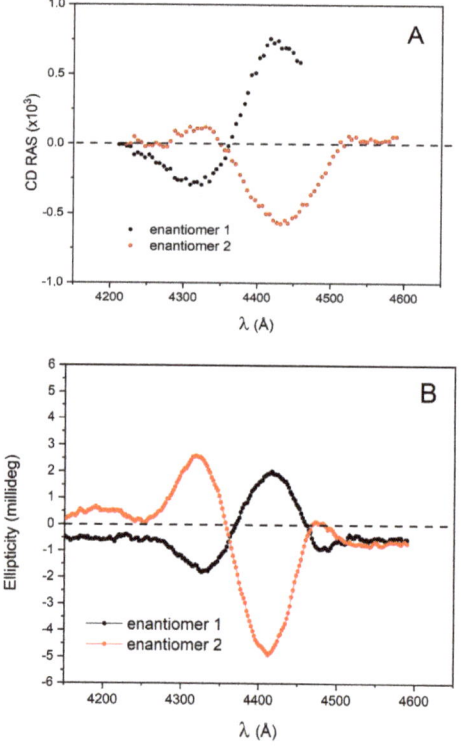

Figure 7. Panel (**A**): CD-RAS experimental spectra for sample C1 (red curve: enantiomer 1) and sample C2 (black curve: enantiomer 2) (see text). The two enantiomers were obtained from a 50 µL droplet of a ZnOEP solution in dichloromethane deposited onto a glass substrate and measured in transmission mode. The resulting film has an average thickness estimated in the range of 10–15 nm from atomic force microscope images (not shown). Panel (**B**): Ellipticity spectra measured on the same samples by a commercial spectrometer for circular dichroism for sample C1 (red curve: enantiomer 1) and sample C2 (black curve: enantiomer 2).

4. Materials and Methods

Details about preparation, kinetics and theoretical explanation of samples A1, A2, and B1 are reported in a recent review about the stereospecific self-assembly processes of porphyrin–proline conjugates [23]. In detail, sample A1 contains II-type aggregates of ZnP(L)Pro(−) in hydroalcoholic solution (EtOH:water 25:75, 5 µM); sample A2 contains aggregates of $H_2P(L)Pro(-)$ in hydroalcoholic solution (EtOH:water 25:75, 10 µM); sample

B contains I-type aggregates of ZnP(D)Pro(−) in hydroalcoholic solution (EtOH:water 25:75, 5 µM). Solid samples were prepared as reported in Ref. [25]. In detail, 50 µL of a 0.1 mM solution in toluene of ZnP(D)Pro(−) or ZnP(L)Pro(−) was cast on a ultraflat glass slide to obtain samples C1 and C2, respectively.

The solutions and casted films were analyzed by a JASCO J-1500 Circular Dichroism Spectrophotometer, equipped with a thermostated cell holder set at 298 K and purged with ultra-pure nitrogen gas. The slit width was set to 2 nm and the scanning speed to 20 mdeg/min in continuous scanning mode. Linear dichroism contribution (LD) was found to be <0.0004 DOD units with respect to the baseline in all the cases examined.

AFM measurements were performed in air using a NanoSurf NaioAFM instrument. Experiments were carried out in tapping mode by using silicon tips with a spring constant of about 48 N/m and a curvature radius of less than 10 nm. The film thickness was estimated as the height differences between the molecule layer and the glass slide, made by excavating across the spot with a needle previously dipped in dichloromethane. The average height of the structures emerging from the glass is approximately 12 ± 3 nm, in a region of 30×30 µm^2.

5. Conclusions

In this article, we have shown that destructuring a spectrometer for circularly polarized light (CPL) by adapting a RAS system—normally used for linearly polarized light—to CPL, a flexible apparatus is now available to measure, with high sensitivity, the circular dichroism of solutions and very thin solid state films. As a consequence, a new class of experiments would then become accessible as an investigation in real time of the chiral thin film growth from solution drops deposited onto a transparent substrate (that on a vertical glass would undoubtedly glide toward the bottom, while the CD-RAS system can be mounted in a vertical plane, with the sample horizontal). The development of different approaches for the characterization of chiral layers is an important opportunity: the application of a modified RAS spectrometer for CD measurements could represent a significant boost for research in this field.

Supplementary Materials: The following are available online at https://www.mdpi.com/article/10.3390/molecules28083471/s1, (A): Optical spectra; (B) Rationalization of the relation between the commercial CD unit and the RAS-CD.

Author Contributions: Conceptualization and design of the study, C.G., R.P. and D.M.; methodology, A.S., M.F. and C.G.; synthesis of the title porphyrins, I.D.F.; preparation of the porphyrin samples and drop-casting on glass, I.D.F. and R.M.; Circular dichroism measurements, R.M. and I.D.F.; CD-RAS experiments, I.T., B.B. and C.G.; AFM experiments, R.M.; writing—original draft preparation, C.G. and R.P.; writing—review and editing, C.G., M.S., G.M. and R.P.; responsible for the financial support to the work, R.P. All authors have read and agreed to the published version of the manuscript.

Funding: This research was funded by MUR PRIN, grant number 2017EKCS35_002, by University of Rome Tor Vergata, Italy, project ASPIRE E84I20000220005.

Data Availability Statement: The data presented in this study and Supplementary Materials are available on request from the corresponding author.

Conflicts of Interest: The authors declare no conflict of interest.

Sample Availability: Samples of the compounds are available on request from the authors.

References

1. Gilp, J.F.M. Epioptics: Linear and non-linear optical spectroscopy of surfaces and interfaces. *J. Phys. Cond. Matt.* **1990**, *2*, 7985.
2. Weightman, P.; Martin, D.S.; Cole, R.J.; Farrell, T. Reflection anisotropy spectroscopy. *Rep. Progr. Phys.* **2005**, *68*, 1251. [CrossRef]
3. Goletti, C. Reflectance Anisotropy Spectroscopy. In *Encyclopedia of Interfacial Chemistry: Surface Science and Electrochemistry*; Wandelt, K., Ed.; Elsevier: Amsterdam, The Netherlands, 2018; pp. 413–420.
4. Kamiya, I.; Aspnes, D.E.; Florez, L.T.; Harbison, J.P. Reflectance-difference spectroscopy of (001) GaAs surfaces in ultrahigh vacuum. *Phys. Rev. B* **1992**, *46*, 15894. [CrossRef] [PubMed]

5. Hofmann, P.; Rose, K.C.; Fernandez, V.; Bradshaw, A.M.; Richter, W. Study of Surface States on Cu(110) Using Optical Reflectance Anisotropy. *Phys. Rev. Lett.* **1995**, *75*, 2039. [CrossRef] [PubMed]
6. Goletti, C.; Bussetti, G.; Arciprete, F.; Chiaradia, P.; Chiarotti, G. Infrared surface absorption in Si(111)2x1 observed with reflectance anisotropy spectroscopy. *Phys. Rev. B* **2002**, *66*, 153307. [CrossRef]
7. Shkrebtii, A.I.; Esser, N.; Richter, W.; Schmidt, W.G.; Bechstedt, F.; Fimland, B.O.; Kley, A.; Del Sole, R. Reflectance anisotropy of GaAs (100): Theory and experiment. *Phys. Rev. Lett.* **1998**, *81*, 721. [CrossRef]
8. Navarro-Quezada, A.; Ghanbari, E.; Wagner, T.; Zeppenfeld, P. Molecular reorientation during the initial growth of perfluoropentacene on Ag (110). *J. Phys. Chem. C* **2018**, *122*, 12704. [CrossRef]
9. Fazi, L.; Raimondo, L.; Bonanni, B.; Fanfoni, M.; Paolesse, R.; Sgarlata, A.; Sassella, A.; Goletti, C. Unveiling the robustness of porphyrin crystalline nanowires toward aggressive chemicals. *Eur. Phys. J. Plus (Open Access)* **2022**, *137*, 300. [CrossRef]
10. Goletti, C.; Bussetti, G.; Chiaradia, P.; Sassella, A.; Borghesi, A. Highly sensitive optical monitoring of molecular film growth by organic molecular beam deposition. *Appl. Phys. Lett.* **2003**, *83*, 4146. [CrossRef]
11. Sassella, A.; Campione, M.; Raimondo, L.; Borghesi, A.; Bussetti, G.; Cirilli, S.; Violante, A.; Goletti, C.; Chiaradia, P. Pseudomorphic growth of organic semiconductor thin films driven by incommensurate epitaxy. *Appl. Phys. Lett.* **2009**, *94*, 073307. [CrossRef]
12. Di Natale, C.; Goletti, C.; Paolesse, R.; Della Sala, F.; Drago, M.; Chiaradia, P.; Lugli, P.; D'Amico, A. Optical anisotropy of Langmuir–Blodgett sapphyrin films. *Appl. Phys. Lett.* **2000**, *77*, 3164. [CrossRef]
13. Goletti, C.; Paolesse, R.; Dalcanale, E.; Berzina, T.; Di Natale, C.; Bussetti, G.; Chiaradia, P.; Froiio; Costa, L.C.M.; D'Amico, A. Thickness Dependence of the Optical Anisotropy for Porphyrin Octaester Langmuir–Schaefer Films. *Langmuir* **2002**, *18*, 6881. [CrossRef]
14. Goletti, C. Reflectance anisotropy at the solid–liquid interface. In *Reference Module in Chemistry, Molecular Sciences and Chemical Engineering*; Wandelt, K., Bussetti, G., Eds.; Elsevier: Amsterdam, The Netherlands, 2023. Available online: https://scitechconnect.elsevier.com/resources/reference-module-chemistry-molecular-sciences-chemical-engineering/ (accessed on 1 March 2023).
15. Goletti, C.; Bussetti, G.; Violante, A.; Bonanni, B.; Di Giovannantonio, M.; Serrano, G.; Breuer, S.; Gentz, K.K.; Wandelt, K. Thickness Dependence of the Optical Anisotropy for Porphyrin Octaester Langmuir–Schaefer Films. *J. Phys. Chem. C* **2015**, *119*, 1782. [CrossRef]
16. Barati, G.; Solokha, V.; Wandelt, K.; Hingerl, K.; Cobet, C. Chloride-induced morphology transformations of the Cu (110) surface in dilute HCl. *Langmuir* **2014**, *30*, 14486. [CrossRef]
17. Lininger, A.; Palermo, G.; Guglielmelli, A.; Nicoletta, G.; Goel, M.; Hinczewski, M.; Strangi, G. Chirality Light–Matter Interaction. *Adv. Mater.* **2022**, 2107325. [CrossRef] [PubMed]
18. IUPAC. *Compendium of Chemical Terminology*, 2nd ed.; McNaught, A.D., Wilkinson, A., Eds.; The "Gold Book"; Blackwell Scientific Publications: Oxford, UK, 1997.
19. Moss, G.P. Basic terminology of stereochemistry (IUPAC Recommendations 1996). *Pure Appl. Chem.* **1996**, *68*, 2193–2222. [CrossRef]
20. Monti, D. Recent Advancements in Chiral Porphyrin Self-Assembly. In *Synthesis and Modification of Porphy-Rinoids—Topics in Heterocyclic Chemistry*; Paolesse, R., Ed.; Springer: Berlin/Heidelberg, Germany, 2014; Volume 33, pp. 231–291.
21. Berova, N.; Nakanishi, K. *Circular Dichroism: Principles and Applications*, 2nd ed.; Berova, N., Nakanishi, K., Woody, R.W., Eds.; Wiley-VCH: New York, NY, USA, 2000; pp. 337–382.
22. Pescitelli, G. ECD exciton chirality method today: A modern tool for determining absolute configurations. *Chirality* **2022**, *34*, 333–363. [CrossRef] [PubMed]
23. Stefanelli, M.; Magna, G.; Di Natale, C.; Paolesse, R.; Monti, D. Stereospecific Self-Assembly Processes of Porphyrin-Proline Conjugates: From the Effect of Structural Features and Bulk Solvent Properties to the Application in Stereoselective Sensor Systems. *Int. J. Mol. Sci.* **2022**, *23*, 15587. [CrossRef]
24. Stefanelli, M.; Magna, G.; Zurlo, F.; Caso, M.F.; Di Bartolomeo, E.; Antonaroli, S.; Venanzi, M.; Paolesse, R.; Di Natale, C.; Monti, D.; et al. Selectivity of Porphyrin-ZnO Nanoparticle conjugates. *ACS Appl. Mater. Interfaces* **2019**, *11*, 12077–12087. [CrossRef]
25. Magna, G.; Traini, T.; Naitana, M.L.; Bussetti, G.; Domenici, F.; Paradossi, G.; Venanzi, M.; Di Natale, C.; Paolesse, R.; Monti, D. Seeding Chiral Ensembles of Prolinated Porphyrin Derivatives on Glass Surface: Simple and Rapid Access to Chiral Porphyrin Films. *Front. Chem.* **2022**, *9*, 804893. [CrossRef]
26. The Manual of the CD Spectrometer Model J-1100/1500, Section "Principles of CD Operation". p. 1. Available online: https://jascoinc.com/products/spectroscopy/circular-dichroism/ (accessed on 1 March 2023).
27. Bussetti, G.; Goletti, C.; Chiaradia, P.; Derry, T. Optical gap between dangling-bond states of a single-domain diamond C (111)-2× 1 by reflectance anisotropy spectroscopy. *Europhys. Lett.* **2007**, *79*, 57002. [CrossRef]
28. Azzam, R.M.A.; Bashara, N.M. *Ellipsometry and Polarized Light*; North Holland: Amsterdam, The Netherlands, 1977.
29. Berkovits, V.L.; Kiselev, V.A.; Safarov, V.I. Optical spectroscopy of (110) surfaces of III–V semiconductors. *Surf. Sci.* **1989**, *211*, 489. [CrossRef]
30. Aspnes, D.E.; Studna, A.A. Anisotropies in the above—Band-gap optical spectra of cubic semiconductors. *Phys. Rev. Lett.* **1985**, *54*, 1956. [CrossRef] [PubMed]

Disclaimer/Publisher's Note: The statements, opinions and data contained in all publications are solely those of the individual author(s) and contributor(s) and not of MDPI and/or the editor(s). MDPI and/or the editor(s) disclaim responsibility for any injury to people or property resulting from any ideas, methods, instructions or products referred to in the content.

Article

Increasing Reaction Rates of Water-Soluble Porphyrins for ^{64}Cu Radiopharmaceutical Labeling

Mateusz Pęgier [1],*, Krzysztof Kilian [1] and Krystyna Pyrzynska [2]

[1] Heavy Ion Laboratory, University of Warsaw, Pasteura 5A, 02-093 Warsaw, Poland
[2] Faculty of Chemistry, University of Warsaw, Pasteura 1, 02-093 Warsaw, Poland
* Correspondence: pegier@slcj.uw.edu.pl

Abstract: Searching for new compounds and synthetic routes for medical applications is a great challenge for modern chemistry. Porphyrins, natural macrocycles able to tightly bind metal ions, can serve as complexing and delivering agents in nuclear medicine diagnostic imaging utilizing radioactive nuclides of copper with particular emphasis on ^{64}Cu. This nuclide can, due to multiple decay modes, serve also as a therapeutic agent. As the complexation reaction of porphyrins suffers from relatively poor kinetics, the aim of this study was to optimize the reaction of copper ions with various water-soluble porphyrins in terms of time and chemical conditions, that would meet pharmaceutical requirements and to develop a method that can be applied for various water-soluble porphyrins. In the first method, reactions were conducted in a presence of a reducing agent (ascorbic acid). Optimal conditions, in which the reaction time was 1 min, comprised borate buffer at pH 9 with a 10-fold excess of ascorbic acid over Cu^{2+}. The second approach involved a microwave-assisted synthesis at 140 °C for 1–2 min. The proposed method with ascorbic acid was applied for radiolabeling of porphyrin with ^{64}Cu. The complex was then subjected to a purification procedure and the final product was identified using high-performance liquid chromatography with radiometric detection.

Keywords: porphyrins; copper-64; positron emission tomography

Citation: Pęgier, M.; Kilian, K.; Pyrzynska, K. Increasing Reaction Rates of Water-Soluble Porphyrins for ^{64}Cu Radiopharmaceutical Labeling. *Molecules* **2023**, *28*, 2350. https://doi.org/10.3390/molecules28052350

Academic Editors: M. Amparo F. Faustino, Carlos J. P. Monteiro and Carlos Serpa

Received: 5 January 2023
Revised: 28 February 2023
Accepted: 1 March 2023
Published: 3 March 2023

Copyright: © 2023 by the authors. Licensee MDPI, Basel, Switzerland. This article is an open access article distributed under the terms and conditions of the Creative Commons Attribution (CC BY) license (https://creativecommons.org/licenses/by/4.0/).

1. Introduction

Porphyrins are natural macrocycles that are able to tightly complex metal ions. The application of porphyrins and their complexes with metal ions in non-invasive diagnostics is gaining increasing importance which allows early detection of disturbances or pathological changes caused by diseases [1,2]. Porphyrins are potent fluorophores, biologically compatible, and are known to preferentially accumulate in tumor tissue [1]. These compounds occurring in nature include the well-known, red-colored heme in hemoglobin, which is responsible for the transport of blood as well as the chlorophylls in some bacteria and plants which are utilized for photosynthesis [3]. Inspired by their role in nature, porphyrins and metalloporphyrins are used not only as agents for tumor diagnostics but also as photosensitizers in photodynamic therapy of cancer [4], contrast agents in magnetic resonance imaging [5], photodynamic antimicrobial chemotherapy [6], or enzyme models in bioinorganic chemistry [7].

Single-photon emission computed tomography (SPECT) and positron emission tomography (PET) are both highly sensitive nuclear imaging techniques, where small amounts of the radioactive tracer as prepared radiopharmaceutical are used to produce images of a given body part. SPECT uses gamma emitters injected into the bloodstream and subsequently taken up by certain tissues [8]. In healthy people, the distribution of the dossed radiopharmaceutical follows the physiological pattern, while in the case of some dysfunction, regions with increased or decreased content can be identified. In PET the short-lived radionuclides with the emission of positron are utilized [9]. Positron, after losing kinetic energy, interacts with an electron in an annihilation reaction, resulting in the emission of two gamma photons at an angle of 180 degrees. The distribution of the radiotracer within different tissues, according to the carrier molecule, can be calculated based on the determination of the point of that annihilation.

Generally, the radiopharmaceutical used in SPECT or PET consists of four parts: radionuclide, bifunctional ligand, linker, and target molecule. The target molecules (small proteins, peptides, fragments of monoclonal antibody) are able to transport complex radionuclides to the diseased tissue containing the appropriate target receptor. The complexes used in nuclear medicine have to exhibit high thermodynamic stability as a strong interaction between the metal and the ligand is necessary to ensure the complete complexation of the radionuclide and to limit transmetalation reaction with competing species [10–12]. Moreover, a radioisotope–ligand complex should exhibit high kinetic inertness to prevent the dissociation of the complex and thus, the release of the radionuclide in the biological medium. The bifunctional ligands are covalently attached to the targeting molecule either directly or through various linkers [3]. The linker can be a simple hydrocarbon chain to increase lipophilicity, a peptide sequence to improve hydrophilicity, or a poly(ethyleneglycol) linker to slow extraction by hepatocytes. The selection of suitable chelators can facilitate the development of an effective PET imaging probe by improving targeting properties. Thus, there is a continuous need for new efficient chelators that can satisfy all the requirements.

The choice of radionuclide for diagnostic radiopharmaceuticals in PET and SPECT depends on its nuclear properties (type of radiation, half-life, energy) as well as radionuclide production, the conditions for radiolabeling, and specific activity. ^{64}Cu, with its half-time of 78.4 h, low positron energy (653 keV), and a short average tissue penetration range (0.7 mm), has received attention for radiopharmaceutical development due to its favorable nuclear decay properties that make it useful in the labeling of antibodies for immuno-PET applications [13]. It undergoes multiple decay paths, allowing not only PET imaging through positron emission but also offering the possibility of treatment due to the emission of β^- radiation (theranostics) [13,14]. DOTA (1,4,7,10-tetra- azacyclododecane-1,4,7,10-tetra-acetic acid) derivatives are mostly used to synthesize bifunctional ligands able to tightly bind the copper ions [15,16]. Porphyrins offer potential for application in nuclear medicine techniques, as ^{64}Cu is a potent radionuclide that can serve as a diagnostic and therapeutic agent. With their natural affinity for tumor tissues, porphyrins can also act as targeting molecules for ^{64}Cu, which can efficiently deliver it to pathological tissues. As potent complexing ligands able to tightly bind ^{64}Cu ions, porphyrins can serve in radiopharmaceutical synthesis [2].

Porphyrins are known for forming very stable complexes with metal ions, which is advantageous in view of possible radiopharmaceutical applications [2,10]. Chelating properties are not significantly affected by the type and number of substituents in the porphyrin ring, which allows the choice of a ligand with appropriate characteristics, such as hydrophilicity/hydrophobicity or partition coefficient octanol/water, for the intended purpose of imaging. The main drawback of their coordination reaction is relatively poor kinetics at room temperature, thus heating under reflux conditions was required over an extended period of time [17]. The rate-limiting steps are determined by the kinetic barrier of the metal insertion step into the planar porphyrin structure and its relative rigidity.

To increase the rate of the formation of metalloporphyrins with medium-sized ions (e.g., Cu(II)) different approaches have been adopted. The rate-limiting step in the metalation of these ions is the deformation of the porphyrin ring. One of the interesting methods is the use of large ions (e.g., Hg(II), Pb(II), Cd(II)) that relatively easily form coordination complexes [18–20]. As these cations are too big to be incorporated completely into a porphyrin ring, they form a complex in which the ion "sits" on the top of the porphyrin macrocycle. It causes much faster distortion of the originally planar ring of the porphyrin ligand and allows relatively easy access for smaller ions (e.g., Cu(II)) from the other side. Yet, as far as the radiopharmaceutical application is concerned, the use of these highly toxic metal ions during synthesis is unfavorable. Another way to implement this approach is to use a reducing agent for Cu(II) such as ascorbic acid, glutathione, or hydroxylamine [18,21].

Using a microwave synthesis unit, higher radiochemical yields can be achieved in shorter reaction time and small quantities of solvents employed [22–24]. Mechanochemistry and ultrasounds have been also recently proposed for the development of a new method for metalloporphyrin synthesis [25–27]. The synthesis of metal complexes under mechanochemical action requires the use of excess metal salt and the addition of NaOH solution to achieve moderate yields. Under ultrasound irradiation using water as a solvent, only an alkaline medium is necessary. The solventless mechanochemical synthesis of copper(II) complex of 5,10,15,20-tetrakis(N-methylpyridinium-4-yl)porphine tetraiodide (Cu-TMPyP) was obtained in 45% yield after 60 min of milling [26]. The synthesized metalloporphyrin required purification to remove the excess metal salt through exclusion chromatography using water as an eluent, followed by water evaporation. The Cu-TMPyP complex under 30 min ultrasound was obtained in 79% yield (determined by HPLC) and 53% isolated yield.

The aim of this work was the optimization of the synthesis of copper complexes with water-soluble porphyrins with cationic and anionic functional groups for potential application in radiopharmaceutical synthesis. Efforts were focused on developing the fastest possible method that would meet the requirements of labeling with short-lived copper isotopes used in PET. An optimized method was applied for the synthesis of the ^{64}Cu–porphyrin complex.

2. Results and Discussion

Studied porphyrins included 5,10,15,20-tetrakis(N-methylpyridinium-4-yl)porphine tetratosylate (TMPyP), 5,10,15,20-tetrakis(4-sulfonato phenyl)porphyrin (TSPP) and for microwave synthesis also 5,10,15,20-tetrakis(4-carboxy phenyl)porphyrin (TCPP) (Figure 1).

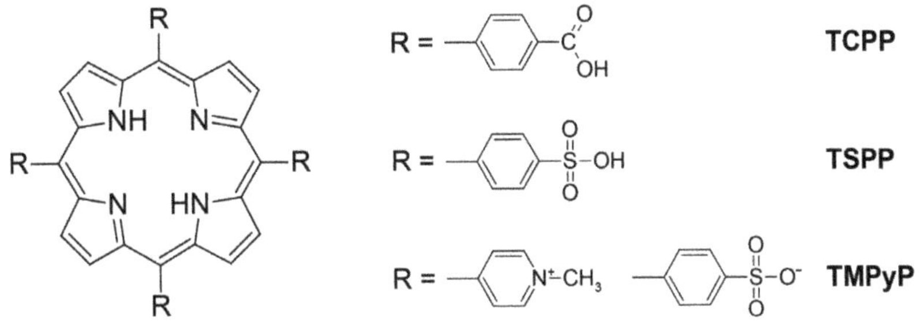

Figure 1. Structures of studied porphyrins.

Electronic spectra of studied porphyrins are typical for this group of compounds [28,29]. The spectrum consists of the most intense band named the Soret band coming from the S_0-S_2 transition and four Q bands corresponding to split S_0-S_1 transitions. The formation of porphyrin complexes (structure presented in Figure 2) causes characteristic changes in electronic spectra. The number of Q bands decreases as symmetry break caused by core nitrogen protons no longer exists. The presence of metal ions also causes shifts of Soret and remaining Q bands. In the case of copper complexes, the shift in the Soret band is relatively small. Shifted Q(0,1) band is characteristic of the Cu complex of all studied porphyrins.

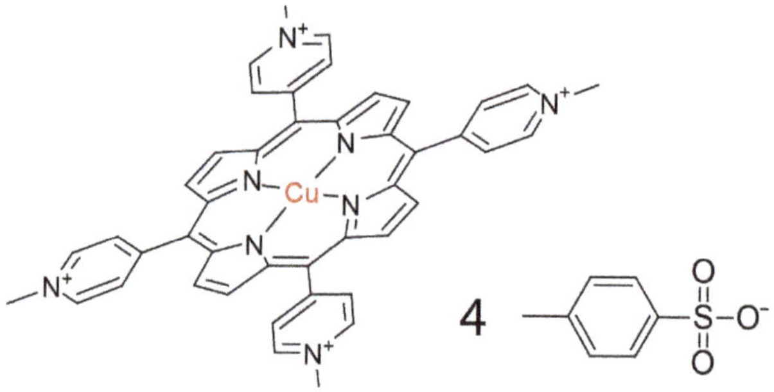

Figure 2. Structure of the copper(II) porphyrin complex on the example of Cu-TMPyP.

Spectra presented in Figure 3 are typical for unprotonated porphyrins (pH 9). They are almost identical to those at pH 7 (with a minimal decrease in absorbance). The change in spectrum in the acidic region refers to the protonation of porphyrin with average pKa~5, depending on functional groups attached to the porphyrin ring. In the case of TMPyP, protonation is not yet visible in the spectrum at pH 4. However, both TSPP and TCPP become protonated at pH 4 as the Soret band is shifted to a higher wavelength and the intensity order of Q bands is inverted. The intensity of all bands is generally higher than in neutral and alkali media. These spectral changes can be observed as a change in color to deep green and it makes spectrophotometry a convenient method for monitoring of porphyrin reaction course. Spectra of highly protonated porphyrins are presented in Figure S1. TCPP tends to precipitate from the solution.

Cu–porphyrin complexes were synthesized in a water environment at three different pH values: acidic pH 4 (acetate buffer), neutral pH 7 (phosphate buffer), and basic pH 9 (borate buffer) to find optimal conditions. Only in the case of TCPP at pH 4 porphyrin tended to precipitate from the solution, especially after heating, so these conditions were not included in studies. Application of a water environment is highly desirable for possible radiopharmaceutical synthesis, as organic solvents generally must be thoroughly separated from the product, which unnecessarily lengthens and complicates the purification step. Different pH values were tested as it has been proven that acid–base equilibria can strongly influence the complexation of porphyrins. The [Cu]:[Porphyrin] ratio for all syntheses was 1:1.

As porphyrins tend to form aggregates in solutions, a study had to be performed to define the appropriate concentration of reagents for spectrophotometric monitoring of complexation reactions. It was completed by recording spectra of porphyrin solutions at different concentrations and analysis of the dependence of absorbance vs. concentration. The deviations from linearity may suggest the formation of aggregates that could influence the identity of the final product, which is important in radiopharmaceutical synthesis and quality control. The high absorbance (well above 1 absorbance unit) of the Soret band made it unavailable to use the most intense porphyrin band, as linearity was quickly violated due to high absorption. For this reason, aggregation was studied by plotting the absorbance of the most intense band in the Q region of the porphyrin spectrum, namely Q1 for all except from TSPP at pH 4, where the Q1 band is virtually invisible and the Q4 band exhibits the highest absorption. Examples of fitted data are presented in Figure S2. Results show that in the studied concentration range, porphyrins show no signs of aggregation, as R^2 for all the plots reaches at least 0.999. The only exception is TSPP at pH 4, where the upper limit is 1.5×10^{-5} mol L^{-1}. Thus, 10^{-5} mol L^{-1} has been chosen as the optimal concentration for all further studies, as it is far enough from the upper limit and made it possible to study even small changes in the spectrum.

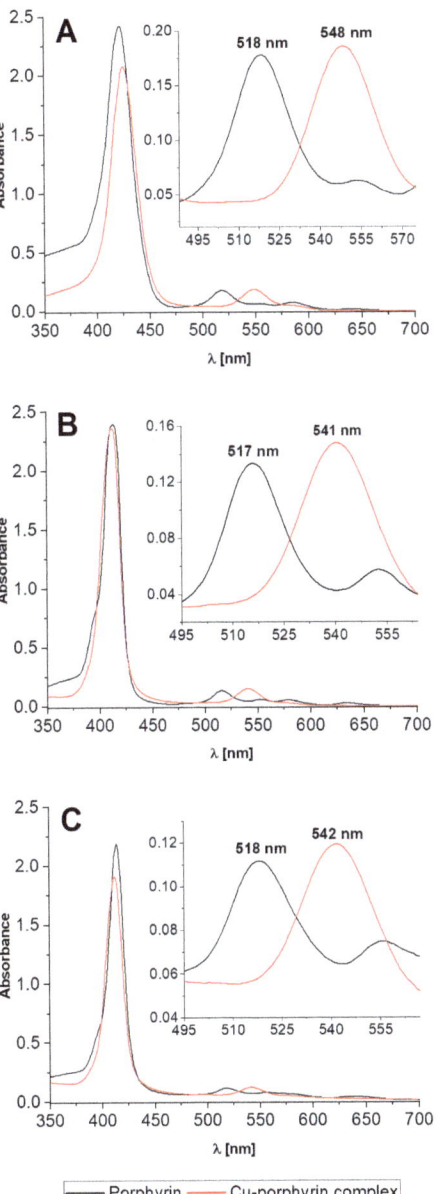

Figure 3. Spectra of studied porphyrins and corresponding Cu complexes: (**A**) TSPP, (**B**) TMPyP, (**C**) TCPP. [Cu-porphyrin] = [Porphyrin] = 10^{-5} mol L^{-1}; pH 9.

Acceleration of the synthesis of Cu(II)–porphyrin complex with the use of large ions can also be realized with the use of a reducing agent [19]. As far as the radiopharmaceutical application is concerned the choice of an acceptable reductant is important. In our study, ascorbic acid (AA) was employed (Figure 4). The reaction was conducted at a sufficiently high pH to maintain a deprotonated porphyrin ring. In the first rapid step, copper(II) is reduced to copper(I). As the ionic radius of Cu$^+$ (96 pm) is larger than that of Cu^{2+} (72 pm), a complex with Cu$^+$ ions remaining out of the ligand plane is formed. This leads to the deformation of the porphyrin ring. Subsequent incorporation of Cu^{2+} ion involves more favorable kinetics leading to almost immediate reaction in optimal conditions.

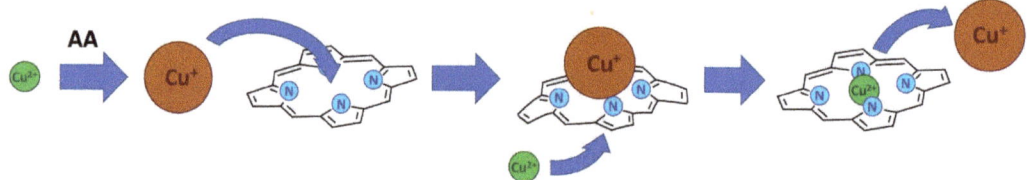

Figure 4. Overview scheme of synthesis of Cu(II)–poprhyrin complex involving formation of complex with large ions generated by reduction of Cu^{2+} ion to Cu$^+$ ion with ascorbic acid (AA).

Synthesis of Cu-TSPP complex in the presence of ascorbic acid (AA) as a reducing agent was running at the highest rate at pH 9 (Figure 5). Since the [AA]:[Cu] ratio equals to 5:1. maximum of complex absorbance is reached in 1 min. In the case of pH 7, a 20-fold excess of AA is needed to reach a maximum in 10 min or 50-fold (maximum is reached in 8 min). Reaction at pH 4 is the slowest and 20 min is not enough for complete synthesis of the Cu-TSPP complex. The reaction rate for TMPyP is generally higher than for TSPP with maximum absorbance reached at all pH values. Similarly to TSPP, at pH 9 the reaction is the fastest with an almost instant process, and maximum absorbance is reached between mixing reagents and recording first spectrum.

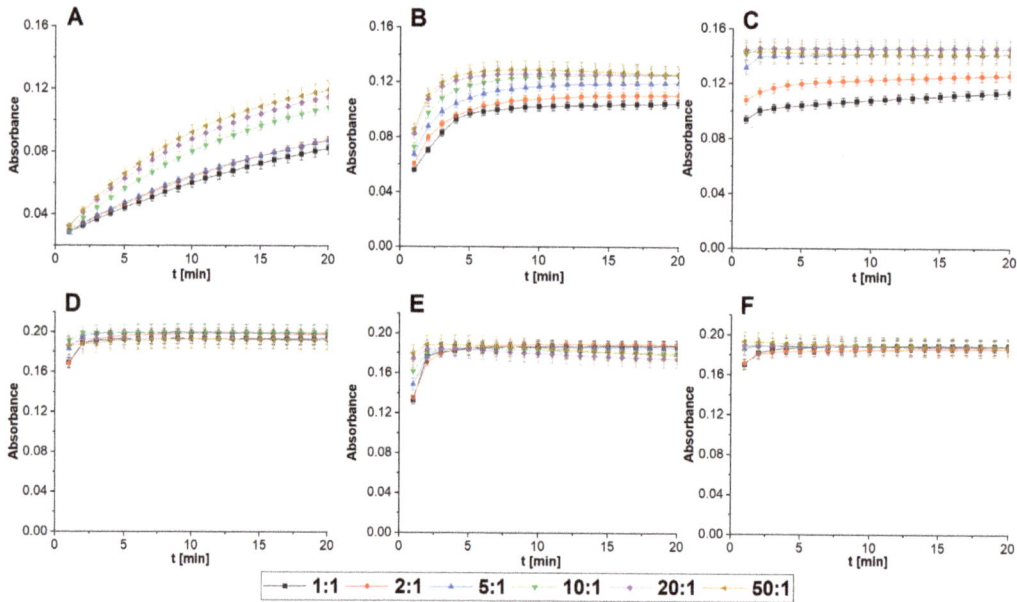

Figure 5. Complexation of Cu with porphyrins in a presence of AA at various [AA]:[Cu] ratios: (**A**) TSPP (pH 4); (**B**) TSPP (pH 7); (**C**) TSPP (pH 9); (**D**) TMPyP (pH 4); (**E**) TMPyP (pH 7); (**F**) TMPyP (pH 9). [Cu^{2+}] = [TSPP] = [TMPyP] = 10^{-5} mol L^{-1}.

Contrary to TSPP, complexation at pH 4 is only slightly inferior to pH 9. From the 10-fold excess of AA upwards the reaction time is about 1 min. For both porphyrins, basic pH led to the deprotonation of pyrrolic nitrogen atoms, which besides the SAT mechanism additionally promoted a high reaction rate of complexation. Strong differences in the behavior of both porphyrins at pH 4 come from the stronger protonation of the porphyrin ring in the case of TSPP. TMPyP remains in a more reactive form with weakly bound protons that can easily be substituted by the incorporation of copper into the π system of porphyrin molecules. As for the other porphyrins, in the case of TCPP, which was the object of the previous study [19], the complexation at pH 9 is the fastest. An almost immediate reaction was achieved with a 10-fold excess of AA. For all studied porphyrins, a significant increase was observed, as the reaction time at room temperature without AA was more than 100 min (Figure S3).

The porphyrin ring is notably resistant to harsh conditions, such as high temperatures. For this reason, to increase the reaction rate microwave-assisted synthesis was applied. Microwave heating provides uniform energy transfer that prevents samples from overheating near reaction vessel walls and the possibility of precise control of reaction parameters. In order to maximize microwave energy, transfer continuous air cooling of the reaction vessel was applied. This allowed for a reduction of the reaction time to a minimum, which is vital for possible radiopharmaceutical synthesis. The obtained results are presented in Figure 6.

Microwave heating drastically reduced the time required for the synthesis of the Cu–porphyrin complex without the addition of any other chemicals, which is important in pharmaceutical preparations. Generally, 1 min was enough to reach the maximum absorbance of the corresponding complex. Only in the case of TMPyP after 2 min small increase was observed. Further elongation of synthesis did not cause any positive change. In the case of Cu-TCPP, a decrease in complex absorbance was noticeable, which shows the limited stability of the product. Compared to conventional heating, where in many cases 60 min was necessary to complete labeling [2], microwave-assisted synthesis showed a significant increase in reaction rate.

To assess the usefulness of the proposed complexation method, the reaction of TCPP with radioactive ^{64}Cu ($t_{1/2}$ = 12.4 h) was performed. As mentioned earlier, the complexation of macrocyclic ligands often requires harsh conditions that can be destructive for the targeting molecules due to irreversible modification of their structure. Thus, the rapid and stable complexation in mild conditions is a scientific challenge for the production of radiopharmaceuticals.

The ^{64}Cu isotope was produced by proton irradiation of ^{64}Ni in a medical cyclotron, then separated and finally dried in the form of ^{64}CuCl$_2$. Before further experiments, it was dissolved in deionized water. This form, without any purification, was used for labeling according to the procedure developed, achieving satisfactory yield with good radiochemical purity. Direct complexation resulted in a 78.7% yield at room temperature immediately. Further purification using solid phase extraction with an anion exchange column eliminated free ^{64}Cu and increased the radiochemical purity up to 99.0%. It was confirmed by radio-HPLC where ^{64}Cu was eluted in 2.3 min, while ^{64}CuTCPP was eluted in 4.0 min referring to the radiometric detector signal (Figure S4).

Peaks are identified according to retention time (Figure 7) and specific spectra (Figure S5). This confirmed that the complexation of Cu(II) with the porphyrin ligand is fast and occurs within a single minute at pH = 9.

An interesting behavior was observed during purification on the anion exchange column. The complex could be eluted only in a specific sequence of eluents. After loading the column, water was used to remove water-soluble and cationic impurities (^{64}Cu^{2+}, buffers, and other metal ions). After that, gentle acidification (2 mL 0.1 mol L^{-1} HCl) of the column was required to remove the complex quantitatively (>98.5%) with 0.5 mL of ethanol. Any other sequences of eluents, including concentrated acids (1 mol L^{-1} HCl), acidification of the solution before the solid phase extraction step, and elution only with organic solvents, resulted in lower recovery or complex decomposition due to high concentration of acids. The explanation could be the mixed sorption mechanism; anion exchange for carboxylic groups combined with π–π interactions between the porphyrin ring and the solid phase of the column. Acidification breaks the ionic interactions and aprotic solvent releases the complex.

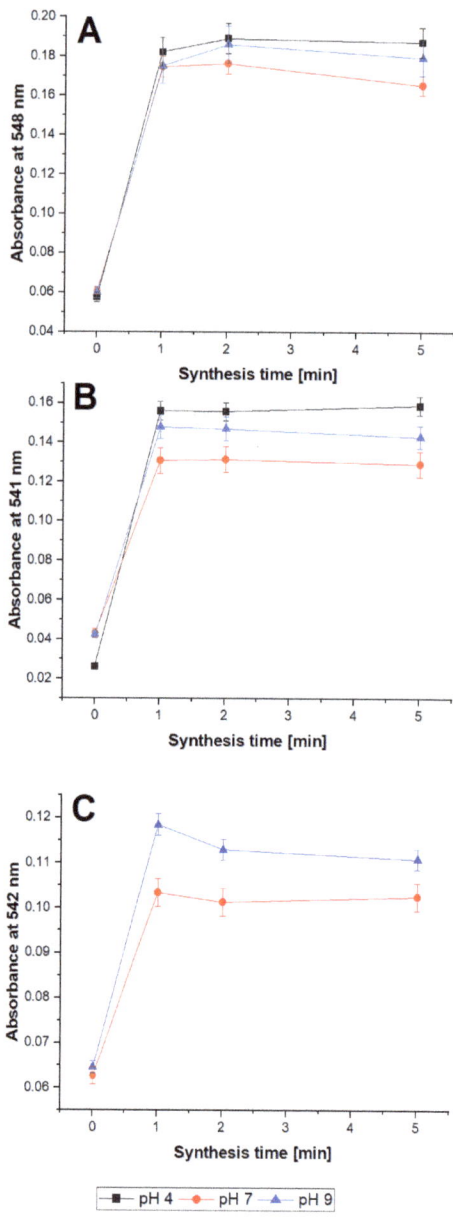

Figure 6. Microwave-assisted synthesis of Cu complex with: (**A**) TMPyP, (**B**) TSPP, (**C**) TCPP. $[Cu^{2+}] = [TSPP] = [TMPyP] = [TCPP] = 10^{-5}$ mol L^{-1}.

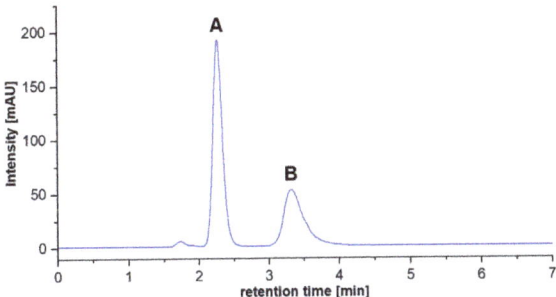

Figure 7. Separation of (**A**) TCPP (retention time 2.2 min) and (**B**) Cu-TCPP complex (retention time 3.3 min). Conditions: Phenomenex Gemini C18 column (150 mm × 4.0 mm i.d., 10 µm), mobile phase: 40:60 acetonitrile:0.05 mol L^{-1} CH$_3$COOH/CH$_3$COONa, pH 4.8, 1 mL min^{-1} flow rate.

Another favorable phenomenon can be observed in the chromatographic separation of the post-reaction mixture. Other divalent metal ions may compete in the complexation process [30,31]. According to the supplier's declaration for ^{64}Cu^{2+}, the solution may contain Ni(II) in a two-fold excess and Zn(II) ions in an approx. 40-fold excess vs. copper-64. As porphyrin ligands are used in excess they are not competitive. However, due to formal requirements in the development of radiopharmaceuticals, it may be necessary to remove these metal ions. The proposed HPLC method also allows for the separation of impurities in the form of Ni-TCPP and ZnTCPP complexes [28].

3. Reagents and Methods

3.1. Reagents

5,10,15,20-tetrakis(4-carboxyphenyl)porphyrin (TCPP) and 5,10,15,20-tetrakis(N-methylpyridinium-4-yl)porphine tetratosylate salt (TMPyP) were obtained from Sigma-Aldrich. 5,10,15,20-tetrakis(4-sulfonatophenyl)porphyrin tetraammonium salt was obtained from PorphyChem. Stock solutions (10^{-3} mol L^{-1}) were prepared by dissolution of the appropriate amount of porphyrin in 5 mL of specific solvent: 3.59 mg of TCPP in 0.02 mol L^{-1} NaOH, 6.82 mg of TMPyP in 0.05 mol L^{-1} HCl and 5.02 mg of TSPP in deionized water.

Amounts of 0.1 mol L^{-1} borate buffer (pH 9) and phosphate buffer (pH 7) were prepared by dissolution of boric acid or sodium dihydrogen phosphate in water. Acetate buffer (pH 4) was prepared by dilution of 99.5% acetic acid with water. Further pH adjustment was performed using concentrated NaOH solution under control of Hanna HI 2210 pH meter (Hanna Instruments Inc., Smithfield, VA, USA) with combined glass electrode.

An amount of 1000 mg L^{-1} standard of Cu(II) ions as nitrates in 2–5% nitric acid (Merck) was further diluted with deionized water.

3.2. Study of Porphyrin Aggregation

Series of 2.5 × 10^{-6}–2 × 10^{-5} mol L^{-1} porphyrin solutions at different pH were prepared and spectra were recorded. Then, linear fitting of the data using least squares method was performed and R^2 factor was calculated to check the linearity of obtained results.

3.3. Direct Synthesis of Cu–porphyrin Complexes

Reactions were monitored with the Perkin Elmer Lambda 20 UV–VIS spectrophotometer (Perkin Elmer Inc., Waltham, MA, USA) with UV WinLab V2.85 software. Spectra were recorded in a range of 350–700 nm in polystyrene cuvettes of 1 cm length. To perform general synthesis at room temperature, 40 µL of stock porphyrin solution (10^{-3} mol L^{-1}) was mixed with pH buffer and then 25 µL of stock Cu solution (100 mg L^{-1} in nitric acid) was added ([Cu] = [Porphyrin] = 10^{-5} mol L^{-1}). Final volume of the sample was 4 mL. Then, spectra were recorded for 4 h.

3.4. Increasing Reaction Rate with the Use of Ascorbic Acid at Room Temperature

The study was conducted at three pH values—4, 7, and 9 at room temperature. Final concentrations of copper and porphyrins were 10^{-5} mol L^{-1}. To study the influence of ascorbic acid (AA) on the complexation the concentration of AA was changed to achieve desired excess in relation to Cu. To carry out the study 40 µL of porphyrin was added to the proper buffer. Then, an appropriate volume of ascorbic acid and 25 µL of copper was added consecutively to reach desired excess of ascorbic acid in relation to Cu. Following excesses of AA were studied: 1, 2, 5, 10, 20, 50. For this reason, two stock solutions of ascorbic acid (AA) in acetate buffer were prepared by dissolution of solid ascorbic acid (Merck KGaA, Darmstadt, Germany): 0.05 mol L^{-1} for 50 and 20-fold excess of AA and 0.001 < for 1, 2, 5 and 10-fold excess of AA. Final concentrations of Cu(II) and porphyrin were 10^{-5} mol L^{-1}. Reaction was monitored spectrophotometrically for 20 min. Due to low stability of AA, stock solutions wereprepared every day.

3.5. Microwave-Assisted Synthesis of Cu–porphyrin Complexes (without Ascorbic Acid)

Microwave-assisted synthesis was conducted without the addition of the AA due to its lack of thermal stability using CEM Discover SP microwave synthesizer (CEM Corporation, Matthews, NC, USA), which heated the sample with automatic real-time control of reaction parameters to prevent overheating and rapid increase in the pressure. Synthesis was performed in sealed 10 mL glass vessels with magnetic stirring. Temperature was set to 140 °C. The amounts of reagents were the same as for direct syntheses. In order to increase microwave power transfer continuous air cooling of the reaction vessel was applied. Reaction time was 1, 2, or 5 min. After cooling step sample was released and a spectrum was recorded.

3.6. Radiosynthesis of ^{64}CuTCPP Complexes

^{64}Cu was purchased from PET radioisotopes manufacturer (Voxel, Cracow, Poland) and was produced in ^{64}Ni(p,n)^{64}Cu reaction conducted on ALCEO solid target system (Comecer, Italy) coupled with GE PETrace 840 cyclotron (GE Healthcare Technologies Inc., Chicago, IL, USA). The requested activity was about 200 MBq in the form of dried ^{64}CuCl$_2$ reconstituted on-site with 1 mL of water. Stock porphyrin solution, buffers, and 0.05 mol L^{-1} ascorbic acid (Merck KGaA, Darmstadt, Germany) were prepared as described above. Water and phosphate-buffered saline (PBS) were of pharmaceutical grade. ^{64}CuTCPP was synthesized at room temperature by mixing 200 µL of ^{64}CuCl$_2$ solution, 400 µL of TCPP, and 200 µL AA stock solutions, filled up with a borate buffer up to 5 mL, and vortexed. For final purification and formulation solution was loaded on anion exchange Waters Sep-Pak Accell Plus QMA column (Waters Corporation, Milford, CT, USA) with an ISM833 Ismatec peristaltic pump (VWR, Radnor, PA, USA) at 1 mL min^{-1}, flushed with 5 mL of water at 2 mL min^{-1}, acidified with 1 mL of 0.1 mol L^{-1} HCl and finally eluted with 0.5 mL of ethanol. After evaporation near dryness, small aliquot of PBS was added to reach expected activity concentration. An Atomlab 500 dose calibrator and wipe tester (Biodex, Shirley, NY, USA) was used for activity determination.

3.7. Determination of Radiochemical Purity of ^{64}CuTCPP Complexes

Radiochemical purity was analyzed with the chromatography system Shimadzu AD20 with SPD-M20A UV–Vis photodiode array and radiometric (GabiStar, Raytest, Germany) detectors. The separation was performed on Phenomenex Gemini C18 column (150 mm × 4.0 mm i.d., 10 µm), with 40:60 acetonitrile: 0.05 M CH$_3$COOH/CH$_3$COONa adjusted to pH 4.8 as a mobile phase and 1 mL min^{-1} flow rate.

4. Conclusions

Studied porphyrins form stable complexes with copper(II) ions. The application of both methods, one with the use of ascorbic acid and the second with microwave heating drastically reduced the time required for the synthesis of Cu(II)–porphyrin complexes.

In the case of the method with AA for all studied porphyrins, optimum conditions for fast and efficient complexation at room temperature comprise at least 10-fold excess of AA over Cu. Reactions with TSPP and TCPP should be conducted at pH 9, while for TMPyP pH 4 also offers a satisfying reaction rate. A method with microwave heating would be applicable only for thermally resistant molecules. If the porphyrin would act as a complexing agent attached to other, thermally sensitive targeting moiety, the method with the use of AA would be more convenient. Both proposed methods are convenient for application in radiopharmaceutical synthesis as they are one step, fast, and no toxic or harsh chemicals are used that would have to be separated. Additionally, they could be potentially adopted for the synthesis of radiopharmaceuticals labeled with short-lived copper isotopes (^{60}Cu ($t_{1/2}$ = 23.7 min), ^{61}Cu ($t_{1/2}$ = 3.3 h), and even ^{62}Cu($t_{1/2}$ = 9.67 min)).

Supplementary Materials: The following supporting information can be downloaded at: https://www.mdpi.com/article/10.3390/molecules28052350/s1, Figure S1. Spectra of protonated porphyrins. [TMPyP] = [TSPP] = [TCPP] = 10^{-5} mol L^{-1}; pH 1.5.; Figure S2: Study of aggregation of porphyrins at different pH and concentrations. Wavelength of peak maximum is given in parentheses.; Figure S3. Complexation of Cu with porphyrins with and without the addition of AA (pH and [AA]:[Cu] ratio in brackets): (A) TSPP (pH 4, 50:1); (B) TSPP (pH 7, 50:1); (C) TSPP (pH 9, 10:1); (D) TMPyP (pH 4, 10:1); (E) TMPyP (pH 7, 20:1); (F) TMPyP (pH 9, 10:1). [Cu2+] = [TSPP] = [TMPyP] = 10^{-5} mol L^{-1}.; Figure S4. Radio-chromatogram of raw product (A) and purified final product (B). Separation: Phenomenex Gemini C18 column (150 mm × 4.0 mm i.d., 10 μm), mobile phase: 40:60 acetonitrile: 0.05 mol L^{-1} CH$_3$COOH/CH$_3$COONa pH 4.8, 1 mL min^{-1} flow rate.; Figure S5. Spectra of (A) TCPP and (B) Cu-TCPP recorded at retention times: 2.2 and 3.3 min, respectively. Separation: Phenomenex Gemini C18 column (150 mm × 4.0 mm i.d., 10 μm), mobile phase: 40:60 acetonitrile: 0.05 mol L^{-1} CH$_3$COOH/CH$_3$COONa pH 4.8, 1 mL min^{-1} flow rate.

Author Contributions: Conceptualization, K.P., K.K. and M.P; supervision: K.P.; data curation, K.K. and M.P.; investigation, K.K. and M.P.; visualization, M.P.; writing—original draft, K.P., K.K. and M.P.; writing—review and editing, K.P., K.K. and M.P. All authors have read and agreed to the published version of the manuscript.

Funding: This research received no external funding.

Institutional Review Board Statement: Not applicable.

Informed Consent Statement: Not applicable.

Data Availability Statement: Data generated during the study are available on request.

Acknowledgments: The authors thank the company Voxel for providing copper-64 used for the experiments.

Conflicts of Interest: The authors declare no conflict of interest.

Sample Availability: Samples of the compounds TCPP, TSPP, TMPyP, and CuTCPP are available from the authors.

References

1. Tsolekile, N.; Nelana, S.; Oluwafemi, O.S. Porphyrin as Diagnostic and Therapeutic Agent. *Molecules* **2019**, *24*, 2669. [CrossRef]
2. Pyrzynska, K.; Kilian, K.; Pęgier, M. Porphyrins as Chelating Agents for Molecular Imaging in Nuclear Medicine. *Molecules* **2022**, *27*, 3311. [CrossRef] [PubMed]
3. Tahoun, M.; Gee, C.T.; McCoy, V.E.; Sander, P.M.; Müller, C.M. Chemistry of porphyrins in fossil plants and Animals. *RSC Adv.* **2021**, *11*, 7552–7563. [CrossRef] [PubMed]
4. Glowacka-Sobotta, A.; Wrotynski, M.; Kryjewski, M.; Sobotta, L.; Mielcarek, J. Porphyrinoids in photodynamic diagnosis and therapy of oral diseases. *J. Porphyr. Phthalocyanines* **2019**, *23*, 1–10. [CrossRef]
5. Imran, M.; Ramzan, M.; Qureshi, A.K.; Khan, M.A.; Tariq, M. Emerging applications of porphyrins and metalloporphyrins in biomedicine and diagnostics magnetic resonance imaging. *Biosensors* **2018**, *8*, 95. [CrossRef]
6. Managa, M.; Nyokong, T. Photodynamic antimicrobial chemotherapy activity of gallium tetra-(4-carboxyphenyl) porphyrin when conjugated to differently shaped platinum nanoparticles. *J. Mol. Struc.* **2015**, *1099*, 432–440. [CrossRef]
7. Woggon, W.D. Metalloporphyrines as Active Site Analogues-Lessons from Enzymes and Enzyme Models. *Acc. Chem. Res.* **2005**, *38*, 127–136. [CrossRef]

8. Israel, O.; Pellet, O.; Biassoni, L.; De Palma, D.; Estrada-Lobato, E.; Gnanasegaran, G.; Kuwert, T.; la Fougère, C.; Mariani, G.; Massalha, S.; et al. Two decades of SPECT/CT-the coming of age of a technology: An updated review of literature evidence. *Eur. J. Nucl. Med. Mol. Imaging* **2019**, *46*, 1990–2912. [CrossRef]
9. Coenen, H.H.; Ermert, J. Expanding PET-applications in life sciences with positron-emitters beyond fluorine-18. *Nucl. Med. Biol.* **2021**, *92*, 241–269. [CrossRef]
10. Amoroso, A.J.; Fallis, I.A.; Pope, S.J.A. Chelating agents for radiolanthanides: Applications to imaging and therapy. *Coord. Chem. Rev.* **2017**, *340*, 198–217. [CrossRef]
11. Price, E.W.; Orvig, C. Matching chelators to radiometals for radiopharmaceuticals. *Chem. Soc. Rev.* **2014**, *43*, 260–290. [CrossRef]
12. Okoye, N.C.; Baumeister, J.E.; Khosroshahi, F.N.; Hennkens, H.M.; Jurisson, S.S. Chelators and metal complex stability for radiopharmaceutical application. *Radiochim. Acta* **2019**, *107*, 1087–1120. [CrossRef]
13. Boschi, A.; Martini, P.; Janevik-Ivanovska, E.; Adriano Duatti, A. The emerging role of copper-64 radiopharmaceuticals as cancer theranostic. *Drug Discov. Today* **2018**, *23*, 1489–1501. [CrossRef]
14. Gutfilen, B.; Souza, S.A.L.; Valentini, G. Copper-64: A real theranostic agent. *Drug Des. Devel. Ther.* **2018**, *12*, 3235–3245. [CrossRef]
15. Mukai, H.; Wada, Y.; Watanabe, Y. The synthesis of 64Cu-chelated porphyrin photosensitizers and their tumor-targeting peptide conjugates for the evaluation of target cell uptake and PET image-based pharmacokinetics of targeted photodynamic therapy agents. *Ann. Nucl. Med.* **2013**, *27*, 625–639. [CrossRef]
16. Cai, Z.; Anderson, C.J. Chelators for copper radionuclides in positron emission tomography radiopharmaceuticals. *J. Labelled Comp. Radiopharm.* **2014**, *57*, 224–230. [CrossRef]
17. Zhou, Y.; Li, J.; Xu, X.; Zhao, M.; Zhang, B.; Deng, S.; Wu, Y. 64Cu-based radiopharmaceuticals in molecular imaging. *Technol. Cancer Res. Treat.* **2019**, *18*, 1533033819830758. [CrossRef]
18. Inamo, M.; Kamiiya, N.; Inada, Y.; Nomura, M.; Funahashi, M. Structural characterization and formation kinetics of sitting-atop (SAT) complexes of some porphyrins with copper(II) ion in aqueous acetonitrile relevant to porphyrin metalation mechanism. structures of aquacopper(II) and Cu(II)−SAT complexes as determined by XAFS spectroscopy. *Inorg. Chem.* **2001**, *40*, 5636–5644. [CrossRef]
19. Kilian, K.; Pęgier, M.; Pyrzynska, K. The fast method of Cu-porphyrin complex synthesis for potential use in positron emission tomography imaging. *Spectrochim. Acta Part A* **2016**, *159*, 123–127. [CrossRef]
20. De Luca, G.; Romero, A.; Scolaro, L.M.; Ricciardi, G.; Rosa, A. Sitting-atop metallo-porphyrin complexes: Experimental and theoretical investigations on such elusive species. *Inorg. Chem.* **2009**, *48*, 8493–8507. [CrossRef]
21. Uzal-Varela, R.; Patinec, V.; Tripier, R.; Valencia, L.; Maneiro, M.; Canle, M.; Platas-Iglesias, C.; Esteban-Gómez, D.; Iglesias, E. On the dissociation pathways of copper complexes relevant as PET imaging agents. *J. Inorg. Biochem.* **2022**, *236*, 111951. [CrossRef] [PubMed]
22. Nascimento, B.F.O.; Pineiro, M.; Rocha Gonsalves, A.M.D.A.; Silva, M.R.; Beja, A.M.; Paixão, J.A. Microwave-assisted synthesis of porphyrins and metalloporphyrins: A rapid and efficient synthetic metod. *J. Porphyr. Phthalocyanines* **2007**, *11*, 77–84. [CrossRef]
23. Pineiro, M. Microwave-assisted synthesis and reactivity of porphyrins. *Curr. Org. Synth.* **2014**, *11*, 89–109. [CrossRef]
24. Matamala-Cea, E.; Valenzuela-Godoy, F.; González, D.; Arancibia, R.; Dorcet, V.; Hamon, J.R.; Néstor Novoa, N. Efficient preparation of 5,10,15,20-tetrakis(4-bromophenyl)porphyrin. Microwave assisted v/s conventional synthetic metod X-ray and hirshfeld surface structural analysis. *J. Mol. Struct.* **2020**, *1201*, 127139. [CrossRef]
25. Ralphs, K.; Zhang, C.; James, S.L. Solventless mechanochemical metallation of porphyrins. *Green Chem.* **2017**, *19*, 102–105. [CrossRef]
26. Gomes, C.; Peixoto, M.; Pineiro, M. Modern metod for the sustainable synthesis of metalloporphyrins. *Molecules* **2021**, *26*, 6652. [CrossRef]
27. Atoyebi, A.O.; Brückner, C. Observations on the mechanochemical insertion of zinc(II), copper(II), magnesium(II), and select other metal(II) ions into porphyrins. *Inorg. Chem.* **2019**, *58*, 9631–9642. [CrossRef]
28. Gouterman, M. Study of the effects of substitution on the absorption spectra of porphin. *J. Chem. Phys.* **1959**, *30*, 1139–1161. [CrossRef]
29. Gouterman, M. Spectra of porphyrins. *J. Mol. Spectrosc.* **1961**, *6*, 138–163. [CrossRef]
30. Kilian, K.; Pyrzynska, K. Spectrophotometric study of Cd(II), Pb(II), Hg(II) and Zn(II) complexes with 5,10,15,20-tetrakis (4-carboxylphenyl)porphyrin. *Talanta* **2003**, *60*, 669–678. [CrossRef]
31. Krężel, A.; Maret, W. The biological inorganic chemistry of zinc ions. *Arch. Biochem. Biophys.* **2016**, *611*, 3–19. [CrossRef]

Disclaimer/Publisher's Note: The statements, opinions and data contained in all publications are solely those of the individual author(s) and contributor(s) and not of MDPI and/or the editor(s). MDPI and/or the editor(s) disclaim responsibility for any injury to people or property resulting from any ideas, methods, instructions or products referred to in the content.

Article

Selective Determination of Glutathione Using a Highly Emissive Fluorescent Probe Based on a Pyrrolidine-Fused Chlorin

Francisco G. Moscoso [1,*], Carla Queirós [2], Paula González [1], Tânia Lopes-Costa [1], Ana M. G. Silva [2] and Jose M. Pedrosa [1,*]

[1] Departamento de Sistemas Físicos, Químicos y Naturales, Universidad Pablo de Olavide, Ctra. Utrera km. 1, 41013 Seville, Spain
[2] LAQV-REQUIMTE, Departamento de Química e Bioquímica, Faculdade de Ciências da Universidade do Porto, 4169-007 Porto, Portugal
* Correspondence: fjgarmos@upo.es (F.G.M.); jmpedpoy@upo.es (J.M.P.)

Abstract: We report the use of a carboxylated pyrrolidine-fused chlorin (TCPC) as a fluorescent probe for the determination of glutathione (GSH) in 7.4 pH phosphate buffer. TCPC is a very stable, highly emissive molecule that has been easily obtained from meso-tetrakis(4-methoxycarbonylphenyl) porphyrin (TCPP) through a 1,3-dipolar cycloaddition approach. First, we describe the coordination of TCPC with Hg(II) ions and the corresponding spectral changes, mainly characterized by a strong quenching of the chlorin emission band. Then, the TCPC-Hg^{2+} complex exhibits a significant fluorescence turn-on in the presence of low concentrations of the target analyte GSH. The efficacy of the sensing molecule was tested by using different TCPC:Hg^{2+} concentration ratios (1:2, 1:5 and 1:10) that gave rise to sigmoidal response curves in all cases with modulating detection limits, being the lowest 40 nM. The experiments were carried out under physiological conditions and the selectivity of the system was demonstrated against a number of potential interferents, including cysteine. Furthermore, the TCPC macrocycle did not showed a significant fluorescent quenching in the presence of other metal ions.

Keywords: carboxylated pyrrolidine-fused chlorin; fluorescence; glutathione detection

Citation: Moscoso, F.G.; Queirós, C.; González, P.; Lopes-Costa, T.; Silva, A.M.G.; Pedrosa, J.M. Selective Determination of Glutathione Using a Highly Emissive Fluorescent Probe Based on a Pyrrolidine-Fused Chlorin. *Molecules* 2023, 28, 568. https://doi.org/10.3390/molecules28020568

Academic Editor: Maged Henary

Received: 22 November 2022
Revised: 16 December 2022
Accepted: 27 December 2022
Published: 5 January 2023

Copyright: © 2023 by the authors. Licensee MDPI, Basel, Switzerland. This article is an open access article distributed under the terms and conditions of the Creative Commons Attribution (CC BY) license (https://creativecommons.org/licenses/by/4.0/).

1. Introduction

Chlorins are an important class of reduced porphyrins (dihydroporphyrins) which, when in the form of Mg^{2+} complexes, can be found as the green photosynthetic pigments (chlorophylls) in plants, microalgae and cyanobacteria [1]. As natural chlorins are often unstable and difficult to handle, the preparation of synthetic analogues has been explored as a very attractive approach to obtain more robust derivatives with functional groups suitable for the desired applications [2]. Due to its structural features, chlorin derivatives show unique spectral properties including: (1) an intense absorption band at ca. 650 nm; (2) high emission quantum yield and (3) upon coordination with metal ions, several metallochlorins can be obtained with more valuable physicochemical properties that enable a broader range of applications, such as in theranostics of cancer [3–6], nonlinear optics [7–9], photoacoustic imaging [10–12] and for dye sensitised solar cells [13–15].

Considering the potential use of chlorin derivatives in biomedical applications, it is important to study their effects in vitro and in vivo systems, including their toxicity, biocompatibility and cellular uptake [16]. Nowadays the great majority of in vitro and in vivo studies are still performed using porphyrin derivatives, nevertheless similar elations can be taken towards chlorin derivatives. For sensing applications in living systems, it is crucial to study and decrease the potential toxicity of the sensors, while maintaining their photophysical properties, like high emission intensity, singlet oxygen (^{1}O$_2$) quantum yields, brightness and luminescence lifetime. In the case of porphyrin and chlorin derivatives, two related factors are extremely important for attaining a sensor with low toxicity, namely the phototoxicity and singlet oxygen generation. A method to decrease the derivatives toxicity is to disable their capacity to penetrate the cellular membranes. For example, in 2011 a dendritic benzoporphyrin was prepared to be used as in vivo oxygen probes and revealed low phototoxicity, this was associated with the bulky structure inability to penetrate the cellular membranes [17]. Another reported method, is the preparation of dyads using dyes that promote the quenching of the triplet state of tetrapyrrolic compounds by energy transfer, disabling the production of singlet oxygen [18]. Another strategy is the preparation of nanoscale structures. Recently, a chlorin-nanoscale-Metal-Organic Framework (TCPCUiO) was reported and revealed applicability both in photodynamic therapy (PDT) and photothermal therapy (PTT). TCPCUiO presented good anticancer activity against H22 tumor-bearing mice in vivo and possessed negligible systematic toxicity—with favorable non thrombogenic and biocompatible properties on blood cells, low systemic toxicities to the function of the liver and kidney and no tissue damage or inflammatory lesions were observed in all major organs [19].

Chlorins can be obtained from porphyrins by a variety of synthetic approches, including through hydrogenation [14], annulation [20] and cycloaddition [21] with diverse entities. Using 1,3-dipolar cycloaddition (1,3-DC) reaction of porphyrins with azomethine ylides [22], we recently reported the synthesis of carboxylated pyrrolidine-fused chlorin (TCPC) and its application in the successful detection of the explosive triacetone triperoxide (TATP) in the gas phase [23].

Glutathione (GSH) is the most abundant thiol in animal cells. It is a tripeptide with a peptide linkage between cysteine (Cys), glycine (Gly) and glutamate (Glu). Cellular GSH exist in the 1–10 mM concentration range whereas it is reduced to 150–200 µM in serum [24] and 1–6 µM in plasma [25–27]. The most important role of GSH in the organism is as an antioxidant agent, preventing damage caused by free radicals or reactive oxygen species (ROS). Moreover, GSH is involved in a vast regulatory process, and an abnormal concentration of GSH is directly related to severe diseases [28]. For instance, some tumours have an extraordinarily high level of GSH [29–31]. Furthermore, a GSH or GSH synthetase deficiency involves massive urinary excretion, metabolic acidosis, and/or a tendency to hemolysis in humans [28]. It has been also associated with mitocondrial disorders [32], Alzheimer's and Parkinson's diseases, among others [33–36].

For these reasons, it is essential to detect GSH to prevent or downplay the effects of the low or high levels of the tripeptide. For the determination of thiols and specifically GSH, some traditional techniques have been applied, such as high-performance liquid chromatography (HPLC) [37–39], surface enhanced Raman scattering (SERS) [39,40] mass spectroscopy (MS) [41] or electrophoresis [42,43]. Although these techniques provide high resolution for low levels of GSH, the cost of equipment, the complexity of sample preparation and extended analysis time means that these analysis methods are not practical for clinical or research purposes.

Electroanalytical methods have also been used to detect GSH [44–46]. However, these techniques are based on reducing the thiol group; the detection mechanism is not specific for GSH, and they require sample pre-treatment to isolate the analyte of interest.

On the other hand, colorimetric and fluorescent probes have become an excellent alternative for GSH detection due to their simple mechanisms, easy sample preparation, and colour changes with the naked eye when the analyte is present [32,47–53]. However, taking advantage of these properties requires selective, sensitive and robust methods. In this work, a highly emissive porphyrin analogue, a carboxylated pyrrolidine-fused chlorin (hereafter TCPC, Figure 1), has been synthesised and used as a fluorescent probe for the turn-on-based sensing of GSH fulfilling all the above requirements.

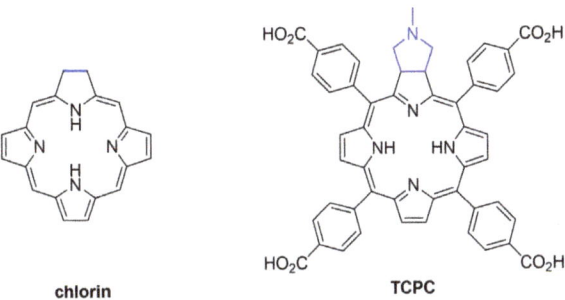

Figure 1. Structure of core chlorin and TCPC.

2. Results and Discussion

2.1. Absorption and Emission Features of TCPC

Chlorins are porphyrin derivatives that suffer a single β,β'-double bond reduction, and their spectral features are consequently similar to those of the porphyrin precursor. Here, we used the carboxylated pyrrolidine-fused chlorin (TCPC) as a fluorescent probe, which was prepared as previously reported [23]. Briefly, its synthesis involved the 1,3-dipolar cycloaddition (1,3-DC) of the porphyrin precursor meso-tetrakis(4-methoxycarbonylphenyl)porphyrin TCPP with azomethine ylide, obtained from sarcosine and paraformaldehyde, followed by methyl ester hydrolysis under alkaline conditions. The absorbance and fluorescence spectra of TCPC in ethanolic solution at different concentrations are depicted in Figure 2. The absorbance spectra show the Soret band with a main peak centred at 418 nm and a shoulder at 404 nm. This splitting of the Soret band is attributed to orbital symmetry breaking after alteration of the macrocycle structure that induced a non-degenerate electronic transition [54–62]. An enhanced intensity of the Q bands at longer wavelengths is also appreciable [63], in contrast to free-base porphyrins [64]. On the other hand, the emission spectra exhibit the characteristic 0-0 peak at 651 nm and the corresponding 0-1 vibronic band at 715 nm typical of porphyrins. However, the emission intensity of the chlorin shows a six-fold increase as compared to its porphyrinic counterpart TCPP in solutions with the same absorbance (Figures S1 and S2). Regarding the concentration effect, the inset in Figure 2 clearly shows that the linearly increase in the absorbance is accompanied by a decrease in the emission intensity for increasing concentrations of TCPC, which is due to an efficient reabsorption of the photoluminescence at 650 nm where the intense Q band is centered.

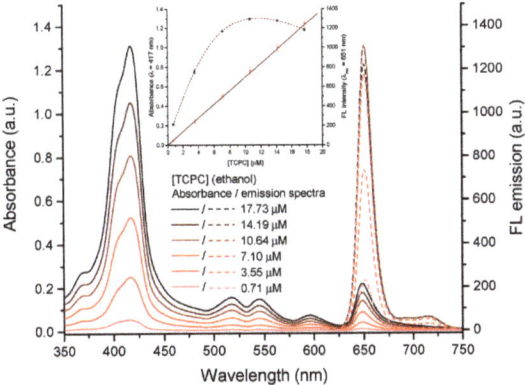

Figure 2. Absorbance (solid lines) and fluorescence (λ_{ex} = 415 nm, dashed lines) spectra of TCPC solutions in ethanol at different concentrations. Inset: Absorbance and FL intensity vs. TCPC concentration. The error bars were calculated from the standard deviation of three independent measurements.

2.2. TCPC-Hg²⁺ Complex

Prior to GSH determination experiments, we formed the TCPC-Hg^{2+} complex by adding an increasing amount of HgCl$_2$ to the chlorin solution, whose colour gradually turned green as the Hg(II) concentration increased from 0 µM to 61.4 µM (solutions a–g in Figure S3). These colour changes are produced by a red shift of the Soret band and the disappearance of the original Q bands, exhibiting a unique absorption at 620 nm, as shown in Figure 3a. Those spectral changes, typical of protonation of the porphyrin ring, are promoted by the Hg^{2+} coordination to the pyrrolic ring nitrogens, adopting a non-planar conformation [65–67]. Moreover, a fluorescence decrease is observed with the formation of the complex, as shown in Figure 3b. The non-planar conformation induced by the mercury ion coordination breaks the π-electrons conjugation of the macrocycle, promoting the excited states of the molecule to relax by non-radiative phenomena, implying a significant quenching of fluorescence emission [68]. The fluorescence quenching follows a linear relationship according to the Stern-Volmer Equation (Figure S4):

$$\frac{I_0}{I} = 1 + K_{SV}[Q] \tag{1}$$

where I_0 is the initial FL intensity at 650 nm, I is the FL intensity at the same wavelength in the presence of the quencher, K_{SV} is the Stern-Volmer constant and [Q] is the quencher concentration. The results are well-fitted to the Stern-Volmer model indicating that the stoichiometry of the complex remains constant in the working concentration range.

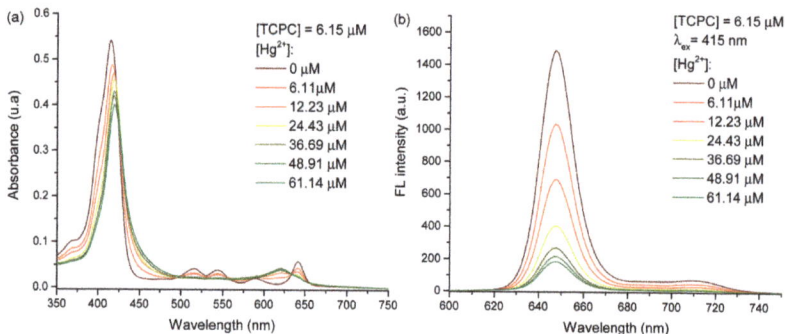

Figure 3. Absorbance (**a**) and fluorescence (λ_{ex} = 415 nm) (**b**) spectra of TCPC aqueous solutions at pH 7.4 with different concentrations of Hg^{2+}.

2.3. GSH Determination

Due to the excellent luminescent properties of TCPC, it was used as a fluorescent probe for GSH determination. For this approach, three different molar ratios of TCPC-Hg^{2+} were fixed, i.e., 1:2, 1:5 and 1:10. After the TCPC-Hg^{2+} complex formation (log K = 5.7 estimated from the absorbance data at 650 nm in Figures 2 and 3, for TCPC:Hg^{2+} = 1:2), increasing amounts of GSH were gradually added to the complex solution. Subsequently, the Hg^{2+} ions are bonded to the thiol groups of two different GSH molecules, forming a [Hg(GSH)$_2$]$^{4-}$ complex [69–71] (log K = 33.4 [72]) (Figure S5). Firstly, the free Hg^{2+} ions coordinate to GSH, causing almost no change in the absorption and fluorescence spectra. Once the mercury ions in excess have been coordinated, the remaining GSH, breaks the TCPC-Hg^{2+} complex due to the higher binding constant of the [Hg(GSH)$_2$]$^{4-}$ complex. This reaction liberates the TCPC molecules triggering the fluorescence turn on, as shown in Figure 4 for a TCPC-Hg^{2+} 1:10 molar ratio. In particular, Figure 4a shows that the absorption spectrum is progressively recovered to its initial state as the analyte is added to the sensing solution, confirming the proposed mechanism. At the same time, the corresponding fluorescence spectra (Figure 4b) show an increase of the emission intensity that is used as the response signal. These changes are also visually observed in Figure S6. Once the reaction is completed and the TCPC molecules are quantitatively in their free form, no additional spectral changes are produced for further addition of GSH. A similar behaviour is found when the experiment is performed using the TCPC:Hg^{2+} ratios of 1:2 and 1:5 (Figures S7 and S8, respectively), with the main difference being the minimum amount of GSH needed to obtain the required spectral changes. Therefore, the TCPC-Hg^{2+} ratio determines the minimum amount of analyte to be detected as explained below.

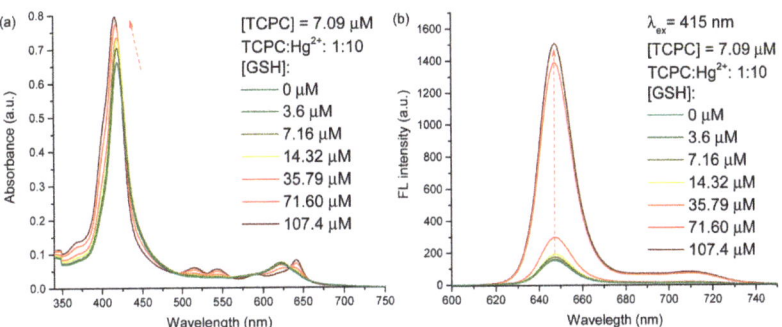

Figure 4. Absorbance (**a**) and fluorescence (λ_{ex} = 415 nm) (**b**) spectra of TCPC-Hg^{2+} complex aqueous solutions (ratio 1:10) at pH 7.4 with different concentrations of GSH.

Figure 5 depicts the normalized fluorescence at the maximum emission wavelength (650 nm, λ_{ex} = 415 nm) of the TCPC molecule vs. the GSH concentration added to the solution for the three-selected TCPC:Hg^{2+} concentration ratios.

As can be seen, the intensity of the signal follows a sigmoidal trend in all cases. Deeper inspection reveals that the position of the curves and the length of the corresponding tails before the intensity takes off are given by the excess amount of Hg^{2+} ions in the media. In this sense, as the TCPC:Hg^{2+} concentration ratio decreases (lower amount of free Hg^{2+} ions in the media) the minimum and maximum concentrations of GSH that can be detected decreases. In all cases, the experimental data can be fitted to the followed equation for calibration purposes:

$$FL = \frac{\alpha}{1 + \beta e^{-kc}} + FL_{res} \qquad (2)$$

where c is the GSH concentration, α, β, and k are empirical parameters, and FL_{res} is the residual fluorescence from non-coordinated TCPC in the absence of GSH. As can be seen in Figure 5, excellent values for the regression coefficient were obtained in all cases. Additionally, the corresponding values for FL_{res} (0.372, 0.179 and 0.108 for TCPC:Hg^{2+} 1:2, 1:5 and 1:10, respectively) increase as the TCPC:Hg^{2+} ratio decreases. This effect is expected for sufficiently low TCPC:Hg^{2+} ratios due to the presence of non-coordinated insensitive TCPC molecules in the media and reduces the GSH concentration range that can be detected, although it corresponds to low concentration regimes that can be more interesting for certain applications.

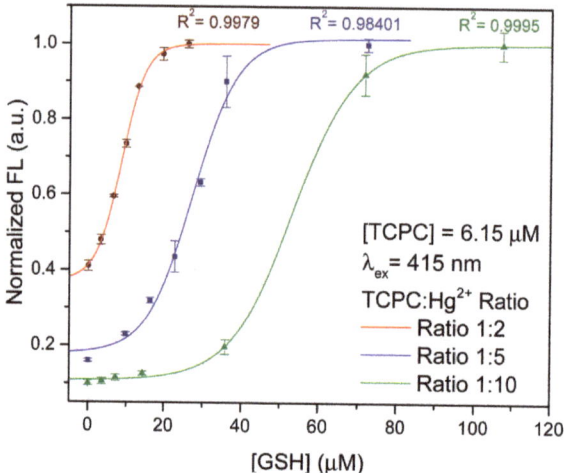

Figure 5. Fluorescence intensity changes when a different amount of GSH is added to (red) 1:2, (blue) 1:5, and (green) 1:10 molar ratio of TCPC-Hg^{2+} complex aqueous solutions at pH 7.4. The error bars were calculated from the standard deviation of three independent measurements.

These considerations are better understood if we focus on an essential feature of any analytical method like the limit of detection (LOD). Traditionally, in the context of simple measurements where the signal varies linearly with the amount of analyte, the linear regression method is used and the LOD is defined as 3σ/slope, where σ is the standard deviation of the intercept. However, the complexity of the data that analytical systems can provide for incoming samples leads to situations where the LOD cannot be calculated as reliably as before. In this way, different strategies to calculate the LOD could be considered for optical sensors with a sigmoidal response. In this case, the LOD can be defined as a quarter of the maximum slope of the curve [73], giving rise to two different values of LOD. The first one is the lower quantity of analyte that can be detected, namely the lower limit of detection (LOD_{low}). The second one is the upper limit of detection, LOD_{up}, and represents the maximum amount of analyte that can be detected being the analyte concentration at which the sensor signal is saturated. The LOD_{low} and LOD_{up} of the different TCPC:Hg^{2+} molar ratios tested in this study are shown in Figure 6. The green shadowed area represents the [GSH] determination region corresponding to the different concentrations of GSH that the sensor can determine reliably. As can be seen, the TCPC:Hg^{2+} ratio to be employed determines both the LOD_{low} and LOD_{up} values and the [GSH] range lying between them. Although these values could be extended to lower and higher GSH concentration regimes by using more extreme TCPC:Hg^{2+} ratios (see Figure S9 for a TCPC:Hg^{2+} 1:100 molar ratio), those used here allow the GSH determination in plasma [24] and serum [25–27] samples. Additionally, Figure S9 demonstrates the effective recovery of the TCPC emission after addition of 10 mM GSH. Moreover, applications where sub-micromolar GSH concentrations need to be determined can be covered by this chlorin-based fluorescent probe, with a LOD_{low} for TCPC:Hg^{2+} 1:2 as low as 40 nM. This value, susceptible to further reduction by reducing the TCPC:Hg^{2+} ratio, is significantly lower than those found in most of the existing literature using fluorescent probes [32,74–81] and very similar to those reporting the lowest values [53,67,75]. These results are summarized in Table S1.

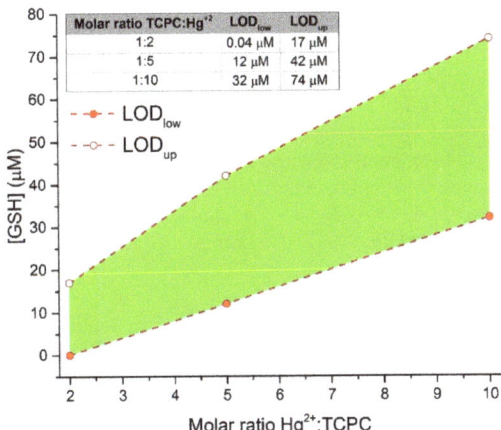

Figure 6. Variation of the LOD_{low} and LOD_{up} again the molar ratio of Hg^{2+}-TCPC. The green region represents the determination range. Inset: Table of values of LOD_{low} and LOD_{up} in µM for the different ratios.

Finally, we carried out the selectivity study selecting two families of potential interferents: metal ions that could compete with Hg^{2+} to quench the chlorin emission, and competing thiols and other common anions (ligands) present in biological samples. Figure 7a shows the response of TCPC against ten equivalents of different metal ions. As can be seen, only Hg^{2+} is able to quench the fluorescence of the TCPC by the mechanism discussed above, even with ten-fold concentrations of the potential interferents. Those results reveal the high selectivity of the TCPC towards mercury ions. On the other hand, Figure 7b shows the results for cysteine (cys), Cl^-, acetate and phosphate dibasic (HPO_4^{2-}). In this experiment, GSH is ten times more concentrated than cys in order to simulate physiological concentrations, while the other ions are in 1 equivalent to the GSH. As can be seen, only cys can produce an important interference in the proposed mechanism. In fact, cys is the most plausible interferent in an intracellular or extracellular medium. As discussed above, this amino acid is part of the structure of glutathione, and the thiol group could coordinate with mercury ions in the media, giving rise to a false positive. However, in the human organism, the concentration of cys is between 30 and 50 times lower than that of GSH [25,82] and therefore, it does not represent real interferent in detecting GSH.

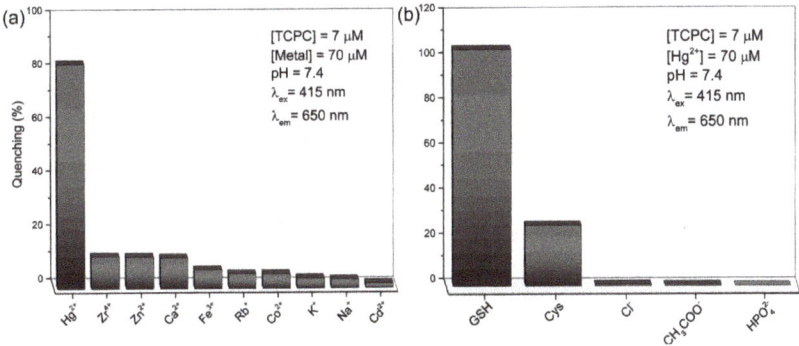

Figure 7. (a) Fluorescence quenching of TCPC in the presence of different metal ions. The experimental conditions are shown in the figure. (b) Recovery of the TCPC fluorescence when GSH, cys and other ions were added. The experimental conditions are shown in the figure. GSH and cys are in physiological concentration. Other ions are in 1 equivalent to the GSH.

3. Materials and Methods

3.1. Chlorin, Reagents and Instrumentals

The synthetic procedure and characterization (including ^{1}H NMR, ^{13}C NMR, and HRMS) for TCPC is described elsewhere [23]. Glutathione and mercury(II) chloride ($HgCl_2 > 99.5\%$) were purchased from Sigma-Aldrich. Other reagents and solvents were purchased as reagent-grade and used without further purification. UV-visible absorbance spectra were recorded using a Cary 100 UV-Vis spectrophotometer (Agilent Technologies, Santa Clara, CA, USA). In addition, fluorescence (FL) emission and excitation spectra were recorded with a Hitachi F-7000 Fluorescence Spectrophotometer (Hitachi, Tokyo, Japan).

3.2. Sensing Experiments

GSH fluorescence determination was carried out by increasing the GSH concentration in a TCPC-Hg^{2+} complex aqueous solution to obtain a calibration curve. To ensure the maximum formation of TCPC-Hg^{2+}, the experiments were performed with different TCPC-Hg^{2+} molar ratios (1:2, 1:5 and 1:10). All experiments were carried out by fixing the pH to 7.4 using a phosphate buffer ($H_2PO_4^-$/HPO_4^{2-}) 0.01 M. All determination assays were carried out at $\lambda_{ex} = 415$ nm, close to the absorption peak of TCPC.

4. Conclusions

The chlorin TCPC exhibits appealing emissive features that prompted us to use it in chemical sensing. In particular, we employed the TCPC molecule as a fluorescent probe to detect different concentrations of GSH with high reliability and sensitivity, being able to detect even 40 nM of GSH in physiological conditions. For this approach, a controlled amount of $HgCl_2$ is added to the solution to form the non-emissive Hg^{2+} complex which, in turn, is broken in the presence of GSH by the formation of the $[Hg(GSH)_2]^{4-}$ complex, releasing free TCPC molecules and triggering the fluorescence turn on. This increase in the fluorescence intensity accurately fits a sigmoidal curve as a function of the GSH concentration, which enables the proposed method to perform quantitative analysis. Furthermore, the specificity of TCPC for the Hg^{2+} coordination and the selectivity of the TCPC-Hg^{2+} complex against GSH was demonstrated.

Supplementary Materials: The following supporting information can be downloaded at: https://www.mdpi.com/article/10.3390/molecules28020568/s1, Figure S1: Absorption spectra of a methanol solution of meso-tetrakis(4-carboxyphenyl)porphyrin (TCPP) at different concentrations. Inset: linear fitting of the maximum absorbance vs. [TCPP].; Figure S2: Fluorescence emission spectra ($\lambda_{ex} = 415$ nm) of a methanol solution of meso-tetrakis(4-carboxyphenyl)porphyrin (TCPP) at different concentrations. Inset: FL emission intensity vs. [TCPP].; Figure S3: TCPC ethanolic solutions with different concentrations of Hg^{2+} (from A to G: 0, 6.11, 12.23, 24.43, 36.69, 48.91, 61.14 µM). See the main text for further details.; Figure S4: TCPC ([TCPC] = 6.15 µM) fluorescence dependence with Hg^{2+} concentration (black dots) and its Stern-Volmer equation linear fitting (red line), $\lambda_{ex} = 415$ nm, $\lambda_{em} = 650$ nm; Figure S5: Schematic representation of the complex $[Hg(GSH)_2]^{4-}$ (see [69] in main article).; Figure S6: Photographs of cuvettes containing a TCPC aqueous solution (left), its corresponding TCPC–Hg^{2+} complex (center) and after addition of an excess of GSH (right) under visible (a) and UV (b) light.; Figure S7: Absorbance (left) and fluorescence ($\lambda_{ex} = 415$ nm) (right) spectra of TCPC–Hg^{2+} complex aqueous solutions (ratio 1:2) at pH 7.4 with different concentrations of GSH.; Figure S8: Absorbance (left) and fluorescence ($\lambda_{ex} = 415$ nm) (right) spectra of TCPC–Hg^{2+} complex aqueous solutions (ratio 1:5) at pH 7.4 with different concentrations of GSH.; Table S1: Limit of detection and response rates of various fluorescent sensors toward GSH.

Author Contributions: Writting—original draft: F.G.M. and C.Q.; Investigation: P.G., F.G.M. and C.Q.; Supervision: J.M.P., T.L.-C. and A.M.G.S.; Formal analysis: F.G.M.; Funding acquisition: A.M.G.S. and J.M.P.; Conceptualization: J.M.P., T.L.-C. and A.M.G.S.; Resources: A.M.G.S. and J.M.P. All authors have read and agreed to the published version of the manuscript.

Funding: This research has been funded by the Spanish AEI/MCIN/10.13039/501100011033 within the NextGenerationEU/PRTR funds through the projects PCI2020-112241 (M-ERA.NET 2019 project7106,

SALMOS) and PID2019-110430 GB-C22 (ADLIGHT). ERDF (80%) and Andalusian CTEICU/JA in the framework of the Operative Programme FEDERAndalucia 2014–2020 through projects P20 01258 (objective 01) and UPO-1381028 (objective 1.2.3.) also contributed to the present research. This work received support from PT national funds (FCT/MCTES, Fundação para a Ciência e Tecnologia and Ministério da Ciência, Tecnologia e Ensino Superior) through the projects UIDB/50006/2020, UIDP/50006/2020 and EXPL/QUI-OUT/1554/ 2021. A. M. G. Silva thanks FCT (Fundação para a Ciência e Tecnologia) for funding through program DL 57/2016–Norma Transitória.

Institutional Review Board Statement: Not applicable.

Informed Consent Statement: Not applicable.

Data Availability Statement: Not applicable.

Conflicts of Interest: The authors declare no conflict of interest.

References

1. Senge, M.O.; Sergeeva, N.N.; Hale, K.J. Classic highlights in porphyrin and porphyrinoid total synthesis and biosynthesis. *Chem. Soc. Rev.* **2021**, *50*, 4730–4789. [CrossRef] [PubMed]
2. Taniguchi, M.; Lindsey, J.S. Synthetic Chlorins, Possible Surrogates for Chlorophylls, Prepared by Derivatization of Porphyrins. *Chem. Rev.* **2017**, *117*, 344–535. [CrossRef] [PubMed]
3. Laranjo, M.; Aguiar, M.C.; Pereira, N.A.; Brites, G.; Nascimento, B.F.; Brito, A.F.; Casalta-Lopes, J.; Gonçalves, A.C.; Sarmento-Ribeiro, A.B.; Pineiro, M.; et al. Platinum(II) ring-fused chlorins as efficient theranostic agents: Dyes for tumor-imaging and photodynamic therapy of cancer. *Eur. J. Med. Chem.* **2020**, *200*, 112468. [CrossRef]
4. Singh, S.; Aggarwal, A.; Bhupathiraju, N.D.K.; Jovanovic, I.R.; Landress, M.; Tuz, M.P.; Gao, R.; Drain, C.M. Comparing a thioglycosylated chlorin and phthalocyanine as potential theranostic agents. *Bioorg. Med. Chem.* **2020**, *28*, 115259. [CrossRef]
5. Srivatsan, A.; Wang, Y.; Joshi, P.; Sajjad, M.; Chen, Y.; Liu, C.; Thankppan, K.; Missert, J.R.; Tracy, E.; Morgan, J.; et al. In Vitro Cellular Uptake and Dimerization of Signal Transducer and Activator of Transcription-3 (STAT3) Identify the Photosensitizing and Imaging-Potential of Isomeric Photosensitizers Derived from Chlorophyll-a and Bacteriochlorophyll-a. *J. Med. Chem.* **2011**, *54*, 6859–6873. [CrossRef] [PubMed]
6. De Annunzio, S.R.; Costa, N.C.S.; Graminha, M.A.; Fontana, C.R.; Mezzina, R.D. Chlorin, phthalocyanine, and porphyrin types derivatives in phototreatment of cutaneous manifestations: A review. *Int. J. Mol. Sci.* **2019**, *20*, 3861. [CrossRef] [PubMed]
7. Zhou, X.; Chen, Y.; Su, J.; Tian, X.; Luo, Y.; Luo, L. In situ second-harmonic generation mediated photodynamic therapy by micelles co-encapsulating coordination nanoparticle and photosensitizer. *RSC Adv.* **2017**, *7*, 52125–52132. [CrossRef]
8. Khadria, A.; de Coene, Y.; Gawel, P.; Roche, C.; Clays, K.; Anderson, H.L. Push–pull pyropheophorbides for nonlinear optical imaging. *Org. Biomol. Chem.* **2017**, *15*, 947–956. [CrossRef]
9. Mrinalini, M.; Naresh, M.; Prasanthkumar, S.; Giribabu, L. Porphyrin-based supramolecular assemblies and their applications in NLO and PDT. *J. Porphyr. Phthalocyanines* **2021**, *25*, 382–395. [CrossRef]
10. Jiao, X.; Zhang, W.; Zhang, L.; Cao, Y.; Xu, Z.; Kang, Y.; Xue, P. Rational design of oxygen deficient TiO_{2-x} nanoparticles conjugated with chlorin e6 (Ce6) for photoacoustic imaging-guided photothermal/photodynamic dual therapy of cancer. *Nanoscale* **2020**, *12*, 1707–1718. [CrossRef]
11. Zhu, Q.H.; Zhang, G.H.; Yuan, W.L.; Wang, S.L.; He, L.; Yong, F.; Tao, G.H. Handy fluorescent paper device based on a curcumin derivative for ultrafast detection of peroxide-based explosives. *Chem. Commun.* **2019**, *55*, 13661–13664. [CrossRef] [PubMed]
12. Zheng, X.; Wang, L.; Liu, S.; Zhang, W.; Liu, F.; Xie, Z. Nanoparticles of Chlorin Dimer with Enhanced Absorbance for Photoacoustic Imaging and Phototherapy. *Adv. Funct. Mater.* **2018**, *28*, 1706507. [CrossRef]
13. Tamiaki, H.; Hagio, N.; Tsuzuki, S.; Cui, Y.; Zouta, T.; Wang, X.F.; Kinoshita, Y. Synthesis of carboxylated chlorophyll derivatives and their activities in dye-sensitized solar cells. *Tetrahedron* **2018**, *74*, 4078–4085. [CrossRef]
14. Pineiro, M.; Gomes, C.; Peixoto, M. Mechanochemical in situ generated gas reactant for the solvent-free hydrogenation of porphyrins. *Green Chem. Lett. Rev.* **2021**, *14*, 339–344. [CrossRef]
15. Wang, X.F.; Kitao, O. Natural Chlorophyll-Related Porphyrins and Chlorins for Dye-Sensitized Solar Cells. *Molecules* **2012**, *17*, 4484–4497. [CrossRef] [PubMed]
16. Ptaszek, M. Rational Design of Fluorophores for In Vivo Applications. *Prog. Mol. Biol. Transl. Sci.* **2013**, *113*, 59–108. [CrossRef] [PubMed]
17. Ceroni, P.; Lebedev, A.Y.; Marchi, E.; Yuan, M.; Esipova, T.V.; Bergamini, G.; Wilson, D.F.; Busch, T.M.; Vinogradov, S.A. Evaluation of phototoxicity of dendritic porphyrin-based phosphorescent oxygen probes: An in vitro study. *Photochem. Photobiol. Sci.* **2011**, *10*, 1056–1065. [CrossRef] [PubMed]
18. Tatman, D.; Liddell, P.A.; Moore, T.A.; Gust, D.; Moore, A.L. Carotenohematoporphyrins as Tumor-Imaging Dyes. Synthesis and In Vitro Photophysical Characterization. *Photochem. Photobiol.* **1998**, *68*, 459–466. [CrossRef]
19. Zheng, X.; Wang, L.; Liu, M.; Lei, P.; Liu, F.; Xie, Z. Nanoscale Mixed-Component Metal–Organic Frameworks with Photosensitizer Spatial-Arrangement-Dependent Photochemistry for Multimodal-Imaging-Guided Photothermal Therapy. *Chem. Mater.* **2018**, *30*, 6867–6876. [CrossRef]

20. Banerjee, S.; Phadte, A.A. β-meso-Annulated meso-Tetraarylchlorins: A Study of the Effect of Ring Fusion on Chlorin Conformation and Optical Spectra. *ChemistrySelect* **2020**, *5*, 11127–11144. [CrossRef]
21. Moura, N.M.M.; Monteiro, C.J.P.; Tomé, A.C.; Neves, M.G.P.; Cavaleiro, J.A. Synthesis of chlorins and bacteriochlorins from cycloaddition reactions with porphyrins. *Arkivoc* **2022**, *2022*, 54–98. [CrossRef]
22. Silva, A.M.G.; Tomé, A.C.; Neves, M.G.P.M.S.; Silva, A.M.S.; Cavaleiro, J.A.S. 1,3-Dipolar Cycloaddition Reactions of Porphyrins with Azomethine Ylides. *J. Org. Chem.* **2005**, *70*, 2306–2314. [CrossRef]
23. Vargas, A.P.; Almeida, J.; Gámez, F.; Roales, J.; Queirós, C.; Rangel, M.; Lopes-Costa, T.; Silva, A.M.; Pedrosa, J.M. Synthesis of a highly emissive carboxylated pyrrolidine-fused chlorin for optical sensing of TATP vapours. *Dye. Pigment.* **2021**, *195*, 109721. [CrossRef]
24. Vidyasagar, M.S.; Kodali, M.; Prakash Saxena, P.; Upadhya, D.; Murali Krishna, C.; Vadhiraja, B.M.; Fernandes, D.J.; Bola Sadashiva, S.R. Predictive and Prognostic Significance of Glutathione Levels and DNA Damage in Cervix Cancer Patients Undergoing Radiotherapy. *Int. J. Radiat. Oncol. Biol. Phys.* **2010**, *78*, 343–349. [CrossRef] [PubMed]
25. Forman, H.J.; Zhang, H.; Rinna, A. Glutathione: Overview of its protective roles, measurement, and biosynthesis. *Mol. Asp. Med.* **2009**, *30*, 1–12. [CrossRef] [PubMed]
26. Meister, A. Glutathione metabolism and its selective modification. *J. Biol. Chem.* **1988**, *263*, 17205–17208. [CrossRef]
27. Camera, E.; Picardo, M. Analytical methods to investigate glutathione and related compounds in biological and pathological processes. *J. Chromatogr. B* **2002**, *781*, 181–206. [CrossRef]
28. Townsend, D.M.; Tew, K.D.; Tapiero, H. The importance of glutathione in human disease. *Biomed. Pharmacother.* **2003**, *57*, 145–155. [CrossRef]
29. Meister, A.; Anderson, M.E. Glutathione. *Annu. Rev. Biochem.* **1983**, *52*, 711–760. [CrossRef]
30. Kennedy, L.; Sandhu, J.K.; Harper, M.E.; Cuperlovic-Culf, M. Role of Glutathione in Cancer: From Mechanisms to Therapies. *Biomolecules* **2020**, *10*, 1429. [CrossRef]
31. Estrela, J.M.; Ortega, A.; Obrador, E. Glutathione in Cancer Biology and Therapy. *Crit. Rev. Clin. Lab. Sci.* **2006**, *43*, 143–181. [CrossRef] [PubMed]
32. Hakuna, L.; Doughan, B.; Escobedo, J.O.; Strongin, R.M. A simple assay for glutathione in whole blood. *Analyst* **2015**, *140*, 3339–3342. [CrossRef] [PubMed]
33. Zeevalk, G.D.; Razmpour, R.; Bernard, L.P. Glutathione and Parkinson's disease: Is this the elephant in the room? *Biomed. Pharmacother.* **2008**, *62*, 236–249. [CrossRef] [PubMed]
34. Pinnen, F.; Sozio, P.; Cacciatore, I.; Cornacchia, C.; Mollica, A.; Iannitelli, A.; DAurizio, E.; Cataldi, A.; Zara, S.; Nasuti, C.; et al. Ibuprofen and Glutathione Conjugate as a Potential Therapeutic Agent for Treating Alzheimer's Disease. *Arch. Pharm.* **2011**, *344*, 139–148. [CrossRef] [PubMed]
35. Dauer, W.; Przedborski, S. Parkinson's Disease. *Neuron* **2003**, *39*, 889–909. [CrossRef]
36. Liu, H.; Wang, H.; Shenvi, S.; Hagen, T.M.; Liu, R.M. Glutathione Metabolism during Aging and in Alzheimer Disease. *Ann. N. Y. Acad. Sci.* **2004**, *1019*, 346–349. [CrossRef]
37. McDermott, G.P.; Francis, P.S.; Holt, K.J.; Scott, K.L.; Martin, S.D.; Stupka, N.; Barnett, N.W.; Conlan, X.A. Determination of intracellular glutathione and glutathione disulfide using high performance liquid chromatography with acidic potassium permanganate chemiluminescence detection. *Analyst* **2011**, *136*, 2578. [CrossRef]
38. Reed, D.; Babson, J.; Beatty, P.; Brodie, A.; Ellis, W.; Potter, D. High-performance liquid chromatography analysis of nanomole levels of glutathione, glutathione disulfide, and related thiols and disulfides. *Anal. Biochem.* **1980**, *106*, 55–62. [CrossRef]
39. Wei, L.; Song, Y.; Liu, P.; Kang, X. Polystyrene nanofibers capped with copper nanoparticles for selective extraction of glutathione prior to its determination by HPLC. *Microchim. Acta* **2018**, *185*, 321. [CrossRef]
40. Huang, G.G.; Hossain, M.K.; Han, X.X.; Ozaki, Y. A novel reversed reporting agent method for surface-enhanced Raman scattering; highly sensitive detection of glutathione in aqueous solutions. *Analyst* **2009**, *134*, 2468. [CrossRef]
41. Ahire, D.S.; Basit, A.; Karasu, M.; Prasad, B. Ultrasensitive Quantification of Drug-metabolizing Enzymes and Transporters in Small Sample Volume by Microflow LC-MS/MS. *J. Pharm. Sci.* **2021**, *110*, 2833–2840. [CrossRef] [PubMed]
42. Zhang, L.Y.; Sun, M.X. Fast determination of glutathione by capillary electrophoresis with fluorescence detection using β-cyclodextrin as modifier. *J. Chromatogr. B* **2009**, *877*, 4051–4054. [CrossRef] [PubMed]
43. Chang, C.W.; Tseng, W.L. Gold Nanoparticle Extraction Followed by Capillary Electrophoresis to Determine the Total, Free, and Protein-Bound Aminothiols in Plasma. *Anal. Chem.* **2010**, *82*, 2696–2702. [CrossRef] [PubMed]
44. Lee, P.T.; Goncalves, L.M.; Compton, R.G. Electrochemical determination of free and total glutathione in human saliva samples. *Sens. Actuators B Chem.* **2015**, *221*, 962–968. [CrossRef]
45. Harfield, J.C.; Batchelor-McAuley, C.; Compton, R.G. Electrochemical determination of glutathione: A review. *Analyst* **2012**, *137*, 2285. [CrossRef] [PubMed]
46. Liu, T.; Zhou, M.; Pu, Y.; Liu, L.; Li, F.; Li, M.; Zhang, M. Silver nanoparticle-functionalized 3D flower-like copper (II)-porphyrin framework nanocomposites as signal enhancers for fabricating a sensitive glutathione electrochemical sensor. *Sens. Actuators B Chem.* **2021**, *342*, 130047. [CrossRef]
47. Chen, X.; Zhou, Y.; Peng, X.; Yoon, J. Fluorescent and colorimetric probes for detection of thiols. *Chem. Soc. Rev.* **2010**, *39*, 2120. [CrossRef]

48. Sun, J.; Chen, N.; Chen, X.; Zhang, Q.; Gao, F. Two-Photon Fluorescent Nanoprobe for Glutathione Sensing and Imaging in Living Cells and Zebrafish Using a Semiconducting Polymer Dots Hybrid with Dopamine and β-Cyclodextrin. *Anal. Chem.* **2019**, *91*, 12414–12421. [CrossRef]
49. Ma, H.; Li, X.; Liu, X.; Deng, M.; Wang, X.; Iqbal, A.; Liu, W.; Qin, W. Fluorescent glutathione probe based on MnO_2–Si quantum dots nanocomposite directly used for intracellular glutathione imaging. *Sens. Actuators B Chem.* **2018**, *255*, 1687–1693. [CrossRef]
50. Jin, P.; Niu, X.; Zhang, F.; Dong, K.; Dai, H.; Zhang, H.; Wang, W.; Chen, H.; Chen, X. Stable and Reusable Light-Responsive Reduced Covalent Organic Framework (COF-300-AR) as a Oxidase-Mimicking Catalyst for GSH Detection in Cell Lysate. *ACS Appl. Mater. Interfaces* **2020**, *12*, 20414–20422. [CrossRef]
51. Huang, M.; Wang, H.; He, D.; Jiang, P.; Zhang, Y. Ultrafine and monodispersed iridium nanoparticles supported on nitrogen-functionalized carbon: An efficient oxidase mimic for glutathione colorimetric detection. *Chem. Commun.* **2019**, *55*, 3634–3637. [CrossRef] [PubMed]
52. Zhang, X.; Zhang, Q.; Yue, D.; Zhang, J.; Wang, J.; Li, B.; Yang, Y.; Cui, Y.; Qian, G. Flexible Metal-Organic Framework-Based Mixed-Matrix Membranes: A New Platform for H_2S Sensors. *Small* **2018**, *14*, 1801563. [CrossRef] [PubMed]
53. Zhang, H.; Wu, S.; Sun, M.; Wang, J.; Gao, M.; Wang, H.B.; Fang, L. In-situ formation of MnO_2 nanoparticles on Ru@SiO_2 nanospheres as a fluorescent probe for sensitive and rapid detection of glutathione. *Spectrochim. Acta Part A Mol. Biomol. Spectrosc.* **2022**, *283*, 121724. [CrossRef] [PubMed]
54. Gouterman, M. Study of the Effects of Substitution on the Absorption Spectra of Porphin. *J. Chem. Phys.* **1959**, *30*, 1139–1161. [CrossRef]
55. Harvey, P.; Stern, C.; Gros, C.; Guilard, R. The photophysics and photochemistry of cofacial free base and metallated bisporphyrins held together by covalent architectures. *Coord. Chem. Rev.* **2007**, *251*, 401–428. [CrossRef]
56. Satake, A.; Kobuke, Y. Artificial photosynthetic systems: Assemblies of slipped cofacial porphyrins and phthalocyanines showing strong electronic coupling. *Org. Biomol. Chem.* **2007**, *5*, 1679–1691. [CrossRef]
57. Durot, S.; Flamigni, L.; Taesch, J.; Dang, T.; Heitz, V.; Ventura, B. Synthesis and solution studies of silver(I)-assembled porphyrin coordination cages. *Chem.-Eur. J.* **2014**, *20*, 9979–9990. [CrossRef]
58. Telfer, S.; McLean, T.; Waterland, M. Exciton coupling in coordination compounds. *Dalton Trans.* **2011**, *40*, 3097–3108. [CrossRef]
59. Yamada, Y.; Nawate, K.; Maeno, T.; Tanaka, K. Intramolecular strong electronic coupling in a discretely H-aggregated phthalocyanine dimer connected with a rigid linker. *Chem. Commun.* **2018**, *54*, 8226–8228. [CrossRef]
60. Ribó, J.; Bofill, J.; Crusats, J.; Rubires, R. Point-dipole approximation of the exciton coupling model versus type of bonding and of excitons in porphyrin supramolecular structures. *Chem.-Eur. J.* **2001**, *7*, 2733–2737. [CrossRef]
61. Hunter, C.; Sanders, J.; Stone, A. Exciton coupling in porphyrin dimers. *Chem. Phys.* **1989**, *133*, 395–404. [CrossRef]
62. Tran-Thi, T.; Lipskier, J.; Maillard, P.; Momenteau, M.; Lopez-Castillo, J.M.; Jay-Gerin, J.P. Effect of the exciton coupling on the optical and photophysical properties of face-to-face porphyrin dimer and trimer. A treatment including the solvent stabilization effect. *J. Phys. Chem.* **1992**, *96*, 1073–1082. [CrossRef]
63. Meehan, E.V. Synthesis of Pyrrole-Modified Porphyrins: Oxachlorins, and the Beckmann Rearrangement of Octaethyl-2-oxachlorin Oxime. Master's Thesis, The University of Connecticut, Storrs, CT, USA, 2014; p. 90.
64. Kim, B.F.; Bohandy, J. Spectroscopy of Porphyrins. *Johns Hopkins APL Tech. Dig.* **1981**, *2*, 153–163.
65. Rayati, S.; Zakavi, S.; Ghaemi, A.; Carroll, P.J. Core protonation of meso-tetraphenylporphyrin with tetrafluoroboric acid: Unusual water-mediated hydrogen bonding of H_4tpp^{2+} to the counterion. *Tetrahedron Lett.* **2008**, *49*, 664–667. [CrossRef]
66. Zakavi, S.; Omidyan, R.; Ebrahimi, L.; Heidarizadi, F. Substitution effects on the UV–vis and 1H NMR spectra of the dications of meso and/or β substituted porphyrins with trifluoroacetic acid: Electron-deficient porphyrins compared to the electron-rich ones. *Inorg. Chem. Commun.* **2011**, *14*, 1827–1832. [CrossRef]
67. Zhao, Y.; Cai, X.; Zhang, Y.; Chen, C.; Wang, J.; Pei, R. Porphyrin-based metal–organic frameworks: Protonation induced Q band absorption. *Nanoscale* **2019**, *11*, 12250–12258. [CrossRef] [PubMed]
68. Shaikh, S.M.; Chakraborty, A.; Alatis, J.; Cai, M.; Danilov, E.; Morris, A.J. Light harvesting and energy transfer in a porphyrin-based metal organic framework. *Faraday Discuss.* **2019**, *216*, 174–190. [CrossRef]
69. McAuliffe, C.; Murray, S. Metal complexes of sulphur-containing amino acids. *Inorg. Chim. Acta Rev.* **1972**, *6*, 103–121. [CrossRef]
70. Mah, V.; Jalilehvand, F. Mercury(II) complex formation with glutathione in alkaline aqueous solution. *JBIC J. Biol. Inorg. Chem.* **2008**, *13*, 541–553. [CrossRef]
71. Fuhr, B.J.; Rabenstein, D.L. Nuclear magnetic resonance studies of the solution chemistry of metal complexes. IX. Binding of cadmium, zinc, lead, and mercury by glutathione. *J. Am. Chem. Soc.* **1973**, *95*, 6944–6950. [CrossRef]
72. Oram, P.D.; Fang, X.; Fernando, Q.; Letkeman, P.; Letkeman, D. The Formation Constants of Mercury(II)-Glutathione Complexes. *Chem. Res. Toxicol.* **1996**, *9*, 709–712. [CrossRef] [PubMed]
73. Fernández-Ramos, M.D.; Cuadros-Rodríguez, L.; Arroyo-Guerrero, E.; Capitán-Vallvey, L.F. An IUPAC-based approach to estimate the detection limit in co-extraction-based optical sensors for anions with sigmoidal response calibration curves. *Anal. Bioanal. Chem.* **2011**, *401*, 2881–2889. [CrossRef] [PubMed]
74. Jia, P.; Hou, J.; Yang, K.; Wang, L. On-off-on fluorescent sensor for glutathione based on bifunctional vanadium oxide quantum dots induced spontaneous formation of MnO_2 nanosheets. *Microchim. Acta* **2021**, *188*, 299. [CrossRef] [PubMed]
75. Qiu, Y.; Huang, J.; Jia, L. A Turn-On Fluorescent Sensor for Glutathione Based on Bovine Serum Albumin-Stabilized Gold Nanoclusters. *Int. J. Anal. Chem.* **2018**, *2018*, 1–5. [CrossRef]

76. Li, X.-Y.; Zhang, Q.; Wang, N.; Liu, J.-J.; Wang, J. Cu^{2+} mediated Fluorescence Switching of Graphene Quantum Dots for Highly Selective Detection of Glutathione. *Chin. J. Anal. Chem.* **2020**, *43*, 339–346. [CrossRef]
77. Cai, L.; Fu, Z.; Cui, F. Synthesis of Carbon Dots and their Application as Turn Off–On Fluorescent Sensor for Mercury (II) and Glutathione. *J. Fluoresc.* **2020**, *30*, 11–20. [CrossRef]
78. Yang, R.; Tang, Y.; Zhu, W. Ratiometric Fluorescent Probe for the Detection of Glutathione in Living Cells. *Chem. J. Chin. Univ.* **2016**, *37*, 643. [CrossRef]
79. Gu, J.; Hu, D.; Wang, W.; Zhang, Q.; Meng, Z.; Jia, X.; Xi, K. Carbon dot cluster as an efficient "off–on" fluorescent probe to detect Au(III) and glutathione. *Biosens. Bioelectron.* **2015**, *68*, 27–33. [CrossRef]
80. Wang, X.; Zhang, Y.; Jin, Y.; Wang, S.; Zhang, Z.; Zhou, T.; Zhang, G.; Wang, F. An Off-Off fluorescence sensor based on ZnS quantum dots for detection of glutathione. *J. Photochem. Photobiol. A Chem.* **2023**, *435*, 114264. [CrossRef]
81. Meng, Y.; Guo, Q.; Jiao, Y.; Lei, P.; Shuang, S.; Dong, C. Smartphone-based label-free ratiometric fluorescence detection of sertraline and glutathione based on the use of orange-emission carbon dots. *Mater. Today Chem.* **2022**, *26*, 101170. [CrossRef]
82. Tian, M.; Guo, F.; Sun, Y.; Zhang, W.; Miao, F.; Liu, Y.; Song, G.; Ho, C.L.; Yu, X.; Sun, J.Z.; et al. A fluorescent probe for intracellular cysteine overcoming the interference by glutathione. *Org. Biomol. Chem.* **2014**, *12*, 6128. [CrossRef] [PubMed]

Disclaimer/Publisher's Note: The statements, opinions and data contained in all publications are solely those of the individual author(s) and contributor(s) and not of MDPI and/or the editor(s). MDPI and/or the editor(s) disclaim responsibility for any injury to people or property resulting from any ideas, methods, instructions or products referred to in the content.

Review

Porphyrin Macrocycles: General Properties and Theranostic Potential

Rica Boscencu [1,*,†], Natalia Radulea [1,†], Gina Manda [2], Isabel Ferreira Machado [3,4], Radu Petre Socoteanu [5,*], Dumitru Lupuliasa [1], Andreea Mihaela Burloiu [1,*], Dragos Paul Mihai [1] and Luis Filipe Vieira Ferreira [4,*]

1 Faculty of Pharmacy, "Carol Davila" University of Medicine and Pharmacy, 6 Traian Vuia, 020956 Bucharest, Romania
2 "Victor Babeș" National Institute of Pathology, 050096 Bucharest, Romania
3 Polytechnic Institute of Portalegre, 7300-110 Portalegre, Portugal
4 BSIRG—Biospectroscopy and Interfaces Research Group, iBB-Institute for Bioengineering and Biosciences, Instituto Superior Técnico and Associate Laboratory i4HB—Institute for Health and Bioeconomy at Instituto Superior Técnico, Universidade de Lisboa, 1049-001 Lisboa, Portugal
5 "Ilie Murgulescu" Institute of Physical Chemistry, Romanian Academy, 060021 Bucharest, Romania
* Correspondence: rica.boscencu@umfcd.ro (R.B.); psradu@yahoo.com (R.P.S.); andreea-mihaela.burloiu@drd.umfcd (A.M.B.); lfvieiraferreira@tecnico.ulisboa.pt (L.F.V.F.)
† These authors contributed equally to this work.

Abstract: Despite specialists' efforts to find the best solutions for cancer diagnosis and therapy, this pathology remains the biggest health threat in the world. Global statistics concerning deaths associated with cancer are alarming; therefore, it is necessary to intensify interdisciplinary research in order to identify efficient strategies for cancer diagnosis and therapy, by using new molecules with optimal therapeutic potential and minimal adverse effects. This review focuses on studies of porphyrin macrocycles with regard to their structural and spectral profiles relevant to their applicability in efficient cancer diagnosis and therapy. Furthermore, we present a critical overview of the main commercial formulations, followed by short descriptions of some strategies approached in the development of third-generation photosensitizers.

Keywords: porphyrin macrocycles; photodynamic therapy; theranostic agents

Citation: Boscencu, R.; Radulea, N.; Manda, G.; Machado, I.F.; Socoteanu, R.P.; Lupuliasa, D.; Burloiu, A.M.; Mihai, D.P.; Ferreira, L.F.V. Porphyrin Macrocycles: General Properties and Theranostic Potential. *Molecules* **2023**, *28*, 1149. https://doi.org/10.3390/molecules28031149

Academic Editors: M. Amparo F. Faustino, Carlos J. P. Monteiro and Carlos Serpa

Received: 19 December 2022
Revised: 16 January 2023
Accepted: 18 January 2023
Published: 23 January 2023

Copyright: © 2023 by the authors. Licensee MDPI, Basel, Switzerland. This article is an open access article distributed under the terms and conditions of the Creative Commons Attribution (CC BY) license (https://creativecommons.org/licenses/by/4.0/).

1. Introduction

Interdisciplinary research on obtaining new pharmaceutical forms to guarantee, in one step, the identification of and therapy for certain tumor formations occupies a special place in the medical field and generates some top themes in terms of objectives targeting improvements in the health of the population. Despite specialists' efforts to find the best solutions in this respect, cancer generally remains the biggest health threat on the planet. Characterized by an uncontrollable growth of cells that are able to invade any part of the organism, cancer is a conglomeration of several neoplastic diseases that can be triggered by a multitude of factors, both endogenous and exogenous. Nowadays, global statistics concerning deaths associated with cancer are alarming [1–3]. At the global level, about 19.3 million new cases were reported in 2020, and it was one of the main causes of death in most countries [4–6]. These statistics justify the need to intensify interdisciplinary research in order to identify efficient strategies for cancer diagnosis and therapy by the use of new molecules with optimal therapeutic potential and minimal adverse effects.

The integration of therapeutic and diagnostic potential into single drug molecules is one of the most important concerns of specialists in the biomedical field [4,7].

Porphyrins possess selectivity for tumor cells, low cytotoxicity in the absence of light, and peculiar photophysical properties (absorption and emission in the visible region, high triplet state quantum yield, high quantum yield of $^1O_2{}^*$ generation), which have drawn research interest for their use as theranostic agents [8–14].

Compared with conventional antitumor therapies, which are known for their highly toxic effects on normal cells, photodynamic therapy (PDT) is noninvasive and has minimal side effects [15].

PDT is a selective method of antitumor treatment based on cell necrosis/apoptosis induced by a reactive species derived from molecular oxygen, generated during the irradiation of photosensitizer (PS) molecules in the presence of oxygen molecules. Tetrapyrrolic molecules such as porphyrins, chlorins, phthalocyanines, and bacteriochlorins are macrocyclic structures frequently used as photosensitizers. This review is focused on porphyrin macrocycles, and summarizes aspects of their structural and spectral profiles relevant to their applicability in cancer theranostics.

2. General Properties of Porphyrin Macrocycle

Porphyrins are among the most examined tetrapyrrolic macrocycle structures with applicability in the biomedical field. Starting at the beginning of the 20th century, it was revealed that the tetrapyrrolic macrocycles present an inherent photosensitive affinity characteristic in relation to tumor cells, proving their potential use in antitumor therapy [16–19]. The structural characteristics, spectral profiles, and great coordinating capacity of tetrapyrrolic molecules are the main factors behind their therapeutic potential.

Structurally, porphyrin molecules include four pyrrolic units bound by methine bridges (the C atoms in the 5, 10, 15, and 20 positions of the macrocycle, also named *meso*-positions) (Figure 1) [20]. The aromatic character of tetrapyrrolic macrocycles is an essential property of these structures, which is important in defining the spectral profile. The π electron system contains 22 electrons, among which 18 are delocalized and follow Hückel's rule concerning the aromatic character. The presence of nitrogen atoms inside the macrocycle provides ligand properties for these molecules, allowing the possibility of coordinating metal ions [21,22].

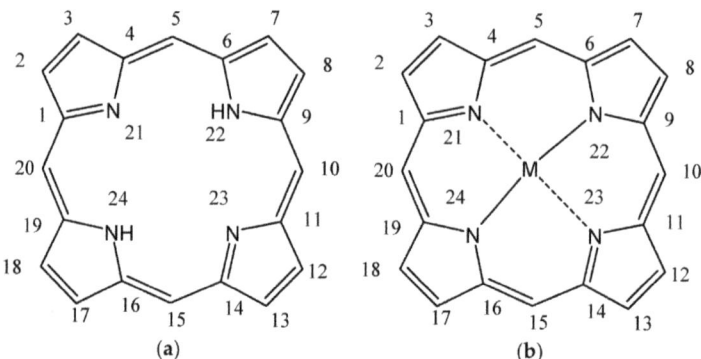

Figure 1. General structures of (**a**) free base porphyrin and (**b**) metalloporphyrin (atoms numbered according to IUPAC (https://doi.org/10.1351/goldbook). (accessed date: 17 January 2023).

Furthermore, the properties of metal ions (ionic volume and electronic configuration) may facilitate reactions of metalloporphyrin with monodentate ligands from the biologic environment [23]. Hemoglobin, cytochromes, catalase, and peroxidases are examples of molecules that contain metalloporphyrin structures and have an important role in biological processes [24–27].

The porphyrinic structures are versatile and can be shaped by attaching peripheral substituents with a polarity degree adequate for internalization in cells.

Their unique structural and spectral properties, especially their excellent photochemical and photophysical properties, define the profile of porphyrins as therapeutic candidates.

Porphyrins have a typical absorption spectrum described by an intense band (Soret band) located between 400 and 440 nm, and four Q bands ($Q_y(1,0)$, $Q_y(0,0)$, $Q_x(1,0)$, $Q_x(0,0)$)

with reduced intensity and placed in the 440–800 nm spectral range (Figure 2). The Soret band is not relevant for photodynamic therapy of deeper tumor tissue; only the 600–800 nm range is useful for antitumor therapy by photosensitization.

In the case of metalloporphyrins, the absorption spectrum is described by the Soret band, placed in the 400–440 nm range, and one or two Q bands located in the 550–600 nm range (Figure 2).

The spectral profile of porphyrins is determined by the molecular symmetry, the nature and position of peripheral macrocycle substituents, and the nature of the metal ions in the case of metalloporphyrins [28–30].

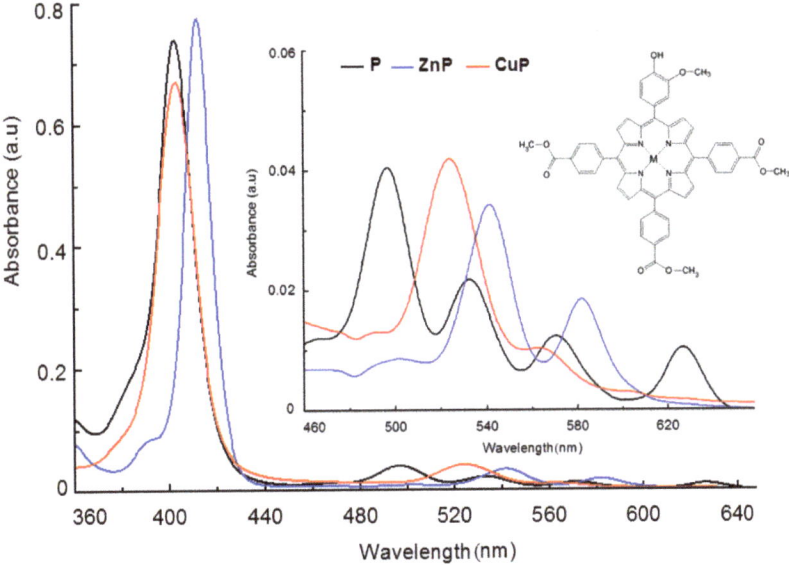

Figure 2. Absorption spectra for: 5-(4-hydroxy-3-methoxyphenyl)-10,15,20-tris(4-carboxymethylphenyl) porphyrin (P), 5-(4-hydroxy-3-methoxyphenyl)-10,15,20-tris(4-carboxymethylphenyl)porphyrinatozinc(II) (ZnP) and 5-(4-hydroxy-3-methoxyphenyl)-10,15,20-tris(4-carboxymethylphenyl)porphyrinatocopper(II) (CuP); (dimethylsulfoxide used as solvent) [31].

A ratio of the relative intensities of the Q bands IV > III > II > I describe, for the porphyrinic compound, a spectrum of absorption of *etio* type and the porphyrin fits the group of etioporphyrins [30]. This type of electronic spectrum is characteristic to the porphyrins, for which six or more of the positions β of the macrocycle have substituents with no π electrons.

Substituents with π electrons attached to the β position of the macrocycle determine spectral modifications by shifting the maximum absorption to a higher wavelength and modifying the ratio of Q band intensities, in the order III > IV > II > I.

Longer shifts to the red side of the absorption maximum are registered when substituents with double or triple bonds occupy the β positions and the ratio of Q band intensity is III > II > IV > I. If the *meso* positions are occupied, the Q band intensity decreases in the order IV > II > III > I [30].

Significant spectral modifications appear by protonation of the nitrogen atoms from the tetrapyrrole core. The acidic environment determines the protonation of the nitrogen atoms, which are inside the tetrapyrrole core, with the transition to the dicationic shape H_4P^{2+} accompanied by the modification of the absorption spectrum, respectively, the reduction of the number of Q bands and red shifts of the absorption peaks [28,29]. Therefore, studies on the spectral influence of pH modifications in biological areas where these tetrapyrroles show activity are needed in terms of their potential therapeutic use.

Another factor that has a major influence on the spectral behavior of porphyrins is molecular aggregation, a phenomenon that depends on the structural characteristics, pH value, polarity, and ionic strength of the solution in which these compounds are activated [32,33]. The 22 π electrons of the porphyrin macrocycle facilitate strong π–π interactions and the formation of J-type or H-type aggregate, each leading to specific spectral features in terms of bathochromic or hypsochromic shifts in the corresponding absorption spectra.

The remarkable absorption and emission properties of tetrapyrrolic chromophores are the most relevant from the point of view of their biomedical applicability [34–36].

Porphyrins have a good fluorescence profile, with emission capacity in the 600–800 nm spectral range, and the potential for reactive oxygen species (ROS) formation by irradiation in the presence of molecular oxygen. The fluorescence spectrum of a porphyrinic compound (Figure 3) includes two spectral bands, as a result of the radiation emission that takes place during the transition of the molecule from the excited singlet state, S_1, to the ground state, S_0; these bands are located at longer wavelengths compared to the excitation wavelength ($\lambda_{emis} > \lambda_{abs}$), due to the loss of energy through vibrational relaxation in the excitation state. The registered difference between λ_{max} of the absorption band and λ_{max} of the emission band is named the Stokes shift.

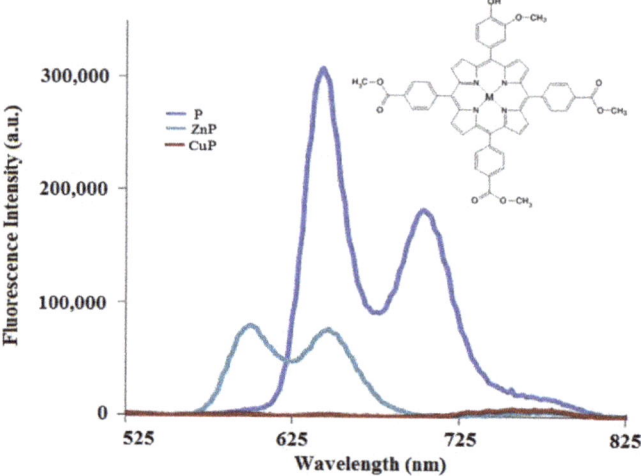

Figure 3. Fluorescence emission spectra for 5-(4-hydroxy-3-methoxyphenyl)-10,15,20-tris(4-carboxymethylphenyl) porphyrin (P), 5-(4-hydroxy-3-methoxyphenyl)-10,15,20-tris(4-carboxymethylphenyl)porphyrinatozinc(II) (ZnP) and 5-(4-hydroxy-3-methoxyphenyl)-10,15,20-tris(4-carboxymethylphenyl)porphyrinatocopper(II) (CuP) (ethanol used as solvent) [23].

The specific parameters for molecular fluorescence evaluation are fluorescence lifetime (τ_s) and fluorescence emission quantum yield (Φ_F). They can be experimentally determined by spectral fluorescence analysis with the use of modern laser equipment; their test excitation time is in the picoseconds, and they have some rapid response photodiodes [37–39].

The fluorescence of the tetrapyrrole structures appears as a result of the aromatic character imprinted on the molecule by the π electrons of the double conjugated bonds. The delocalized electrons can be involved in π→π* transitions and go through singlet excitation states.

The fluorescence of porphyrins is influenced by the nature of the peripheral substituents of the macrocycle; some substituents, such as –NH$_2$, –OH, and –O–CH$_3$, have the effect of delocalizing the π electrons, with an increased probability of transitions between the excited singlet state and the ground state, and they have an effect on increasing the quantum yields. On the other hand, electron withdrawing groups (e.g., –NO$_2$ or CO$_2$H) lower or eliminate the fluorescence [40]. The values of the parameters that define the fluorescent profile of a tetrapyrrole (lifetime of fluorescence (τ_s) and fluorescence efficiency (Φ_F)) are modified by complexation with metal ions, and the modifications are dependent on the electronic structure of the cation. With the complexation of metal ions with high volume (e.g., Pd^{2+}), the values of Φ_F and τ_s decrease. The efficiency of singlet oxygen species formation also decreases [39].

Porphyrinic compounds with paramagnetic metal ions (Fe^{2+}, Cu^{2+}, Co^{2+}) present phosphorescent properties, while the lifetime of the excited singlet state is very short [41,42].

3. Aspects regarding Porphyrin Macrocycles as Photosensitizers

3.1. Short Background regarding to the Porphyrinoid Photosensitizers

Among the first clinical observations on the photosensitization effect of a chemical compound (eosin) on the tumor cells are those made from 1903 by the pharmacologist Hermann von Tappeiner and the dermatologist Albert Jesionek. By topical application of the eosin (photosensitive dye) and exposure of the treated area to light, the two researchers realized for the first time the photodynamic therapy and they introduced the term of "photodynamic reaction" [43].

Subsequent studies have shown that hematoporphyrin (Hp), combined with ultraviolet light, has a therapeutic potential on the skin diseases, including psoriasis [44–47].

The purification of hematoporphyrin led to the HpD, a combination of monomers and oligomers with tetrapyrrole structures, with remarkable fluorescent properties and affinity for the tumor cells. HpD was used for the first time by R. L. Lipson as an imagistic imaging agent to visualize the tumor lesion during surgery [45,48]. Later, the team led by Thomas J. Dougherty purifiedHpD and obtained a heterogenous mixture of porphyrins, which was marketed as Photofrin [49,50].

The clinical use of Photofrin was limited by its difficult accumulation at cellular level, weak absorption at λ = 630 nm, and the patients' photosensitivity to the sun for a long period of time after the treatment [51,52].

The deficiencies appearing in the clinical usage of HpD and Photofrin led to the development of new porphyrinoid photosensitizers with unique structure, high absorption in the visible-near infrared spectral domain and high singlet oxygen quantum yield. They were classified as second generation photosensitizers, and include the porphyrin and porphyrin analogue photosensitizers (chlorin, (iso)bacteriochlorin, phthalocyanines) such as Hemoporfin®, Foscan®, Visudyne®, Tookad®, Radachlorin®, Photochlor®, and Photocyanine® [53,54].

However, after years of development, the second-generation porphyrin photosensitizers had some disadvantages which include reduced solubility in water and biological fluids, difficult location at the cell level, the molecular aggregation tendency, and the effect of reducing the photodynamic efficiency [55–57].

In recent years, researchers have obtained and clinically investigated a new generation of photosensitizers, the so-called third-generation photosensitizers, characterized by good solubility in biologic fluids, great capacity for generating reactive oxygen species (ROS) and good selectivity in relation to tumor cells [58–60].

The synthesis of third generation PS is carried out in a chemo- and regioselective manner, by the attachment of fragments of bioactive molecules (folic acids, peptides, sterols, proteins, anti-tumor monoclonal antibodies, sugars), polyethyleneglycol or nanoparticles to the structures from the second-generation porphyrin PS [55–57,61–66]. These porphyrin nanophotosensitizers are currently in the development stage and have not been used in clinical trials.

Due to their structural shape and unique affinity for tumor cells, porphyrins can be used as excellent carriers to transport anticancer drugs into tumor tissues, with synergistic effects in the identification and destruction of tumors. Therefore, current research based on the development of porphyrins as theranostic agents mainly only includes studies on porphyrins as anticancer drugs and porphyrins-bonded anticancer drugs.

3.2. Basic Photochemical Principles in PDT

Establishing a framework protocol to completely cure cancer by using optimal pharmacological molecules with minimal adverse effects is one of the prime objectives of research studies in the pharmaceutical and oncological fields. Tetrapyrrolic macrocycles are among the most studied families of compounds with regard to their therapeutic potential, especially their antitumor potential. Through their structural profile and electronic load, these structures can efficiently generate reactive oxygen species (ROS), such as singlet oxygen, in the presence of molecular oxygen and light.

Clinical experiments of irradiating porphyrinic structures (at the cell level) in the presence of molecular oxygen laid the foundation for a modern noninvasive treatment method: photodynamic therapy (PDT) [16,17,67].

The clinical procedure is initiated by administering a photosensitizer (PS), followed by light irradiation at a certain wavelength, usually a red light laser, after the accumulation of PS in the tumor tissue. Light irradiation has an important role in photodynamic therapy, because it offers the necessary energy to produce reactive species, which destroy tumor cells. Moreover, molecular oxygen has a determinant role in initiating and propagating the reactions of destroying tumor formations [68].

The activation of PS molecules by irradiation generates a series of competitive reactions of ROS, which induce necrosis and/or apoptosis of tumor cells. Apoptosis is a form of programmed cell death (the cells start their own destruction) as a response to the physiological key index and intracellular injury. During apoptosis, cell collapse is characterized by key morphological and chemical modifications, followed by phagocytosis. These modifications take place while maintaining the levels of adenosine triphosphate, activating caspase (an enzyme with proteolytic activity and a predominant role in the process of self-destruction), chromatin condensation, and the fragmentation of DNA and formation of apoptotic bodies. During apoptosis, an associated inflammatory response is not identified [69].

Necrosis represents a form of uncontrolled cell death in response to a series of factors, such as physical injury, ischemia, excessive ROS accumulation, or the presence of infectious agents. Unlike apoptosis, necrosis is a local inflammatory response seen as a result of cells being liberated directly in the surrounding tissue [69].

Schematically, the photophysical processes that form the basis of generating ROS are rendered in Figure 4 (Jablonsky diagram).

In the initial stage, by absorbing light radiation, the molecules of the photosensitizer go from the ground state (S_0) to a superior energetic state, the excited singlet state (S_1); the transformation takes place rapidly (~10–15 s) through the transition of electrons on the S_0 level to the S_1 excited singlet state. The duration of the S_1 state is in nanoseconds. Stabilization of the S_1 state can be reached by the emission of photons (fluorescence) or by a process called intersystem crossing (ISC), by which the electronic spin is reversed and the electron passes onto an energetic level (T_1) of the excited triplet state [70–72]. The energy of T_1 species can be dissipated by the passage to a fundamental state with weak light emission [70–72]. The lifetime of the T_1 species is in micro- to milliseconds, much longer than the lifetime of the excited singlet species; consequently, it is more likely that the PS found in the excited triplet state participates in photodynamic reactions. Thus, the photodynamic effect is the result of the energetic or electronic transfer of the photosensitizer found in the T_1 state on the way to the organic substratum or molecular oxygen. The stabilization reaction of the excited triplet state (T_1) of the photosensitizer can occur by two types of mechanisms, classified as reactions of type I and II [69].

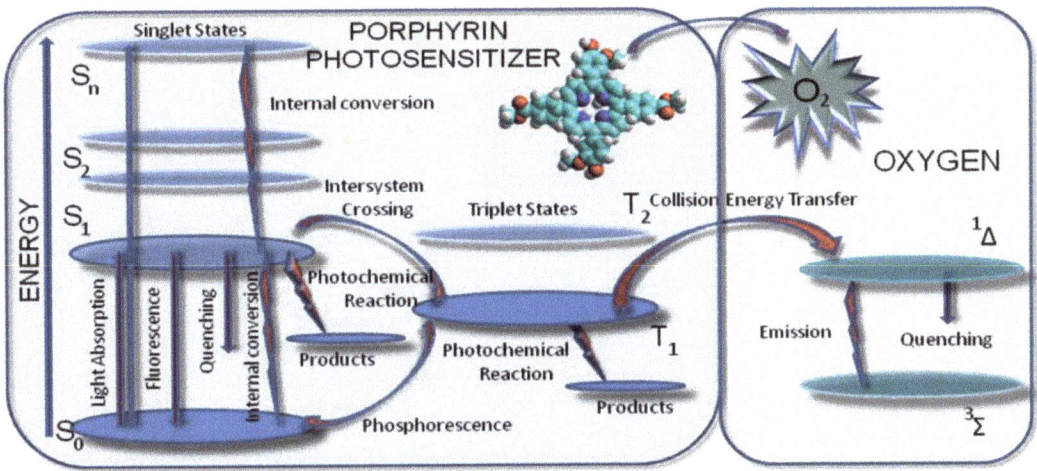

Figure 4. Jablonsky diagram (S_0—ground state of photosensitizer; S_1, S_2, excited singlet states of photosensitizer; T_1, excited triplet state of photosensitizer).

Type I reactions (free radical mechanism) are reactions of oxidation and reduction and imply the transfer of electrons and hydrogen atoms between the photosensitizer (found in the excited singlet or triplet state) and the biomolecules, resulting in free radical formation; these radicals interact with oxygen molecules, resulting in the formation of superoxide anion radicals, which can generate forward hydroxyl radicals and may determine the formation of cytotoxic species.

Type II reactions (singlet oxygen mechanism) take place through the interaction of photosensitizer molecules in the excited triplet state and molecular oxygen (triplet state) with the formation of singlet oxygen, a very reactive species in relation to the cell components (lipids, proteins, and nucleic acids) with a large potential to destroy tumor cells [73–79]. Moreover, type II reactions proceed without the chemical transformation of PS molecules, which is why the set of photochemical reactions can be remade with a new activation. As the half-life of the reactive species is very short, fewer than 40 ns, the photodegradation reactions unfold next to the areas of initiation of the photochemical reactions, which is why the PS location at the cell level is determined through its photodynamic reactions [15,80,81].

3.3. Chemical and Pharmaceutical Aspects Concerning Photosensitizers with Tetrapyrrolic Structures Used in Diagnosis and Antitumor Therapy

The theranostic agent profile assigned to an active substance defines its potential to act simultaneously as a marker for the detection of tumor cells and as a therapeutic agent [8].

Photodynamic therapy (PDT) uses porphyrins as photosensitizers, which, by their structural and spectral properties, have superior pharmacological potential compared to classic antitumor molecules [9,82,83]. At present, PDT is an efficient therapeutic method successfully applied for the treatment of tumor formation in the skin, breast, lung, throat and brain [84–87]. Photodynamic therapy is a selective method of antitumor treatment based on the cell necrosis/apoptosis induced by ROS generated during the irradiation of photosensitizer molecules in the presence of oxygen molecules. The phototoxic reactions induced by PDT are limited to the tumor area, where the photosensitizer accumulates and the irradiation takes place, and for this reason, PDT has limited side effects compared to the standard treatment methods [88]. Lately, the scientific community has made significant investments into developing and implementing new active molecules in oncology that are both therapeutic and diagnostic [89–92]. Regardless of the therapeutic purpose, ensuring optimal biomedical efficiency requires that the photosensitizer meets the following mandatory requirements [72,82]:

- It has a unique, well-defined structure with maximum purity and can be obtained by modern ecological methods;
- It has a structural profile that allows optimal internalization at the tumor level and is described by a well-defined distribution of lipophillic/hydrophilic groups at the periphery of tetrapyrrolic macrocycles;
- It has a well-defined spectral profile with maximum absorption in the therapeutic range (600–850 nm) associated with good efficiency in generating singlet oxygen ($\Phi_\Delta \geq 0.5$) and a triplet state with a lifetime in the microsecond range;
- It is soluble in nontoxic solvents accepted for pharmaceutical formulations;
- It is nontoxic in the absence of light;
- It provides rapid clearance from the organism;
- There is no toxic effect with the necessary dose for a therapeutic effect;
- There is an absence of toxicity for the photosensitizer metabolites.

The molecular structure of the photosensitizer is the predominant factor in its pharmacokinetic and pharmacodynamic evolution, no matter the type of therapeutic approach. Biodistribution at the tumor mass level has direct consequences in terms of the pharmacodynamic efficiency of any active substance [53]. Among the factors that influence the intracellular uptake of PS are its solubility in biological fluids and the amphiphilic character of the molecule, which determines the penetration of the double-lipid stratum of the cell membrane.

By evaluating the relationship between the PS structure and the biodistribution capacity at the tumor tissue level, Boyle and Dolphin classified PS into three main groups [53]:
- Hydrophobic photosensitizers, which have peripheral substituent hydrophobic functional groups and very reduced solubility in polar solvents, alcohol, or water, at a physiological pH;
- Hydrophilic photosensitizers, which have three or more peripheral functional substituent hydrophilic groups and slight water solubility at physiological pH;
- Amphiphilic photosensitizers, which have in their structure hydrophobic and hydrophilic functional groups and water solubility at physiological pH.

The solubility of tetrapyrrolic compounds meant for biomedical applications represent a current problem in the field of pharmaceutical chemistry, because the clinical efficiency of the active substance is directly influenced by its solubility in the biological fluid and membrane environment [93]. For example, chlorin-type macrocycles, although characterized by stronger absorption capacity in the red area of the visible spectrum, present the disadvantage of limited water and alcohol solubility. Another disadvantage of tetrapyrrole-type photosensitizers is the molecular aggregation tendency, with the effect of reducing the photodynamic efficiency, specifically the efficiency of producing reactive singlet oxygen and the fluorescence lifetime [58,76].

At present, the efforts of specialists who work in medicinal chemistry are oriented toward synthesizing photosensitizers with an amphiphilic structure by introducing peripheral substituents (lipophillic and hydrophilic) at the periphery of the macrocycle.

Some modern synthesis methods introduce, as substitutes, residues of polyethylene glycol in order to obtain a soluble structure in water with excellent photophysical properties [94–98].

A series of photosensitizers with macrocycle structures approved or in clinical trials for diagnosis and photodynamic therapy of cancer, are presented in Table 1.

Table 1. General presentation of photosensitizers with macrocycle structures approved or in clinical trials for diagnosis and photodynamic therapy of cancer [99–104].

Photosensitizer	Active Substance, Activation Wavelength λ (nm)	Clinical Approval
PHOTOFRIN® (Axcan Pharma Inc., Mont-Saint-Hilaire, QC, Canada)	Porfimer sodium 630 nm	Theranostic for tumor formations in lung, brain, cervix; therapy for bladder and esophagus cancer; in clinical trial for brain cancer diagnosis
FOSCAN® (Biolitec Pharma Ltd., Jena, Germany)	5,10,15,20-tetrakis(3-hydroxyphenyl)chlorin (temoporfin) 652 nm	Photodynamic therapy for carcinoma with squamous cells at late stage in head and throat level; diagnosis of brain, bladder, and ovarian tumors; in clinical studies for therapy for breast, pancreas, prostate, lung, stomach, and skin cancer
LASERPHYRIN® (Meiji Seika Pharma Co., Ltd., Tokyo, Japan)	Mono-L-aspartyl chlorin e-6 (talaporfin) 654 nm	Photodynamic therapy for lung and brain cancer; intraoperative photodiagnosis of malignant brain formations; clinical studies for treatment of colorectal and liver neoplasms
VISUDYNE® (Novartis Pharmaceuticals, East Hanover, NJ, USA)	Verteporfin 690 nm	Photodynamic therapy for macular exudative degeneration with subfoveal choroidal neovascularisation of melanoma and psoriasis; clinical trial for diagnosis of ovarian tumor formation

Table 1. Cont.

Photosensitizer	Active Substance, Activation Wavelength λ (nm)	Clinical Approval
PHOTOCHLOR® (Roswell Park Cancer Institute, Buffalo, NY, USA)	2-(1-hexyloxyethyl)-2-devinyl pyropheophorbide-a (HPPH) 664 nm	Clinical trials for therapy (first and second stage) of cervical intraepithelial neoplasia, esophagus tumor, skin, lung, and oral cavity cancer; clinical trials of marker potential in identifying different types of cancer
PURLYTIN® (Miravant Medica Technologies, Santa Barbara, CA, USA)	Sn(II) ethyl-ethiopurpurin 660 nm	Clinical trials for photodynamic therapy of breast adenocarcinoma, Kaposi's sarcoma, prostate cancer, cerebral metastasis, and psoriasis
TOOKAD® (WST09) (Negma Lerads/Steba Biotech, Toussous le Noble, France)	Palladium bacteriopheophorbide 762 nm	Stage II/III clinical trials targeting photodynamic therapy for prostate cancer
LUTRIN® (Pharmacyclics Inc., Sunnyvale, CA, USA)	Motexafin lutetium (Lu-Tex) 732 nm	Clinical trials for prostate, breast, cervical, and brain cancer, melanoma, and Kaposi's sarcoma

Table 1. *Cont.*

Photosensitizer	Active Substance, Activation Wavelength λ (nm)	Clinical Approval
PHOTOSENS® (SSC NIOPIK, Moscow, Russia)	Aluminum phthalocyanine tetrasulfonate chloride R = H or -SO₃H 676 nm	Clinical trials for gastric, oral, skin and breast cancer

3.3.1. PHOTOFRIN®

Photofrin® (Axcan Pharma, Inc.) is a first-generation photosensitizer used in photodynamic therapy. Chemically, it is a mixture of monomers, dimers, and oligomers of hematoporphyrin, as well as esters and ethers corresponding to these compounds. The preparation of this mixture from derivatives of hematoporphyrin was reported for the first time by Dougherty in 1983 [49].

Spectrally, phosphate-buffered saline solution containing Photofrin® presents in the spectrum of molecular absorption at a maximum 630 nm ($\varepsilon = 3.0 \times 10^3$ M^{-1} cm^{-1}) and has lower efficiency of singlet oxygen generation ($\Phi_\Delta = 0.01$). In clinical applications, Photofrin® is given as an intravenous injection at a dose of 2–5 mg/kg, and after 24–48 h it is activated by light irradiation at $\lambda = 630$ nm (light dose 100–200 J cm^{-2}).

Photofrin® was approved by the Canadian Health Agency in 1993 for antitumor therapy against bladder cancer [49]. The therapeutic applicability of Photofrin® was later extended for the treatment of lung, esophageal, bladder, and early stages of cervical cancer [73,74,76,105]. Photofrin® is used to identify and treat tumor formations at the brain and lung level, and cervical dysplasia [9,102]. Wilson et al. used Photofrin® as a marker to identify tumor formations at the brain level and as an antitumor agent for the treatment of waste tumor cells after surgery [106–109]. The results obtained by using PDT with Photofrin® were explained by a series of disadvantages, mainly determined by the chemical composition of the PS (mixture of chemical compounds, some of them inactive PDT). The non-homogeneous composition of PS determines a difficult location at the tumor mass level, with weak radiation absorption at long radiation time and activation.

The most important disadvantage of Photofrin® is the long photosensitization after treatment, due to which patients must be protected against bright light for 6–12 weeks. The ability to repeat the treatment is limited by these disadvantages [17,51,52].

Other pharmaceutical forms that contain, as active substances, mixtures of monomers, dimers, and oligomer structures derived from hematoporphyrin are Photogem® and Photosan®. Photogem® was approved for clinical use in Russia and Brazil, and Photogem® was clinically approved in the European Union [110–112].

3.3.2. FOSCAN®

The tetrapyrrole compound 5,10,15,20-tetrakis(3-hydroxyphenyl)chlorin, known as temoporfin or Foscan® (the commercial name), is one of the strongest photosensitizers used in photodynamic therapy against cancer. Temoporfin has a symmetric structure defined by identical substitutes of the four *meso* positions of the tetrapyrrole macrocycle, and it was obtained and characterized in terms of its toxicological and physico-chemical aspects by

Bonnet et al. [113]. It is a dark violet crystalline powder, insoluble in water but soluble in alcohol, acetone, and ethyl acetate.

The structure of 5,10,15,20-tetrakis(3-hydroxyphenyl)chlorin is characterized by a maximum absorption capacity at 652 nm ($\varepsilon \sim 35,000$ M^{-1} cm^{-1}) and good distribution at the tumor tissue level. The photophysical properties of temoporfin include a quantum yield of singlet oxygen generation $\Phi_\Delta = 0.4$ (measured in dimethylsulfoxide) and high efficiency in initiating phototoxic reactions.

Foscan® (Biolitec Pharma Ltd., Jena, Germany) is a solution for injection that contains, as an active substance, temoporfin 4 mg/mL with ethanol solvent and propylene glycol co-solvent. The commercial formula does not contain water, because the stability of temoporfin decreases in an aqueous environment.

In clinical applications, Foscan® is intravenously administered at a dose of 0.15 mg/kg and then activated by laser light irradiation at a wavelength of 652 nm about 98 h later. An energy dose of 10–20 J cm^{-2} is used for irradiation. Foscan® is administered only at oncology centers under the medical supervision of a photodynamic therapy specialist. The therapeutic effect is mediated by generating reactive oxygen species, resulting in intracellular interactions between temoporfin, light, and molecular oxygen. By restrictive exposure of tumor formations to laser radiation, the cell destruction by reactive oxygen species is limited to the components of the tumor cells. Foscan is an antitumor agent that was approved in 2001 for photodynamic therapy of advanced squamous cell carcinoma at the brain and throat level (a form of cancer that starts at the nasal, throat, and ear membrane level), and is administered to patients who do not respond to chemotherapy or radiotherapy.

Foscan® has applicability in antitumor diagnosis, as a marker in identifying the modifications that appear in bladder cancer, and in the diagnosis of ovarian cancer and tumors at the brain level [100,114–116]. It was clinically investigated for use in breast, pancreas, lung, stomach, and skin cancer [63,99].

Temoporfin is a second-generation photosensitizer and, under the same experimental conditions (photosensitizer dose, energy dose), presents superior clinical efficiency (by approximately 100 times) to the first-generation Photofrin® [117].

A disadvantage of temoporfin is the long removal time from plasma (up to 4–6 weeks after intravenous administration), which induces photosensitivity in patients, similar to Photofrin® [118]. Further disadvantages include side effects such as constipation, difficulty digesting food, necrotizing stomatitis, and facial edema.

To obtain a complete response to treatment, homogeneous distribution of the PS molecule at the tumor formation level is required, which is dependent on the pharmaceutical formula in which the agent is administered.

The problems of solubility and transport to the cells are topics under investigation in pharmaceutical research, constituting a major objective for introducing agents to pharmacokinetically improve the controlled and selective delivery of PS to tumor cells while reducing toxicity to healthy tissue [119].

That is why, in order to optimize the response to antitumor therapy, other medicines that contain temoporfin were included: Foslip® and Fospeg® (Biolitec Pharma Ltd., Jena, Germany), which have demonstrated a better ability to internalize the PS in the tumor mass, with better clinical effects compared to Foscan® [120]. The experimental data prove that Foslip® and Fospeg® have efficient photosensitivity similar to Foscan®, but reduced cytotoxicity [121,122].

Foslip® contains temoporfin conditioned in conventional liposomes. Fospeg® has temoporfin expressed in liposomes covered with polyethylene glycol (PEG). It presents pharmaceutical properties superior to Foscan® (demonstrated by studies carried out on rats with tumors), and allows the administration of a third Foscan dose in order to induce necrosis in the whole tumor volume [123].

Recent in vivo comparative studies by Reshtov et al. reported better clinical efficiency for the liposomal formula with PEG, with reduced time between administration and irradiation of PS at the skin level, which significantly reduces the side effects at the skin level [124]. Moreover, biological evaluation of the two liposome formulae (Foslip® and Fospeg®) has followed the internalizing potential, with the distribution of PS at the tumor mass level correlated with photosensitizing efficiency [119,123–125].

3.3.3. LASERPHYRIN®

Mono-L-aspartyl chlorin e-6 is a second-generation photosensitizer obtained by the interaction of dicyclohexylcarbodiimide of chlorin e-6acid and di-*tert*-butylaspartate [126]. It is a dark green, water-soluble crystalline powder. The absorption spectrum of mono-L-aspartyl chlorin e-6 is typical of chlorine type compounds, and it has generated interest for its clinical applications (with a maximum absorption band at 654 nm; $\varepsilon = 4.0 \times 10^4$ M^{-1} cm^{-1}). The molecule's good efficiency in generating singlet oxygen is shown by the parameter of quantum efficiency, $\Phi_\Delta = 0.77$ (Φ_Δ determined by photoirradiation in a buffer solution of phosphate) [63,113].

Talaporfin has a shorter accumulation time at the tumor level, more rapid clearance, and better biodisponibility than Photofrin®.

Laserphyrin® (ME2906; Meiji Seika Pharma Co., Ltd., Tokyo, Japan) was approved in 2004 for clinical use in photodynamic therapy for lung cancer. At present, it is under clinical investigation for the treatment of colorectal neoplasm and brain, throat, and liver tumors. In clinical applications, PDT with Laserphyrin® follows a procedure similar to the one used for Photofrin®, but the intravenous administration uses a smaller photosensitizer dose (0.5–3.5 mg/kg) and irradiation is performed at 664 nm 4 h after that dose. The primary action mechanism of mono-L-aspartyl chlorin e-6, by the effect of photosensitization, implies the degradation of tumor formation by vascular stasis and the direct effects of tumor cytotoxicity [127].

Based on the fluorescence properties and the observed internalization ability at the tumor level, the latest studies have demonstrated the clinical usefulness of Laserphyrin® as an intraoperative photodiagnostic agent for brain tumors. The histological degree of malignity was correlated with the intensity of fluorescence and the concentration of photosensitizer in the tumor tissue.

Intravenous administration of 40 mg/m^2 Laserphyrin® 24 h before surgery and irradiation with laser light at 664 nm along with the surgery have allowed the guided resection of malignant tumors, and the results obtained have shown the theranostic agent profile of Talaporfin [104].

3.3.4. VISUDYNE®

Chemically, verteporfin is a benzoporphyrin derivative, which was obtained in 1982 by Dolphin et al. by the reaction of the protoporphyrin dimethylester with dimethyl acetylenedicarboxylate [128]. The two regiomer forms (±) have similar pharmacokinetic properties concerning accumulation and clearance, which is why verteporfin has been considered from the pharmacologic point of view.

Verteporfin is a dark green, water-insoluble powder; in clinical applications, it is used in some formulae that contain lactose, dimyristoylphosphatidylcholine, sodium phosphatidylglycerol, ascorbyl palmitate, and butylhydroxytoluene. AsVisudyne® precipitates in sodium chloride or other parenteral solutions, when preparing perfusion solutions saline is not used; instead, it is perfused with 5% dextrose solution and standard lines of perfusion, with hydrophilic membrane filters with pore size of at least 1.2 μm.

Visudyne® is a second-generation photosensitizer. It was developed by QLT Phototherapeutics (Vancouver, BC, Canada) and was clinically tested in collaboration with Ciba Vision Corporation (Duluth, GA, USA) for the photodynamic treatment of exudative macular degeneration (wet) with subfoveal choroidal neovascularization, cutaneous membranes, and psoriasis [63]. Since 2001, it has been used as therapy in over 70 countries. Presently, Visudyne® is manufactured by Novartis Pharmaceuticals (East Hanover, NJ, USA) in the form of a sterile powder for intravenous perfusion.

From the spectral point of view, Verteporfin has maximum absorption at 686 nm, with a molar absorption coefficient of $\varepsilon \sim 3.4 \times 10^4$ M^{-1} cm^{-1}. The singlet oxygen generating efficiency is $\Phi_\Delta = 0.7$ (methanol solvent).

The intravenous dose is 6 mg/kg, 30 min before irradiation, with an energy dose of 100 J/cm^2. Activation of the photosensitizer takes place at λ = 686 nm. This photosensitizer rapidly accumulates in tumors (30–150 min after intravenous administration), while the remaining photosensitivity lasts only for several days, minimizing the photosensitivity effect on the patient [129].

Verteporfin registered accumulation and a clearance rate in the tumor mass 20 times greater than Photofrin®. The plasma half-life by the clearance of Verteporfin is approximately 6 h. Moreover, the red shift of the maximum absorption compared to Photofrin® gives the advantage of better irradiation of tumor tissue. At approximately 690 nm, the tissue penetration power of light radiation is 50 times bigger compared to Photofrin® [130–132]. Visudyne® presents the advantage of spectral superiority to photosensitizers with second-generation porphyrinic structures (Foscan®, Photochlor®, and Laserphyrin®).

The latest studies have demonstrated the clinical efficiency of Visudyne® in pancreatic tumor treatment, even in advanced stages of the disease [133]. Visudyne® is a photosensitizer with potential as a theranostic agent, under investigation with promising results in ovarian cancer [134].

Among the adverse reactions to Visudyne® administration, the most frequently reported are vision disorders, sensitivity in the region where the perfusion is administered (inflammation, pain, edema), and, rarely, allergic reactions.

3.3.5. PHOTOCHLOR®

HPPH is a second-generation photosensitizer with a macrocycle structure, lipophilic character, and better internalization at the cell level. Pandey et al. reported on the preparation of HPPH from methyl pheophorbide, a derivative extracted from spirulina [135,136].

The spectral behavior of HPPH is relevant for biomedical applications. In micellar solution with 1% Tween-80, HPPH registers an absorption maximum at 665 nm, an extinction molar coefficient $\varepsilon \sim 4.75 \times 10^4$ M^{-1} cm^{-1}, and singlet oxygen formation quantum yield $\Phi_\Delta = 0.48$ (determined in dichloromethane). Through its molecular structure, Photochlor® was proved to have some spectral characteristics and a certain amount of tumor accumulation superior to Photofrin® or Foscan® [117,137]. After intravenous administration, HPPH is selectively located in the tumor cell cytoplasm and presents a pharmacokinetic and pharmacotoxicological profile superior to Photofrin® [135,136].

In clinical applications, Photochlor® is administered intravenously at a dose of 0.15 mg/kg. Irradiation takes place at 665 nm, within 24–48 h from administration. Similar to Photofrin®, Photochlor® is not metabolized, but is cleared from human plasma and slowly excreted [138].

Photochlor® was investigated at the Roswell Park Cancer Institute (Buffalo, NY, USA) in stage I/II clinical studies of tumor formations at the esophagus level, in the mouth cavity, at the lung level, and for cervical intraepithelial neoplasm.

The use of Photochlor® in PDT did not show any photosensitivity in patients after treatment, but it is recommended to avoid solar exposure for 7–10 days after administration [139]. The capacity to internalize at the tumor mass level and the fluorescence properties associated with the HPPH structure allow the use of this photosensitizer as a marker for the identification of different types of cancer [140].

3.3.6. PURLYTIN®

Sn(II) ethyl-ethiopurpurin is a macrocyclic photosensitizer with Sn(II) metal ion. The compound was obtained and characterized by Morgan et al. and formulated and commercialized by Miravant Medica Technologies (Santa Barbara, CA, USA) [101–103,141].

The Sn(II) ethyl-ethiopurpurin complex occurs in the form of a water-insoluble powder, and for clinical application as a photosensitizer, a liposomal formulation is required.

Spectrally, the complex presents maximum absorption at 660 nm in the therapeutic field, thus assuring deep penetration in the tissues during irradiation. Its molecular absorption coefficient is $\varepsilon \sim 26{,}400\ M^{-1}\ cm^{-1}$, and the singlet oxygen formation quantum yield, measured in acetonitrile, is $\Phi_\Delta = 0.70$ [142].

In clinical applications, Purlytin® is administered intravenously as a liposomal solution at a dose of 1.2 mg/kg body weight. Activation by irradiation is carried out 24 h after administration, with laser light of 660 nm.

Purlytin® is under investigation in stage I/II clinical studies for PDT applied to breast cancer, Kaposi's sarcoma in AIDS patients, prostate cancer and brain metastases, and treatment of psoriasis. There have been no reported adverse side effects correlated with systemic toxicity with the administration of Purlytin®, but induced photosensitivity was observed in patients treated for a period of two weeks [101,102].

3.3.7. TOOKAD®

Padoporfin has a complex combination structure of a macrocycle bacteriochlorophyll ligand with Pd(II) as the metal ion. It is manufactured by Negma Lerads/Steba Biotech, Toussous le Noble, France, and is used in stage II/III clinical studies of photodynamic therapy against prostate cancer [143–146].

The absorption spectrum of Padaporfin presents a band in the therapeutic range with a maximum at 762 nm and a molar extinction coefficient of $\varepsilon \sim 8.85 \times 10^4\ M^{-1}\ cm^{-1}$. Consequently, it can be activated with light rays at 762 nm, i.e., deep tissue penetrating radiation. From the photophysical point of view, Padaporfin generates reactive species with good quantic efficiency of $\Phi_\Delta = 0.50$ (measured in dimethylsulfoxide) [101,102].

The clinical procedure for PDT mediated by Tookad® is similar to that described for Photofrin®; it is administered intravenously at a dose of 2–4 mg/kg, and the irradiation takes place half an hour after administration.

The photosensitizer molecule has a strong lipophilic character, and the advantage of rapid clearance from the organism (less than 20 min) without inducing cutaneous phototoxicity compared to Visudyne® and Lutrin® [101,102].

3.3.8. LUTRIN®

Texaphyrin lutetium or Motexafin lutetium (Lu-Tex) (commercial name Lutrin®) is a second-generation photosensitizer that was synthesized in 1994 by Sessler et al. by a condensation reaction of the derivatives of diformyltripyrrane and phenylenediamine [147]. The structure of the complex combination is characterized by water solubility and high selectivity in relation to tumor formation.

The spectral profile of the photosensitizer is described by absorption in the visible field at a maximum of 732 nm and a molar absorption coefficient of $\varepsilon = 4.2 \times 10^4\ M^{-1}\ cm^{-1}$ (determined in methanol solution). From the photophysical point of view, the capacity of Lutrin® to generate reactive species of singlet oxygen is lower compared to other second-generation photosensitizers; in methanol solution, the singlet oxygen quantum yield is $\Phi_\Delta = 0.11$.

The clinical photodynamic approach for antitumor therapy follows a procedure similar to that for Photofrin®, with a compound dose of 0.6–7.2 mg/kg and activation at 732 nm, 3 h after administration.

Lutrin® was approved for clinical use in therapy for recurrent prostate cancer and uterine and cervical cancer, and is in stage I/II/III clinical studies for breast cancer, melanoma, and Kaposi's sarcoma [101,102]. The results obtained in the clinical evaluation of the therapeutic effect of Lutrin® according to the compound dose, wavelength, and energy used in photochemical activation have confirmed good clinical efficacy by using a larger dose of the medicine (2 mg/kg) [148]. Clinical studies have also revealed some effects of photosensitivity induced by photodynamic therapy with Lutrin® in patients with prostate cancer. The ligand of the macrocycle texaphyrin structure, together with Gd (III) ion, generates a complex combination (Xcytrin™) with applicability in tumor diagnosis and as an agent with radiotherapy potential and a chemosensitizer [149,150].

Xcytrin™ was investigated in stage III clinical studies for the treatment of primary tumor formation at the brain level. Xcytrin™ is characterized by an excellent potential for locating tumor cells compared to healthy ones, and through photoactivation it generates reactive species in the presence of intracellular oxygen and determines cell death by apoptosis [101,102,151].

3.3.9. PHOTOSENS®

Photosens® is a mixture of complex structures with photosensitizing ability. Chemically, the basic structure is represented by a complex combination of a central ion Al(III) and a macrocycle phthalocyanine ligand; the peripheral substitutes of the macrocycle are $-SO_3H$ groups. In the absorption spectrum, it presents a Q band with a maximum 676 nm and a molar absorption coefficient of $\varepsilon \sim 20 \times 10^4$ M^{-1} cm^{-1}.

Photosens® has the largest molar absorption coefficient among all second-generation photosensitizers. The protocol for the clinical application of photodynamic therapy is similar to that applied for Photofrin®, with a reduced active substance dose. Photosens® is administered intravenously at a dose of 0.5–0.8 mg/kg, and is activated by light at 150 J/cm² at 676 nm 24–72h after administration.

Photosens® is produced by the Niopik pharmaceutical company (Moscow, Russia) and was evaluated in stage III clinical studies for the treatment of skin cancer with squamous cells, breast tumors, and oropharyngeal and lung cancer. In order to improve the degree of internalization at the tumor level, Swiss clinicians have developed a liposomal formulation of phthalocyanine with Zn(II) (CGP55847) that they evaluated in stage III clinical studies for the therapy of carcinoma with squamous cells [101,152].

4. Strategies Aimed at Improving the Therapeutic Potential of Porphyrin Photosensitizers

With all of the advantages that second-generation photosensitizers have, including their possible double role as markers and therapeutic agents, it is a priority for specialists in oncology and medicinal chemistry to identify new structures with optimal pharmacological potential to assure the most advantageous cost/benefit ratio with regard to improving the health of the population.

Simultaneously with the use of second-generation photosensitizers in therapy, in recent years, through structural functional modifications of tetrapyrrolic macrocycles, researchers have obtained and clinically investigated structures characterized by good solubility in biologic fluids, great capacity for generating reactive oxygen species (ROS), good selectivity in relation to tumor cells, good clearance, and the absence of side effects associated with administration. These structures define a new generation of photosensitizers, the so-called third-generation photosensitizers, with pharmacological profiles superior to those of the second generation. The design and synthesis of third generation PS are carried out in a chemo- and regioselective manner, taking into account the set of mechanisms that govern the identification and destruction of tumor cells.

The main disadvantages of second-generation photosensitizers include reduced solubility in water and biological fluids, difficult location at the cell level, reduced clearance rate from the organism, molecular aggregation tendency, and the effect of reducing the photodynamic efficiency [102–104].

Presently, research based on development porphyrins in the oncological field, mainly includes two type studies: porphyrins as theranostic agents, and porphyrins-bonded anticancer drugs. Regarding the first type of study, in this section we will briefly describe some strategies aimed at improving the therapeutic potential of porphyrin as theranostic agents.

4.1. Design and Synthesis of New Unsymmetrical Porphyrins for Theranostics Applications

Numerous studies have reported that structural modifications by attaching functional groups with different polarities to the macrocycle, or introducing metal ions into the tetrapyrrole macrocycles, increase the stability and diminish the aggregation tendency of PS molecules. On the other hand, attaching functional groups with rich π electron content to peripheral macrocycles determines the red shift of maximum PS absorption together with spectral optimization [153–159].

Among the macrocycle structures, at present, porphyrins are the most studied in terms of architectural modifications with improvements in the spectral and pharmacologic profiles.

In recent years, our research group synthesized a panel of new amphiphilic porphyrins by ecological and versatile technological approaches in the framework of *"green chemistry"* (solvent-free reactions activated by microwave irradiation), as efficient strategies in the synthesis of new photosensitizers [31,160–178].

In the present study, we reviewed some recently reported unsymmetrical porphyrins as potential theranostics photosensitizers for cancer. These structures, by their advantages (good cellular uptake and appreciable biocompatibility in relation to tumor cells, appreciable fluorescence properties, and good singlet oxygen generation capacity), could provide a useful reference for the development of theranostics tools.

The particular profile determined by the presence of polar groups such as –OH, –OCOCH$_3$ and –COCH$_3$, enhances their ability to act as weak intermolecular physical bond generators, with increasing solubility of PS in polar environments.

By implementing drug design strategies, research has been aimed at obtaining unsymmetrical porphyrins with functional groups with acceptable volume that do not overly increase the molecular mass of the photosensitizers. Taking into account that these compounds must cross the cell membrane, the hydrophilic/lipophillic character must not be lost, and this was why we selected A3B and A2B2 porphyrins (Table 2) [31,160–178].

Table 2. Unsymmetrical A3B and A2B2 type *meso*-substituted porphyrins.

A3B type *meso*-substituted porphyrins	R_1	$R_2 = R_3 = R_4$	Ref.
TCMPMOHp M(II)TCMPMOHp	2-OCH$_3$, 4-OH phenyl	4-(C(=O)OCH$_3$) phenyl	[31]
TMAPMOHp M(II)TMAPMOHp	2-OCH$_3$, 4-OH phenyl	2-OCH$_3$, 4-OC(=O)CH$_3$ phenyl	[160]
TMAPDOH	2-OH, 4-OH phenyl	2-OCH$_3$, 4-OC(=O)CH$_3$ phenyl	[161]

M = Zn(II), Cu(II)

Table 2. Cont.

	R₁ = R₃	R₂ = R₄	Ref.
TCMPOHo M(II)TCMPOHo	2-hydroxyphenyl	4-(methoxycarbonyl)phenyl	[162]
TRMOPP M(II)TRMOPP	3,4-methylenedioxyphenyl	4-(methoxycarbonyl)phenyl	[163,172]
TCMPOHp M(II)TCMPOHp	4-hydroxyphenyl	4-(methoxycarbonyl)phenyl	[164,170]
TCMPOMo M(II)TCMPOMo	2-acetoxy-3-methoxyphenyl	4-(methoxycarbonyl)phenyl	[165,172]
TCMPOHm M(II)TCMPOHm	3-hydroxyphenyl	4-(methoxycarbonyl)phenyl	[166,175]
TMAPOHm M(II)TMAPOHm	3-hydroxyphenyl	3-methoxy-4-acetoxyphenyl	[167,178]
TMAPOHo M(II)TMAPOHo	2-hydroxyphenyl	3-methoxy-4-acetoxyphenyl	[167,178]
M(II)TMAPOHp	4-hydroxyphenyl	3-methoxy-4-acetoxyphenyl	[171]
A2B2 type *meso*-substituted porphyrins	R₁ = R₃	R₂ = R₄	
TCMPDMOH	2-hydroxy-3-methoxyphenyl	4-(methoxycarbonyl)phenyl	[168]

Among the most important advantages offered by these structural configurations is the excellent solubility of PS in PEG 200, a nontoxic and pharmacologically accepted solvent. In fact, the choice of A3B and A2B2 isomers for our studies was justified because they ensure a balance between good cellular localization, the generation of good singlet oxygen yields for efficient PDT, and good solubility in solvents accepted for pharmaceutical formulation.

Regarding the relationship between the structural and photophysical profiles of the A3B type porphyrins, studies show that structural asymmetry induces slight changes in the values of singlet oxygen formation quantum yield, fluorescence emission quantum yield, and the fluorescence lifetime of PS relative to symmetrical structures (Table 3) [162,172,179].

Furthermore, a comparative study on the photophysical behavior of 5,10,15,20-tetrakis(3-hydroxyphenyl) porphyrin and the corresponding reduced form 5,10,15,20-tetrakis(3-

hydroxyphenyl) chlorin (active substance of Foscan®) highlighted very small differences between the photophysical parameter values (Φ_Δ, Φ_F, and τ_F) (Table 3) [180]. The date presented in Table 3 confirmed that the photophysical properties of PS molecules can be modified by the coordination of metal ion at the porphyrinic ring, and changes related to electronic structure of the metal ion [159,160,162,172,179].

Regarding the potential for the internalization of A3B and A2B2 porphyrins at the cellular level, in vitro studies on various cell lines confirmed their localization in the cellular environment depending on the cell type, dose of the compound, and loading time [160,161,168]. As example, a laser scanning microscopy image of 5-(4-hydroxy-3-methoxyphenyl)-10,15,20–tris(4-acetoxy-3-methoxyphenyl)porphyrin (10 µM) uptake by human HT-29 colon carcinoma cells, is presented in Figure 5 [160].

Table 3. Singlet oxygen formation quantum yield (Φ_Δ), fluorescence emission quantum yield (Φ_F), and fluorescence lifetime (τ_F) for a series of tetrapyrrolic compounds (solvent CHCl$_3$, except *marked structures were CH$_3$OH was used).

Tetrapyrrolic Compound		Ref.
TPP $\Phi_\Delta = 0.4$, $\Phi_F = 0.11$, $\tau_F = 9.5$ ns	Zn(II)TPP $\Phi_\Delta = 0.23$, $\Phi_F = 0.04$, $\tau_F = 2.2$ ns	[162]
TPPOHo $\Phi_\Delta = 0.32$, $\Phi_F = 0.10$, $\tau_F = 9.4$ ns	Zn(II)TPPOHo $\Phi_\Delta = 0.20$, $\Phi_F = 0.04$, $\tau_F = 3.7$ ns	[179]
TCMPOHo $\Phi_\Delta = 0.39$, $\Phi_F = 0.07$, $\tau_F = 9.2$ ns	Zn(II)TCMPOHo $\Phi_\Delta = 0.27$, $\Phi_F = 0.03$, $\tau_F = 2$ ns	[162]

Table 3. *Cont.*

Tetrapyrrolic Compound		Ref.
TCMPOMo Φ_Δ = 0.42, Φ_F = 0.09, τ_F = 7.89 ns	**Zn(II)TCMPOMo** Φ_Δ = 0.16, Φ_F = 0.06, τ_F = 1.75 ns	[172]
mTHPP* Φ_Δ = 0.46, Φ_F = 0.10, τ_F = 9.2 ns	***m*THPC*** Φ_Δ = 0.43, Φ_F = 0.09, τ_F = 8.6 ns	[180]

In vitro biological studies performed on various cell lines (human colon carcinoma HT-29 cells, mouse L929 fibroblasts, peripheral blood mononuclear cells, human normal dermal HS27 cells, HaCaT keratinocytes, human peripheral blood SC monocytes) confirmed good biocompatibility for these asymmetrical structural configurations [160,161,168]. Furthermore, new amphiphilic porphyrins generated important amounts of singlet oxygen when activated with light at 635 nm [161,181].

In a recent study, one of the A3B-type asymmetric structures synthesized by us was investigated in vitro for its ability to reduce the number of tumor cells by PDT. In addition, the profile of stress gene expression changes triggered in vitro by porphyrin PDT in HT-29 human colon carcinoma cells was established [181].

By in vitro experimental PDT performed with TMAPMOHp, we observed the concomitant activation of particular cellular responses to oxidative stress, hypoxia, DNA damage, proteotoxic stress, and inflammation. This web of interconnected stressors underlies cell death, but can also trigger protective mechanisms that may delay tumor cell death or even defend cells against the deleterious effects of PDT. This stress-related molecular profile allowed us to identify potential therapeutic targets, such as the cytoprotective transcription factor NRF2, which could be used to develop co-therapies aimed at increasing PDT efficacy in cancer cells or protecting normal tissues [181].

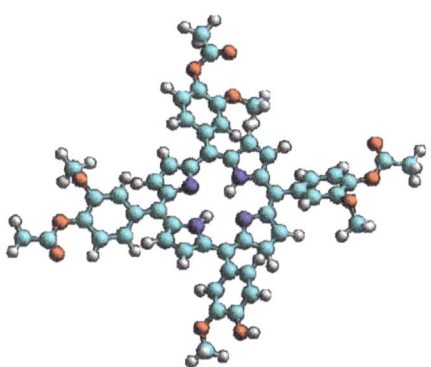

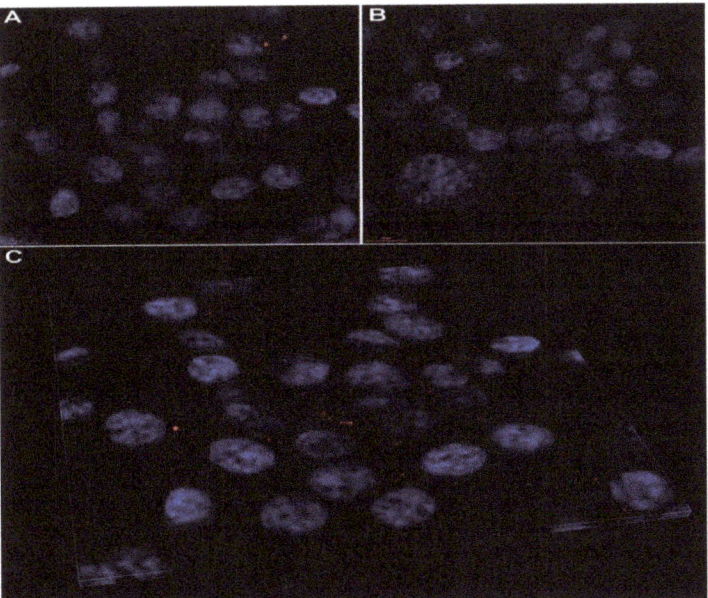

Figure 5. Representative laser scanning microscopy image of 5-(4-hydroxy-3-methoxyphenyl)-10,15,20–tris(4-acetoxy-3-methoxyphenyl) porphyrin (10 μM) uptake by human HT-29 colon carcinoma cells: (**A**): porphyrin (red) scattered throughout cytosol compared to control (**B**); (**C**): 3D volume rendering of 2.36 μm z-stack from A; nuclei were stained with DAPI (blue); scale bar 10 μm in (**A**,**B**) and 20 μm in (**C**) [160].

4.2. Functionalizing of Porphyrin Type Macrocycles with Fragments of Bioactive Molecules

There are several strategies for improving the clinical performance of porphyrins, such as using a modified derivative design or specific delivery vehicles. The functionalizing of porphyrin macrocycles with moieties of carbohydrate, amino acids, antibody, and peptide, or their encapsulation into liposomes, micelles and nanoparticles, have been highlighted as a main strategy in development of third-generation PS.

The most interesting systems with photosensitizing properties nowadays are developed using natural resources, such as chlorophyll derivatives (chlorin compounds), which have a similar structure to hemoglobin and can be preferentially located in tumor cells [55–57]. The research group led by Li synthesized and evaluated a new photosensi-

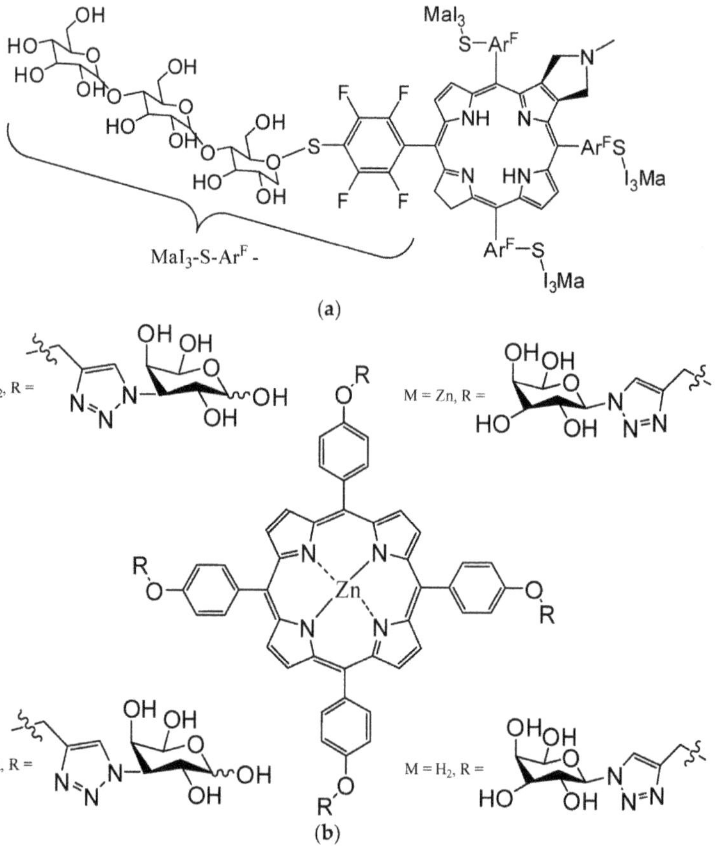

Figure 8. Structures of macrocycles photosensitizers functionalized with maltotriose (a) [185] or galactose (b) [185] fragments.

4.3. Functionalizing of Porphyrin with Metal-Based Nanoparticles

An alternative to using structural modifications to improve the biodistribution of PS and increase its accumulation at tumor sites is to incorporate PS in the transport agents (liposomal systems, micelles, or nanoparticles [163,168].

Nanomedicine is an ever-growing field with huge potential for future discoveries and substantial progress in medicine, especially in cancer theranostics. In recent years, interdisciplinary studies have pointed out that translational nanotheranostics are a promising way forward in medicine. Biotech, pharmaceutical, and medical sciences companies have been active in its evolution, and are dynamic collaborators with researchers, governments, and educational institutions in developing and translating cancer nanomedicine. Two iron oxide contrast agents are now on the market, Ferridex and Gastro MARK from Advanced Magnetics, Inc., Cambridge, MA., and according to the analyses performed on several metal-based NPs (iron oxide, magnetic NPs, gold nanorods, colloidal gold) for imaging/theranostics, most clinical trials in nanomedicine are confined to polymer/lipid NPs, probably due to concerns regarding the adverse effects and/or toxicity of metal-based NPs [188].Therefore, there is a need to develop nanotheranostic tools and associated strategies, complemented by comprehensive toxicological studies, in order to find solutions for reducing the counteracting of clearly defined toxicity issues, while exploiting their huge potential as theranostic agents.

Polymer-coated NPs can act as carriers for porphyrins. Metal-based NPs carry an additional theranostic property: irradiation with near infrared (NIR) light will activate the covering photosensitizer in PDT and trigger a concurrent photothermal effect, mediated by the associated metal. This will lead to the enhanced power of the nanosystems for tumor cell ablation and/or a vasodilatation-mediated increase in NP accumulation in tumors. Moreover, the same NIR radiation used to activate porphyrinic photosensitizers can induce additional ROS production by direct interaction with the metal core of NPs. By their excellent photophysical and photochemical properties, porphyrins can be good generators of tumor-destructive ROS in PDT. Additionally, the fluorescence of the compounds allows the monitoring of NS uptake by cancer cells, thus improving their imaging power (combined magnetic and fluorescence characteristics). As porphyrinic compounds preferentially accumulate in tumors, a concentration of NS in the diseased tissue can be achieved, the therapeutic dose can consequently be lowered, and unwanted side effects could be partially avoided.

To overcome the drawbacks related to the aggregation and low solubility of PS molecules in biological media, in recent studies, functionalized metal-based nanocarriers were used to deliver porphyrin to tumors [189–192]. Moreover, coating inorganic NPs with biocompatible polymers (PEG, chitosan, alginate, heparin, cellulose, gelatin, etc.) can reduce particle uptake by phagocytes, promoting the stabilization and biocompatibility of the system and increasing the blood circulation time. Hydrophilic polymeric chains can be physically linked to porphyrins in order to enhance porphyrin attachment and the biocompatibility of NP/NS. For metal NP synthesis, several methodologies can be used, including co-precipitation. Porphyrins can be attached to NPs by means of appropriate linkers. Applying a thin silica layer on the surface of NPs greatly improves the conjugation of porphyrin on the functionalized NP surface.

Recently, Vieira Ferreira and collab. [189,190] developed superparamagnetic iron NPs coated with a porphyrin derivative. The surface photochemistry of two tetrapyrrolic structures was reported as TCMP (Figure 9) and TCMPMOH$_P$, covalently bound to silica-coated magnetite nanoparticles (Fe$_3$O$_4$@SiO$_2$) and imbibed into a polyethylene glycol (PEG) matrix, in the form of a fine powder.

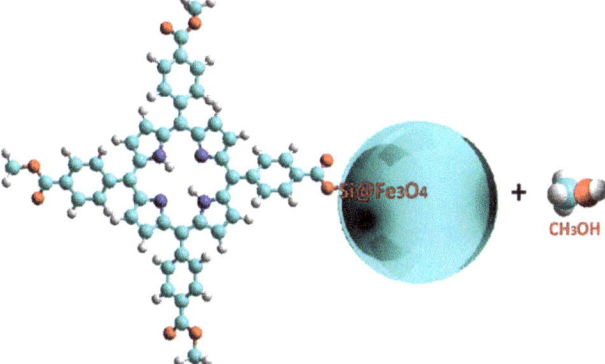

Figure 9. New covalent bond formed after transesterification reaction of TCMP with silanol groups of the surface of silica-coated magnetite nanoparticles [190].

Two populations of porphyrin molecules exist, free porphyrin (unquenched) and porphyrin with Fe$_3$O$_4$@SiO$_2$ NPs attached (quenched). In this way, the amount of available excited porphyrin for singlet oxygen formation is reduced, thus the presence of Fe$_3$O$_4$@SiO$_2$ NPs may not favor this specific PDT action by the type II mechanism, but it prevents aggregation of the porphyrins, even when a PEG matrix is used, and type I mechanisms remain active. In conclusion, the possibility of loading porphyrinic molecules covalently bound to ferromagnetic nanoparticles represents a great advantage in nanotechnology, because it avoids the major problems related to the aggregation and low solubility of PS in biological media; moreover, this approach allows the establishment of an accurate diagnosis and the optimization of the therapeutic outcome.

5. Conclusions

Tetrapyrrolic compounds are among the most studied with regard to their antitumor potential, because their versatile structure and spectral profile make them applicable as photosensitizers.

At present, one of the primary objectives of research in the pharmaceutical and oncological fields is to establish an optimum protocol for the identification and treatment of tumor formations by using molecules with exceptional pharmacological properties and minimal side effects.

In this review, we summarized the theoretical aspects regarding the structural, spectral, and biomedical profiles of porphyrin macrocycles. The bibliographical data concerning the properties and theranostic potential of this tetrapyrrole type are complemented by some ideas related to strategies for improving the biodistribution of PS at tumor sites. Our research group synthesized and characterized a panel of new amphiphilic porphyrins exhibiting structural and spectral profiles suitable for theranostic applications. "Green chemistry" (solvent-free reactions activated by microwave irradiation) was applied for the synthesis of amphiphilic porphyrins. We can appreciate the fact that extensive interdisciplinary research, accompanied by detailed toxicological assessments, can guide researchers in finding the most appropriate variants for the development of new tetrapyrroles as theranostic agents. Furthermore, the mechanisms of action at the cellular level should be thoroughly investigated for scientific documentation of the balance between benefits and side effects. Finally, we anticipate that the information presented in this review will encourage the design and development of new porphyrin macrocycles as theranostic agents.

Author Contributions: Conceptualization, R.B., N.R. and G.M.; writing—original draft preparation, N.R., R.B. and G.M.; writing—review and editing, R.B., N.R., G.M., I.F.M., R.P.S., D.L., A.M.B., D.P.M. and L.F.V.F. All authors have read and agreed to the published version of the manuscript.

Funding: This research received no external funding.

Institutional Review Board Statement: Not applicable.

Informed Consent Statement: Not applicable.

Data Availability Statement: Not applicable.

Acknowledgments: The research was supported by the Ministry of Research, Innovation and Digitalization, Romania, through PORPHYDERM project (ctr. no. 637PED/2022) and the Nucleu project PN 23.16.02.01/2022.

Conflicts of Interest: The authors declare no conflict of interest.

Abbreviations

TCMPMOH$_P$	5-(4-hydroxy-3-methoxyphenyl)-10,15,20-tris(4-carboxymethylphenyl)porphyrin
Zn(II)TCMPMOH$_P$	5-(4-hydroxy-3-methoxyphenyl)-10,15,20-tris(4-carboxymethylphenyl)porphyrinatozinc(II)
Cu(II)TCMPMOH$_P$	5-(4-hydroxy-3-methoxyphenyl)-10,15,20-tris(4-carboxymethylphenyl)porphyrinatocopper(II)
TMAPMOHp	5-(4-hydroxy-3-methoxyphenyl)-10,15,20-tris(4-acetoxy-3-methoxyphenyl)porphyrin
Zn(II)TMAPMOHp	5-(4-hydroxy-3-methoxyphenyl)-10,15,20-tris(4-acetoxy-3-methoxyphenyl)porphyrinatozinc(II)
Cu(II)TMAPMOHp	5-(4-hydroxy-3-methoxyphenyl)-10,15,20-tris(4-acetoxy-3-methoxyphenyl)porphyrinatocopper(II)
TMAPDOH	5-(2,4-dihydroxyphenyl)-10,15,20-tris(4-acetoxy-3-methoxyphenyl)porphyrin
TCMPOHo	5-(2-hydroxyphenyl)-10,15,20-tris(4-carboxymethylphenyl)porphyrin
Zn(II)TCMPOHo	5-(2-hydroxyphenyl)-10,15,20-tris(4-carboxymethylphenyl)porphyrinatozinc(II)
Cu(II)TCMPOHo	5-(2-hydroxyphenyl)-10,15,20-tris(4-carboxymethylphenyl)porphyrinatocopper(II)
TRMOPP	5-[(3,4-methylenedioxy)phenyl]-10,15,20-tris(4-carboxymethylphenyl)porphyrin

Zn(II)TRMOPP	5-[(3,4-methylenedioxy)phenyl]-10,15,20-tris(4-carboxymethylphenyl)porphyrinatozinc(II)
Cu(II)TRMOPP	5-[(3,4-methylenedioxy)phenyl]-10,15,20-tris(4-carboxymethylphenyl)porphyrinatocopper(II)
TCMPOHp	5-(4-hydroxyphenyl)-10,15,20-tris(4-carboxymethylphenyl)porphyrin
Zn(II)TCMPOHp	5-(4-hydroxyphenyl)-10,15,20-tris(4-carboxymethylphenyl)porphyrinatozinc(II)
Cu(II)TCMPOHp	5-(4-hydroxyphenyl)-10,15,20-tris(4-carboxymethylphenyl)porphyrinatocopper(II)
TCMPOMo	5-(4-acetoxy-3-methoxyphenyl)-10,15,20-tris(4-carboxymethylphenyl)porphyrin
Zn(II)TCMPOMo	5-(4-acetoxy-3-methoxyphenyl)-10,15,20-tris(4-carboxymethylphenyl)porphyrinatozinc(II)
Cu(II)TCMPOMo	5-(4-acetoxy-3-methoxyphenyl)-10,15,20-tris(4-carboxymethylphenyl)porphyrinatocopper(II)
TCMPOHm	5-(3-hydroxyphenyl)-10,15,20-tris(4-carboxymethylphenyl)porphyrin
Zn(II)TCMPOHm	5-(3-hydroxyphenyl)-10,15,20-tris(4-carboxymethylphenyl)porphyrinatozinc(II)
Cu(II)TCMPOHm	5-(3-hydroxyphenyl)-10,15,20-tris(4-carboxymethylphenyl)porphyrinatocopper(II)
TMAPOHm	5-(3-hydroxyphenyl)-10,15,20-tris(4-acetoxy-3-methoxyphenyl)porphyrin
Zn(II)TMAPOHm	5-(3-hydroxyphenyl)-10,15,20-tris(4-acetoxy-3-methoxyphenyl)porphyrinatozinc(II)
Cu(II)TMAPOHm	5-(3-hydroxyphenyl)-10,15,20-tris(4-acetoxy-3-methoxyphenyl)porphyrinatocopper(II)
TMAPOHo	5-(2-hydroxyphenyl)-10,15,20-tris(4-acetoxy-3-methoxyphenyl)porphyrin
Zn(II)TMAPOHo	5-(2-hydroxyphenyl)-10,15,20-tris(4-acetoxy-3-methoxyphenyl)porphyrinatozinc(II)
Cu(II)TMAPOHo	5-(2-hydroxyphenyl)-10,15,20-tris(4-acetoxy-3-methoxyphenyl)porphyrinatocopper(II)
Zn(II)TMAPOHp	5-(4-hydroxyphenyl)-10,15,20-tris(4-acetoxy-3-methoxyphenyl)porphyrinatozinc(II)
Cu(II)TMAPOHp	5-(4-hydroxyphenyl)-10,15,20-tris(4-acetoxy-3-methoxyphenyl) porphyrinatocopper(II)
TCMPDMOH	5,15-bis-(4-hydroxy-3-methoxyphenyl)-10,20-bis(4-carboxymethylphenyl)porphyrin
TPP	5,10,15,20-tetrakis-phenyl porphyrin
Zn(II)TPP	5,10,15,20-tetrakis-phenyl porphyrinatozinc(II)
TPPOHo	5-(2-hydroxyphenyl)-10,15,20-triphenyl porphyrin
Zn(II)TPPOHo	5-(2-hydroxyphenyl)-10,15,20-triphenyl porphyrinatozinc(II)
mTHPP	5,10,15,20-tetrakis(3-hydroxyphenyl)porphyrin
mTHPC	5,10,15,20-tetrakis(3-hydroxyphenyl)chlorin
TCMP	5,10,15,20-tetrakis(4-carboxymethylphenyl)porphyrin

References

1. Siegel, R.L.; Miller, K.D.; Jemal, A. Cancer statistics. *CA Cancer J. Clin.* **2020**, *70*, 7–30. [CrossRef] [PubMed]
2. Siegel, R.L.; Miller, K.D.; Fuchs, H.E.; Jemal, A. Cancer statistics. *CA Cancer J. Clin.* **2021**, *71*, 7–33. [CrossRef]
3. Siegel, R.L.; Miller, K.D.; Fuchs, H.E.; Jemal, A. Cancer statistics. *CA Cancer J. Clin.* **2022**, *72*, 7–33. [CrossRef]
4. Sarbadhikary, P.; George, B.P.; Abrahamse, H. Recentadvances in photosensitizers as multifunctional theranostic agents forimaging-guided photodynamictherapy of cancer. *Theranostics* **2021**, *11*, 9054–9088. [CrossRef] [PubMed]
5. Sung, H.; Ferlay, J.; Siegel, R.L.; Laversanne, M.; Soerjomataram, I.; Jemal, A.; Bray, F. Global cancer statistics 2020: Globocan estimates of incidence and mortality worldwide for 36 cancers in 185 countries. *CA Cancer J. Clin.* **2021**, *71*, 1–41. [CrossRef] [PubMed]
6. Ferlay, J.; Colombet, M.; Soerjomataram, I.; Parkin, D.M.; Piñeros, M.; Znaor, A.; Bray, F. Cancer statistics for the year 2020: An overview. *Int. J. Cancer* **2021**, *149*, 778–789. [CrossRef]
7. Kelkar, S.S.; Reineke, T.M. Theranostics: Combining imaging and therapy. *Bioconjug. Chem.* **2011**, *22*, 1879–1903. [CrossRef]
8. Josefsen, L.B.; Boyle, R.W. Unique diagnostic and therapeutic roles of porphyrins and phthalocyanines in photodynamic therapy, imaging and theranostics. *Theranostics* **2012**, *2*, 916–966. [CrossRef] [PubMed]
9. Ethirajan, M.; Chen, Y.; Joshi, P.; Pandey, R.K. The role of porphyrin chemistry in tumor imaging and photodynamic therapy. *Chem. Soc. Rev.* **2011**, *40*, 340–362. [CrossRef]
10. Pandey, R.K.; Zheng, G. Porphyrins as photosensitizers in photodynamic therapy. In *The Porphyrin Handbook*; Kadish, K.M., Guilard, R., Smith, K.M., Eds.; Academic Press: New York, NY, USA, 2000; Volume 6, pp. 157–230.
11. Simpson, M.C.; Novikova, I.N. Porphyrins: Electronic structure and ultraviolet/visible absorption spectroscopy. In *Fundamentals of Porphyrin Chemistry: A 21st Century Approach*; Brothers, P.J., Senge, O.M., Eds.; John Wiley & Sons Ltd.: New Jersey, NJ, USA, 2022; Volume 1, pp. 505–586.
12. Tsolekile, N.; Nelana, S.; Oluwafemi, O.S. Porphyrin as Diagnostic and Therapeutic Agent. *Molecules* **2019**, *24*, 2669. [CrossRef] [PubMed]
13. Plekhova, N.; Shevchenko, O.; Korshunova, O.; Stepanyugina, A.; Tananaev, I.; Apanasevich, V. Development of Novel Tetrapyrrole Structure Photosensitizers for Cancer Photodynamic Therapy. *Bioengineering* **2022**, *9*, 82. [CrossRef] [PubMed]
14. Zhao, X.; Liu, J.; Fan, J.; Chao, H.; Peng, X. Recent progress in photosensitizers for overcoming the challenges of photodynamic therapy: From molecular design to application. *Chem. Soc. Rev.* **2021**, *50*, 4185–4219. [CrossRef] [PubMed]
15. Henderson, B.W.; Dougherty, T.J. How does photodynamic therapy work? *Photochem. Photobiol.* **1992**, *55*, 145–157. [CrossRef] [PubMed]
16. Dolmans, D.E.; Fukumura, D.; Jain, R.K. Photodynamic therapy for cancer. *Nat. Rev. Cancer* **2003**, *3*, 380–387. [CrossRef] [PubMed]
17. Juzeniene, A.; Peng, Q.; Moan, J. Milestones in the development of photodynamic therapy and fluorescence diagnosis. *Photochem. Photobiol. Sci.* **2007**, *6*, 1234–1245. [CrossRef] [PubMed]

18. Meyer-Betz, F. Untersuchung uber die biologische (photodynamische) wirkung des hamatoporphyrins und anderer derivate des blutundgallenfarbstoffs. *Dtsch. Arch. Klin. Med.* **1913**, *112*, 476–503.
19. Policard, A. Etudes sur les aspects offerts par des tumeursexperimentales examinees a la lumiere des Woods. *Compt. Rend. Soc. Biol.* **1924**, *91*, 1423–1426.
20. Battersby, A.R.; Fookes, C.J.; Matcham, G.W.; McDonald, E. Biosynthesis of the pigments of life: Formation of the macrocycle. *Nature* **1980**, *285*, 17–21. [CrossRef]
21. Brothers, P.J.; Gosh, A. Coordination chemistry. In *Fundamentals of Porphyrin Chemistry: A 21st Century Approach*; Brothers, P.J., Senge, O.M., Eds.; John Wiley & Sons Ltd.: New Jersey, NJ, USA, 2022; Volume 4, pp. 141–146.
22. Sessler, J.L.; Tomat, E. Transition-metal complexes of expanded porphyrins. *Acc. Chem. Res.* **2007**, *40*, 371–379. [CrossRef]
23. Nelson, J.S.; Roberts, W.G.; Berns, M.W. In vivo studies on the utilization of Mono-L-aspartyl chlorin (NPe6) for photodynamic therapy. *Cancer Res.* **1987**, *47*, 4681–4685.
24. Mauzerall, D. *Porphyrins, Chlorophyll, and Photosynthesis*; Springer: Berlin/Heidelberg, Germany, 1977; pp. 117–124. [CrossRef]
25. Armstrong, D.; Stratton, R.D. *Oxidative Stress and Antioxidant Protection: The Science of Free Radical Biology and Disease*; John Wiley &Sons: Hoboken, NJ, USA, 2016; pp. 415–470.
26. Nantes, I.L.; Crespilho, F.N.; Mugnol, K.C.U.; Araujo-Chaves, J.C.; Nascimento, O.R.; Pinto, S.M.S. Magnetic circular dichroism applied in the study of symmetry and functional properties of porphyrinoids. In *Circular Dichroism: Theory and Spectroscopy*, 1st ed.; David, S.R., Ed.; Nova Science Publishers: New York, NY, USA, 2010; pp. 321–344.
27. Van Santen, R. Catalysis in perspective: Historic review. In *Catalysis: From Principles to Applications*, 1st ed.; Beller, M., Renken, A., Van Santen, R., Eds.; Wiley-VCH: Weinheim, Germany, 2012; pp. 3–19.
28. Gouterman, M. Optical spectra and electronic structure of porphyrins and related rings. In *The Porphyrins*; Dolphin, D., Ed.; Academic Press: New York, NY, USA, 1978; Volume 3, pp. 11–87.
29. Gouterman, M.; Wagniere, G.H.; Snyder, L.C. Spectra of porphyrins: Part II. Four orbital model. *J. Mol. Spectrosc.* **1963**, *11*, 108–127. [CrossRef]
30. Milgrom, L.R. What porphyrins are and what they do. In *The Colours of Life. An Introduction to the Chemistry of Porphyrins and Related Compounds*; Oxford University Press: Oxford, UK, 1997; Volume 1, pp. 1–85.
31. Boscencu, R.; Socoteanu, R.P.; Manda, G.; Radulea, N.; Anastasescu, M.; Gama, A.; Ferreira Machado, I.; Vieira Ferreira, L.F. New A$_3$B porphyrins as potential candidates for theranostic. Synthesis and photochemical behaviour. *Dyes Pigments* **2019**, *160*, 410–417. [CrossRef]
32. Dar, U.A.; Shah, S.A. UV–visible and fluorescence spectroscopic assessment of meso-tetrakis-(4-halophenyl) porphyrin; H2TXPP (X=F, Cl, Br, I) in THF and THF-water system: Effect of pH and aggregation behaviour. *Spectrochim. Acta Part A Mol. Biomol. Spectrosc.* **2020**, *240*, 118570. [CrossRef] [PubMed]
33. Wu, Q.; Xia, R.; Li, C.; Li, Y.; Sun, T.; Xie, Z.; Jing, X. Nanoscale aggregates of porphyrins: Red-shifted absorption, enhanced absorbance and phototherapeutic activity. *Mater. Chem. Front.* **2021**, *5*, 8333–8340. [CrossRef]
34. Dabrowski, J.M.; Pucelik, B.; Regiel-Futyra, A.; Brindell, M.; Mazuryk, O.; Kyzioł, A.; Arnaut, L.G. Engineering of relevant photodynamic processes through structural modifications of metallotetrapyrrolic photosensitizers. *Coord. Chem. Rev.* **2016**, *325*, 67–101. [CrossRef]
35. Meshkov, I.N.; Bulach, V.; Gorbunova, Y.G.; Gostev, F.E.; Nadtochenko, V.A.; Tsivadze, A.Y.; Hosseini, M.W. Tuning photochemical properties of phosphorus (V) porphyrin photosensitizers. *Chem. Commun.* **2017**, *53*, 9918–9921. [CrossRef]
36. Horváth, O.; Valicsek, Z.; Fodor, M.A.; Major, M.M.; Imran, M.; Grampp, G.; Wankmüller, A. Visible light-driven photophysics and photochemistry of water-soluble metalloporphyrins. *Coord. Chem. Rev.* **2016**, *325*, 59–66. [CrossRef]
37. Vieira Ferreira, L.F.; Ferreira Machado, I.L. Surface photochemistry: Organic molecules within nanocavities of calixarenes. *Curr. Drug Discov. Technol.* **2007**, *4*, 229–245. [CrossRef]
38. Strickler, S.J.; Berg, R.A. Relationship between Absorption Intensity and Fluorescence Lifetime of Molecules. *J. Chem. Phys.* **1962**, *37*, 814–822. [CrossRef]
39. Harriman, A.; Hosie, R.J. Luminescence of porphyrins and metalloporphyrins. Fluorescence of substituted tetraphenylporphyrins. *J. Photochem.* **1981**, *15*, 163–167. [CrossRef]
40. Nifiatis, F.; Athas, J.C.K.; Gunaratne, D.; Gurung, Y.; Monette, K.; Shivokevich, P.J. Substituent effects of porphyrin on singlet oxygen generation quantum yields. *Open Spectrosc. J.* **2011**, *5*, 1–12. [CrossRef]
41. Harriman, A. Luminescence of porphyrins and metalloporphyrins, Zinc(II), nickel(II) and manganese(II) porphyrins. *J. Chem. Soc. Faraday Trans. 1 Phys. Chem. Condens. Phases* **1980**, *76*, 1978–1985. [CrossRef]
42. Chen, C.Y.; Sun, E.; Fan, D. Synthesis and physicochemical properties of metallobacteriochlorins. *Inorg. Chem.* **2012**, *51*, 9443–9464. [CrossRef] [PubMed]
43. Von Tapiener, H.; Jesionek, A. TherapeutischeVersuchemitfluoreszierendenStoffen. *Münchener Med. Wochenschr.* **1903**, *50*, 2042–2051.
44. Silver, H. Psoriasis vulgaris treated with hemathoporphyrin. *Arch Dermatol. Physiol.* **1937**, *36*, 1118–1119.
45. Dougherty, T.J.; Kaufman, J.E.; Goldfarb, A.; Weishaupt, K.R.; Boyle, D.; Mittleman, A. Photoradiation therapy for the treatment of malignant tumors. *Cancer Res.* **1978**, *38*, 2628–2635. [PubMed]
46. Dahlman, A.; Wile, A.G.; Burns, R.G.; Johnson, F.; Berns, M.W. Laser photoradiation therapy of cancer. *Cancer Res.* **1982**, *43*, 430–434.
47. Diezel, W.; Meffert, H.; Sonnichsen, N. Stability of nystatin in an o/w-cream. *Dermatol. Monatsschr.* **1980**, *166*, 75–79.

48. Dougherty, T.J.; Henderson, B.; Schwartz, S.; Winkelman, J.W.; Lipson, R.L. Historical perspective Photodynamic therapy. In *Basic Principles and Clinical Application*; Henderson, B.W., Dougherty, T.J., Eds.; Marcel Dekker Inc.: New York, NY, USA, 1992; pp. 1–15.
49. Dougherty, T.J. Hematoporphyrin as a photosensitizer of tumors. *Photochem. Photobiol.* **1983**, *38*, 377–379. [CrossRef]
50. Dougherty, T.J. Photodynamic therapy. *Photochem.Photobiol.* **1993**, *58*, 895–900. [CrossRef]
51. Jori, G. Tumour photosensitizers: Approaches to enhance the selectivity and efficiency of photodynamic therapy. *J. Photochem. Photobiol B Biol.* **1996**, *36*, 87–93. [CrossRef]
52. Detty, M.R.; Gibson, S.L.; Wagner, S.J. Current clinical and preclinical photosensitizers for use in photodynamic therapy. *J. Med. Chem.* **2004**, *47*, 3897–3915. [CrossRef] [PubMed]
53. Boyle, R.B.; Dolphin, D. Structure and biodistribution relationships of photodynamic sensitizers. *Photochem. Photobiol.* **1996**, *64*, 469–485. [CrossRef] [PubMed]
54. Kudinova, N.V.; Berezov, T.T. Photodynamic therapy of cancer: Search for ideal photosensitizer. *Biochem. Suppl. Ser. B Biomed. Chem.* **2010**, *4*, 95–103. [CrossRef]
55. Myrzakhmetov, B.; Arnoux, P.; Acherar, S.; Vanderesse, R.; Frochot, C. Folic acid conjugates with photosensitizers for cancer targeting in photodynamic therapy: Synthesis and photophysical properties. *Bioorg. Med. Chem.* **2017**, *25*, 1–10.
56. Clark, P.A.; Alahmad, A.J.; Qian, T.; Zhang, R.R.; Wilson, H.K.; Weichert, J.P.; Palecek, S.P.; Kuo, J.S.; Shusta, E.V. Analysis of cancer-targeting alkylphosphocholine analog permeability characteristics using a human induced pluripotent stem cell blood-brain barrier model. *Mol. Pharm.* **2016**, *13*, 3341–3349. [CrossRef] [PubMed]
57. Zhang, R.R.; Swanson, K.I.; Hall, L.T.; Weichert, J.P.; Kuo, J.S. Diapeutic cancer-targeting alkylphosphocholine analogs may advance management of brain malignancies. *CNS Oncol.* **2016**, *5*, 223–231. [CrossRef]
58. Yoon, I.; Li, J.Z.; Shim, Y.K. Advance in photosensitizers and light delivery forphotodynamic therapy. *Clin. Endosc.* **2013**, *46*, 7–23. [CrossRef]
59. Van Straten, D.; Mashayekhi, V.; de Bruijn, H.S.; Oliveira, S.; Robinson, D.J. Oncologic photodynamic therapy: Basic principles, current clinical status and future directions. *Cancers* **2017**, *9*, 19. [CrossRef]
60. Tanaka, M.; Kataoka, H.; Mabuchi, M.; Sakuma, S.; Takahashi, S.; Tujii, R.; Akashi, H.; Ohi, H.; Yano, S.; Morita, A.; et al. Anticancer effects of novel photodynamic therapy with glycoconjugated chlorin for gastric and colon cancer. *Anticancer Res.* **2011**, *31*, 763–769.
61. Lapa, C.; Schreder, M.; Schirbel, A.; Samnick, S.; Kortum, K.M.; Hermann, H.; Kropf, S.; Einsele, H.; Buck, A.K.; Wester, H.J. [68Ga] Penixafor-PET/CT for imaging of chemokine receptor CXCR4 expression in multiple myeloma–Comparison to [18F] FDG and laboratory values. *Theranostics* **2017**, *7*, 205–212. [CrossRef] [PubMed]
62. Broughton, L.J.; Giuntini, F.; Savoie, H.; Bryden, F.; Boyle, R.W.; Maraveyas, A.; Madden, L.A. Duramycin-porphyrin conjugates for targeting of tumour cells using photodynamic therapy. *J. Photochem. Photobiol. B.* **2016**, *163*, 374–384. [CrossRef] [PubMed]
63. Yan, G.; Li, Z.; Xu, W.; Zhou, C.; Yang, L.; Zhang, Q.; Li, L.; Liu, F.; Han, L.; Ge, Y. Porphyrin-containing polyaspartamide gadolinium complexes as potential magnetic resonance imaging contrast agents. *Int. J. Pharm.* **2011**, *407*, 119–125. [CrossRef] [PubMed]
64. Mokwena, M.G.; Kruger, C.A.; Ivan, M.; Heidi, A. A review of nanoparticle photosensitizer drug delivery uptake systems for photodynamic treatment of lung cancers. *Photodiagn. Photodyn. Ther.* **2018**, *22*, 147–154. [CrossRef] [PubMed]
65. Fakayode, O.J.; Kruger, C.A.; Songca, S.P.; Abrahamse, H.; Oluwafemi, O.S. Photodynamic therapy evaluation of methoxy-polyethylene glycol-thiol-SPIONs-gold-meso-tetrakis(4-hydroxyphenyl) porphyrin conjugate against breast cancer cells. *Mater. Sci. Eng. C* **2018**, *92*, 737–744. [CrossRef] [PubMed]
66. Penon, O.; Marín, M.J.; Russell, D.A.; Pérez-García, L. Water soluble, multifunctional antibody-porphyrin gold nanoparticles for targeted photodynamic therapy. *J. Colloid Interface Sci.* **2017**, *496*, 100–110. [CrossRef]
67. Robertson, C.A.; Evans, D.H.; Abrahamse, H. Photodynamic therapy (PDT): A short review on cellular mechanisms and cancer research applications for PDT. *J. Photochem. Photobiol. B* **2009**, *96*, 1–8. [CrossRef]
68. Baron, E.D.; Suggs, A.K. Introduction to photobiology. *Dermatol. Clin.* **2014**, *32*, 255–266. [CrossRef]
69. Josefsen, L.B.; Boyle, R.W. Photodynamic therapy: Novel third-generation photosensitizers one step closer? *Br. J. Pharmacol.* **2008**, *154*, 1–3. [CrossRef]
70. Wayne, C.E.; Wayne, R.P. Photochemical principles. In *Photochemistry, of Oxford Chemistry Primers, Oxford Science*; Oxford University Press: Oxford, UK, 1996; Volume 39, pp. 11–12.
71. Schweitzer, C.; Schmidt, R. Physical mechanisms of generation and deactivation of singlet oxygen. *Chem. Rev.* **2003**, *103*, 1685–1757. [CrossRef]
72. Gilbert, A.; Baggott, J. *Essentials of Molecular Photochemistry*; Blackwell Scientific: Oxford, UK, 1991; Chapter 4; pp. 92–99.
73. Dougherty, T.J.; Gomer, C.J.; Henderson, B.W.; Jori, G.; Kessel, D.; Korbelik, M.; Moan, J.; Peng, Q. Photodynamic therapy. *J. Natl. Cancer Inst.* **1998**, *90*, 889–905. [CrossRef] [PubMed]
74. Castano, A.P.; Demidova, T.N.; Hamblin, M.R. Mechanisms in photodynamic therapy: Part one–Photosensitizers, photochemistry and cellular localization. *Photodiagn. Photodyn. Ther.* **2004**, *1*, 279–293. [CrossRef] [PubMed]
75. Silva, J.N.; Filipe, P.; Morlière, P.; Mazière, J.C.; Freitas, J.P.; de Castro, C.J.L.; Santus, R. Photodynamic therapies: Principles and present medical ap-plications. *Bio Med. Mater. Eng.* **2006**, *16*, 147–154.
76. Nyman, E.S.; Hynninen, P.H. Research advances in the use of tetrapyrrolic photosensitisers for photodynamic therapy. *J. Photochem. Photobiol. B Biol.* **2004**, *73*, 1–28. [CrossRef]
77. MacDonald, I.J.; Dougherty, T.J. Basic principles of photodynamic therapy. *J. Porphyr. Phthalocyanines* **2001**, *5*, 105–129. [CrossRef]

78. Bonnett, R. Photosensitizers of the porphyrin and phthalocyanine series for photodynamic therapy. *Chem. Soc. Rev.* **1995**, *24*, 19–33. [CrossRef]
79. Sharman, W.M.; Allen, C.M.; van Lier, J.E. Photodynamic therapeutics: Basic principles and clinical applications. *Drug Discov. Today* **1999**, *4*, 507–517. [CrossRef]
80. Moan, J.; Berg, K. The photodegradation of porphyrins in cells can be used to estimate the lifetime of singlet oxygen. *Photochem. Photobiol.* **1991**, *53*, 549–553. [CrossRef]
81. Hirth, A.; Michelsen, U.; Wohrle, D. PhotodynamischeTumortherapie. *Chem. Unserer Zeit* **1999**, *33*, 84–94. [CrossRef]
82. Sandland, J.; Malatesti, N.; Boyle, R. Porphyrins and related macrocycles: Combining photosensitization with radio- or optical-imaging for next generation theranostic agents. *Photodiagn. Photodyn. Ther.* **2018**, *23*, 281–294. [CrossRef]
83. O'Connor, A.E.; Gallagher, W.M.; Byrne, A.T. Porphyrin and nonporphyrin photosensitizers in oncology: Preclinical and clinical advances in photodynamic therapy. *Photochem. Photobiol.* **2009**, *85*, 1053–1074. [CrossRef] [PubMed]
84. Allison, R.R.; Cuenca, R.E.; Downie, G.H.; Camnitz, P.; Brodish, B.; Sibata, C.H. Clinical photodynamic therapy of head and neck cancers–A review of applications and outcomes. *Photodiagn. Photodyn. Ther.* **2005**, *2*, 205–222. [CrossRef] [PubMed]
85. Huang, Z. A review of progress in clinical photodynamic therapy. *Technol. Cancer Res. Treat.* **2005**, *4*, 283–293. [CrossRef] [PubMed]
86. Allison, R.; Moghissi, K.; Downie, G.; Dixon, K. Photodynamic therapy (PDT) for lung cancer. *Photodiagn. Photodyn. Ther.* **2011**, *8*, 231–239. [CrossRef] [PubMed]
87. Banerjee, S.M.; Mac Robert, A.J.; Mosse, C.A.; Periera, B.; Bown, S.G.; Keshtgar, M.R.S. Photodynamic therapy: Inception to application in breast cancer. *Breast* **2017**, *31*, 105–113. [CrossRef] [PubMed]
88. Agostinis, P.; Berg, K.; Cengel, K.A.; Foster, T.H.; Girotti, A.W.; Gollnick, S.O.; Hahn, S.M.; Hamblin, M.R.; Juzeniene, A.; Kessel, D.; et al. Photodynamic therapy of cancer: An update. *CA Cancer J. Clin.* **2011**, *61*, 250–281. [CrossRef]
89. Stegh, A.H. Toward personalized cancer nanomedicine–past, present, and future. *Integr. Biol.* **2012**, *5*, 48–65. [CrossRef]
90. Pineiro, M.; Pereira, M.M.; d' Rocha Gonsalves, A.M.; Arnaut, L.G.; Formosinho, S.J. Singlet oxygen quantum yields from halogenated chlorins: Potential new photodynamic therapy agents. *J. Photochem. Photobiol. A Chem.* **2001**, *138*, 147–157. [CrossRef]
91. Spagnul, C.; Turner, L.C.; Boyle, R.W. Immobilized photosensitizers for antimicrobial applications. *J. Photochem. Photobiol. B Biol.* **2015**, *150*, 11–30. [CrossRef]
92. Orlandi, V.T.; Caruso, E.; Tettamanti, G.; Banfi, S.; Barbieri, P. Photoinduced antibacterial activity of two dicationic 5,15-diarylporphyrins. *J. Photochem. Photobiol. B Biol.* **2013**, *127*, 123–132. [CrossRef]
93. Moritz, M.N.O.; Gonçalves, J.L.S.; Linares, I.A.P.; Perussi, J.R.; de Oliveira, K.T. Semi-synthesis and PDT activities of a new amphiphilic chlorin derivative. *Photodiagn. Photodyn. Ther.* **2017**, *17*, 39–47. [CrossRef] [PubMed]
94. Mandal, A.K.; Sahin, T.; Liu, M.; Lindsey, J.S.; Bocian, D.F.; Holten, D. Photophysical comparisons of PEGylated porphyrins, chlorins and bacteriochlorins. *New J. Chem.* **2016**, *40*, 9648–9656. [CrossRef]
95. Pisarek, S.; Maximova, K.; Gryko, D. Strategies toward the synthesis of amphiphilic porphyrins. *Tetrahedron* **2014**, *38*, 6685–6715. [CrossRef]
96. Liu, M.; Chen, C.Y.; Mandal, A.K.; Chandrashaker, V.; Evans-Storms, R.B.; Pitner, J.B.; Bocian, D.F.; Holten, D.; Lindsey, J.S. Bioconjugatable, PEGylated Hydroporphyrins for Photochemistry and Photomedicine. Narrow-Band, Red-Emitting Chlorins. *New J. Chem.* **2016**, *40*, 7721–7740. [CrossRef] [PubMed]
97. Zhang, N.; Jiang, J.; Liu, M.; Taniguchi, M.; Mandal, A.K.; Evans-Storms, R.B.; Pitner, J.B.; Bocian, D.F.; Holten, D.; Lindsey, J.S. Bioconjugatable, PEGylated Hydroporphyrins for Photochemistry and Photomedicine. Narrow-Band, Near-Infrared-Emitting Bacteriochlorins. *New J. Chem.* **2016**, *40*, 7750–7767. [CrossRef]
98. Roy, A.; Magdaong, N.C.M.; Jing, H.; Rong, J.; Diers, J.R.; Kang, H.S.; Niedzwiedzki, D.M.; Taniguchi, M.; Kirmaier, C.; Lindsey, J.S.; et al. Balancing Panchromatic Absorption and Multistep Charge Separation in a Compact Molecular Architecture. *J. Phys. Chem. A* **2022**, *126*, 9353–9365. [CrossRef]
99. Allison, R.R.; Sibata, C.H. Oncologic photodynamic therapy photosensitizers: A clinical review. *Photodiagn. Photodyn. Ther.* **2010**, *7*, 61–75. [CrossRef] [PubMed]
100. Celli, J.P.; Spring, B.Q.; Rizvi, I.; Evans, C.L.; Samkoe, K.S.; Verma, S.; Pogue, B.W.; Hasan, T. Imaging and Photodynamic Therapy: Mechanisms, Monitoring and Optimization. *Chem. Rev.* **2010**, *110*, 2795–2838. [CrossRef]
101. Yano, S.; Hirohara, S.; Obata, M.; Hagiya, Y.; Ogura, S.-I.; Ikeda, A.; Kataoka, H.; Tanaka, M.; Joh, T. Current states and future views in photodynamic therapy. *J. Photochem. Photobiol. C Photochem. Rev.* **2011**, *12*, 46–67. [CrossRef]
102. Baskaran, R.; Lee, J.; Yang, S.G. Clinical development of photodynamic agents and therapeutic applications. *Biomater. Res.* **2018**, *22*, 1–8. [CrossRef]
103. Monro, S.; Colón, L.K.; Yin, H.; Roque, J.; Konda, P.; Gujar, S.; Thummel, R.; Lilge, L.; Cameron, G.C.; McFarland, A.S. Ttransition metal complexes and photodynamic therapy from a tumor-centered approach: Challenges, opportunities, and highlights from the development of TLD. *Chem. Rev.* **2019**, *119*, 797–828. [CrossRef] [PubMed]
104. Akimoto, J.; Fukami, S.; Ichikawa, M.; Mohamed, A.; Kohno, M. Intraoperative photodiagnosis for malignant glioma using photosensitizer talaporfin sodium. *Front. Surg.* **2019**, *6*, 12. [CrossRef] [PubMed]
105. Allison, R.R.; Mota, H.C.; Sibata, C.H. Clinical PD/PDT in North America: An historical review. *Photodiagn. Photodyn. Ther.* **2004**, *1*, 263–277. [CrossRef] [PubMed]
106. Yang, V.X.; Muller, P.J.; Herman, P.; Wilson, B.C. A multispectral fluorescence imaging system: Design and initial clinical tests in intra-operative Photofrin photodynamic therapy of brain tumors. *Lasers Surg. Med.* **2003**, *32*, 224–232. [CrossRef]

107. Bogaards, A.; Varma, A.; Zhang, K.; Zach, D.; Bisland, S.K.; Moriyama, E.H.; Lilge, L.; Muller, P.J.; Wilson, B.C. Fluorescence image guided brain tumour resection with adjuvant metronomic photodynamic therapy: Preclinical model and technology development. *Photochem. Photobiol. Sci.* **2005**, *4*, 438–442. [CrossRef]
108. Bogaards, A.; Varma, A.; Collens, S.P.; Lin, A.H.; Giles, A.; Yang, V.X.D.; Bilbao, J.M.; Lilge, L.D.; Muller, P.J.; Wilson, B.C. Increased brain tumor resection using fluorescence image guidance in a preclinical model. *Lasers Surg. Med.* **2004**, *35*, 181–190. [CrossRef]
109. Olivo, M.; Wilson, B.C. Mapping ALA-induced PPIX fluorescence in normal brain and brain tumour using confocal fluorescence microscopy. *Int. J. Oncol.* **2004**, *25*, 37–45. [CrossRef]
110. Hage, R.; Ferreira, J.; Bagnato, V.S.; Vollet-Filho, J.D.; Plapler, H. Pharmacokinetics of photogem using fluorescence spectroscopy in dimethylhydrazine-inducedmurine colorectal carcinoma. *Int. J. Photoenergy* **2012**, *20*, 1–8. [CrossRef]
111. Trindade, F.Z.; Pavarina, A.C.; Ribeiro, A.P.D.; Bagnato, V.S.; Vergani, C.E.; Souza Costa, C.A. Toxicity of photodynamic therapy with LED associated to Photogem®: An in vivo study. *Lasers Med. Sci.* **2012**, *27*, 403–411. [CrossRef]
112. Benes, J.; Pouckova, P.; Zeman, J.; Zadinova, M.; Sunka, P.; Lukes, P.; Kolarova, H. Effects of tandem shock waves combined with photosan and cytostatics on the growth of tumours. *Folia Biol.* **2011**, *57*, 255–260.
113. Bonnett, R.; Martinez, G. Photobleaching of photosensitisers used in photodynamic therapy. *Tetrahedron* **2001**, *57*, 9513–9547. [CrossRef]
114. Lovell, J.F.; Liu, T.W.B.; Chen, J.; Zheng, G. Activatable Photosensitizers for Imaging and Therapy. *Chem. Rev.* **2010**, *110*, 2839–2857. [CrossRef] [PubMed]
115. Zimmermann, A.; Ritsch-Marte, M.; Kostron, H. mTHPC-mediated photodynamic diagnosis of malignant brain tumors. *Photochem. Photobiol.* **2001**, *74*, 611–616. [CrossRef] [PubMed]
116. Kostron, H.; Rossler, K. Surgical intervention in patients with malignant glioma. *Wien Med. Wochenschr.* **2006**, *156*, 338–341. [CrossRef]
117. Yow, C.M.N.; Chen, J.Y.; Mak, N.K.; Cheung, N.H.; Leung, A.W.N. Cellular uptake, subcellular localization and photodamaging effect of Temoporfin (Mthpc) in nasopharyngeal carcinoma cells: Comparison with hematoporphyrin derivative. *Cancer Lett.* **2000**, *157*, 123–131. [CrossRef]
118. Ronn, A.M.; Nouri, M.; Lofgren, L.A.; Steinberg, B.M.; Westerborn, A.; Windal, T.; Shikowitz, M.J. Human tissue levels and plasma pharmacokinetics of temoporfin (Foscan®, mTHPC). *Lasers Med. Sci.* **1996**, *11*, 267–272. [CrossRef]
119. Allen, T.M.; Cullis, P.R. Liposomal drug delivery systems: From concept to clinical applications. *Adv. Drug Deliv. Rev.* **2013**, *65*, 36–48. [CrossRef]
120. Gaio, E.; Scheglmann, D.; Reddi, E.; Moret, F. Uptake and photo-toxicity of Foscan®, Foslip® and Fospeg® in multicellular tumor spheroids. *J. Photochem. Photobiol. B Biol.* **2016**, *161*, 244–252. [CrossRef]
121. Compagnin, C.; Moret, F.; Celotti, L.; Miotto, G.; Woodhams, J.H.; MacRobert, A.J.; Scheglmann, D.; Iratni, S.; Reddi, E. Meta-tetra (hydroxyphenyl)chlorin-loaded liposomes sterically stabilised with poly (ethylene glycol) of different lenght and density: Characterisation, in vitro cellular uptake and phototoxicity. *Photochem. Photobiol. Sci.* **2011**, *10*, 1751–1759. [CrossRef]
122. Kiesselich, T.; Berlanda., J.; Plaetzer., K.; Krammer., B.; Berr., F. Comparative characterisation of the efficiency and cellular pharmacokinetics of Foscan® and Foslip® based photodynamic treatment in human biliary tract cancer cell lines. *Photochem. Photobiol. Sci.* **2007**, *6*, 619–627. [CrossRef]
123. Bovis, M.J.; Woodhams, J.H.; Loizidou, M.; Scheglmann, D.; Bown, S.G.; MacRobert, A.J. Improved in vivo delivery of m-THPC via pegylated liposomes for use in photodynamic therapy. *J. Control. Release* **2012**, *30*, 196–205. [CrossRef] [PubMed]
124. Reshetov, V.; Lassalle, H.P.; François, A.; Dumas, D.; Hupont, S.; Gräfe, S.; Filipe, V.; Jiskoot, W.; Guillemin, F.; Zorin, V.; et al. Photodynamic therapy with conventional and PEGylated liposomal formulations of m-THPC (temoporfin): Comparison of treatment efficacy and distribution characteristics in vivo. *Int. J. Nanomed.* **2013**, *8*, 3817–3831. [CrossRef] [PubMed]
125. Peng, W.; Samplonius, D.F.; de Visscher, S.; Roodenburg, J.L.; Helfrich, W.; Witjes, M.J. Photochemical internalization (PCI)-mediated enhancement of bleomycin cytotoxicity by liposomal mTHPC formulations in human head and neck cancer cells. *Lasers Surg. Med.* **2014**, *46*, 650–658. [CrossRef]
126. Chen, Y.; Zheng, X.; Dobhal, M.P.; Gryshuk, A.; Morgan, J.; Dougherty, T.J.; Oseroff, A.; Pandey, R.K. Methyl pyropheophorbide-a analogues: Potential fluorescent probes for the peripheral-type benzodiazepine receptor. Effect of central metal in photosensitizing efficacy. *J. Med. Chem.* **2005**, *48*, 3692–3695. [CrossRef]
127. Spikes, J.D.; Bommer, J.C. Photosensitizing properties of Mono-L-aspartyl chlorin e6 (NPe6): A candidate sensitizer for the photodynamic therapy of tumors. *J. Photochem. Photobiol. B.* **1993**, *17*, 135–143. [CrossRef]
128. Figueiredo, T.L.C.; Dolphin, D. Mesoarylporphyrins as dienophiles in diels-alder reactions: A novel approach to the synthesis of chlorins, bacteriochlorins, and naphthoporphyrins. *Org. Chem.* **1998**, *11*, 297–300.
129. Blant, S.A.; Ballini, J.P.; van den Bergh, H.; Fontolliet, C.; Wagnières, G.; Monnier, P. Time-dependent biodistribution of tetra(m-hydroxyphenyl) chlorin and benzoporphyrin derivatives monoacid ring A in the hamster model; comparative fluorescence microscopy study. *Photochem. Photobiol.* **2000**, *71*, 333–340. [CrossRef]
130. Levy, J.G.; Jones, C.A.; Pilson, L.A. The preclinical and clinical development and potential application of benzoporphyrin derivative. *Int. Photodyn.* **1994**, *1*, 3–5.

131. Aveline, B.M.; Hasan, T.; Redmond, R.W. The effects of aggregation, protein binding and cellular incorporation on the photophysical properties of benzoporphyrin derivative monoacid ring A (BPDMA). *J. Photochem. Photobiol. B Biol.* **1995**, *30*, 161–169. [CrossRef]
132. Stables, G.I.; Ash, D.V. Photodynamic therapy. *Cancer Treat Rev.* **1995**, *21*, 311–323. [CrossRef]
133. Huggett, M.T.; Jermyn, M.; Gillams, A.; Illing, R.; Mosse, S.; Novelli, M. Phase I/II study of verteporfin photodynamic therapy in locally advanced pancreatic cancer. *Br. J. Cancer* **2014**, *110*, 1698–1704. [CrossRef]
134. Zhong, W.; Celli, J.P.; Rizvi, I.; Mai, Z.; Spring, B.Q.; Yun, S.H.; Hasan, T. In vivo highresolution fluorescence microendoscopy for ovarian cancer detection and treatment monitoring. *Br. J. Cancer.* **2009**, *101*, 2015–2022. [CrossRef] [PubMed]
135. Pandey, R.K.; Bellnier, D.A.; Smith, K.M.; Dougherty, T.J. Chlor and porphyrin derivatives as potential photosensitizers in photodynamic therapy. *Photochem. Photobiol.* **1991**, *53*, 65–72. [CrossRef]
136. Pandey, R.K.; Sumlin, A.B.; Constantine, S.; Aoudia, M.; Potter, W.R.; Belinier, D.A.; Henderson, B.W.; Roders, M.A.; Smith, K.M.; Dougherty, T.J. Alkyl ether analogs of chlorophyll-a derivatives, PartSynthesis, photophysical properties and photodynamic efficacy. *Photochem. Photobiol.* **1996**, *64*, 194–204. [CrossRef] [PubMed]
137. Marchal, S.; Francois, A.; Dumas, D.; Guillemin, F.; Bezdetnaya, L. Relationship between subcellular localization of Foscan and caspase activation in photosensitized MCF-7 cells. *Br. J. Cancer* **2007**, *96*, 944–951. [CrossRef] [PubMed]
138. Bellnier, D.A.; Greco, W.R.; Loewen, G.M.; Nava, H.; Oseroff, A.; Pandey, R.K.; Dougherty, T.J. Population pharmacokinetics of the photodynamic therapy agent 2-[1-hexyloxyethyl]-2-devinyl pyropheophorbide-a in cancer patients. *Cancer Res.* **2003**, *63*, 1806–1813. [PubMed]
139. Bellnier, D.A.; Greco, W.R.; Nave, H.; Loewen, G.M.; Oseroff, A.R.; Pandey, R.K.; Dougherty, T.J. Mild skin photosensitivity in cancer patients following injection of Photochlor (2-[1-hexyloxyethyl]-2-devinyl pyropheophorbide-a; HPPH) for photodynamic therapy. *Cancer Chemother. Pharmacol.* **2006**, *57*, 40–45. [CrossRef] [PubMed]
140. Slansky, G.; Li, A.; Dobhal, M.P.; Goswami, L.; Graham, A.; Chen, Y.; Kanter, P.; Alberico, R.A.; Spernyak, J.; Morgan, J.; et al. Chlorophyll-a analogues conjugated with aminobenzyl-DTPA as potential bifunctional agents for magnetic resonance imaging and photodynamic therapy. *Bioconjug. Chem.* **2005**, *16*, 32–42.
141. Morgan, A.R.; Garbo, G.M.; Keck, R.; Selman, S.H. New photosensitizers for photodynamic therapy: Combined effect of metallopurpurin derivatives and light on transplantable bladder tumors. *Cancer Res.* **1988**, *48*, 194–198.
142. Pogue, B.W.; Redmond, R.W.; Trivedi, N.; Hasan, T. Photophysical properties of tin ethyl etiopurpurin I (SnET2) and tinoctaethylbenzochlorin (SnOEBC) in solution and bound to albumin. *Photochem. Photobiol.* **1998**, *68*, 809–815. [CrossRef]
143. Kawczyk-Krupka, A.; Wawrzyniec, K.; Musiol, S.K.; Potempa, M.; Bugaj, A.M.; Sieron, A. Treatment of localized prostate cancer using WST-09 and WST-11 mediated vascular targeted photodynamic therapy-a review. *Photodiagn. Photodyn. Ther.* **2015**, *12*, 567–574. [CrossRef] [PubMed]
144. Azzouzi, A.R.; Lebdai, S.; Benzaghou, F.; Stief, C. Vascular-targeted photodynamic therapy with TOOKAD(R) soluble in localized prostate cancer: Standardization of the procedure. *World J. Urol.* **2015**, *33*, 937–944. [CrossRef] [PubMed]
145. Azzouzi, A.R.; Vincendeau, S.; Barret, E.; Cicco, A.; Kleinclauss, F.; van der Poel, H.G. Padeliporfin vascular-targeted photodynamic therapy versus active surveillance in men with low-risk prostate cancer (CLIN1001 PCM301): An open-label, phase 3, randomised controlled trial. *Lancet Oncol.* **2017**, *18*, 181–191. [CrossRef] [PubMed]
146. Betrouni, N.; Boukris, S.; Benzaghou, F. Vascular targeted photodynamic therapy with TOOKAD(R) soluble (WST11) in localized prostate cancer: Efficiency of automatic pre-treatment planning. *Lasers Med. Sci.* **2017**, *32*, 1301–1307. [CrossRef] [PubMed]
147. Sessler, J.L.; Hemmi, G.; Mody, T.D.; Murai, T.; Burrell, A.; Young, S.W. Texaphyrins: Synthesis and applications. *Acc. Chem. Res.* **1994**, *27*, 43–50. [CrossRef]
148. Patel, H.; Mick, R.; Finlay, J.; Zhu, T.C.; Rickter, E.; Cengel, K.A. Motexafin lutetium-photodynamic therapy of prostate cancer: Short- and long-term effects on prostate-specific antigen. *Clin. Cancer Res.* **2008**, *14*, 4869–4876. [CrossRef]
149. Bradley, K.A.; Pollack, I.F.; Reid, J.M.; Adamson, P.C.; Ames, M.M.; Vezina, G. Motexafin gadolinium and involved field radiation therapy for intrinsic pontine glioma of childhood: A Children's oncology group phase I study. *Neuro-Oncology* **2008**, *10*, 752–758. [CrossRef]
150. Bradley, K.A.; Zhou, T.; McNall-Knapp, R.Y.; Jakacki, R.I.; Levy, A.S.; Vezina, G. Motexafin-gadolinium and involved field radiation therapy for intrinsic pontine glioma of childhood: A children's oncology group phase 2 study. *Int. J. Radiat. Oncol. Biol. Phys.* **2013**, *85*, 55–60. [CrossRef]
151. Thomas, S.R.; Khuntia, D. Motexafin gadolinium: A promising radiation sensitizer in brain metastasis. *Expert Opin. Drug Discov.* **2011**, *6*, 195–203. [CrossRef]
152. Weersink, R.A.; Bogaards, A.; Gertner, M.; Davidson, S.R.H.; Zhang, K.; Netchev, G.; Trachtenberg, J.; Wilson, B.C. Techniques for delivery and monitoring of TOOKAD (WST09)-mediated photodynamic therapy of the prostate: Clinical experience and praticalities. *J. Photochem. Photobiol. B* **2005**, *79*, 211–222. [CrossRef]
153. Zhang, J.; Jiang, C.; Longo, J.P.F.; Azevedo, R.B.; Zhang, H.; Muehlmann, L.A. An updated overview on the development of new photosensitizers for anticancer photodynamic therapy. *Acta Pharm. Sin. B* **2018**, *8*, 137–146. [CrossRef] [PubMed]
154. Li, J.W.; Wu, Z.M.; Magetic, D.; Zhang, L.J.; Chen, Z.L. Antitumor effects evaluation of a novel porphyrin derivative in photodynamic therapy. *Tumor Biol.* **2015**, *36*, 9685–9692. [CrossRef] [PubMed]
155. Liao, P.Y.; Wang, X.R.; Gao, Y.H.; Zhang, X.H.; Zhang, L.J.; Song, C.H. Synthesis, photophysical properties and biological evaluation of β- alkylamino porphyrin for photodynamic therapy. *Bioorg. Med. Chem.* **2016**, *24*, 6040–6047. [CrossRef]

156. Chen, J.J.; Hong, G.; Gao, L.J.; Liu, T.J.; Cao, W.J. In vitro and in vivo antitumor activity of a novel porphyrin-based photosensitizer for photodynamic therapy. *J. Cancer Res. Clin. Oncol.* **2015**, *141*, 1553–1561. [CrossRef] [PubMed]
157. Nakai, M.; Maeda, T.; Mashima, T.; Yano, S.; Sakuma, S.; Otake, E. Synthesis and photodynamic properties of glucopyanoside conjugated indium (III) porphyrins as a bifunctionalagent. *J. Porphyr. Phthalocyanies* **2013**, *17*, 1173–1182. [CrossRef]
158. Lucky, S.S.; Soo, K.C.; Zhang, Y. Nanoparticles in photodynamic therapy. *Chem. Rev.* **2015**, *115*, 1990–2042. [CrossRef]
159. Xodo, L.E.; Cogoi, S.; Rapozzi, V. Photosensitizers binding to nucleic acids as anticancer agents. *Future Med. Chem.* **2016**, *8*, 179–194. [CrossRef] [PubMed]
160. Boscencu, R.; Manda, G.; Radulea, N.; Socoteanu, R.P.; Ceafalan, L.C.; Neagoe, I.V.; Ferreira Machado, I.; Basaga, S.H.; Vieira Ferreira, L.F. Studies on the synthesis, photophysical and biological evaluation of some unsymmetrical meso-tetrasubstituted phenyl porphyrins. *Molecules* **2017**, *22*, 1815. [CrossRef]
161. Boscencu, R.; Manda, G.; Socoteanu, R.P.; Hinescu, M.E.; Neagoe, I.V.; Olariu, L.; Dumitriu, B. Porphyrin Derivative for Theranostic Use. Patent Application No. 2017 01030/2017; published in RO-BOPI 8, 30 August 2018.
162. Ferreira, L.F.V.; Ferreira, D.P.; Oliveira, A.S.; Boscencu, R.; Socoteanu, R.; Ilie, M.; Constantin, C.; Neagu, M. Synthesis, photophysical and cytotoxicity evaluation of A3B type mesoporphyrinic compounds. *Dyes Pigments* **2012**, *95*, 296–303. [CrossRef]
163. Boscencu, R.; Socoteanu, R.P.; Nacea, V.; Constantin, C.; Manda, G.; Neagu, M.; Ilie, M.; Baconi, D.L.; Oliveira, A.S.; Vieira Ferreira, L.F. Asymmetrically Substituted Tetrapyrrolic Compound, Obtaining Process and Cell Level Biological Evaluation. Patent No. 123419 B1, 30 March 2012.
164. Socoteanu, R.; Boscencu, R.; Nacea, V.; Constantin, C.; Manda, G.; Neagu, M.; Ilie, M.; Oliveira, A.S.; Vieira Ferreira, L.F. Biofunctionalized Porphyrinic Compound. Patent No. 125018 B1, 30 April 2015.
165. Boscencu, R.; Socoteanu, R.; Constantin, C.; Neagu, M.; Manda, G.; Nacea, V.; Ilie, M.; Gird, C.E.; Oliveira, A.S.; Vieira Ferreira, L.F. Asymmetric Substituted Porphyrinic Compound as A Photosensitizing Agent and Process for Its Preparation. Patent No. 126761B1, 30 December 2014.
166. Socoteanu, R.; Boscencu, R.; Nacea, V.; Sousa Oliveira, A.; Vieira Ferreira, L.F. Microwave assisted synthesis of unsymmetrical tetrapyrrolic compounds. *Rev. Chim.* **2008**, *59*, 969–972. [CrossRef]
167. Boscencu, R.; Socoteanu, R.; Vasiliu, G.; Nacea, V. Synthesis under solvent free conditions of some unsymmetrically substituted porphyrinic compounds. *Rev. Chim.* **2014**, *65*, 888–891.
168. Boscencu, R.; Manda, G.; Socoteanu, R.P.; Hinescu, M.E.; Radulea, N.; Neagoe, I.V.; Vieira Ferreira, L.F. Tetrapyrrolic Compound with Theranostic Applications and Obtaining Process. Patent No. 131946, 29 March 2019.
169. Socoteanu, R.; Manda, G.; Boscencu, R.; Vasiliu, G.; Oliveira, A.S. Synthesis, Spectral Analysis and Preliminary in Vitro Evaluation of Some Tetrapyrrolic Complexes with 3d Metal Ions. *Molecules* **2015**, *20*, 15488–15499. [CrossRef] [PubMed]
170. Boscencu, R. Unsymmetrical Mesoporphyrinic Complexes of Copper(II) and Zinc(II). Microwave-Assisted Synthesis, Spectral Characterization and Cytotoxicity Evaluation. *Molecules* **2011**, *16*, 5604–5617. [CrossRef]
171. Boscencu, R. Microwave Synthesis under Solvent-Free Conditions and Spectral Studies of Some Mesoporphyrinic Complexes. *Molecules* **2012**, *17*, 5592–5603. [CrossRef]
172. Boscencu, R.; Oliveira, A.S.; Ferreira, D.P.; Vieira Ferreira, L.F. Synthesis and spectral evaluation of some unsymmetrical mesoporphyrinic complexes. *Int. J. Mol. Sci.* **2012**, *13*, 8112–8125. [CrossRef]
173. Boscencu, R.; Licsandru, D.; Socoteanu, R.; Oliveira, A.S.; Vieira Ferreira, L.F. Synthesis and characterization of some mesoporphyrinic compounds unsymetricaly substituted. *Rev. Chim.* **2007**, *58*, 498–502.
174. Boscencu, R.; Socoteanu, R.; Ilie, M.; Oliveira, A.S.; Constantin, C.; Vieira Ferreira, L.F. Synthesis, spectraland biological evaluation of some mesoporphyrinic complexes of Zn(II). *Rev. Chim.* **2009**, *60*, 1006–1011.
175. Boscencu, R.; Ilie, M.; Socoteanu, R.; Oliveira, A.S.; Constantin, C.; Neagu, M.; Manda, M.; Vieira Ferreira, L.F. Microwave Synthesis, Basic Spectral and Biological Evaluation of Some Copper (II) Mesoporphyrinic Complexes. *Molecules* **2010**, *15*, 3731–3743. [CrossRef]
176. Boscencu, R.; Socoteanu, R.; Oliveira, A.S.; Vieira Ferreira, L.F.; Nacea, V.; Patrinoiu, G. Synthesis and characterization of some unsymmetrically-substituted mesoporphyrinic mono-hydroxyphenyl complexes of Copper(II). *Pol. J. Chem.* **2008**, *82*, 509–522.
177. Boscencu, R.; Socoteanu, R.; Oliveira, A.S.; Vieira Ferreira, L.F. Studies on Zn(II) monohydroxyphenylmesoporphyrinic complexes. Synthesis and characterization. *J. Serb. Chem. Soc.* **2008**, *73*, 713–726. [CrossRef]
178. Vasiliu, G.; Boscencu, R.; Socoteanu, R.; Nacea, V. Complex combinations of some transition metals with new unsymmetrical porphyrins. *Rev. Chim.* **2014**, *65*, 998–1001.
179. Oliveira, A.S.; Licsandru, D.; Boscencu, R.; Socoteanu, R.; Nacea, V.; Vieira Ferreira, L.F. A Singlet Oxygen Photogeneration and luminescence study of unsymmetrically-substituted meso-porphyrinic compounds. *Int. J. Photoenergy* **2009**, *2009*, 10. [CrossRef]
180. Bonnett, R.; Charlesworth, P.; Djelal, B.; Foley, S.; McGarvey, D.J.; Truscott, T. Photophysical properties of 5,10,15, 20–tetrakis (*m*-hydroxyphenyl)porphyrin (m-THPP), 5,10,15,20–tetrakis (*m*-hydroxyphenyl)chlorin (m-THPC) and 5,10,15,20–tetrakis (*m*-hydroxyphenyl)bacteriochlorin (m-THPBC): A comparative study. *J. Chem. Soc. Perkin Trans. 2* **1999**, *2*, 325–328. [CrossRef]
181. Dobre, M.; Boscencu, R.; Neagoe, I.V.; Surcel, M.; Milanesi, E.; Manda, G. Insight into the web of stress responses triggered at gene expression level by porphyrin-PDT in HT29 human colon carcinoma cells. *Pharmaceutics* **2021**, *13*, 1032. [CrossRef] [PubMed]
182. Costa, L.D.; Silva, J.; Fonseca, S.M.; Arranja, C.T.; Urbano, A.M.; Sobral, A. Photophysical characterization and in vitro phototoxicity evaluation of 5,10,15,20-tetra(quinolin-2-yl) porphyrin as a potential sensitizer forphotodynamic therapy. *Molecules* **2016**, *21*, 439. [CrossRef]

183. Jinadasa, R.G.W.; Zhou, Z.H.; Vicente, M.G.; Smith, K.M. Syntheses and cellular investigations of diaspartateandaspartate-lysine chlorin e6 conjugates. *Org. Biomol. Chem.* **2016**, *14*, 1049–1064. [CrossRef]
184. Gushchina, O.I.; Larkina, E.A.; Nikolskaya, T.A.; Mironov, A.F. Synthesis of aminoderivatives of chlorin e6 and investigation of their biological activity. *J. Photochem. Photobiol. B* **2015**, *153*, 76–81. [CrossRef]
185. Narumi, A.; Tsuji, T.; Shinohara, K.; Yamazaki, H.; Kikuchi, M.; Kawaguchi, S. Maltotriose-conjugation to a fluorinated chlorin derivative generatinga PDT photosensitizer with improved water solubility. *Org. Biomol. Chem.* **2016**, *14*, 3608–3613. [CrossRef]
186. Pereira, P.; Rizvi, W.; Bhupathiraju, N.; Berisha, N.; Fernandes, R.; Tomé, J.; Drain, C.M. Carbon-1 versus Carbon-3 linkage of d-Galactose to porphyrins: Synthesis, uptake, and photodynamic efficiency. *Bioconjug. Chem.* **2018**, *29*, 306–315. [CrossRef]
187. Lin, T.-Y.; Li, Y.; Liu, Q.; Chen, J.-L.; Zhang, H.; Lac, D.; Zhang, H.; Ferrara, K.W.; Hogiu, W.S.; Lam, K.S.; et al. Novel theranosticnanoporphyrins for photodynamic diagnosis and trimodal therapy for bladder cancer. *Biomaterials* **2016**, *104*, 339–351. [CrossRef]
188. Kim, B.H.; Shin, K.; Kwon, S.G.; Jang, Y.; Lee, H.S.; Lee, H.; Jun, S.W.; Lee, I.; Han, S.Y.; Yim, Y.H.; et al. Sizing by weighing: Characterizing sizes of ultrasmall-sized iron oxide nanocrystals using MALDI-TOF mass spectrometry. *J. Am. Chem. Soc.* **2013**, *135*, 2407–2410. [CrossRef] [PubMed]
189. Vieira Ferreira, L.F.; Ferreira Machado, I.; Gama, A.; Socoteanu, R.P.; Boscencu, R.; Manda, G.; Calhelha, R.C.; Ferreira, I.C.F. RPhotochemical/Photocytotoxicity studies of new tetrapyrrolic structures as potential candidates for cancer theranostics. *Curr. Drug Discov. Technol.* **2020**, *17*, 661–669. [CrossRef] [PubMed]
190. Vieira Ferreira, L.F.; Ferreira Machado, I.; Gama, A.; Lochte, F.; Socoteanu, R.P.; Boscencu, R. Surface photochemical studies of nano-hybrids of A3B porphyrins and Fe3O4 silica-coated nanoparticles. *J. Photochem. Photobiol. A Chem.* **2020**, *387*, 1–8. [CrossRef]
191. Li, S.; Shen, X.; Xu, Q.-H.; Cao, Y. Gold nanorod enhanced conjugated polymer/photosensitizer composite nanoparticles for simultaneous two-photon excitation fluorescence imaging and photodynamic therapy. *Nanoscale* **2019**, *11*, 19551–19560. [CrossRef]
192. Nowostawska, M.; Corr, S.A.; Byrne, S.J.; Conroy, J.; Volkov, Y.; Gun'ko, Y.K. Porphyrin-magnetite nanoconjugates for biological imaging. *J. Nanobiotechnol.* **2011**, *9*, 1–12. [CrossRef] [PubMed]

Disclaimer/Publisher's Note: The statements, opinions and data contained in all publications are solely those of the individual author(s) and contributor(s) and not of MDPI and/or the editor(s). MDPI and/or the editor(s) disclaim responsibility for any injury to people or property resulting from any ideas, methods, instructions or products referred to in the content.

MDPI
St. Alban-Anlage 66
4052 Basel
Switzerland
www.mdpi.com

Molecules Editorial Office
E-mail: molecules@mdpi.com
www.mdpi.com/journal/molecules

Disclaimer/Publisher's Note: The statements, opinions and data contained in all publications are solely those of the individual author(s) and contributor(s) and not of MDPI and/or the editor(s). MDPI and/or the editor(s) disclaim responsibility for any injury to people or property resulting from any ideas, methods, instructions or products referred to in the content.

www.ingramcontent.com/pod-product-compliance
Lightning Source LLC
LaVergne TN
LVHW070152100526
838202LV00015B/1936